D1300449

ECOLOGICAL ENGINEERING

Principles and Practice

ECOLOGICAL ENGINEERING

Principles and Practice

Patrick C. Kangas

LEWIS PUBLISHERS

A CRC Press Company
Boca Raton London New York Washington, D.C.

Library of Congress Cataloging-in-Publication Data

Kangas, Patrick C.
Ecological engineering: principles and practice / Patrick Kangas.
p. cm.
Includes bibliographical references and index.
ISBN 1-56670-599-1 (alk. paper)
1. Ecological engineering. I. Title.

GE350.K36 2003
628—dc21 2003051689

Direct all inquiries to CRC Press LLC, 2000 N.W. Corporate Blvd., Boca Raton, Florida 33431.

Lewis Publishers is an imprint of CRC Press LLC

International Standard Book Number 1-56670-599-1
Library of Congress Card Number 2003051689
Printed in the United States of America 1 2 3 4 5 6 7 8 9 0
Printed on acid-free paper

Dedication

I would like to dedicate this book to my ecology professors at Kent State University: G.D. Cooke, R. Mack, L.P. Orr, and D. Waller; at the University of Oklahoma: M. Chartock, M. Gilliland, P.G. Risser, and F. Sonlietner; and at the University of Florida: E.S. Deevey, J. Ewel, K. Ewel, L.D. Harris, A.E. Lugo, and H.T. Odum.

Preface

This text is intended as a graduate level introduction to the new field of ecological engineering. It is really a book about ecosystems and how they can be engineered to solve various environmental problems. The Earth's biosphere contains a tremendous variety of existing ecosystems, and ecosystems that never existed before are being created by mixing species and geochemical processes together in new ways. Many different applications are utilizing these old and new ecosystems but with little unity, yet. Ecological engineering is emerging as the discipline that offers unification with principles for understanding and for designing all ecosystem-scale applications. In this text three major principles (the energy signature, self-organization, and preadaptation) are suggested as the foundation for the new discipline.

H. T. Odum, the founder of ecological engineering, directly inspired the writing of this book through his teaching. An important goal was to review and summarize his research, which provides a conceptual framework for the discipline. Odum's ideas are found throughout the book because of their originality, their explanatory power, and their generality.

Acknowledgments

This book benefited greatly from the direct and indirect influences of the author's colleagues in the Biological Resources Engineering Department at the University of Maryland. They helped teach an ecologist some engineering. Art Johnson and Fred Wheaton, in particular, offered models in the form of their own bioengineering texts.

Strong credit for the book goes to the editors at CRC Press, especially Sara Kreisman, Samar Haddad, Matthew Wolff, and Brian Kenet, whose direction brought the book to completion. Kimberly Monahan assisted through managing correspondence and computer processing. Joan Breeze produced the original energy circuit diagrams. David Tilley completed the diagrams and provided important insights on industrial ecology, indoor air treatment, and other topics. Special acknowledgment is due to the author's students who shared research efforts in ecological engineering. Their work is included throughout the text. David Blersch went beyond this contribution in drafting many of the figures. Finally, sincere appreciation goes to the author's wife, Melissa Kangas, for her patience and help during the years of work needed to complete the book.

Author

Patrick Kangas, Ph.D. is a systems ecologist with interests in ecological engineering and tropical sustainable development. He received his B.S. degree from Kent State University in biology, his M.S. from the University of Oklahoma in botany and ecology, and his Ph.D. degree in environmental engineering sciences from the University of Florida. After graduating, Dr. Kangas took a position in the biology department of Eastern Michigan University and taught there for 11 years. In 1990 he moved to the University of Maryland where he is coordinator of the Natural Resources Management Program and associate professor in the Biological Resources Engineering Department. He has conducted research in Puerto Rico, Brazil, and Belize and has led travel–study programs throughout the neotropics. Dr. Kangas has published more than 50 papers, book chapters, and contract reports on a variety of environmental subjects.

Table of Contents

1 Introduction

Ecological engineering combines the disciplines of ecology and engineering in order to solve environmental problems. The approach is to interface ecosystems with technology to create new, hybrid systems. Designs are evolving in this field for wastewater treatment, erosion control, ecological restoration, and many other applications. The goal of ecological engineering is to generate cost effective alternatives to conventional solutions. Some designs are inspired by ancient human management practices such as the multipurpose rice paddy system, while others rely on highly sophisticated technology such as closed life support systems. Because of the extreme range of designs that are being considered and because of the combination of two fields traditionally thought to have opposing directions, ecological engineering offers an exciting, new intellectual approach to problems of man and nature. The purpose of this book is to review the emerging discipline and to illustrate some of the range of designs that have been practically implemented in the present or conceptually imagined for the future.

A CONTROVERSIAL NAME

A simple definition of ecological engineering is "to use ecological processes within natural or constructed imitations of natural systems to achieve engineering goals" (Teal, 1991). Thus, ecosystems are designed, constructed, and operated to solve environmental problems otherwise addressed by conventional technology. The contention is that ecological engineering is a new approach to both ecology and engineering which justifies a new name. However, because these are old, established disciplines, some controversy has arisen from both directions. On one hand, the term *ecological engineering* is controversial to ecologists who are suspicious of the engineering method, which sometimes generates as many problems as it solves. Examples of this concern can be seen in the titles of books that have critiqued the U.S. Army Corps of Engineers' water management projects: *Muddy Water* (Maass, 1951), *Dams and Other Disasters* (Morgan, 1971), *The River Killers* (Heuvelmans, 1974), *The Flood Control Controversy* (Leopold and Maddock, 1954), and *The Corps and the Shore* (Pilkey and Dixon, 1996). In the past, ecologists and engineers have not always shared a common view of nature and, because of this situation, an adversarial relationship has evolved. Ecologists have sometimes been said to be afflicted with "physics envy" (Cohen, 1971; Egler, 1986), because of their desire to elevate the powers of explanation and prediction about ecosystems to a level comparable to that achieved by physicists for the nonliving, physical world. However, even though engineers, like physicists, have achieved great powers of physical explanation and prediction, no ecologist has ever been said to have exhibited "engineering envy."

On the other hand, the name of *ecological engineering* is controversial to engineers who are hesitant about creating a new engineering profession based on an approach that relies so heavily on the "soft" science of ecology and that lacks the quantitative rigor, precision, and control characteristic of most engineering. Some engineers might also dismiss ecological engineering as a kind of subset of the existing field of environmental engineering, which largely uses conventional technology to solve environmental problems. Hall (1995a) described the situation presented by ecological engineering as follows: "This is a very different attitude from that of most conventional engineering, which seeks to force its design onto nature, and from much of conventional ecology, which seeks to protect nature from any human impact." Finally, M. G. Wolman may have summed up the controversy best, during a plenary presentation to a stream restoration conference, by suggesting that ecological engineering is a kind of oxymoron in combining two disciplines that are somewhat contradictory.

The challenge for ecologists and engineers alike is to break down the stereotypes of ecology and engineering and to combine the strengths of both disciplines. By using a "design with nature" philosophy and by taking the best of both worlds, ecological engineering seeks to develop a new paradigm for environmental problem solving. Many activities are already well developed in restoration ecology, appropriate technology, and bioengineering which are creating new designs for the benefit of man and nature. Ecological engineering unites many of these applications into one discipline with similar principles and methods.

The idea of ecological engineering was introduced by H. T. Odum. He first used the term *community engineering*, where *community* referred to the ecological community or set of interacting species in an ecosystem, in an early paper on microcosms (H. T. Odum and Hoskin, 1957). This reference dealt with the design of new sets of species for specific purposes. The best early summary of his ideas was presented as a chapter in his first book on energy systems theory (H. T. Odum, 1971). This chapter outlines many of the agendas of ecological engineering that are suggested by the headings used to organize the writing (Table 1.1). Thirty years later, this chapter is perhaps still the best single source on principles of ecological engineering. H. T. Odum pioneered ecological engineering by adapting ecological theory for applied purposes. He carried out major ecosystem design experiments at Port Aransas, Texas (H. T. Odum et al., 1963); Morehead City, North Carolina (H. T. Odum, 1985, 1989); and Gainesville, Florida (Ewel and H. T. Odum, 1984), the latter two of which involved introduction of domestic sewage into wetlands. He synthesized the use of microcosms (Beyers and H. T. Odum, 1993) and developed an accounting system for environmental decision making (H. T. Odum, 1996). Models of ecologically engineered systems are included throughout this book in the "energy circuit language" which H. T. Odum developed. This is a symbolic modeling language (Figure 1.1) that embodies thermodynamic constraints and mathematical equivalents for simulation (Gilliland and Risser, 1977; Hall et al., 1977; H. T. Odum, 1972, 1983; H. T. Odum and E. C. Odum, 2000).

William Mitsch, one of H. T. Odum's students, is now leading the development of ecological engineering. He has strived to outline the dimensions of the field

TABLE 1.1
Headings from Chapter 10 in *Environment, Power and Society* That Hint at Important Features of Ecological Engineering

The network nightmare
Steady states of planetary cycles
Ecological engineering of new systems
Multiple seeding and invasions
The implementation of a pulse
Energy channeling by the addition of an extreme
Microbial diversification operators
Ecological engineering through control species
The cross-continent transplant principle
Man and the complex closed systems for space
Compatible living with fossil fuel
How to pay the natural networks
The city sewer feedback to food production
Specialization of waste flows
Problem for the ecosystem task forces
Energy-based value decisions
Replacement value of ecosystems
Life-support values of diversity
Constitutional right to life support
Power density
Summary

Source: From Odum, H. T. 1971. *Environment, Power, and Society.* John Wiley & Sons, New York.

(Mitsch, 1993, 1996; Mitsch and Jorgenson, 1989), and he has established a model field laboratory on the Ohio State University campus for the study of alternative wetland designs (see Chapter 9).

Thus, although ecological engineering is presented here as a new field, it has been developing for the last 30 years. The ideas initiated by H. T. Odum are now appearing with greater frequency in the literature (Berryman et al., 1992; Schulze, 1996). Of note, a journal called *Ecological Engineering* was started in 1992, with Mitsch as editor-in-chief, and two professional societies have been formed (the International Ecological Engineering Society founded in 1993 and the American Ecological Engineering Society founded in 2001).

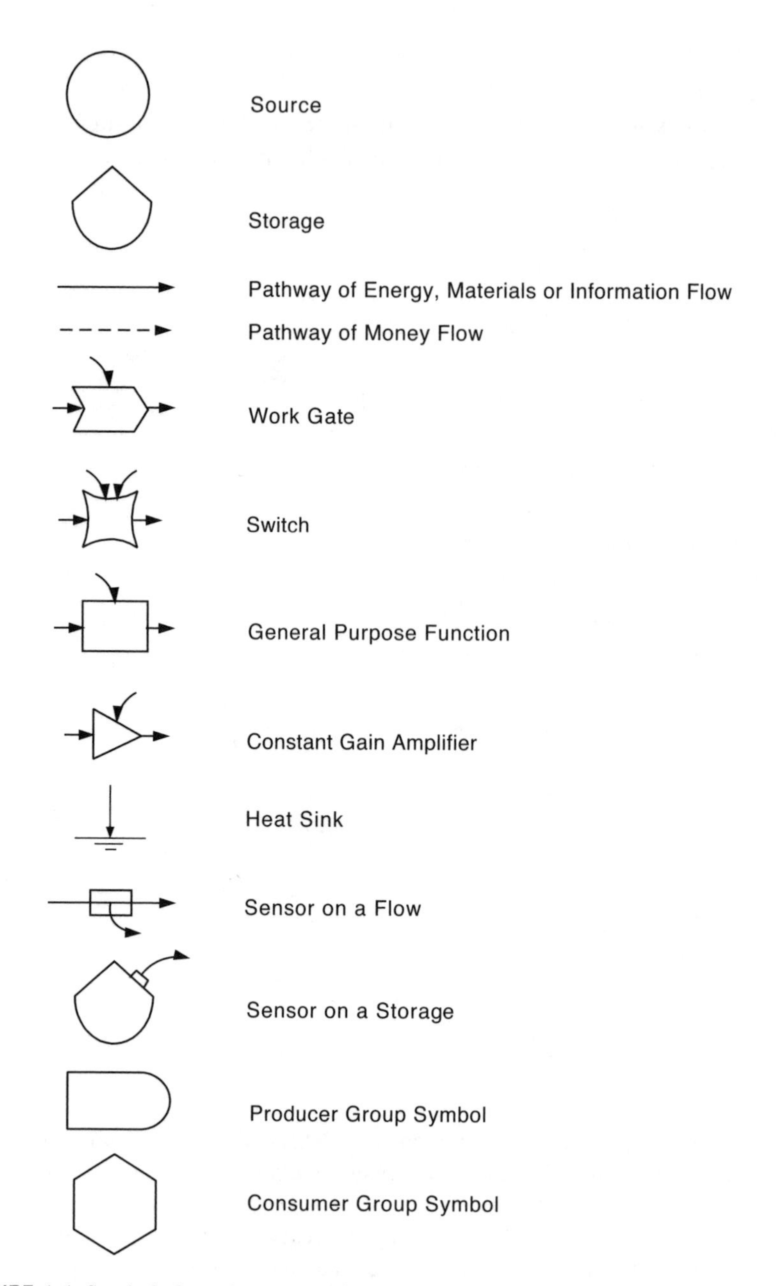

FIGURE 1.1 Symbols from the energy circuit language. (Adapted from Odum, H. T. 1983. *Systems Ecology: An Introduction*. John Wiley & Sons, New York. With permission.)

RELATIONSHIP TO ECOLOGY

Because ecological engineering uses ecosystems to solve problems, it draws directly on the science of ecology. This is consistent with other engineering fields which

TABLE 1.2
The Matching of Disciplines from the Sciences with Disciplines of Engineering, Showing the Correspondence between the Two Activities

Scientific Field or Topic	Engineering Field
Chemistry	Chemical engineering
Mechanics	Mechanical engineering
Electricity	Electrical engineering
Ecology	Ecological engineering

also are based on particular scientific disciplines or topics (Table 1.2). The principles and theories of ecology are fundamental for understanding natural ecosystems and, therefore, also for the design, construction, and operation of new ecosystems for human purposes. The ecosystem is the network of biotic (species populations) and abiotic (nutrients, soil, water, etc.) components found at a particular location that function together as a whole through primary production, community respiration, and biogeochemical cycling. The ecosystem is considered by some to be the fundamental unit of ecology (Evans, 1956, 1976; Jørgensen and Muller, 2000; E. P. Odum, 1971), though other units such as the species population are equally important, depending on the scale of reference. The fundamental nature of the ecosystem concept has been demonstrated by its choice as the most important topic within the science in a survey of the British Ecological Society (Cherrett, 1988), and E. P. Odum chose it as the number one concept in his list of "Great Ideas in Ecology for the 1990s" (E. P. Odum, 1992). Reviews by Golley (1993) and Hagen (1992) trace the history of the concept and provide further perspective.

Functions within ecosystems include (1) energy capture and transformation, (2) mineral retention and cycling, and (3) rate regulation and control (E. P. Odum, 1962, 1972, 1986; O'Neill, 1976). These aspects are depicted in the highly aggregated P–R model of Figure 1.2. In this model energy from the sun interacts with nutrients for the production (P) of biomass of the system's community of species populations. Respiration (R) of the community of species releases nutrients back to abiotic storage, where they are available for uptake again. Thus, energy from sunlight is transformed and dissipated into heat while nutrients cycle internally between compartments. Control is represented by the external energy sources and by the coefficients associated with the pathways. Rates of production and respiration are used as measures of ecosystem performance, and they are regulated by external abiotic conditions such as temperature and precipitation and by the actions of keystone species populations within the system, which are not shown in this highly aggregated model. Concepts and theories about control are as important in ecology as they are in engineering, and a review of the topic is included in Chapter 7.

Ecosystems can be extremely complex with many interconnections between species, as shown in Figure 1.3 (see also more complex networks: figure 6 in Winemiller, 1990 and figure 18.4 in Yodzis, 1996). Boyce (1991) has even suggested that ecosystems "are possibly the most complex structures in the universe." Charles

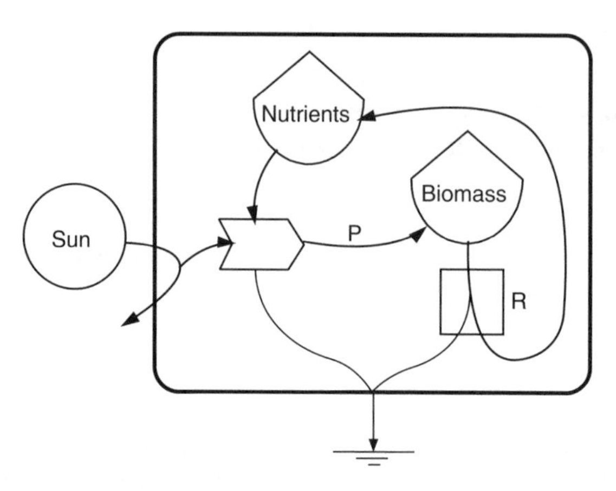

FIGURE 1.2 Basic P–R model of the ecosystem. "P" stands for primary production and "R" stands for community respiration.

Elton, one of the founders of modern ecology, described this complexity for one of his study sites in England with a chess analogy below (Elton, 1966; see also Kangas, 1988, for another chess analogy for understanding ecological complexity):

> In the game of chess, counted by most people as capable of stretching parts of the intellect pretty thoroughly, there are only two sorts of squares, each replicated thirty-two times, on which only twelve species of players having among them six different forms of movement and two colours perform in populations of not more than eight of any one sort. On Wytham Hill, described in the last chapter as a small sample of midland England on mostly calcareous soils but with a full range of wetness, there are something like a hundred kinds of "habitat squares" (even taken on a rather broad classification, and ignoring the individual habitat units provided by hundreds of separate species of plants) most of which are replicated inexactly thousands of times, though some only once or twice, and inhabited altogether by up to 5000 species of animals, perhaps even more, and with populations running into very many millions. Even the Emperor Akbar might have felt hesitation in playing a living chess game on the great courtyard of his palace near Agra, if each square had contained upwards of two hundred different kinds of chessmen. What are we to do with a situation of this magnitude and complexity? It seems, indeed it certainly is, a formidable operation to prepare a blueprint of its organization that can be used scientifically.

A variety of different measures have been used to evaluate ecological complexity, depending on the qualities of the ecosystem (Table 1.3). The most commonly used measure is the number of species in the ecosystem or some index relating the number of species and their relative abundances. Complexity can be overwhelming and it can inhibit the ability of ecologists to understand ecosystems. Therefore, very simple ecosystems are sometimes important and useful for study, such as those found in the hypersaline conditions of the Dead Sea or Great Salt Lake in Utah, where high salinity stress dissects away all but the very basic essence of ecological structure

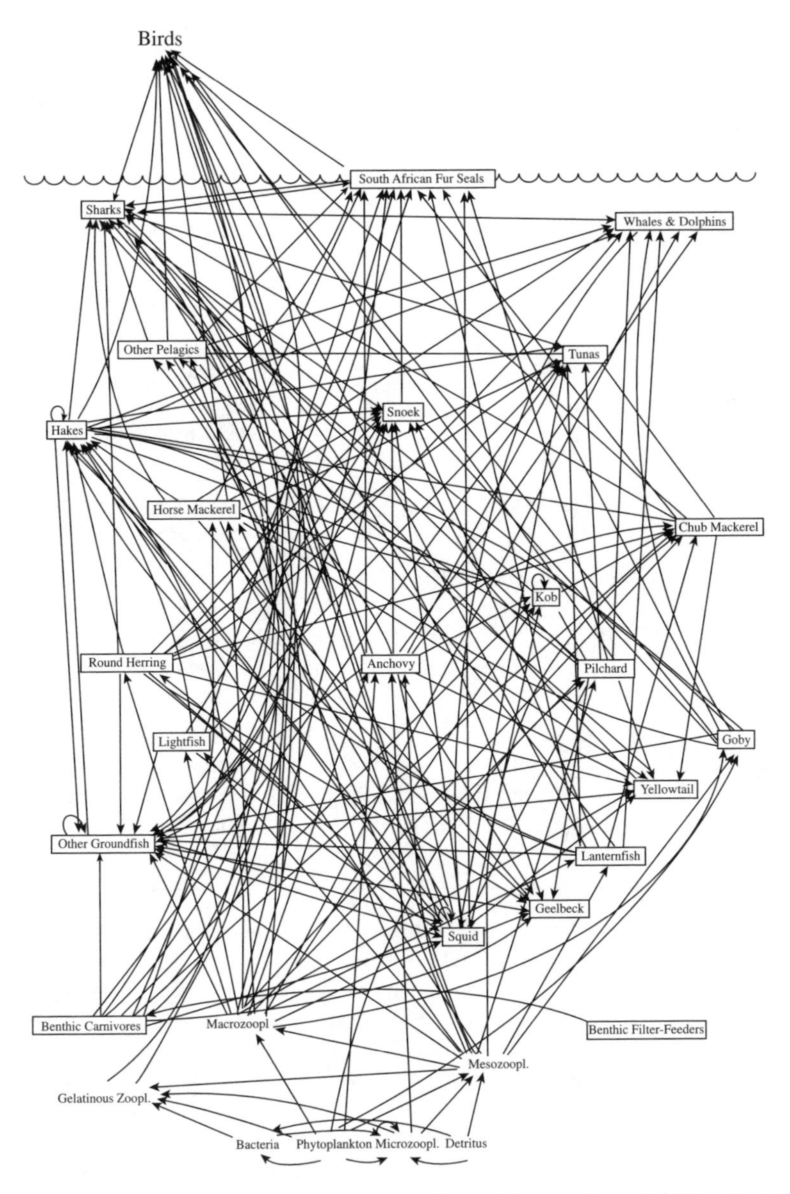

FIGURE 1.3 Diagram of a complex ecosystem. (From Abrams, P. et al. 1996. *Food Webs: Integration of Patterns and Dynamics*. Chapman & Hall, New York. With permission.)

and function. E. P. Odum (1959) described the qualities of simplicity in the following quote about his study site in the Georgia saltmarshes:

> The saltmarshes immediately struck us as being a beautiful ecosystem to study functionally, because over vast areas there is only one kind of higher plant in it and a relatively few kinds of macroscopic animals. Such an area would scarcely interest the

TABLE 1.3
Selected Indices for Estimating Different Conceptions of Complexity of Ecosystems

Index	Description	
Richness diversity (E. P. Odum, 1971)	S	where S = number of species
Shannon–Weaver diversity (E. P. Odum, 1971)	$-\Sigma (n_i/N) \log (n_i/N)$	where n_i = importance value for each species N = total of importance values
Pigment diversity (Margalef, 1968)	D430/D665	where D430 = optical absorption at 430 millimicrons D665 = optical absorption at 665 millimicrons
Food web connectance (Pimm, 1982)	L/[S(S–1)/2]	where L = actual number of links in a food web S = number of species in a food web
Forest complexity (Holdridge, 1967)	(S)(BA)(D)(H)/1000	where S = number of tree species BA = basal area of trees (m2/ha) D = density of trees (number of stems/ha) H = maximum tree height (m)
Ascendency (Ulanowicz, 1997)	$T\sum_{i,j}\left(\frac{T_{ij}}{T}\right)\log\left[\frac{T_{ij}T}{\sum_k T_{kj}\sum_m T_{im}}\right]$	where T = total system flow T_{ij} = flow of energy or materials from trophic category i to j T_{kj} = flow from k to j T_{im} = flow from i to m

field botanist; he would be through with his work in one minute; he would quickly identify the plant as *Spartina alterniflora*, press it, and be gone. Even the number of species of insects seems to be small enough so that one has hopes of knowing them all, something very difficult to do in most vegetation. … The strong tidal fluctuations and salinity variations cut down on the kinds of organisms which can tolerate the environment, yet the marshes are very rich. Lots of energy and nutrients are available and lots of photosynthesis is going on so that the few species able to occupy the habitat are very abundant. There are great masses of snails, fiddler crabs, mussels, grasshoppers and marsh wrens in this kind of marsh. One can include a large part of the ecosystem in the study of single populations. Consequently, fewer and more intensive sampling and other methods can be used. … In other words the saltmarsh is potentially to the ecologist what the fruit fly, *Drosophila*, is to the geneticist, that is to say, a system lending itself to study and experimentation as a whole. The geneticist would not select elephants to study laws and principles, for obvious reasons; yet ecologists have often attempted to work out principles on natural systems whose size, taxonomic complexity, or ecological life span presents great handicaps.

The science of ecology covers several hierarchical levels: individual organisms, species populations, communities, ecosystems, landscapes, and even the global scale. To some extent the science is fragmented because of this wide spectrum of hierarchical levels (Hedgpeth, 1978; McIntosh, 1985), and antagonistic attitudes arise sometimes between ecologists who specialize on one level. This situation is often the case between those studying the population and ecosystem levels. For example, some population ecologists do not even believe ecosystems exist because of their narrow focus on the importance of species to the exclusion of higher levels of organization. These kinds of antagonistic attitudes are counterproductive, and conscious efforts are being made to unify the science (Jones and Lawton, 1995; Vitousek, 1990). Ulanowicz (1981) likens the need for unification in ecology to the search for a unified force theory in physics (for gravitational, electromagnetic, and intranuclear forces), and he suggests network flow analysis as a solution. However, as noted by O'Neill et al. (1986): "Ecology cannot set up a single spatiotemporal scale that will be adequate for all investigations." In this regard, scale and hierarchy theories have been suggested as the key to a unified ecology (Allen and Hoekstra, 1992), but even this approach does not fully cover the discipline. Clearly, ecological engineers need more than just information on energy flow and nutrient cycles. Knowledge from all hierarchical levels of nature is required, and a flexible concept of the ecosystem is advocated in this book (Levin, 1994; O'Neill et al., 1986; Patten and Jørgensen, 1995; Pace and Groffman, 1998). Ecosystem science has become highly quantitative with the development of generalized models and relationships (DeAngelis, 1992; Fitz et al., 1996). Although not completely field tested and verified, this body of knowledge provides a basis for rational design of new, constructed ecosystems. Using analogies from physics, perhaps these models will fill the role of the "ideal gases" (Mead, 1971) or the "perfect crystals" that May (1973, 1974a) indicated in the following quote: "... in the long run, once the 'perfect crystals' of ecology are established, it is likely that a future 'ecological engineering' will draw upon the entire spectrum of theoretical models, from the very abstract to the very particular, just as the more conventional branches of science and engineering do today." In this text several well-known ecological models (such as the logistic population growth equation and the species equilibrium from island biogeography) are used throughout to provide a quantitative framework for ecological engineering design.

As a final aside to the discussion of the relationship of ecology to ecological engineering, an interesting situation has arisen with terminology. Lawton and others have begun referring to some organisms such as earthworms and beavers (Gurney and Lawton, 1996; Jones et al., 1994; Lawton, 1994; Lawton and Jones, 1995) as being "ecosystem engineers" because they have significant roles in structuring their ecosystems. While this is an evocative and perhaps even appropriate description, confusion should be avoided between the human ecological engineers and the organisms ascribed to similar function. In fact, this is an example of the fragmentation of ecology since none of the authors who discuss animals as ecosystem engineers seem to be aware of the field of human ecological engineering.

RELATIONSHIP TO ENGINEERING

The relation of ecological engineering to the overall discipline of engineering is not well developed, probably because most of the originators of the field have been primarily ecologists rather than engineers. This situation is changing rapidly but to a large extent the early work has been dominated by ecology. Ecological engineering

TABLE 1.4
Comparisons of Definitions of Engineering

Definition	Reference
The art and science of applying the laws of the natural sciences to the transformation of materials for the benefit of mankind	Futrell, 1961
The art of directing the great sources of power in nature for the use and convenience of man	1828 definition cited in Ferguson, 1992
The art and science by which the properties of matter and the energies of nature are made useful to man	Burke, 1970
The art of applying the principles of mathematics and science, experience, judgment, and common sense to make things which benefit people	Landis, 1992
The art and science concerned with the practical application of scientific knowledge, as in the design, construction, and operation of roads, bridges, harbors, buildings, machinery, lighting and communication systems, etc.	Funk & Wagnalls, 1973
The art or science of making practical application of the knowledge of pure sciences	Florman, 1976

draws on the traditional engineering method but, surprisingly, this method is relatively undefined, at least as compared with the scientific method. The contrast between science and engineering may be instructive for understanding the method used by engineers:

> "Scientists primarily produce knowledge. Engineers primarily produce things." (Kemper, 1982)
> "Science strives to understand how things work; engineering strives to make things work." (Drexler, 1992)
> "The scientist describes what is; the engineer creates what never was." (T. von Karrsan, seen in Jackson, 2001)

Thus, engineering as a method involves procedures for making useful things. This is confirmed by a comparison of definitions (Table 1.4). It is interesting to note that most of these definitions refer to engineering as an art and, to many observers, engineering can best be described as what engineers do, rather than by some formal set of operations arranged in a standard routine. McCabe and Eckenfelder (1958) outline the development of a hybrid "engineering science" in the following quote:

> Engineering, historically, originates as an art based on experience. Empiricism is gradually replaced by engineering science developed through research, the use of mathematical analysis, and the application of scientific principles. Today's emphasis in engineering, and in engineering education, is, and should be, on the development and use of the engineering science underlying the solution of engineering problems.

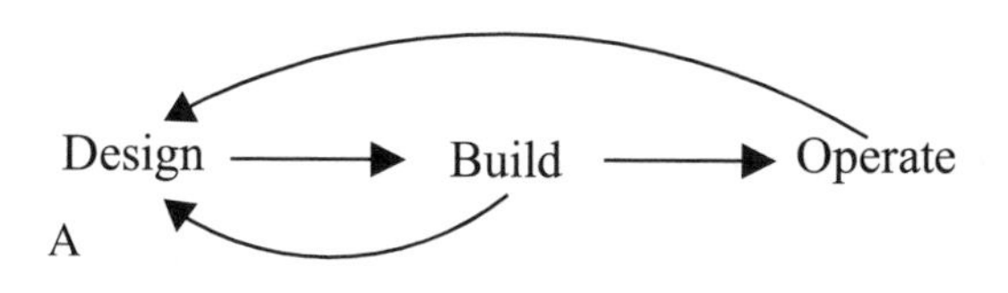

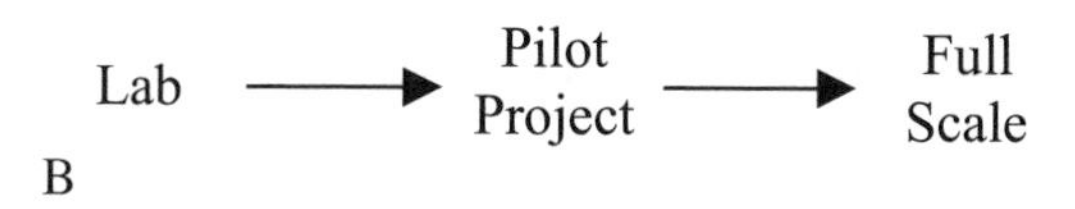

FIGURE 1.4 Views of the role of design in engineering. (A) The sequence of actions in engineering. Design is continually evaluated by comparison of performance in relation to design criteria. (B) Increasing scales of testing required for development of a successful design.

The critical work of engineering is to design, build, and operate useful things. Although different people are usually involved with each phase of this sequence, there is a constant feedback to the design activity (Figure 1.4A). Thus, it may be said that design is the essential element in engineering (Florman, 1976; Layton, 1976; Mikkola, 1993). Design is a creative process for making a plan to solve a problem or to build something. It involves rational, usually quantitatively based, decision making that utilizes knowledge derived from science and from past experience. A protocol is often used to test a design against a previously established set of criteria before full implementation. This protocol is composed of a set of tests of increasing scale (Figure 1.4B), which builds confidence in the choice of design alternatives. Horenstein (1999) provides a comparison of qualities of good vs. bad design that indicates the basic concerns in any engineering project (Table 1.5). A number of books have been written that describe the engineering method with a focus on design (Adams, 1991; Bucciarelli, 1994; Ferguson, 1992; Vincenti, 1990), and the work of Henry Petroski (1982, 1992, 1994, 1996, 1997a) is particularly extensive, including his regular column in the journal *American Scientist.*

Although design may be the essential element of engineering, other professions related to ecological engineering also rely on this activity as a basis. Obviously, architecture utilizes design intimately to construct buildings and to organize landscapes. As an example, Ian McHarg's (1969) classic book entitled *Design with Nature* has inspired a generation of landscape architects to utilize environmental sciences as a basis for design. *Design with Nature* is now a philosophical stance that describes how to interface man and nature into sustainable systems with applications which range from no-till agriculture to urban planning. Another important precursor for ecological engineering is Buckminster Fuller's "Comprehensive Anticipatory Design Science," which prescribes a holistic approach to meeting the needs of humanity by "doing more with less" (Baldwin, 1996; Edmondson, 1992; Fuller, 1963). Finally, many hybrid architect-scientist-engineers have written about ecolog-

TABLE 1.5
Dimensions of Engineering Design

Good Design	Bad Design
Works all the time	Works initially, but stops working after a short time
Meets all technical requirements	Meets only some technical requirements
Meets cost requirements	Costs more than it should
Requires little or no maintenance	Requires frequent maintenance
Is safe	Poses a hazard to user
Creates no ethical dilemma	Fulfills a need that is questionable

Source: Horenstein, M. N. 1999. *Design Concepts for Engineers.* Prentice Hall, Upper Saddle River, NJ. With permission.

ically based design which is fundamentally relevant for ecological engineering (Orr, 2002; Papanek, 1971; Todd and Todd, 1984, 1994; Van Der Ryn and Cowan, 1996; Wann, 1990, 1996; Zelov and Cousineau, 1997). These works on ecological design are perhaps not sufficiently quantitative to strictly qualify as engineering, but they contain important insights necessary for sound engineering practice.

The relationship between ecological engineering and several specific engineering fields also needs to be clarified. Of most importance is the established discipline of environmental engineering. This specialization developed from sanitary engineering (Okun, 1991), which dealt with the problem of treatment of domestic sewage and has traditionally been associated with civil engineering. The field has broadened from its initial start and now deals with all aspects of environment (Corbitt, 1990; Salvato, 1992). Ecological engineering is related to environmental engineering in sharing a concern for the environment but differs from the latter fundamentally in emphasis. There is a commitment to using ecological complexity and living ecosystems with technology to solve environmental problems in ecological engineering, whereas environmental engineering relies on new chemical, mechanical, or material technologies in problem solving. A series of joint editorials published in the journal *Ecological Engineering* and the *Journal of Environmental Engineering* provide further discussion on this relationship (McCutcheon and Mitsch, 1994; McCutcheon and Walski, 1994; Mitsch, 1994). Hopefully, ecological and environmental engineering can evolve on parallel tracks with supportive rather than competitive interactions. In practice, closer ties may exist between ecological engineering and the established discipline of agricultural engineering. As noted by Johnson and Phillips (1995), "agricultural engineers have always dealt with elements of biology in their practices." Because ecology as a science developed from biology, a natural connection can be made between ecological and agricultural engineering, using biology as a unifying theme. At the university level, this relationship is being strengthened as many agricultural engineering departments are broadening in perspective and converting into biological engineering departments.

DESIGN OF NEW ECOSYSTEMS

Ecological engineers design, build, and operate new ecosystems for human purposes. Engineering contributes to all of these phases but, as noted above, the design phase is critical. While the designs in ecological engineering use sets of species that have evolved in natural systems, the ecosystems created are new and have never existed before. Some names have been coined for the new ecosystems including "domestic ecosystems" (H. T. Odum, 1978a), "interface ecosystems" (H. T. Odum, 1983), and "living machines" (Todd, 1991). The new systems of ecological engineering are the product of the creative imagination of the human designers, as is true of any engineering field, but in this case the self-organization properties of living systems also make a contribution. This entails a natural selection of species appropriate for the boundary conditions of the design provided by the designer. Thus, ecologically engineered systems are the product of input from the human designer and from the system being designed, through the feedback of natural selection. This quality of the design makes ecological engineering a unique kind of engineering and an intellectually exciting new kind of applied ecology.

Many practical applications of ecological engineering exist, though often with different names (Table 1.6). The applications are often quite specific, and only time will tell if they will eventually fall under the general heading of ecological engineering. All of the applications in Table 1.6 combine a traditional engineering contribution to a greater or lesser extent, such as land grading, mechanical pump systems, or material support structures, with an ecological system consisting of an interacting set of loosely managed species populations. The best known examples of ecological engineering are those which require an even balance of the design between the engineering and the ecological aspects.

Environmental problem solving is a goal of ecological engineering, but only a subset of the environmental problems that face humanity can be dealt with by constructed ecosystem designs. Most amenable to ecological engineering may be various forms of pollution cleanup or treatment. In these cases, ecosystems are sought that will use the polluted substances as resources. Thus, the normal growth of the ecosystem breaks down or stabilizes the pollutants, sometimes with the generation of useful byproducts. This is a case of turning problems into solutions, which is an overall strategy of ecological engineering. Many examples of useful byproducts from ecologically engineered systems are described in this book.

An ecological engineering design relies on a network of species to perform a given function, such as wastewater treatment or erosion control. The function is usually a consequence of normal growth and behavior of the species. Therefore, finding the best mix of species for the design of a constructed ecosystem is a challenge. The ecological engineer must understand diversity to meet this challenge. Diversity is one of the most important concepts in the discipline of ecology (Huston, 1994; Patrick, 1983; Rosenzweig, 1995). Table 1.7 compares two ecosystems in order to illustrate the relative magnitudes of local species diversity. Globally, there are over a million species known to science, and estimates of undescribed species (mostly tropical rainforest insects) range up to 30 million (May, 1988; Wilson, 1988). Knowledge of taxonomy is critical for understanding diversity. This is the field of

TABLE 1.6
Listing of Applications of New Ecosystems in Ecological Engineering

Activity	Type of Constructed Ecosystem
Soil bioengineering	Fast growing riparian tree species for bank stabilization and erosion control
Bioremediation	Mixes of microbial species and/or nutrient additions for enhanced biodegradation of toxic chemicals
Phytoremediation	Hyperaccumulator plant species for metal and other pollutant uptake
Reclamation of disturbed lands	Communities of plants, animals, and microbes that colonize and restore ecological values
Compost engineering	Mechanical and microbial systems for breakdown of organic solid wastes and generation of soil amendments
Ecotoxicology	Ecosystems in microcosms and mesocosms for evaluating the effects of toxins
Food production	Facilities and species for intensive food production including greenhouses, hydroponics, aquaculture, etc.
Wetland mitigation	Wetland ecosystems that legally compensate for damage done to natural wetlands
Environmental education	Exhibits and/or experiments involving living ecosystems in aquaria or zoos
Wastewater treatment	Wetlands and other aquatic systems for degradation of municipal, industrial, or storm wastewaters

biology that systematically describes the relationships between species, including a logical system of naming species so that they can be distinguished.

Biodiversity is a property of nature that has been conceptually revised recently and is the main focus of conservation efforts. It has grown from the old concept of species diversity which has long been an important component of ecological theory. With the advent of the term, sometime in the 1980s, the old concept has been broadened to include other forms of diversity, ranging from the gene level to the landscape. This broadening was necessary to bring attention to all forms of ecological and evolutionary diversity, especially in relation to forces which reduce or threaten to reduce diversity in living systems. In a somewhat similar fashion, the term *biocomplexity* has recently been introduced (Cottingham, 2002; Michener et al., 2001), which relates to the old concept of complexity (see Table 1.3). To some extent

TABLE 1.7
Comparisons of Species Diversity of Two Ecosystems

Taxa	Mirror Lake, NH	Linesville Creek, PA
Algae	> 188	157
Macrophytes	37	"several"
Bacteria	> 150	> 8 ("not well-studied")
Fungi	> 20	32
Zooplankton and Protozoa	> 50	55
Macroinvertebrates	> 400	171
Fish	6	10
Reptiles and Amphibians	4–7	"several"
Birds	4–5	"several"
Mammals	2–5	1
TOTAL	> 850	> 434

Note: Mirror Lake data is from Likens (1992) and Linesville Creek data is from Coffman et al. (1971).

there is a shallowness to the trend of adding the prefix *bio* to established concepts that have existed for a relatively long time in ecology. However, the trend is positive because it indicates the growing importance of these concepts beyond the boundaries of the academic discipline. Biodiversity prospecting is the name given to the search for species useful to humans (Reid, 1993; Reid et al., 1993) and ecological engineers might join in this effort. The search for plant species that accumulate metals for phytoremediation is one example and others can be imagined.

Design of new ecosystems requires the creation of networks of energy flow (food chains and webs) and biogeochemical cycling (uptake, storage, and release of nutrients, minerals, pollutants) that are developed through time in successional changes of species populations. H. T. Odum (1971) described this design process in the following words:

> The millions of species of plants, animals, and microorganisms are the functional units of the existing network of nature, but the exciting possibilities for great future progress lie in manipulating natural systems into entirely new designs for the good of man and nature. The inventory of the species of the earth is really an immense bin of parts available to the ecological engineer. A species evolved to play one role may be used for a different purpose in a different kind of network as long as its maintenance flows are satisfied. The design of manmade ecological networks is still in its infancy, and the properties of the species pertinent to network design, such as storage capacity, conductivity, and time lag in reproduction, have not yet been tabulated. Because organisms may self-design their relationships once an approximately workable seeding

has been made, ecological network design is already possible even before all the principles are all known.

Species populations are the tools of ecological engineering, along with conventional technology. These are living tools whose roles and performance specifications are still little known. Yet these are the primary components used in ecological engineering, and designers must learn to use them like traditional tools described by Baldwin (1997): "A whole group of tools is like an extension of your mind in that it enables you to bring your ideas into physical form." Perhaps ecological engineers need the equivalent of the *Whole Earth Catalogs* which described useful tools and practices for people interested in environment and social quality (Brand, 1997). Of course, it is the functions and interactions of the species that are important. Ecosystems are made up of invisible networks of interactions (Janzen, 1988) and species act as circuit elements to be combined together in ecological engineering design.

An exciting prospect is to develop techniques of reverse engineering (Ingle, 1994) in order to add to the design capabilities of ecological engineering. This approach would involve study of natural ecosystems to guide the design of new, constructed ecosystems that more closely meet human needs. Reverse engineering is fairly well developed at the organismal level as noted by Griffin (1974):

> Modern biologists, who take it for granted that living and nonliving processes can be understood in the same basic terms, are keenly aware that the performances of many animals exceed the current capabilities of engineering, in the sense that we cannot build an exact copy of any living animal or functioning organ. Technical admiration is therefore coupled with perplexity as to how a living cell or animal can accomplish operations that biologists observe and analyze. It is quite clear that some "engineering" problems were elegantly solved in the course of biological evolution long before they were even tentatively formulated by our own species .... Practical engineering problems are not likely to be solved by directly copying living machinery, primarily because the "design criteria" of natural selection are quite different from those appropriate for our special needs. Nevertheless, the basic principles and the multifaceted ingenuity displayed in living mechanisms can supply us with invaluable challenge and inspiration.

This process has been termed either bionics (Halacy, 1965; Offner, 1995) or variations on biomimesis (McCulloch, 1962) such as biomimicry (Benyus, 1997) and biomimetics (Sarikaya and Aksay, 1995), and it is the subject of several texts (French, 1988; Vogel, 1998; Willis, 1995). Walter Adey's development of algal turf scrubber technology based on coral reef algal systems, which is described in Chapter 2, is a prime example of this kind of activity at the ecosystem level of organization, as is the new field of industrial ecology described in Chapter 6.

PRINCIPLES OF ECOLOGICAL ENGINEERING

As with all engineering disciplines, ecological engineering draws on traditional technology for parts of designs. These aspects are not covered in this book in order to focus more on the special aspects of the discipline which deal with ecological systems. Depending on the application, traditional technology can contribute up to about one half of the

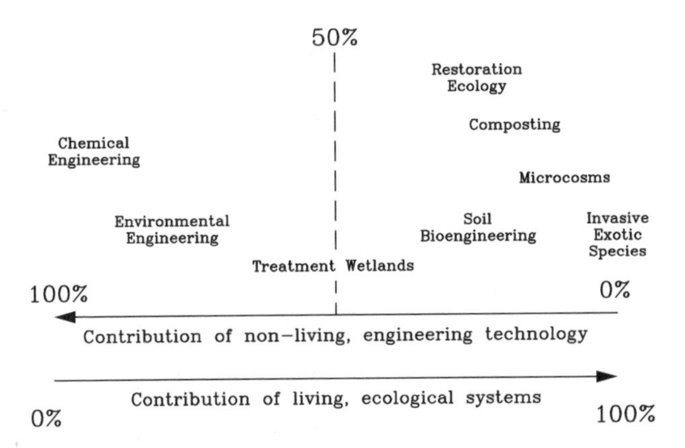

FIGURE 1.5 The realm of ecological engineering as defined by relative design contributions from traditional technology vs. ecological systems. Ecological engineering applications occur to the right of the 50% line. The six examples of ecological engineering applications covered in chapters of this book are shown with hypothetical locations in the design space. See also Mitsch (1998b).

design with the other portion contributed by the ecological system itself (Figure 1.5). Other types of engineering applications address environmental problems but with less contribution from nature. For example, conventional wastewater treatment options from environmental engineering use microbial systems but little other biodiversity, and chemical engineering solutions use no living populations at all. Case study applications of ecological engineering described in this book are shown in Figure 1.5 with overlapping ranges of design contributions extending from treatment wetlands, which can have a relatively even balance of traditional technology and ecosystem, to exotic species, which involve no traditional technology input. Three principles of ecological engineering design, common to all of the applications shown in Figure 1.5 and inherent in ecological systems, are described in Table 1.8.

TABLE 1.8
Principles for Ecological Engineering

Energy signature	The set of energy sources or forcing functions which determine ecosystem structure and function
Self-organization	The selection process through which ecosystems emerge in response to environmental conditions by a filtering of genetic inputs (seed dispersal, recruitment, animal migrations, etc.)
Preadaptation	The phenomenon, which occurs entirely fortuitously, whereby adaptations that arise through natural selection for one set of environmental conditions just happen also to be adaptive for a new set of environmental conditions that the organism had not been previously exposed to

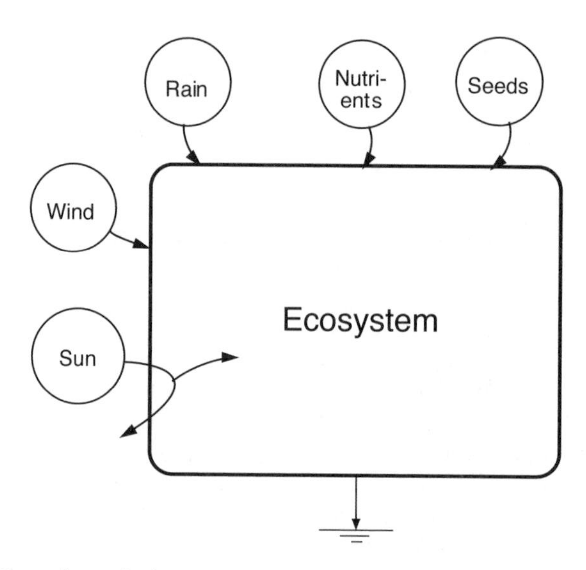

FIGURE 1.6 View of a typical energy signature of an ecosystem.

Energy Signature

The energy signature of an ecosystem is the set of energy sources that affects it (Figure 1.6). Another term used for this concept is *forcing functions*: those outside causal forces that influence system behavior and performance. H. T. Odum (1971) suggested the use of the energy signature as a way of classifying ecosystems based on a physical theory of energy as a source of causation in a general systems sense. A fundamental aspect of the energy signature approach is the recognition that a number of different energy sources affect ecosystems. Kangas (1990) briefly reviewed the history of this idea in ecology. Basically, sunlight was recognized early in the history of ecology as the primary energy source of ecosystems because of its role in photosynthesis at the level of the organism and, by extrapolation, in primary production at the level of the ecosystem. Organic inputs were formally recognized as energy sources for ecosystems in the 1960s with the development of the detritus concept, primarily in stream ecology (Minshall, 1967; Nelson and Scott, 1962) and in estuaries (Darnell, 1961, 1964; E. P. Odum and de la Cruz, 1963). The terms *autochthonous* (sunlight-driven primary production from within the system) vs. *allochthonous* (detrital inputs from outside the system) were coined in the 1960s to distinguish between the main energy sources in ecosystems. Finally, in the late 1960s H. T. Odum introduced the concept of *auxiliary energies* to account for influences on ecosystems from sources other than sunlight and organic matter. E. P. Odum (1971) provided a simple definition of this concept: "Any energy source that reduces the cost of internal self-maintenance of the ecosystem, and thereby increases the amount of other energy that can be converted to production, is called an auxiliary energy flow or an energy subsidy." H. T. Odum (1970) calculated the first energy signature for the rain forest in the Luquillo Mountains of Puerto Rico, which included values for 10 auxiliary energies.

From a thermodynamic perspective, energy has *the* ability to do work or to cause things to happen. Work caused by the utilization of the energy signature creates organization as the energy is dissipated or, in other words, as it is used by the system that receives it. Different energies (sun, wind, rain, tide, waves, etc.) do different kinds of work, and they interact in systems to create different forms of organization. Thus, each energy signature causes a unique kind of system to develop. The wide variety of ecosystems scattered across the biosphere reflect the many kinds of energy sources that exist. Although this concept is easily imagined in a qualitative sense, H. T. Odum (1996) developed an accounting system to quantify different kinds of energy in the same units so that comparisons can be made and metrics can be used for describing the energetics of systems. Other conceptions of ecology and thermodynamics are given by Weigert (1976) and Jørgensen (2001).

The one-to-one matching of energy signature to ecosystem is important in ecological engineering, where the goal is the design, construction, and operation of useful ecosystems. The ecological engineer must ensure that an appropriate energy signature exists to support the ecosystem that is being created. In most cases the existing energy signature at a site is augmented through design. Many options are available. Subsidies can be added, such as water, fertilizer, aeration, or turbulence, to direct the ecosystem to develop in a certain way (i.e., encourage wetland species by adding a source of water). Also, stressors can be added, such as pesticides, to limit development of the ecosystem (i.e., adding herbicides to control invasive, exotic plant species).

Self-Organization

Many kinds of systems exhibit self-organization but living systems are probably the best examples. In fact, self-organization in various forms is so characteristic of living systems that it has been largely taken for granted by biologists (though see Camazine et al., 2001) and is being "rediscovered" and articulated by physical scientists and chemists. Table 1.9 lists some of the major general systems themes emerging on self-organization. These are exciting ideas that are revolutionizing and unifying the understanding of both living and nonliving systems.

Self-organization has been discussed since the 1960s in ecosystem science (Margalef, 1968; H. T. Odum, 1967). It applies to the process by which species composition, relative abundance distributions, and network connections develop over time. This is commonly known as succession within ecology, but those scientists with a general systems perspective recognize it as an example of the larger phenomenon of self-organization. The mechanism of self-organization within ecosystems is a form of natural selection of those species that reach a site through dispersal. The species that successfully colonize and come to make up the ecosystem at a site have survived this selection process by finding a set of resources and favorable environmental conditions that support a population of sufficient size for reproduction. Thus, it is somewhat similar to Darwinian evolution (i.e., descent with modification of species) but at a different scale (see Figure 5.11). In fact, Darwinian evolution occurs within all populations while self-organization occurs between the populations within the ecosystem (Whittaker and Woodwell, 1972). Margalef (1984) has succinctly

TABLE 1.9
Comparison of Emerging Ideas on Self-Organization

Proponent	Conceptual Basis	System of Study
Stuart Kauffman (1995)	Systems evolve to the "edge of chaos," which allows the most flexibility; studied with adaptive "landscapes"	General systems with emphasis on biochemical systems
Per Bak (1996)	Self-organized criticality; studied with sand pile models	General systems with emphasis on physical systems
Mitchel Resnick (1994)	Emergence of order from decentralized processes; studied with an individual-based computer program called STAR LOGO	General systems
Manfred Eigen (Eigen and Schuster, 1979)	Hypercycles or networks of autocatalyzed reactions; studied with chemistry	Origin of life; biochemical systems
Ilya Prigogine (1980)	Dissipative structures; studied with nonequilibrium thermodynamics	General systems with emphasis on chemical systems
Francisco Varela (Varela et al., 1974)	Autopoiesis; studied with chemistry	Origin of life; biochemical systems

described this phenomenon: "Ecosystems are the workshops of evolution; any ecosystem is a selection machine working continuously on a set of populations."

H. T. Odum has gone beyond this explanation to build an energy theory of self-organization from the ideas of Alfred Lotka (1925). He suggests that selection is based on the relative contribution of the species to the overall energetics of the ecosystem. Successful species, therefore, are those that establish feedback pathways which reinforce processes contributing to the overall energy flow. H. T. Odum's theory is not limited to traditional ecological energetics since it allows all species contributions, such as primary production, nutrient cycling, and population regulation of predators on prey, to be converted into energy equivalent units. This is called the maximum power principle or Lotka's principle, and H. T. Odum has even suggested that it might ultimately come to be known as another law of thermodynamics if it stands the test of time as the first and second laws have. The maximum power principle is a general systems theory indicating forms of organization that will develop to dissipate energy, such as the autocatalytic structures of storages and interactions, hierarchies, and pulsing programs, which characterize all kinds of systems (H. T. Odum, 1975, 1982, 1995; H. T. Odum and Pinkerton, 1955). Belief in this theory is not necessary for acceptance of the importance of self-organization

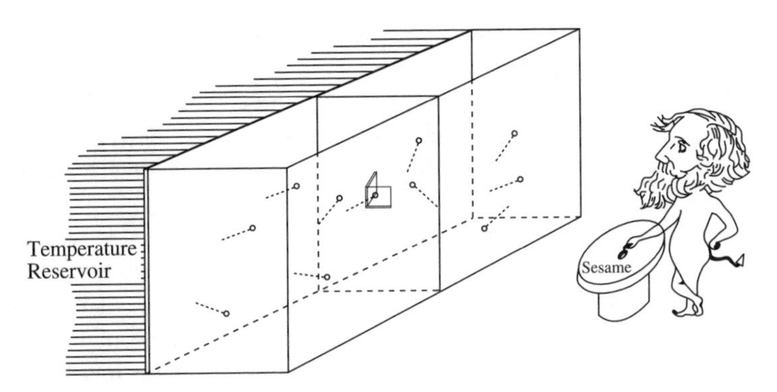

FIGURE 1.7 Maxwell's demon controls the movement of gas molecules in a closed chamber. (From Morowitz, H. J. 1970. *Entropy for Biologists, An Introduction to Thermodynamics.* Academic Press, New York. With permission.)

in ecosystems, and the new systems designed, built, and operated in ecological engineering will be tests of the theory.

According to H. T. Odum (1989a) "the essence of ecological engineering is managing self-organization" which takes advantage of natural energies processed by ecosystems. Mitsch (1992, 1996, 1998a, 2000) has focused on this idea by referring to self-organization as self-design (see also H.T. Odum, 1994a). With this emphasis he draws attention to the design element that is so important in engineering. Utilizing ecosystems, which self-design themselves, the ecological engineer helps to guide design but allows natural selection to organize the systems. This is a way to harness the biodiversity available to a design. For some purposes the best species may be known and they can be preferentially seeded into a particular design. However, in other situations self-organization may be used to let nature choose the appropriate species. In this case the ecological engineer provides excess seeding of many species and self-design occurs automatically. For example, if the goal is to create a wetland for treatment of a waste stream, the ecological engineer would design a traditional containment structure with appropriate inflow and outflow plumbing and then seed the structure with populations from other systems to facilitate self-organization of the living part of the overall design. Interaction of the waste stream with the species pool provides conditions for the selection of species best able to process and transform the waste flow.

The selection force in ecological self-organization may be analogous to an old paradox from thermodynamics (Figure 1.7). Maxwell's demon was the central actor of an imaginary experiment devised by J. Clerk Maxwell in the early days of the development of the field of thermodynamics (Harman, 1998; Klein, 1970). The tiny demon could sense the energy level of gas molecules around him in a closed chamber and operate a door between two partitions. He allowed fast-moving gas molecules to pass through the door and accumulate on one side of the chamber while keeping slow-moving molecules on the other side by closing the door whenever they came nearby. In this way he created order (the final gradient in fast and slow molecules) from disorder (the initial even distribution of fast and slow molecules) and cheated the second law of thermodynamics. In an analogous fashion, the force causing selection of species in self-organization may be thought to be the ecological equivalent of Maxwell's demon (H. T. Odum 1983). The ecological demon operates a metaphorical door through which species pass during succession, creating the orderly networks of ecosystems from the disorderly mass of species that reach a site through dispersal.

Self-organization is a remarkable property of ecosystems that is well known to ecologists (Jørgensen et al., 1998; Kay, 2000; Perry, 1995; Straskraba, 1999), but it is a new tool for engineers to use along with the other, more familiar tools of traditional technology. It will be very interesting to observe how engineers react to and come to assimilate the self-designing property of ecosystems into the engineering method as the discipline of ecological engineering develops over time. Control over designs is fundamental in traditional engineering as noted by Petroski (1995): "… the objective of engineering is control — getting things to function as we want them to in a particular situation or use." However, control over nature is not always possible or desirable (Ehrenfeld, 1981; McPhee, 1989). As noted by Orr (2002): "A rising tide of unanticipated consequences and 'normal accidents' mock the idea that experts are in control or that technologies do only what they are intended to do." Ecological engineering requires that some control over design be given up to nature's self-organization and this will require a new mind-set among engineers. Some positive aspects of systems that are "out of control" are discussed in Chapter 7.

PREADAPTATION

Self-organization can be accelerated by seeding with species that are preadapted to the special conditions of the intended system. This requires knowledge of both the design conditions of the ecosystem to be constructed and the adaptations of species. As an example, when designing an aquatic ecosystem to treat acid drainage from coal mines, seeding from a naturally acidic bog ecosystem should speed up self-design since the bog species are already adapted to acid conditions. Thus, the bog species can be said to be preadapted to fit into the design for acid mine drainage treatment because of their adaptations for acidity. Adaptation by species occurs through Darwinian evolution along environmental gradients (Figure 1.8) and in relation to interactions with other species (i.e., competition and predation). The adaptation curve in Figure 1.8 is bell-shaped since performance can only be optimized over a small portion of an environmental gradient. The biological mechanisms of adaptation include physiological, morphological, and behavioral features. One sense of a species' ecological niche is as the sum total of its adaptations. Hutchinson (1957, 1965, 1978) envisioned this concept as a hypervolume of space along environmental gradients on which a species can exist and reproduce. The niche is an important concept in ecology and reviews are given by MacArthur (1968), Schoener (1988), Vandermeer (1972), and Whittaker and Levin (1975). The concept covers all of the resources required by a species including food, cover, and space (see also the related concept of habitat discussed in Chapter 5). Each species has its own niche and only one species can occupy a niche according to the competitive exclusion principle (Hardin, 1960). As an aside, Pianka (1983) suggested that ecologists might

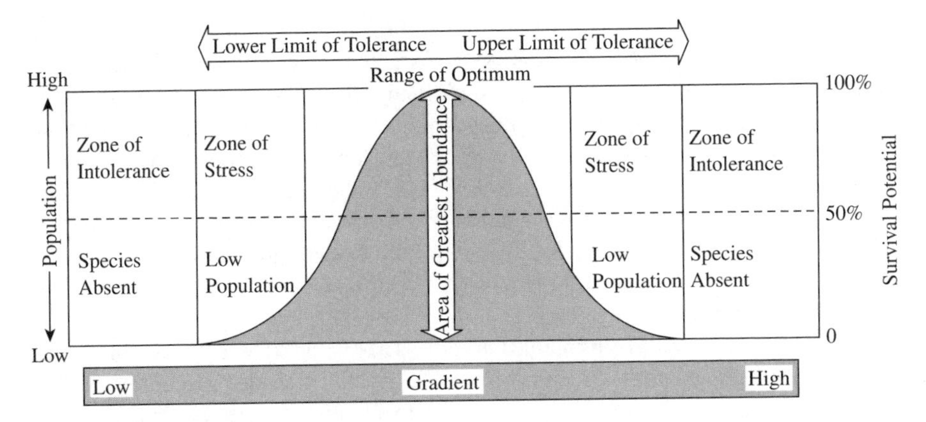

FIGURE 1.8 A performance curve for adaptation of a species along an environmental gradient. (From Furley, P. A. and W. W. Newey. 1988. *Geography of the Biosphere: An Introduction to the Nature, Distribution and Evolution of the World's Life Zones*. Butterworth & Co., London. With permission.)

develop periodic tables of niches, using Dimitri Mendeleev's periodic table of the chemical elements as a model. This creative idea provides a novel approach for dealing with ecological complexity but it has not been developed.

In contrast to the concept of adaptation, preadaptation is a relatively minor concept of evolutionary biology (Futuyma, 1979; Grant, 1991; Shelley, 1999). Wilson and Bossert (1971) describe it in terms of mutations which initially occur at random:

> In other words, within a population with a certain genetic constitution, a mutant is no more likely to appear in an environment in which it would be favored than one in which it would be selected against. When a favored mutation appears, we can therefore speak of it as exhibiting true preadaptation to that particular environment. That is, it did not arise as an adaptive response to the environment but rather proves fortuitously to be adapative after it arises. … Abundant experimental evidence exists to document the preadaptive nature of some mutants.

Preadaptations are then "preexisting features that make organisms suitable for new situations" (Vogel, 1998). E.P. Odum (1971) cited Thienemann (1926) who termed this the "taking-advantage principle," whereby a species in one habitat can take advantage of an adaptation that developed in a different habitat. Gould (1988) has criticized the name preadaptation as "being a dreadful and confusing term" because "it suggested foresight or planning in the evolutionary process" (Brandon, 1990). However, no such foresight or planning is implied and preadaptation is an apparently random phenomenon in nature. Gould suggests the term *exaptation* in place of preadaptation, but in this book the old term is retained.

Vogel (1998) has noted "preadaptation may be so common in human technology that no one pays it much attention." As an example, he notes that waterwheels in mills used to extract power from streams were preadapted for use as paddle wheels in the first generation of steamboats. Similarly, the use of preadapted species may become common in ecological engineering designs of the future. These species will accelerate the development of useful systems and lead to improved performance. Biodiversity prospecting and a knowledge of the niche concept will be needed to take advantage of these species. Rapport et al. (1985) give a table of preadaptations to stress in natural ecosystems. New systems developing with pollution are sources of preadapted species for treatment ecosystems. Likewise, invasive, exotic species often are successful due to preadaptation to human disturbance and can be seed sources for ecological engineering if permissible. Greater attention to the phenomenon of preadaptation can lead to new ways of thinking about biodiversity that may enrich both ecology and engineering.

In conclusion, the three principles described above provide a foundation for the new discipline of ecological engineering. The overall design procedure is (1) to provide an appropriate energy signature, (2) to identify species that may be preadapted to the design conditions and use them as a seed source, and (3) if preadaptated species cannot be identified, to introduce a diversity of species through multiple seeding into the system to facilitate self-organization.

STRATEGY OF THE BOOK

This book is intended to be a survey of the discipline of ecological engineering, rather than a design manual. One theme is to review examples of the new, ecologically engineered systems and to put them in the context of ecological concepts and theory. In this sense the book is an introduction to ecology for engineers. It is hoped that the science of ecology will provide suggestions for ways to improve the design of the wide range of ecologically engineered systems that are being built and tested. The book also should be relevant to ecologists as an introduction to the special, new ecosystems that are appearing with increasing frequency in many applications. While it is true that these are "artificial ecologies," the suggestion is made that ecology as an academic discipline can advance through their study.

The following six chapters focus on case study applications in ecological engineering. Examples of designs are described along with ecological details for each case study. A chapter also is included on economics which is critical for real-world implementation of the new designs of ecological engineering. Finally, a conclusion is presented with a theory of new ecosystems and prospects for the future of the discipline.

2 Treatment Wetlands

INTRODUCTION

The use of wetlands for treating wastewater is probably the best example of ecological engineering because the mix of ecology and engineering is nearly even. The idea is to use an ecosystem type (wetlands) to address a specific human need that ordinarily requires a great deal of engineering (wastewater treatment). This application of ecological engineering emerged in the early 1970s from a number of experimental trials and is today a growing industry based on a tremendous amount of experience as reflected by a large published literature. Although there is, of course, still much to be learned, the use of wetlands for wastewater treatment is no longer a novel, experimental idea, but rather an accepted technology that is beginning to mature and to diffuse throughout the U.S. and elsewhere. The focus of the chapter is on treatment of domestic sewage with wetlands, which was the first application of the technology, but many other kinds of wastewaters (urban stormwater runoff, agricultural and industrial pollution, and acid mine drainage) are now treated with wetlands.

Domestic sewage probably is the least toxic wastewater produced by humans and, in hindsight, it was logical that ecologists would choose it as the first type of wastewater to test for treatment with wetlands. The dominant parameters of sewage that require treatment are total suspended solids (TSS), organic materials measured by biological oxygen demand (BOD), nutrients (primarily nitrogen and phosphorus), and pathogenic microbes (primarily viruses and fecal coliform bacteria). In a sense wetlands are preadapted to treat these parameters in a wastewater flow because they normally receive runoff waters from surrounding terrestrial systems in natural landscapes. Wetlands are sometimes said to act as a "sponge" in absorbing and slowly releasing water flow and as a "filter" in removing materials from water flow; these qualities preadapt them for use in wastewater treatment.

STRATEGY OF THE CHAPTER

A principal purpose of this chapter is to review the history of the treatment wetland technology. This effort will search for the kinds of thinking that went on during the development of the technology and, thus, it will provide perspective on the nature of ecological engineering. This is important since ecological engineering is a new field with a unique approach that combines ecology and engineering. Hopefully, a careful examination of the history of this example will reveal aspects of the whole field. The chapter will not attempt to describe the state-of-the-art in wetland wastewater treatment, especially since this has been done so well by Kadlec and Knight (1996) and others. Rather, the emphasis will be on the early studies. Examination of these studies, which were conducted in the 1970s and which are the "ancestors"

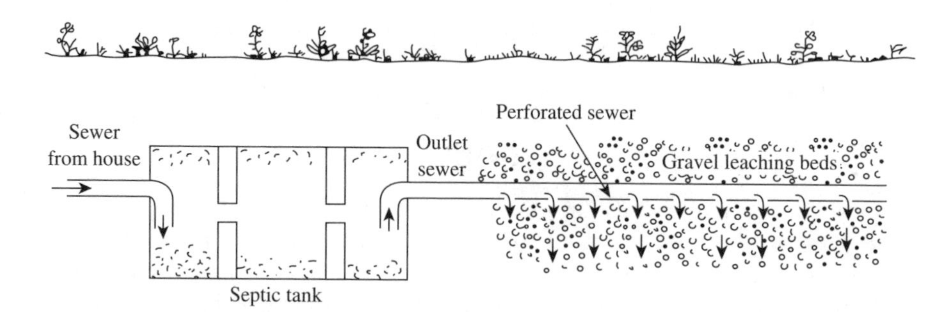

FIGURE 2.1 View of a septic tank and leaching bed. (From Clapham, W. B., Jr. 1981. *Human Ecosystems*. MacMillan, New York. With permission.)

of the present technology, should yield insight into the thought processes of ecological engineering.

A summary of the old field of sanitary engineering from which conventional sewage treatment technologies have evolved is described first. This is followed by a discussion of the history of use of wetlands for sewage treatment, including the proposal of hypotheses about where the original ideas came from and who had them. It is suggested that ecologists played the critical role in the development of treatment wetland technology and that engineering followed the ecology. The conceptual basis of treatment wetlands is covered and the role of biodiversity is discussed with emphasis on several important taxa. A comparison is made of mathematical equations used to describe analogous decay processes in ecology and sanitary engineering, which indicates similarities between the fields. Finally, two variations of treatment ecosystems are examined in detail to demonstrate the design process: Walter Adey's algal turf scrubbers and John Todd's living machines.

SANITARY ENGINEERING

Modern conventional methods of treating domestic sewage use a sequence of subsystems in which different treatment processes are employed. At the scale of the individual home, septic tanks with drain fields are used (Figure 2.1). This is a simple but remarkably effective system that is used widely (Kahn et al., 2000; Kaplan, 1991). Physical sedimentation occurs in the septic tank itself and the solid sludge must be removed periodically. Anaerobic metabolism by microbes occurs inside the tank, which initiates the breakdown of organic matter in the sewage. Liquids eventually flow out from the tank into a drain field of gravel and then into the surrounding soil where microbes continue to consume the organic matter and physical/chemical processes filter out pathogens and nutrients. The larger-scale sewage treatment plants (Figure 2.2) use similar processes for primary treatment (sedimentation of sludge) and secondary treatment (microbial breakdown of organic matter) in a more highly engineered manner. Processes can be aerobic or anaerobic depending on basic design features. Not shown in Figure 2.2 is a final treatment step, usually chlorination in most plants or use of an ultraviolet light filter, which eliminates pathogens. Note

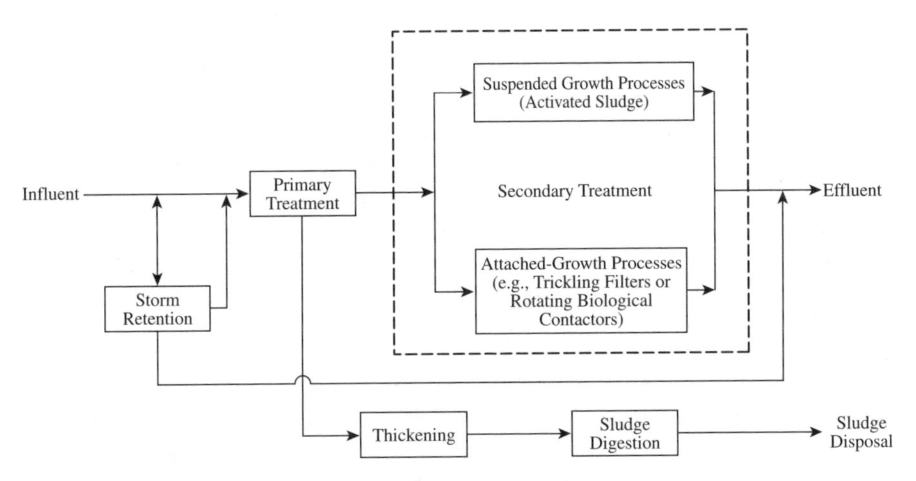

FIGURE 2.2 Processes that take place in a conventional wastewater treatment plant. (Adapted from Lessard, P. and M. B. Beck. 1991. *Environ. Sci. Technol.* 25:30–39.)

also that nutrients are not removed and are usually discharged in the effluent unless some form of tertiary treatment is employed.

The technologies discussed above are used throughout the world to treat human sewage and are the products of a long history of sanitary engineering design. Sawyer (1944), in an interesting paper which represents one of the first uses of the term *biological engineering*, traces the origins of the conventional technologies back to 19th century England and the industrial revolution, but the formal origin of the field of sanitary engineering seems to be the early 20th century United States. In his classic work on stream sanitation, Phelps (1944) places the origin at the research station of the U.S. Public Health Service, opened in 1913 in Cincinnati, Ohio. He calls this station an "exceptional example of the coordinated work of men trained in medicine, engineering, chemistry, bacteriology, and biology" which gives an indication of the interdisciplinary nature of this old field. The station was later named the Robert A. Taft Sanitary Engineering Center and it housed a number of important figures in the field.

Sanitary engineering developed the kinetic and hydraulic aspects of moving and treating sewage with characteristic engineering quantification. The field also involved a great deal of biology and even some ecology, which is particularly relevant in the context of the history of ecological engineering. Admittedly most of the biology has involved only microbes and, in particular, only bacteria (Cheremisinoff, 1994; Gaudy and Gaudy, 1966; Gray, 1989; James, 1964; Kountz and Nesbitt, 1958; Parker, 1962; la Riviere, 1977). Moreover, sanitary engineers seemed to have their own particular way of looking at biology as witnessed by their use of terms such as *slimes* (see Gray and Hunter, 1985; Reid and Assenzo, 1963). Even though this term is quite descriptive, a conventional biologist might think of it as too informal. Another example of their view of biology (see Finstein, 1972; Hickey, 1988 as examples) is the use of the name *sewage fungus* to describe not a fungus but a filamentous bacterium (*Sphaerotilus*) with a gelatinous sheath. Ecologists usually tend to be a bit more precise with biological taxonomy than this [though Hynes (1960) used the term *sewage fungus* in his seminal text on the biology of pollution]. These semantic issues are easily outweighed by the contributions of sanitary engineers to the biology and ecology of sewage treatment. It is significant that sanitary engineers were viewing sewage treatment much differently compared with conventional ecologists. To them sewage was an energy source and their challenge was to design an engineered ecosystem to consume it. This attitude is reflected in a humorous quote attributed to an "anonymous environmental engineer" that was used to introduce an engineering text (Pfafflin and Ziegler, 1979): "It may be sewage to you, but it is bread and butter to me." Meanwhile, more conventional ecologists wrote only on the negative effects of sewage on ecosystems as a form of pollution (Hynes, 1960; Warren and Doudoroff, 1971; Welch, 1980). Because of the negative perspective, this form of applied ecology was not a precursor to the treatment wetland technology.

One important example of classic sanitary engineering is the understanding of what happens when untreated sewage is discharged into a river. This was the state-of-the-art in treatment technology up to the 20th century throughout the world and it is still found in many lesser-developed countries. The problem was worked out by Streeter and Phelps (1925) and is the subject of Phelps' (1944) classic book. The river changes dramatically downstream from the sewage outfall with very predictable consequences in the temperate zone (Figure 2.3), in a pattern of longitudinal succession. Here succession takes the form of a pattern of species replacement in space along a gradient, rather than the usual case of species replacement in one location over time (see Sheldon, 1968 and Talling, 1958 for other examples of longitudinal succession). Streeter and Phelps developed a simple model that shows how the stream ecosystem treats the sewage (Figure 2.4). In the model, sewage waste creates BOD, which is broken down by microbial consumers. The action of the consumers draws down the dissolved oxygen in the river water resulting in the oxygen sag curve seen in both Figure 2.3 and Figure 2.4. Sewage is treated when BOD is completely consumed and when dissolved oxygen returns. This process has been referred to as natural purification or self-purification by a number of authors (McCoy, 1971; Velz, 1970; Wuhrmann, 1972). It is important because it conceptualizes how a natural ecosystem can be used to treat sewage wastewater and is a precursor to the use of wetland ecosystems for wastewater treatment.

Other early sanitary engineers contributed ecological perspectives to their field. A. F. Bartsch, who worked at the Taft Sanitary Engineering Center, wrote widely on ecology (Bartsch 1948, 1970; Bartsch and Allum, 1957). H. A. Hawkes was another author who contributed important early writings on ecology and sewage treatment (Hawkes, 1963, 1965). Many of the important early papers written by sanitary engineers were compiled by Keup et al. (1967), and Chase (1964) provides a brief review of the field.

Unlike most sanitary engineering systems, which focused solely on microbes, the trickling filter component of conventional sewage treatment plants has a high diversity of species and a complex food web. The trickling filter (Figure 2.5) is a large open tank filled with gravel or other materials over which sewage is sprayed. As noted by Rich (1963),

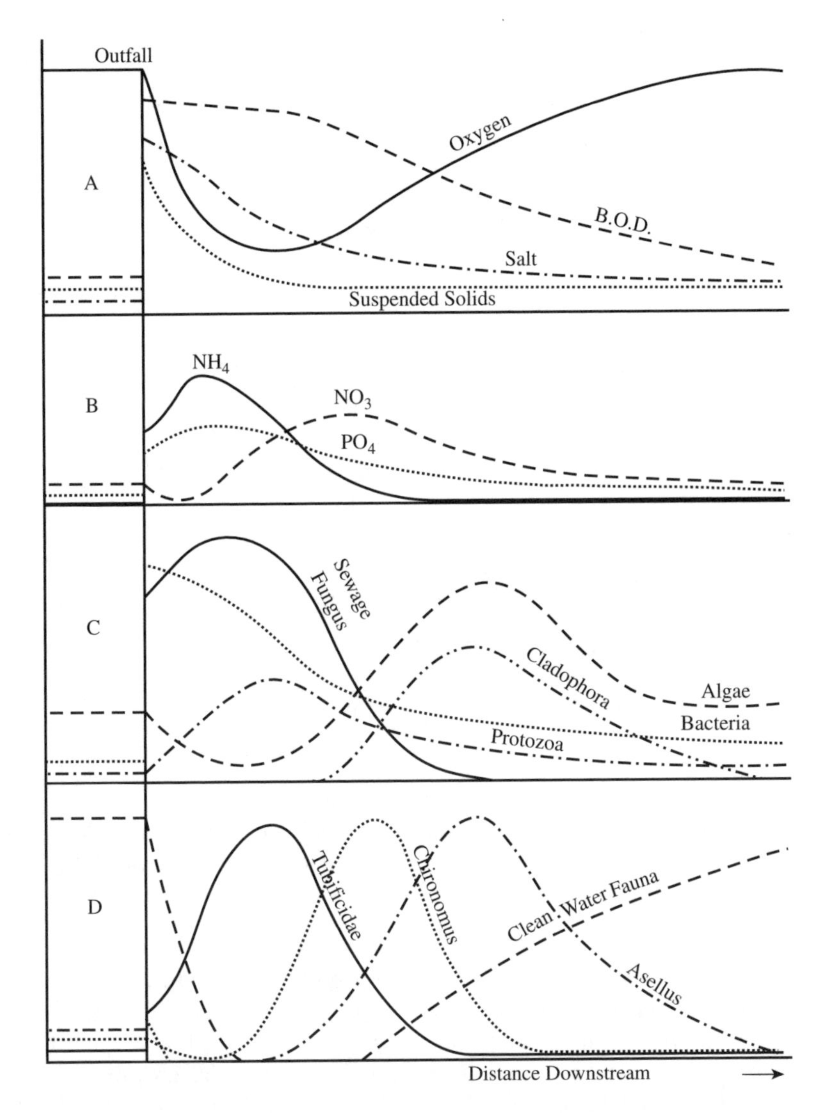

FIGURE 2.3 The longitudinal succession of various ecological parameters caused by the discharge of sewage into a river. A and B: physical and chemical changes; C: changes in microorganisms; D: changes in larger animals. (From Hynes, H. B. N. 1960. *The Biology of Polluted Waters*. Liverpool University Press, Liverpool, U.K. With permission.)

> The term "filter" is a misnomer, because the removal of organic material is not accomplished with a filtering or straining operation. Removal is the result of an adsorption process which occurs at the surfaces of biological slimes covering the filter media. Subsequent to their absorption, the organics are utilized by the slimes for growth and energy.

The gravel or other materials provide a surface for microbes that consume the organic material in sewage. The bed of gravel also provides an open structure that allows a

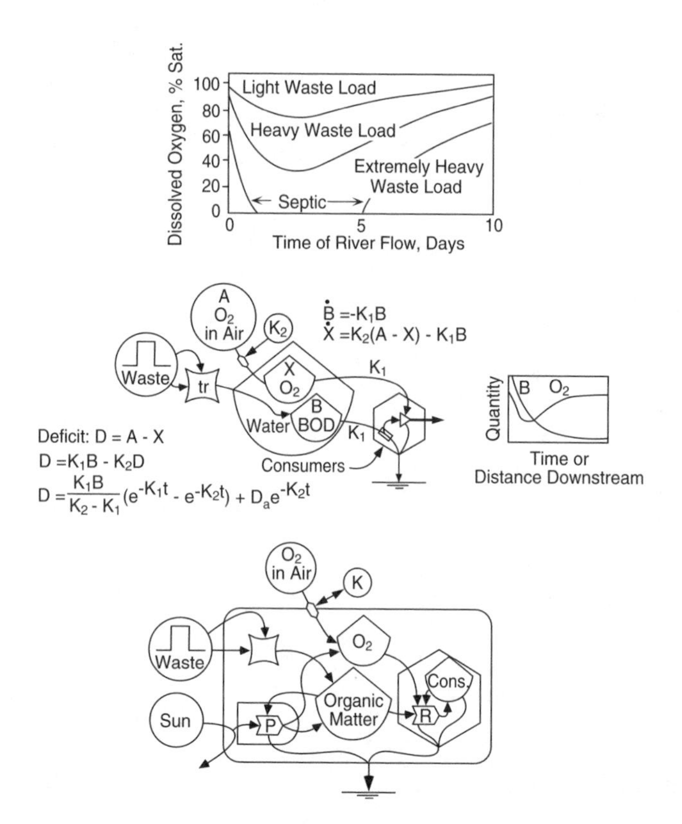

$$\dot{B} = -K_1B$$
$$\dot{X} = K_2(A - X) - K_1B$$
$$D = \frac{K_1B}{K_2 - K_1}(e^{-K_1t} - e^{-K_2t}) + D_ae^{-K_2t}$$

FIGURE 2.4 Several views of the Streeter–Phelps model of biodegradation of sewage in a river ecosystem. (From Odum, H. T. 1983. *Systems Ecology: An Introduction*. John Wiley & Sons, New York. With permission.)

free circulation of air for the aerobic metabolism of microbes, which is more efficient than anaerobic metabolism. A relatively high diversity of organisms colonizes the tank because it is open to the air. Insects, especially filter flies (*Pschodidae*), are important as grazers on the "biological slimes" (Sarai, 1975; Usinger and Kellen, 1955). For optimal aerobic metabolism the film of microbial growth should not exceed 2 or 3 mm, and the invertebrate animals in the trickling filter help to maintain this thickness through their feeding. The overall diversity of trickling filters is depicted with traditional alternative views of ecological energy flow in Figure 2.6 and Figure 2.7. The food web (Figure 2.6) describes the network of direct, trophic (i.e., feeding) interactions within the ecosystem. Both the topology of the food web networks (Cohen, 1978; Cohen et al., 1990; Pimm, 1982) and the flows within the networks (Higashi and Burns, 1991; Wulff et al., 1989) are important subjects in ecological theory. The trophic pyramid (Figure 2.7) describes the pattern of amounts of biomass or energy storage at different aggregated levels (i.e., trophic levels) within the ecosystem. Methods for aggregation of components, such as with trophic levels, are necessary in ecology in order to simplify the complexity of ecosystems. For example, a trophic level consists of all of the organisms in an ecosystem that feed at the same level of energy transformation (i.e., primary producers, herbivores,

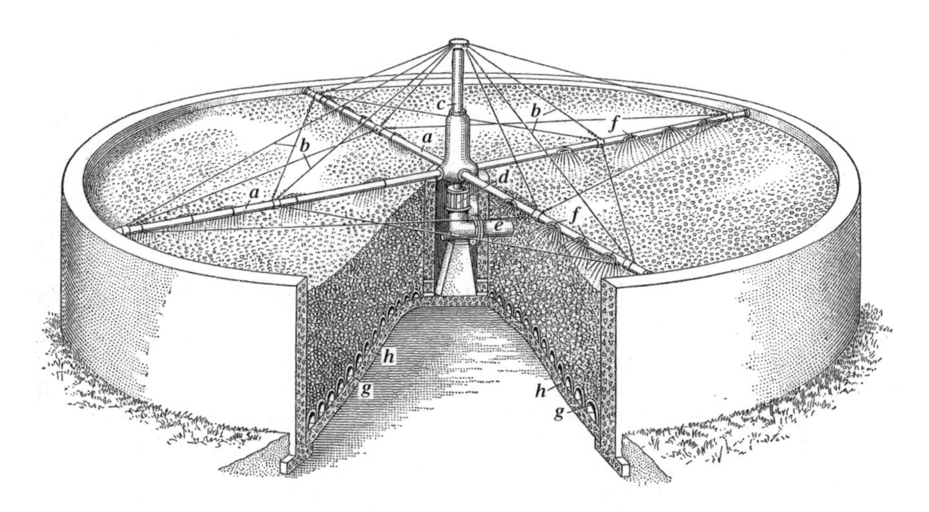

FIGURE 2.5 View of a typical trickling filter system. The distributor arms, a, are supported by diagronal rods, b, which are fastened to the vertical column. c. This column rotates on the base, d, that is connected to the inflow pipe. e. The sewage flows through the distributor arms and from there to the trickling filter by means of a series of flat spray nozzles, f, from which the liquid is discharged in thin sheets. The nozzles are staggered on adjacent distributor arms in order for the sprays to cover overlapping areas as the mechanism rotates. The bottom of the filter is underdrained by means of special blocks or half-tiles, g, which are laid on the concrete floor, h. (From Hardenbergh, W. A. 1942. *Sewerage and Sewage Treatment* (2nd ed.). International Textbook Co., Scranton, PA.)

primary carnivores, etc.). Magnitudes are shown visually on the trophic pyramid by the relative sizes of the different levels. A pyramid shape results because of the progressive energy loss at each level due to the second law of thermodynamics. Energy flow is an important topic in ecology though the concept of "flow" is an abstraction of the complex process that actually takes place. Colinvaux (1993) labels the abstraction of the complex process that actually takes place. Colinvaux (1993) labels the concept as a hydraulic analogy in reference to the simpler dynamics of water movements implied by the term, flow. McCullough (1979) articulated the abstraction more fully as follows,

> The problem concerns energy flux through the system; because it is unidirectional, and perhaps because of a poor choice of terminology, an erroneous impression has developed. Ecologists speak so glibly about energy flow that it is necessary to emphasize that energy does not "flow" in natural ecosystems. It is located, captured or cropped, masticated, and digested by organisms at the expense of considerable performance of work. Far from flowing, it is moved forcibly (and sometimes even screamingly) from one trophic level to the next.

Studies of energy flow, while imperfect in method, provide empirical measurements of ecological systems for making synthetic comparisons and for quantifying magnitudes of contributions of component parts to the whole ecosystem.

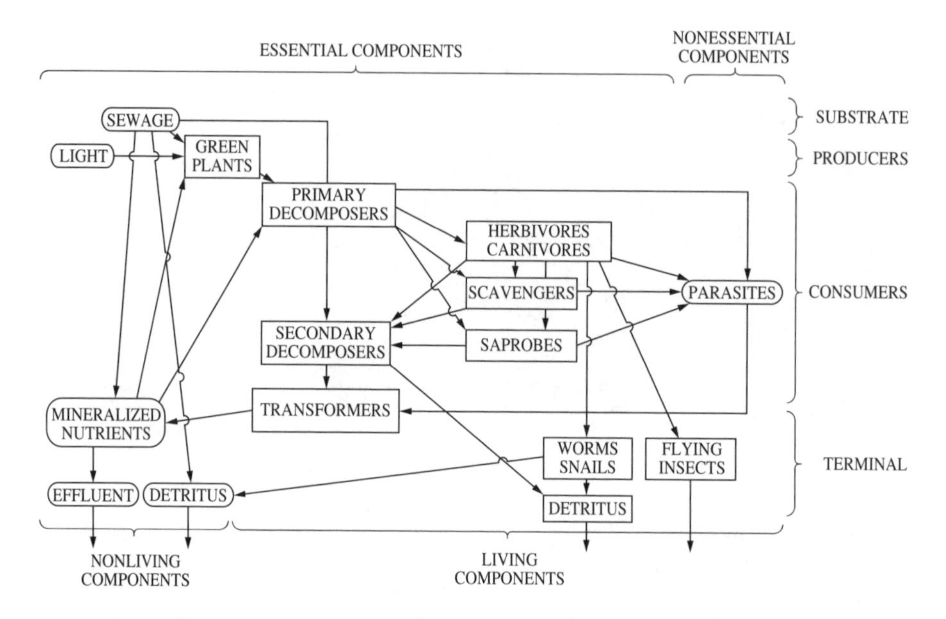

FIGURE 2.6 Food web diagram of a trickling filter ecosystem. (From Cooke, W.G. 1959. *Ecology.* 40:273–291. With permission.)

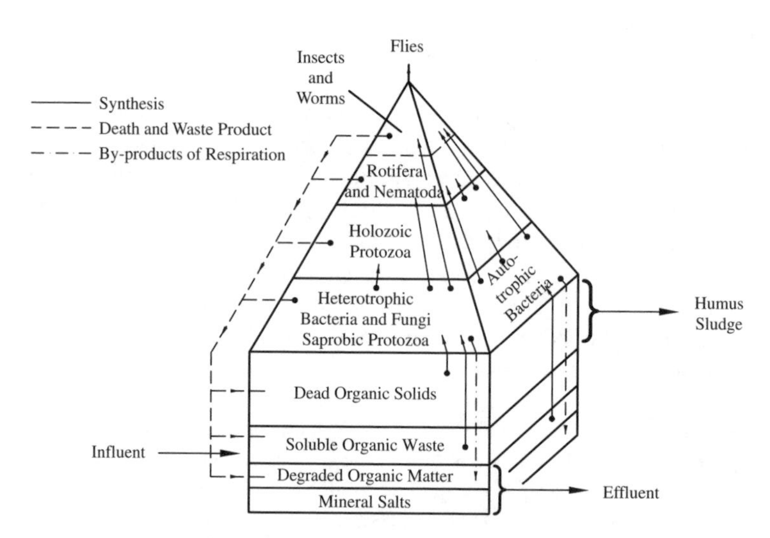

FIGURE 2.7 Trophic pyramid diagram of a trickling filter ecosystem. (From Hawkes, H.A. 1963. The Ecology of Waste Water Treatment. Macmillan, New York. With permission.)

The trickling filter is a fascinating ecosystem because of its ecological complexity and its well-known engineering details. Interestingly, Mitsch (1990), in a passing reference, suggested that some of the new constructed treatment wetlands have many characteristics of "horizontal trickling filters." Perhaps a detailed study of the old

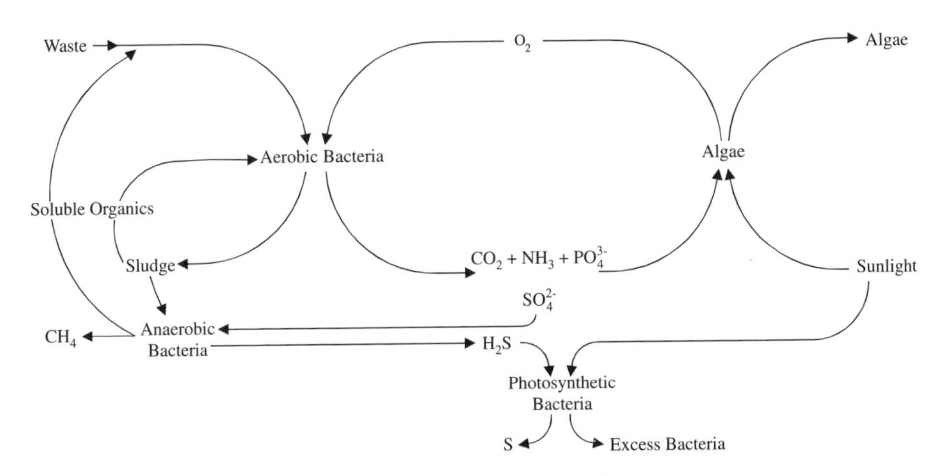

FIGURE 2.8 Metabolic cycling that takes place in oxidation stabilization ponds during wastewater treatment. (Adapted from Oswald, W. J. 1963. *Advances in Biological Waste Treatment.* W. W. Eckenfelder, Jr. and J. McCabe (eds.). MacMillan, New York.)

trickling filter literature will provide useful design information for future work on treatment wetlands.

Other treatment systems have evolved that have more direct similarity to wetlands (Dinges, 1982). Oxidation or waste stabilization lagoons are simply shallow pools in which sewage is broken down with long retention times (Gloyna et al., 1976; Mandt and Bell, 1982; Middlebrooks et al., 1982). This is a very effective technique that relies on biotic metabolism for wastewater treatment (Figure 2.8). Perhaps even closer to the wetland option is land treatment in which sewage is simply sprayed over soil in a grassland or forest (Sanks and Asano, 1976; Sopper and Kardos, 1973; Sopper and Kerr, 1979). In this system sewage is treated as it filters through the soil by physical, chemical, and biological processes.

AN AUDACIOUS IDEA

The use of wetlands for wastewater treatment was begun in the early 1970s. Whose idea was this? It is important to understand the origin of this application since it will reveal information on the nature of ecological engineering. One hypothesis is that the origin of treatment wetlands was a result of the technological progress of sanitary engineering systems (Figure 2.9). This is a reasonable hypothesis in that the pathways require no especially dramatic technical jumps and in each case ecosystems are used to consume the sewage. Of course, sewage was originally just released into streams as Streeter and Phelps had studied in the early 1900s. This is exactly the same approach taken with wetlands in the 1970s but with one treatment ecosystem (the river) being changed for another (the wetland). Although this hypothesis is reasonable, there is much more to the history.

Rather than a gradual progression of technological steps, there was an explosion of ideas, all at about the same time, for combining wetlands and sewage for waste-

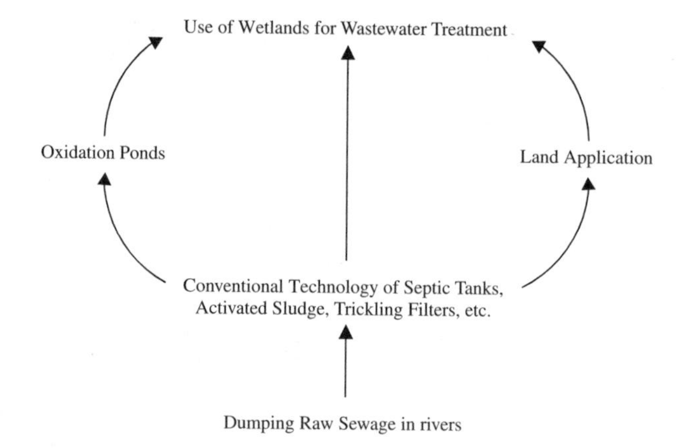

FIGURE 2.9 Hypothetical pathways of technological evolution of the use of wetlands for wastewater treatment from sanitary engineering systems.

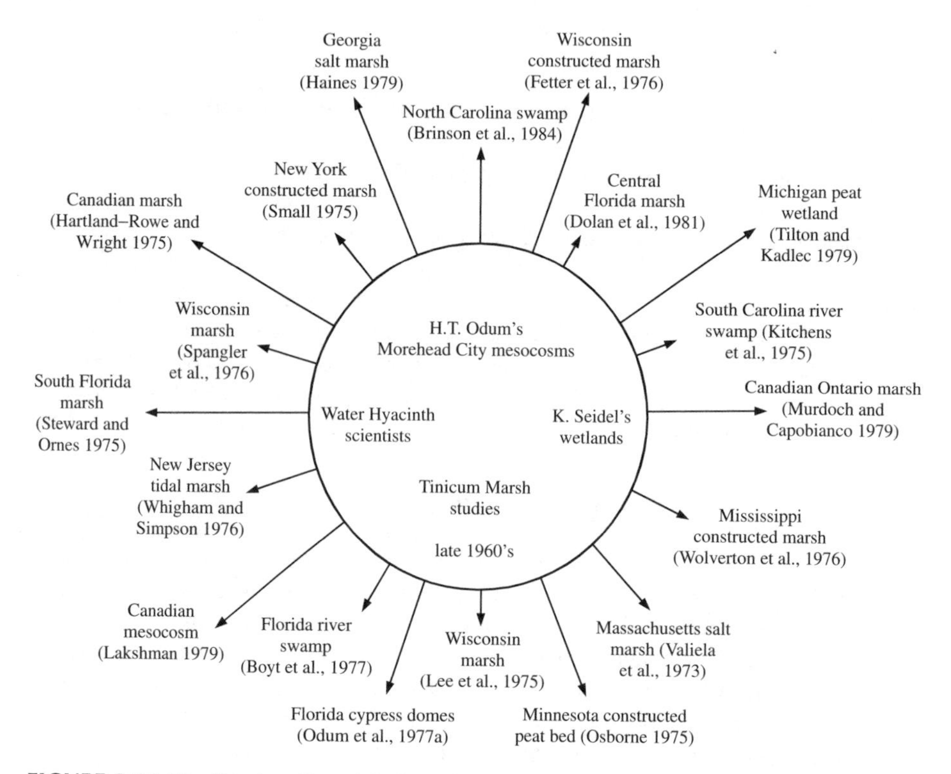

FIGURE 2.10 The "big-bang" model of a technological explosion of early treatment wetland projects.

water treatment (Figure 2.10). An examination of the literature shows that, starting in the early 1970s and extending through the decade, a large number of studies were conducted over a relatively short period of time to test wetlands as a system for wastewater treatment. This is shown in Figure 2.10 with references scattered around a central core of possible antecedent studies. The model that is represented in this figure is a kind of "big bang" explosion of creative trials of the idea of using wetlands for wastewater treatment. This kind of model has been proposed by Kauffman (1995) for technological jumps. He uses an analogy with the evolutionary explosion that took place at the start of the Cambrian era when many of the modern taxonomic groups of organisms appeared suddenly in a kind of creative explosion of biodiversity. In the same sense there was an explosion of studies on wetlands for wastewater treatment in the 1970s and the present state of the art in this technology traces back to this creative time.

What might have triggered this explosion of studies? Several authors have proposed that the Clean Water Act, which was passed in 1972, may have been an important influence (Knight, 1995; Reed et al., 1995). The most significant aspect of this legislation may have been the shifting emphasis in research funding towards alternative treatment technologies. However, the general intention of the Act was to reduce pollutant loads to natural systems, not to increase them as occurs when treating wastewater with wetlands. It seems unlikely, moreover, that either an act of legislation or even increased research funding were the actual triggers to the explosion of studies, because these are not strong motivators of scientific advancement. In fact, there must have been a kind of sociopolitical resistance against putting wastewater into natural wetlands from several sources in the early 1970s. First, the environmental movement was growing, and environmentalists sought to preserve wilderness and to oppose any changes in natural systems caused by human actions. This movement took definite form with the first Earth Day celebration in April 1970, almost at the exact beginning of trials of wastewater treatment with wetlands. Second, society as a whole in the U.S. had just come to recognize cultural eutrophication as a significant issue (Bartsch, 1971; Beeton and Edmondson, 1972; Hutchinson, 1973; Likens 1972). Eutrophication, or the aging of an aquatic ecosystem through filling in with inorganic and organic sediments, is a natural phenomenon (actually a form of ecological succession). However, humans can accelerate this process through additions of nitrogen and phosphorus found in various kinds of wastewater (i.e., cultural eutrophication). Finally, in addition to the obstacles mentioned above, there was a normal resistance to the idea of using wetlands to treat wastewater, resistance that always occurs when a new technology is introduced. This was led by sanitary engineers who utilized conventional treatment technologies and by government officials who regulate the industry, and it continues in the present. Thus, the use of wetlands to treat domestic sewage was an audacious idea in the early 1970s, which faced many hurdles (Figure 2.11). The only positive influence may have been the first energy crisis in 1973, which provided the incentive for reducing costs in many sectors of the economy (K. Ewel, personal communication). In retrospect, it seems somewhat amazing that the idea was allowed to be tested at all.

The use of wetlands to treat wastewater came from an intellectually courageous group of ecologists who saw the positive dimension of the idea (as a form of ecological engineering) and who were not held back by the negative dimension (that it represented intentional pollution of a natural ecosystem type in order to treat wastewater). The concept seems to have arisen from at least four specific antecedent

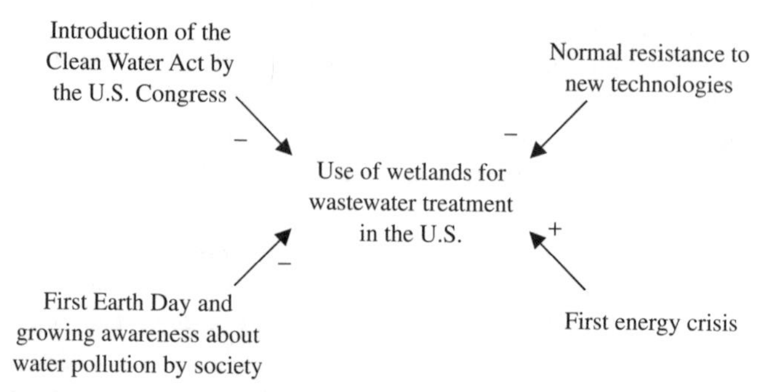

FIGURE 2.11 Causal diagram of sociopolitical influences on the development of the treatment wetland technology in the U.S. during the early 1970s.

activities that appeared in the late 1960s, as shown in the center of Figure 2.10. Bastian and Hammer (1993), Kadlec and Knight (1996), and Knight (1995) provide some discussion of the history of the treatment wetland technology, and they note the possible early influence of several of these antecedent works. These early initiatives are especially important because they predate the early 1970s explosion of studies. Short descriptions of these are given below:

1. Tinicum Marsh is a natural, freshwater tidal marsh near Philadelphia, PA. It is dominated by wild rice (*Zizania aquatica*) and common reed (*Phragmites australis*) and has been highly altered by a variety of human impacts. In the late 1960s the marsh became the focus of a conservation struggle over its value as open space within the urban setting and several studies were conducted on its ecology. One study by Ruth Patrick reviewed the marsh's ability to improve water quality. The findings showed significant reductions in BOD and in nitrogen and phosphorus from the effluent discharge of a nearby sewage treatment plant. The data on water quality improvement owing to the marsh became one of the political arguments for preserving it as urban open space. This example of an inadvertent discharge was the first of many similar studies made in the 1970s. Information on Tinicum Marsh is given by McCormick (1971), by Goodwin and Niering (1975), and in an original contract report by Grant and Patrick (1970).
2. Water hyacinths (*Eichhornia crassipes*) are floating plants of tropical origin that have very high productivity. This quality causes them to act as weeds in clogging waterways and much research has gone into developing methods for controlling their growth. In the late 1960s and early 1970s a number of workers sought to take advantage of the water hyacinth's fast growth rates by testing out possible wastewater treatment designs (Boyd, 1970; Rogers and Davis, 1972; Scarsbrook and Davis, 1971; Sheffield, 1967; Steward, 1970). The concept is to grow water hyacinths on sewage effluent and periodically harvest their biomass. Large amounts of nutrient could be stripped from the water as a result of uptake

driven by the high productivity. These early studies were continued through the 1970s (Cornwell et al., 1977; Taylor and Steward, 1978; Wooten and Dodd, 1976), and they also led to modifications such as by Wolverton and McDonald (1979a, 1979b).

3. Professor Kathe Seidel was a German scientist who started experimenting with the use of wetland plants for various kinds of wastewater treatment in the 1950s at the Max Planck Institute. Seidel seems to have been the first worker to test the concept of treatment wetlands and she published extensively in German (Seidel, 1966). Unfortunately, her work did not become widely known to western scientists until a publication appeared in English in the early 1970s (Seidel, 1976).
4. H. T. Odum ran a large project, which began in 1968, on testing the effects of domestic sewage on estuarine ecosystems at Morehead City, NC (H. T. Odum, 1985, 1989b). Experimental ponds that received sewage were compared with control ponds that received fresh water. The results indicated that sewage ponds had lower diversity of species and other characteristics of cultural eutrophication (algal blooms, extremes in oxygen concentrations) relative to controls, but both systems self-organized ecological structure and function with available species. This experiment did not deal with treating sewage specifically but rather with sewage effects as a pollutant. This focus is indicated by H. T. Odum's placement of the study in his text on microcosms (Beyers and H. T. Odum, 1993) not under the "wastes" chapter but under the chapter on "ponds and pools." However, H. T. Odum's later project on cypress swamps for wastewater treatment in the 1970s (Ewel and H. T. Odum, 1984) clearly traces back to the Morehead City project, as noted by Knight (1995), who served as a young research assistant studying the estuarine ponds. H. T. Odum seems to have had even earlier premonitions on the treatment wetland idea while working on the Texas coast in the 1950s, as indicated by the following quote from Montague and H. T. Odum (1997):

> A serendipitous example one of us (HTO) has observed over some years is the sewage waste outflow from a small treatment plant at Port Aransas, Texas. Wastes were released to a bare sand flat starting about 1950. As the population grew, wastes increased. Now there is an expansive marsh with a zonation of species outward from the outfall. Freshwater cattail marsh occurs immediately around the outfall. Beyond that is a saltmarsh of Spartina and Juncus through which the wastewaters drain before reaching adjacent coastal waters.

These four projects or lines of research seem to have set the stage for or actually triggered the explosion of studies in the 1970s. Apparently, the idea arose in scientists' minds to try wetlands for wastewater treatment and then positive feedback occurred as other scientists got caught up in trying the approach with different kinds of designs. Table 2.1 summarizes the early published studies according to their basic research design. Although there is a balanced representation between types of studies, the inadvertent experiment was the most common kind of study. In this approach

TABLE 2.1
Classification of Early Treatment Wetland Studies

(A) Natural Wetlands

(1) Inadvertent Experiment

Wisconsin marsh (Spangler et al., 1976)
Wisconsin marsh (Lee et al., 1975)
Canadian Northwest Territories (NWT) marsh (Hartland-Rowe and Wright, 1975)
Canadian Ontario marsh (Murdoch and Capobianco, 1979)
South Carolina river swamp (Kitchens et al., 1975)
Florida river swamp (Boyt et al., 1977)

(2) Purposeful Additions of Actual Sewage

New Jersey tidal marsh (Whigham and Simpson, 1976)
Florida cypress dome (Odum et al., 1977a)
Michigan peat wetland (Tilton and Kadlec, 1979)
Central Florida marsh (Dolan et al., 1981)
North Carolina swamp (Brinson et al., 1984)
Georgia saltmarsh (Haines, 1979)

(3) Addition of Simulated Sewage

Massachusetts saltmarsh (Valiela et al., 1973)
South Florida marsh (Steward and Ornes, 1975)

(B) Constructed Wetlands

(4) Pilot Scale System

New York constructed marsh (Small, 1975)
Minnesota constructed peat bed (Osborne, 1975)
Mississippi constructed marsh (Wolverton et al., 1976)
Wisconsin constructed marsh (Fetter et al., 1976)

(5) Mesocosm

Canadian Saskatchewan marsh (Lakshman, 1979)

Note: References are from Figure 2.10.

a study was made of the performance of a natural wetland that had been receiving sewage for a number of years. The situation arises when sewage is discharged inadvertently (and illegally) into a natural wetland. This kind of study has advantages of showing long-term performance, but there is no experimental control and no replication. All of the other kinds of studies listed in Table 2.1 have various degrees of experimental design, though complications often arose. Particularly interesting are the studies that used simulated sewage. The studies listed in Table 2.1 were field studies, which is ecology at its best. Problems occur in such experiments but they are views of how nature responds in the real world. In each case the systems of wetlands and sewage that emerged were new systems with altered biogeochemistry, different relative abundances of plants, animals and microbes, and new food web structures. The ecosystems self-organize from available components into new systems that are partly engineered and partly natural. The engineered subsystems range from simple deployments of pipes and pumps that discharge sewage into an existing wetland to complicated constructed wetlands that are actually hybrids of machine and ecosystem with multiple units in series and parallel connections and with sophisticated flow regulation devices. Some of the studies, such as the cypress project in Florida, were well funded and resulted in many publications about various aspects of the treatment wetland system. Other studies were represented by only a single publication with little system description except some water quality data. Most of the studies were short term and "died out" while a few continued to develop and are represented in the present-day technology. This seems reminiscent of the early automobile industry in Detroit, Michigan, around the turn of the twentieth century when many new auto designs were built and tested by small and large companies (Clymer, 1960). The innovators in the early automobile industry were mechanics who were able to coevolve with entrepreneurs and who in turn could mold and adapt existing technology (such as bicycles). The innovators of the treatment wetland technology were ecologists who were able to coevolve with engineers and regulators and who could mold and adapt wetland ecosystems with existing conventional wastewater treatment technology. An important exception is Robert Kadlec, who is one of the few early workers trained as an engineer rather than as an ecologist. Kadlec has continued his study of sewage treatment by a natural Michigan peatland for three decades, and he is a leader in creating quantitative design knowledge on treatment wetlands (Kadlec and Knight, 1996).

A kind of modest industry has evolved out of the early wetlands for wastewater treatment studies of the 1970s. Table 2.2 offers a hypothetical description of this evolution with speculations for the future. After the period of "optimism and enthusiasm" of the 1970s, problems with the technology began to appear. The best example may be problems with the capacity for long-term phosphorus uptake that have been reviewed extensively by Curtis Richardson (1985, 1989; Richardson and Craft, 1993). These kinds of problems are being addressed and the field is moving forward. It appears the technology will continue to grow into a viable commercial scale industry that will rival conventional treatment technologies, especially for rural or other relatively specialized situations.

THE TREATMENT WETLAND CONCEPT

Basically, the same physical/chemical/biological processes are used to treat domestic sewage in both conventional wastewater treatment plants and treatment wetland systems. The differences occur mainly in dimensions of space and time: wetlands

TABLE 2.2
Stages in the Evolution of the Treatment Wetland Technology

1970s	"Optimism and Enthusiasm"
	An explosion of ideas takes place; tests are performed in a variety of wetland types using different experimental strategies.
1980s	"Caution and Skepticism"
	Many of the original studies are discontinued; long-term treatment ability (especially for phosphorus removal) is questioned (see Richardson's many papers and Kadlec's "aging" concept); many review papers are written.
1990s	"Maturation"
	An almost exclusive emphasis emerges on the use of constructed wetlands rather than natural wetlands for wastewater treatment; Kadlec and Knight's book entitled *Treatment Wetlands* is published; management ideas evolve to address limitations brought up in the 1980s.
2000s	"Commercialization"
	The technology of treatment wetlands expands, especially in less developed countries throughout the world; constructed wetlands become a widely accepted alternative technology for certain scenarios of wastewater treatment.

need significantly more space and more time than conventional plants to provide treatment. The trade-off is economic with the wetlands option being cheaper in utilizing a higher ratio of natural vs. purchased inputs (Figure 2.12), at least conceptually.

A key factor in wastewater treatment is hydraulic residence time, as noted by Knight (1995):

> The typical hydraulic residence time in a modern AWT (advanced wastewater treatment) plant is about 12 hr., and solids residence time might be only about 1–2 days. In a typical treatment wetland, the minimum hydraulic residence time is greater than 5 days and in some is over 100 days. Solids residence time is typically much longer as organic material slowly spirals through the system undergoing numerous transformations.

Knight's use of the verb *spiral* is significant in the above quote. Spiralling is a metaphor used to describe material processing in stream ecosystems that combines cycling and transport. In the classic sense, materials cycle through an ecosystem along transformation pathways between abiotic and biotic compartments (Pomeroy, 1974a). The study of these cycles is termed variously biogeochemistry (Schlesinger, 1997), mineral cycling (Deevey, 1970), or nutrient cycling (Bormann and Likens,

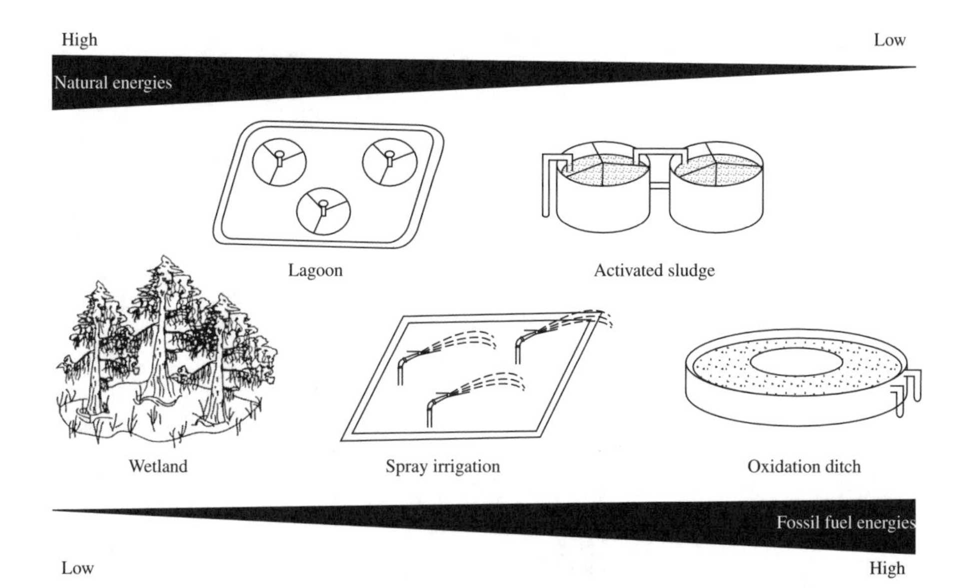

FIGURE 2.12 Locations of various wastewater treatment technologies along gradients of energy input. (From Knight, R. L. 1995. *Maximum Power: The Ideas and Applications of H. T. Odum*. C. A. S. Hall (ed.). University Press of Colorado, Niwot, CO. With permission.)

1967). In terms of abiotic compartments, some elements, such as carbon, nitrogen, and sulfur, have gaseous phases while others, such as phosphorus, potassium, and calcium, are primarily limited to soil and sediment phases. Most elements are taken up by plants for use in the organic matter production of photosynthesis and are released either from living tissue or after deposition as detritus (i.e., storage of nonliving organic matter) through respiration. Thus, each element has its own cycle through the ecosystem, though they are all coupled. Traditionally, cycling was essentially considered to occur at one point in space. This conception makes sense for an aggregated view of a forest or lake ecosystem where internal cycling quantitatively dominates amounts flowing in or out at any point. However, in stream and river ecosystems internal cycling is less important because of the constant movements due to water flow. Stream ecologists developed the spiraling concept (Figure 2.13) to account for both internal cycling and longitudinal transport of materials in a two- or three-dimensional sense as opposed to the one-dimensional sense of internal cycling as a point process (Elwood et al., 1983; Newbold 1992; Newbold et al., 1981, 1982). Wagener et al. (1998) have extended the spiraling concept to soils, and as indicated by Knight's quote, this may be the appropriate perspective for material processing in treatment wetlands. It is such complex system functioning that characterizes treatment of sewage in wetlands.

Sewage is discharged in a treatment wetland usually at a series of points (often along a perforated pipe) rather than at a single point, and it moves by gravity as a thin sheet-flow through the wetland. This kind of flow, either at or below the surface, allows adequate contact with all ecosystem components involved in the treatment process. Channel flows, with depths greater than about 30 cm, will not allow adequate treatment because they reduce residence time.

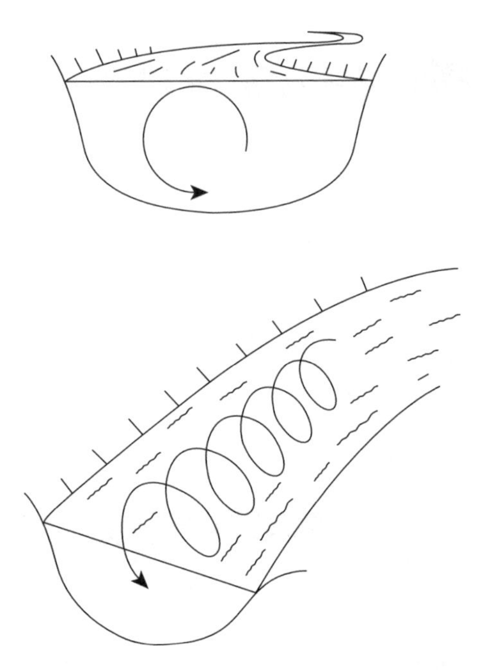

FIGURE 2.13 The spiraling concept of material recycling in stream ecosystems. (From Newbold, J. D. 1992. *The Rivers Handbook: Hydrological and Ecological Principles*. Vol. 1. P. Calow and G. E. Petts (eds.). Blackwell Scientific, Oxford, UK. With permission.)

The efficiency of treatment wetlands is evaluated by input–output methods which quantify assimilatory capacity. A mass balance approach is most useful, which demonstrates percent removal of TSS, BOD, nutrients, and pathogens. Usually this is done by measuring water flow rates (for example, million gallons/day) and concentrations of sewage parameters (usually mg/l for TSS, BOD, and nutrients and numbers of individual organisms per unit volume for pathogens). When water flow rates are multiplied by concentrations, along with suitable conversion factors, the total mass of input can be compared with the total mass of output and uptake efficiencies calculated. If water flow rates cannot be quantified, comparisons between inputs and outputs can be made with concentration data alone, but this approach is not as complete as the full mass balance approach.

The dominant processes that remove the physical–chemical parameters of sewage in wetlands are shown in Figure 2.14 and highlighted in Table 2.3. Many kinds of transformations are involved in these treatment processes and much is known about their kinetics. In general, treatment efficiencies are variable but high enough for the technology to be considered competitive.

The treatment wetland technology works best in tropical or subtropical climates where biological processes are active throughout the annual cycle. An open question still exists about year-round use of treatment wetlands in colder climates where biological processes are reduced during the winter season, but some workers believe that the technology can be utilized in these regions (Lakshman, 1994; Werker et al., 2002). It also is most appropriate for rural areas where waste volumes to be treated

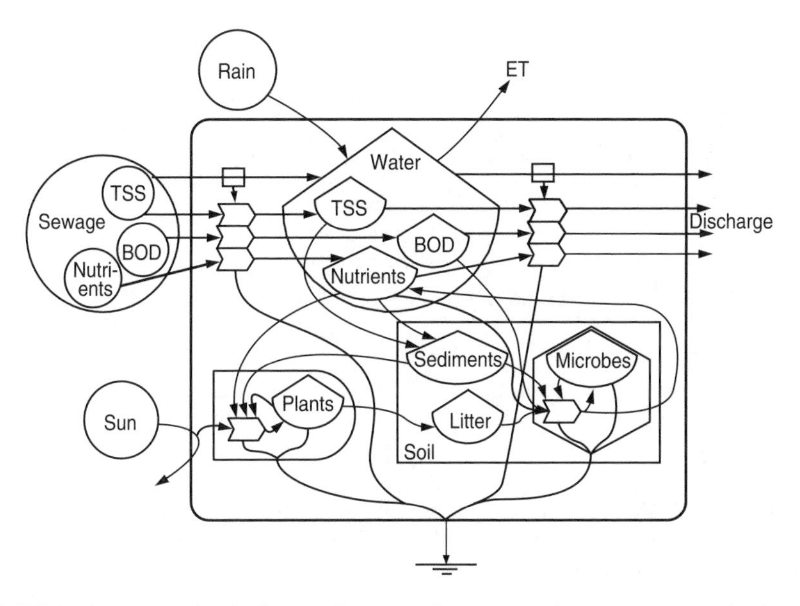

FIGURE 2.14 Energy circuit diagram for the main processes in a treatment wetland.

TABLE 2.3
Listing of the Dominant Processes of Water Quality Dynamics in Treatment Wetlands

Process	Pathway within treatment wetland
Sedimentation	TSS in water to sediments
Adsorption, Precipitation	Nutrients in water to sediments
Biodegradation	BOD in water to microbes
Chemical transformation	Nutrients in water to microbes and microbes to nutrients in water
Metabolic uptake	Nutrients in water to plants
Overall input	TSS, BOD, nutrients in sewage source to water storage
Overall output	TSS, BOD, nutrients in water storage to discharge

Note: Pathways are from Figure 2.14.

are small to moderate. In urban settings, where waste volumes are high, conventional treatment plants are more appropriate than treatment wetlands because they handle large flows with small area requirements.

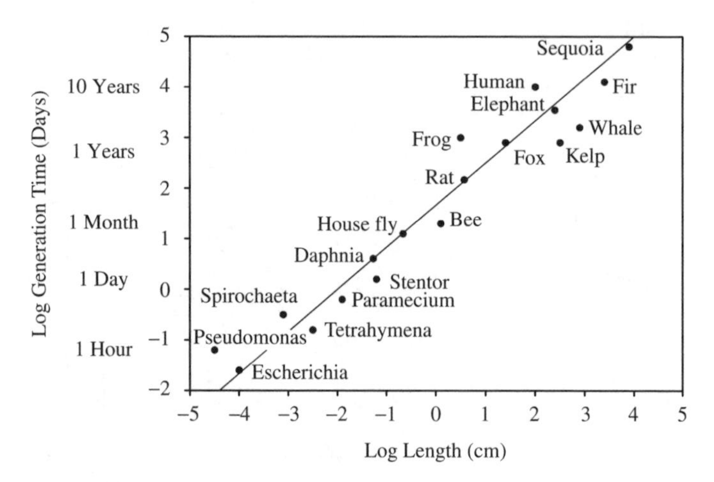

FIGURE 2.15 A scale graph of biodiversity. (From Pedros-Alio, C. and R. Guerrero. 1994. *Limnology: A Paradigm of Planetary Problems*. R. Margalef (ed.). Elsevier, Amsterdam, the Netherlands. With permission.)

Wetlands that are specially constructed for wastewater treatment are the most common form of the technology today. A few types of natural wetlands are used (Breaux and Day, 1994; Knight, 1992), but these are special case situations. The two main classes of constructed treatment wetlands differ in having either surface or subsurface water flows. The state of the art is given in book-length surveys by Campbell and Ogden (1999), Kadlec and Knight (1996), Reed et al. (1995), and Wolverton and Wolverton (2001), and in a number of edited volumes (Etnier and Guterstam, 1991; Godfrey et al., 1985; Hammer, 1989; Moshiri, 1993; Reddy and Smith, 1987). Other useful reviews are given by Bastian (1993), Brown and Reed in a series of papers (Brown and Reed, 1994; Reed and Brown, 1992; Reed, 1991), Cole (1998), Ewel (1997), and Tchobanolous (1991).

BIODIVERSITY AND TREATMENT WETLANDS

Most engineering-oriented discussions of treatment wetlands focus on microbiology, but other forms of biodiversity are, or can be designed to be, involved. Microbes occupy the smallest and fastest (in terms of generation time) realm of biodiversity, making up about the lower quarter of the graph in Figure 2.15. Do other realms of biodiversity have roles to play in existing or possible treatment wetlands? The consensus from many engineers and treatment plant operators seems to be that these roles, to the extent that they even exist, are minor. Another perspective is that the use of biodiversity in treatment wetlands is in the early stage of development and broader roles may be self-organizing or may be designed in the future for more effective performance. For example, Cowan (1998) found more species of frogs and toads in a treatment wetland in central Maryland as compared with a nearby reference wetland. Is this high amphibian diversity playing a functional role in treatment wetland performance? Ecology as a science may be able to lead the design of biodiversity in treatment wetlands through ecological engineering. Several examples of important taxa are discussed below.

MICROBES

The term *microbe* includes a number of different types of organisms that occur at the microscopic range of scale. The ecology and physiology of microbes is much different from macroscopic organisms, because of their small size and because their surface-to-volume ratios are so much larger (Allen, 1977). Thus, the methods of study for microbes are almost completely different from methods used for larger organisms. These qualities separate microbial ecologists from other ecologists and, to some extent, limit interaction between the two groups. The ecology of microbes in general is introduced by Margulis et al. (1986), Allsopp et al. (1993), and Hawksworth (1996), while references focusing on bacteria are given by Fenchel and Blackburn (1979), Pedros-Alio and Guerrero (1994), and Boon (2000). Microbes perform the main biological work of waste treatment in their metabolism. This is especially true for carbon and nitrogen, though less applicable for phosphorus. Organic materials, such as BOD, are consumed through aerobic or anaerobic respiration reactions, and nitrogenous compounds are ultimately converted to nitrogen gas through nitrification and denitrification reactions. Thus, microbes may be thought of as the principal functional forms of biodiversity in treatment wetlands. The basic theory in wastewater treatment engineering considers the dynamics of pollutants, such as BOD, and microbial communities within bioreactors (Figure 2.16), and this approach is used as a starting point for understanding the behavior of treatment wetlands.

Microbes can be either attached to surfaces or suspended in the wastewater. Attached microbes form biofilms (Characklis and Marshall, 1990; Flemming, 1993; Lappin-Scott and Costerton, 1995). These are the "slimes" mentioned earlier (Ben-Ari, 1999). Suspended microbes are important where artificial turbulence is applied as in fluidized beds and activated sludge units.

In natural ecosystems microbes are usually found attached to particles of detritus. Two historic views of the relationship are shown in Figure 2.17. In practice, it is difficult or impossible to separate the living microbial organisms from the nonliving detritus particles, and they are often treated as a complex in ecological field work. From the perspective of detritivores who consume detritus, Cummins (1974) suggested that the complex is like a peanut butter cracker. In this anthropocentric metaphor, the microbes are the nutritious peanut butter because of their low carbon to nitrogen ratio, while the detritus particle is the nutritionally poor cracker because of its high carbon to nitrogen ratio (see the composting section in Chapter 6 for more discussion of the carbon to nitrogen ratio). Thus, a detritivore obtains more nutrition from the microbe than from the detritus particle itself, but both must be ingested because they form a unit. The detritus complex is an important part of most ecosystems. It is associated with soils and sediments but it can be suspended, as in oceanic systems where it is termed *marine snow* (Silver et al., 1978). General reviews of the ecology of detritus are given by Melchiorri-Santolini and Hopton (1972),

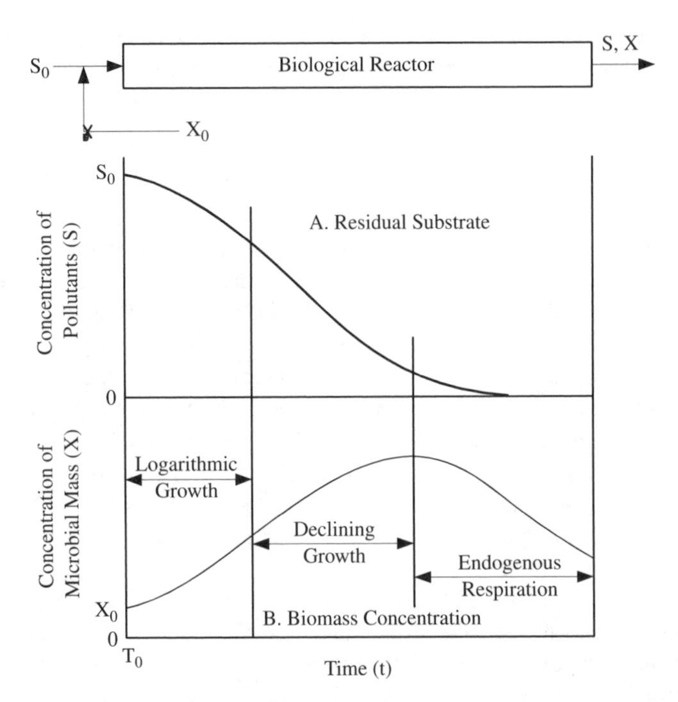

FIGURE 2.16 Views of the basic theory of biological reactor functioning. (From Tenney, M. W. et al. 1972. *Nutrients in Natural Waters*. H. E. Allen and J. R. Kramer (eds.). John Wiley & Sons, New York. With permission.)

Pomeroy (1980), Rich and Wetzel (1978), Schlesinger (1977), Sibert and Naiman (1980), and Vogt et al. (1986).

Higher Plants

Higher plants, especially flowering plants, are an obvious feature of wetlands including treatment wetlands (Cronk and Fennessy, 2001). Although wetlands can be defined broadly (Cowardin et al., 1979), a general definition is that a wetland is an ecosystem with rooted, higher plants where the water table is at or near the soil surface for at least part of the annual cycle. Lower plants, such as algae, mosses, and ferns, can be important but they are usually less dominant than the flowering plants. A variety of life forms fall under the category of higher plants in wetlands, including trees, emergent macrophytes (grasses, sedges, rushes), and floating leafed and submerged macrophytes. Although their function in treatment wetlands is secondary to microbes, they do play significant roles (Gersberg et al., 1986; Peterson and Teal, 1996; Pullin and Hammer, 1991).

Figure 2.18 depicts a general model that covers many of the higher plant life-forms and illustrates several important functions. The plants themselves are composed of aboveground (stems, shoots, and leaves) and belowground (roots and rhizomes, which are underground stems) components which interact in the central process of primary production. Belowground components physically support the

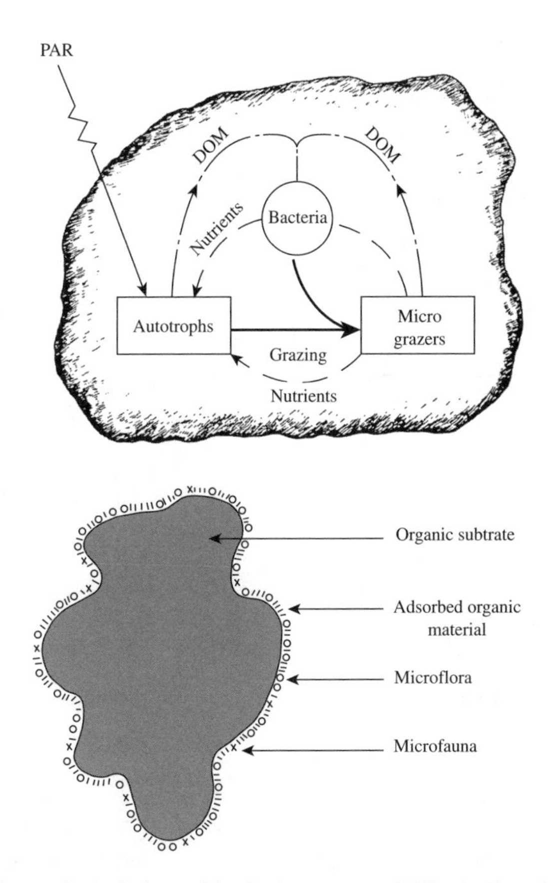

FIGURE 2.17 Two early depictions of the detritus concept. PAR: photosynthetically available radiation; DOM: dissolved organic matter. (The top part of the figure is from Goldman, J. C. 1984. *Flows of Energy and Materials in Marine Ecosystems: Theory and Practice*. M. J. R. Fasham (ed.). Plenum Press, New York. With permission. The bottom part of the figure is from Darnell, R. M. 1967. *Estuaries*. G. H. Lauff (ed.). Publ. No. 83, American Association for the Advancement of Science, Washington, DC. With permission.)

shoots and facilitate uptake of nutrients. Photosynthesis occurs in shoots and leaves that are exposed to sunlight during a portion of the year when the temperature is above freezing (i.e., the growing season). Aboveground and belowground components die and transform into litter and soil organic matter, respectively (both forms of detritus), where decomposition and recycle by detritivores takes place.

Probably the most obvious contribution of higher plants to the treatment process is uptake of nutrients. If biomass is harvested and removed from the system, uptake can play a role in removing nutrients. However, without harvest, nutrients eventually recycle providing no net treatment. This situation leads to a description of treatment wetlands as alternating sinks and sources for nutrients. They are sinks during the growing season when uptake dominates the mass balance, and they are sources during the winter and early spring when decomposition and seasonal flushing dominate the mass balance. Harvest can cause treatment wetlands to be primarily sinks,

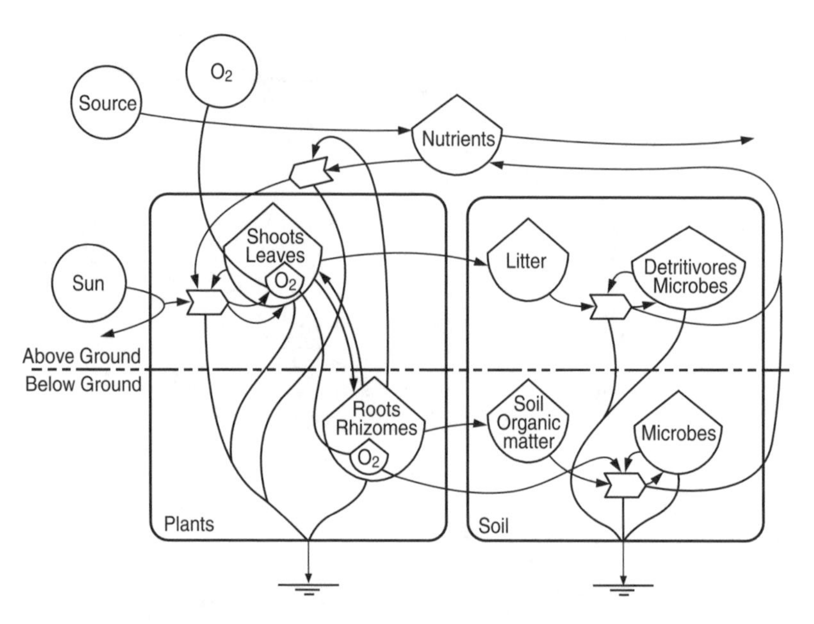

FIGURE 2.18 Energy circuit diagram of an aquatic plant-based system.

but this option is often expensive. Also, only small amounts of nutrients can be removed by harvesting because plant biomass contains only small percentages of nutrients (about 5% by mass).

A more important contribution of higher plants to the treatment process is their support of microbes within the rhizosphere, which is the zone adjacent to living roots in soils and sediments. Roots provide surfaces that are colonized by biofilms, and they leak organic molecules and oxygen that directly support microbes. These kinds of flows are shown in Figure 2.18 in the belowground zone. Oxygenation of sediments through air spaces that connect shoots and roots is a very important function of many macrophytes (Dacey, 1981; Gunnison and Barko, 1989; Jaynes and Carpenter, 1986; Kautsky, 1988). Wetland sediments are normally anaerobic and oxygen leakage from roots supports more efficient aerobic metabolism by microbes in the rhizosphere. The contributions of roots in creating microzones within sediments may be a critical role in supporting microbes that transform nitrogenous materials in wastewater to nitrogen gas. Removal of nitrogen through denitrification is a reliable function in treatment wetlands that is a definite tool to be used by ecological engineers.

A final note on higher plants in treatment wetlands deals with the special features of individual species. Paradoxically, some of the most useful species in treatment wetlands, such as water hyacinth (*Eichhornia crassipes*), common reed (*Phragmites* sp.) and cattail (*Typha* sp.), are considered pests that sometime require control when they occur in natural wetlands. These species are characterized by fast growth, which is a positive feature in treatment wetlands but a negative feature in natural wetlands where they outcompete other plant species. Several of these species are nonnative or "exotic" in North America and the ecology of these kinds of species is covered in Chapter 7. Duckweed (family *Lemnaceae*) is another type of plant with fast growth that has been used for wastewater treatment (Culley and Epps, 1973; Harvey and Fox, 1973). Species from this plant family are small, floating-leaved plants that can completely cover the water surface of a pond or small lake. In an early reference Hillman and Culley (1978) envision a dairy farm system in which 10 acres (4 ha) of duckweek lagoons could treat the wastewater of a herd of 100 cattle. One more recent design for domestic wastewater treatment uses this species exclusively (Buddhavarapu and Hancock, 1991), which was developed by a company named appropriately the Lemna Corporation. Other species, not now used in treatment wetlands (*Symplocarpus foetidus*), may have special qualities that preadapt them for this use. Skunk cabbage is one example that has adaptation for growth during early spring when temperatures are too low for other species. Heat generated by enhanced respiration (Knutson, 1974; Raskin et al., 1987) supports the early growth. Perhaps this species could be manipulated and managed to extend the growing season of treatment wetlands, which is now a limiting factor for their implementation in cold climates.

Protozoans

Protozoans are microscopic animals found primarily in soils and sediments. A variety of groups are known, roughly separated by locomotion type: amoebae, flagellates, and ciliates, along with foraminifera. Their primary role in treatment wetlands is as predators on the bacteria. This predation controls or regulates bacteria populations by selecting for fast growth. Predation is always selective, with the predators choosing among alternative prey individuals. This is true for all organisms from protozoans to killer whales and has important consequences. Predators affect and improve the genetic basis of prey populations by selecting against individuals that are easy to catch (i.e., the sick, the dumb, the weak, and the very young or old) and selecting for those individuals that are hard to catch (i.e., the healthy, the smart, the strong, and the middle-aged). As an example, Fenchel (1982) simulates a predator–prey model for protozoans and bacteria which generate a classic oscillating pattern over time (see also Figure 4.5). The interaction between predators and prey is an important topic in ecological theory (Berryman, 1992; Kerfoot and Sih, 1987), and knowledge of the subject will provide ecological engineers with an important design tool (see also the discussion of top-down control in Chapter 7).

Because bacteria and other microbes metabolize the organic matter in wastewater, protozoans indirectly control treatment effectiveness through their predation. Treatment of BOD is thus a "bacterial–protozoan partnership" (Sieburth, 1976), and this interaction is illustrated in Figure 2.19. This highly organized food web also has been called the microbial loop because of the strong and fast interconnections between components in the system. The microbial loop was first discussed for the oceanic plankton (Azam et al., 1983; Pomeroy, 1974b), but it may well apply to treatment wetlands as well. Numbers of organisms are very high in these mixed microbial systems — on the order of 10^5–10^6 individuals/gram for bacteria and 10^3–10^4 individuals/gram for protozoans (Chapman, 1931; Spotte, 1974; Waksman, 1952). The importance of protozoans is well known in both natural ecosystems (Bick

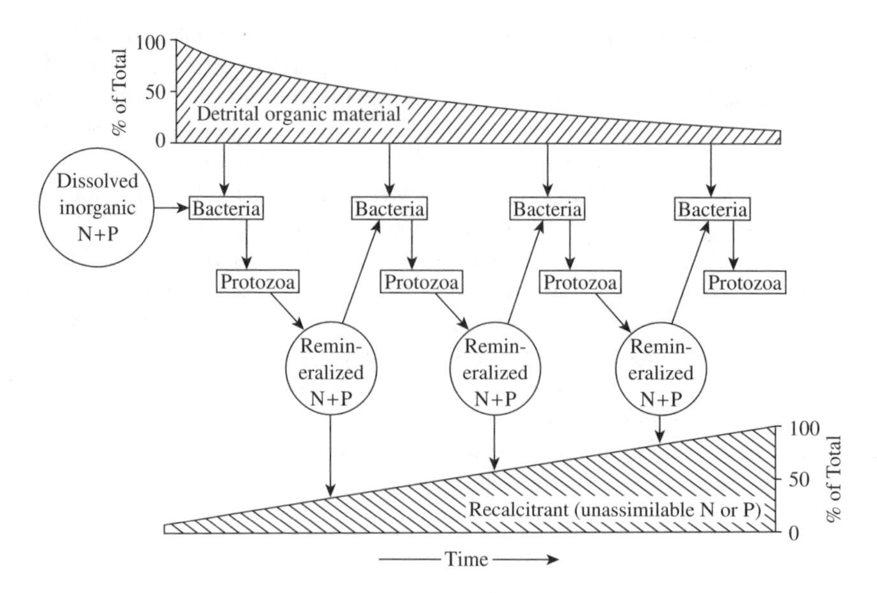

FIGURE 2.19 The role of protozoans in regulating decomposition processes. (From Caron, D. A. and J. C. Goldman. 1990. *Ecology of Marine Protozoa*. G. M. Capriulo (ed.). Oxford University Press, New York. With permission.)

and Muller, 1973; Stout, 1980) and conventional wastewater treatment systems (Barker, 1946; Bhatla and Gaudy, 1965; Curds, 1975; Kinner and Curds, 1987). Experiments with seeding protozoans into wastewater treatment systems have been shown to improve treatment (Curds et al., 1968; McKinney and Gram, 1956), and it is possible that protozoans may be able to be manipulated by ecological engineers in treatment wetlands.

Mosquitoes

Mosquitoes are biting flies of the insect family *Culicidae*. They are well known to be associated with wetlands since their larvae are aquatic. For many people, mosquitoes are a negative form of biodiversity because of their biting behavior and because some species can transmit diseases. Worldwide there are nearly 3000 species of mosquitoes and they range from the arctic to the tropics. They are remarkable animals with a very short breeding cycle and with an adult stage characterized by acute sensory perception and strong flight capability. Of course, the main problem with mosquitoes is that in certain species the females prey on human blood. They do this in order to acquire a concentrated dose of protein needed for the development of eggs in the reproductive cycle. The main genera are *Anopheles*, *Aedes*, and *Culex* and they are considered to be “man’s worst enemy” (Gillett, 1973) because of their association with diseases. The most important diseases are mainly tropical and include malaria, yellow fever, dengue, encephalitis, and filariasis (Foote and Cook, 1959). The latest concern in the U.S. is the West Nile strain of encephalitis which was first recorded in New York City but which has recently spread across the country.

Large-scale, organized control of mosquitoes has been evolving for more than 100 years (Hardenburg, 1922) and it involves many of the issues now associated with invasive exotic species (see Chapter 7). The main methods of control are to restrict breeding habitats and to use chemical pesticides and biological control agents. In general, control of mosquito populations is difficult to achieve because of their dispersal abilities and short breeding cycle. Also, some species such as the Asian tiger mosquito (*Aedes albopictus*) are preadapted to live in human habitats by breeding in manmade containers (old tires, flower pots, clogged gutters, children's swimming pools, etc.). Of special interest, mosquitoes provided one of the first examples of the development of resistance to pesticides when DDT use became widespread after World War II (King, 1952). Because of the importance of the problem, organized mosquito control districts have been formed in many parts of the U.S., which are supported by local taxes. Lichtenberg and Getz (1985) provide a thorough review of the economics of mosquito control for one particular situation that is probably applicable in a more general sense.

Mosquitoes often are associated with treatment wetlands when ponded water provides habitat for larvae. They can participate in the treatment process because the larvae feed on organic matter, but their contribution probably is quite minor. The most important issue is whether or not treatment wetlands are a significant source of mosquito production. Kadlec and Knight (1996) have reviewed the subject and indicate that this is generally not a major concern. However, others are not convinced, as noted by Martin and Eldridge (1989): "We have the basic knowledge to design and operate created wetlands systems today. The major drawback is mosquito problems, which must be solved before created wetlands can be universally accepted by public health officials and the general public." Designs for controlling mosquito production in treatment wetlands are discussed by Anonymous (1995a), Russell (1999), and Stowell et al. (1985).

As a final aside, a significant amount of ecological engineering has been involved in mosquito control efforts in natural saltmarshes, especially along the U.S. East Coast (Carlson et al., 1994; Dale and Hulsman, 1991; Resh, 2001). In this habitat mosquito species lay eggs on moist soil surfaces rather than in standing water. Larvae develop in the eggs but will hatch only when covered by tidewater or rainwater. Historically, control efforts in this case dealt with water level manipulations, either drainage with ditches or canals or impoundment where water is contained behind dikes (Clark, 1977; Provost, 1974). Thus, construction activities often were necessary for control of saltmarsh mosquitoes. This work represented an interesting example of ecological engineering since it involved knowledge of mosquito biology, saltmarsh hydrology, and the design of impoundment and drainage systems (Figure 2.20). Skill was required to combine these areas into new saltmarshes with altered hydrology and fewer mosquitoes, and many efforts failed by actually producing more mosquitoes. These practices are no longer undertaken, partly because many systems have already been constructed and partly because of the environmental impacts caused by the changes to natural saltmarshes. In fact, some old mosquito control systems are currently being restored, which is another ecological engineering challenge (Axelson et al., 2000). A related problem is the design of irrigation systems in regard to pest populations (Jobin and Ippen, 1964).

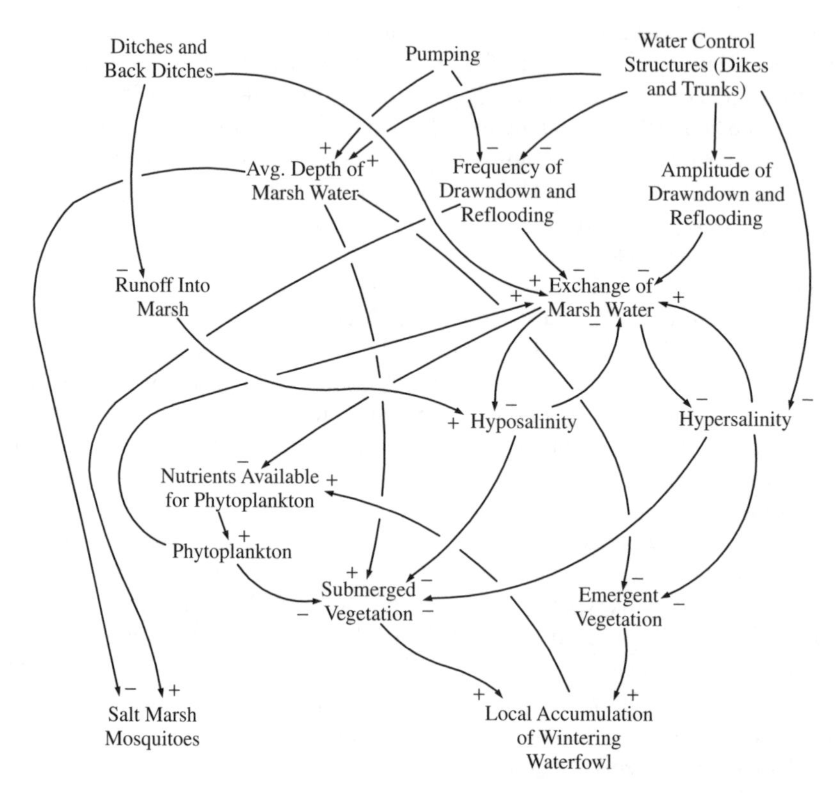

FIGURE 2.20 Causal diagram of interactions involved in mosquito control through water level manipulations in tidal marshes. (Adapted from Montague, C. L., A. V. Zale, and H. F. Percival. 1985. Technical Report No. 17. Florida Cooperative Fish and Wildlife Research Unit, University of Florida, Gainesville, FL.)

Muskrats

Muskrats (*Ondatra zibethicus*) are large, semiaquatic rodents that are distributed throughout most of temperate North America and, as exotic species, in Europe. They are primarily herbivorous, feeding on rhizomes of emergent macrophytes in marshes. However, their indirect effects on marsh ecosystems are probably more significant than their direct effect of grazing (Kangas, 1988). These indirect effects include construction activities that result in mounds, which they use for overwintering, and in burrows and paths, which they use to facilitate movements within the often dense vegetation of marshes. The ecology and natural history of muskrats are well known (Errington, 1963; Johnson, 1925; O'Neil, 1949) and include complex patterns of population oscillations (Elton and Nicholson, 1942), which occurred at least historically before landscapes became fragmented by human development. The network of direct and indirect effects performed by muskrats makes them keystone species in natural marsh ecosystems because of their important control functions (see Chapter 7 for discussion of the keystone species concept).

Muskrats can occur in treatment wetlands, which provide ideal habitats due to a dominance of preferred food plants and stable water levels. In fact, Latchum (1996) found the highest density of muskrat mounds ever reported at 62.5/ha in a treatment wetland in central Maryland. This is particularly significant because mounds are often used as an index of population size for muskrats (Danell, 1982). High densities of muskrats can build up in natural marshes and cause "eat-outs" where a large amount of plant biomass is harvested over a short period of time, leading to a crash in the muskrat population. This is reflected in the oscillatory dynamics commonly reported for muskrats. However, muskrat populations may be stabilized at high levels in treatment wetlands due to the steady supply of nutrients and optimal habitat conditions, creating a kind of "sustainable eat-out." This result does not match with the "paradox of enrichment" described in theoretical ecology (Rosenzweig, 1971), where enrichment of a predator–prey system causes it to become unstable or, in extreme cases, to collapse. In treatment wetlands the marsh vegetation (as prey)–muskrat herbivore (as predator) system is enriched with sewage nutrients, but it becomes stabilized at higher levels rather than destabilized. This may be an exception to the paradox, like other counter examples (Abrams and Walters, 1996; McAllister et al., 1972) due to additional complexities in a real-world example. Muskrats are close relatives of the lemmings (*Lemmus* sp. and *Dicrostonyx* sp.), which have been suggested to be components in homeostatic networks of the tundra (Schultz, 1964, 1969). Enrichment in treatment marshes may push the marsh vegetation–muskrat herbivore system into an alternative stable state with a new homeostatic structure.

A small amount of literature exists on the effect of muskrats on treatment wetlands, though most seem to feel it is negative (Table 2.4). There is an obvious negative effect due to their burrowing into dikes or berms that enclose constructed wetlands (Figure 2.21), which is part of the natural role of the muskrat in spreading water over the landscape. Latchum (1996) traced out many more possible impacts as shown in the causal diagram in Figure 2.22. This diagram illustrates a network of positive and negative and direct and indirect effects that muskrats may have in a treatment wetland. The most significant effects seem to come from the construction activities that cascade through a number of processes to influence treatment capacity. Mounds are the most obvious construction feature of muskrats, and they act like compost piles (see Chapter 6) in accelerating the decomposition rate (Berg and Kangas, 1989; Wainscott et al., 1990). Overall, mounds may have a direct negative effect on primary production since plant materials are used in construction, but they have several indirect positive effects through increasing decomposition.

Paths and burrows constructed by muskrats can be extensive in marshes. At high water levels, they can act as channels or macropores (Beven and Germann, 1982) and they probably increase water flow. This can have a negative impact on treatment capacity if some of the pollutant load passes through the wetland without treatment. In general, treatment effectiveness is directly related to retention time. The preferential flow in channels or macropores reduces retention time and therefore also treatment effectiveness. Muskrat burrows increase aeration of the sediments and they probably have many similar effects as burrowing crustaceans (Montague, 1980,

TABLE 2.4
Comments from the Literature about Muskrats in Treatment Wetlands

Quote	Reference
The proponents of reeds argue for a monoculture of reeds, while others argue that bulrushes are superior. Survivability of each species will depend on other factors, such as plant pests. Muskrats love cattails and bulrushes, while reeds apparently are inedible …	Campbell and Ogden, 1999
Muskrats can damage dikes by burrowing into them. Although muskrats generally prefer to start their burrows in water that is more than 3 ft deep, they can be a problem in shallower waters. Muskrats can be excluded by installing an electric fence low to the ground or by burying muskrat-proof wire mats in the dikes during construction.	Davis, no date
I should point out that for pest control, we do muskrat trapping to prevent destruction of berms …	Wile et al., 1985
Muskrats are also problematic in constructed wetlands because they burrow into dikes, creating operational headaches and potential for system failure …	Kadlec and Knight, 1996

1982; Richardson, 1983; Ringold, 1979). Aeration generally has a positive effect on metabolism and biodiversity, though it inhibits denitrification, which requires anaerobic conditions.

Perhaps the most important general contribution of muskrats in marsh ecosystems is generating spatial heterogeneity, which is not shown in Figure 2.22. Through their various construction activities muskrats create a mosaic pattern of open areas within dense marsh vegetation. Spatial heterogeneity increases diversity through a number of mechanisms (Hutchings et al., 2000; Kolasa and Pickett, 1991; Shorrocks and Swingland, 1990; Smith, 1972). This quality provides redundancy in system design, which is similar to a safety factor in engineering. The spatial heterogeneity caused by muskrats in treatment marshes also may help explain why this system does not match with the "paradox of enrichment" (Scheffer and DeBoer, 1995), as noted above.

The cumulative impact of muskrats on treatment wetlands is unknown, though both positive and negative effects have been noted. Although it is not completely clear, the obvious negative effects seem to dominate over the less obvious positive effects. For example, at the treatment wetland studied by Latchum (1996), muskrats were judged to be negative because they became trapped in some of the mechanical parts of the system. Overall, the fact that muskrats can act as positive, keystone species in natural marshes but as negative, pest species in treatment marshes is a paradox. However, active design and management through ecological engineering

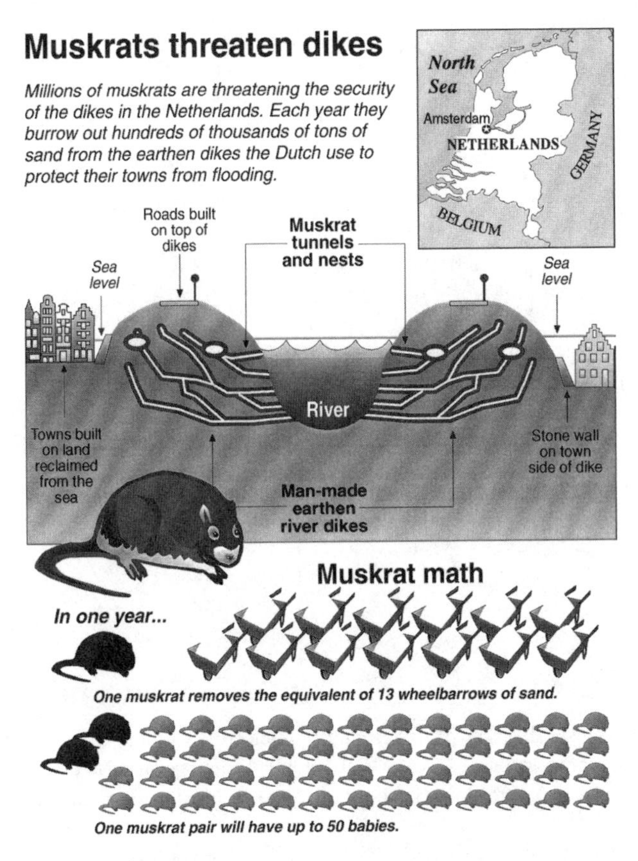

FIGURE 2.21 Views of the problems that muskrats cause by burrowing. (From World Wide Photos. New York, NY. With permission.)

may shift this balance. Perhaps their ecological role can be used to improve treatment capacity. One strategy might be to take advantage of their concentration of biomass in mounds by harvesting the mounds in the spring to remove nutrients. This might reduce the cost of harvesting because the muskrats would be doing some of the collection work for free, and the mound material might be used, like compost, as a soil amendment. In a sense the muskrat is a basic element in the ecological and hydrologic self-organization of temperate, humid landscapes. They have evolved to spread water around and regulate wetland processes in marsh ecosystems. It would be a significant accomplishment of ecological engineering if their adaptations could be used productively. Ultimately, a treatment marsh without muskrats is an incomplete ecosystem.

Aquaculture Species

The aquacultural production of useful species from domestic wastewater is related to the topic of treatment wetlands. Allen (1973) called these systems "sewage farming" and many examples exist (Allen and Carpenter, 1977; Costa-Pierce, 1998;

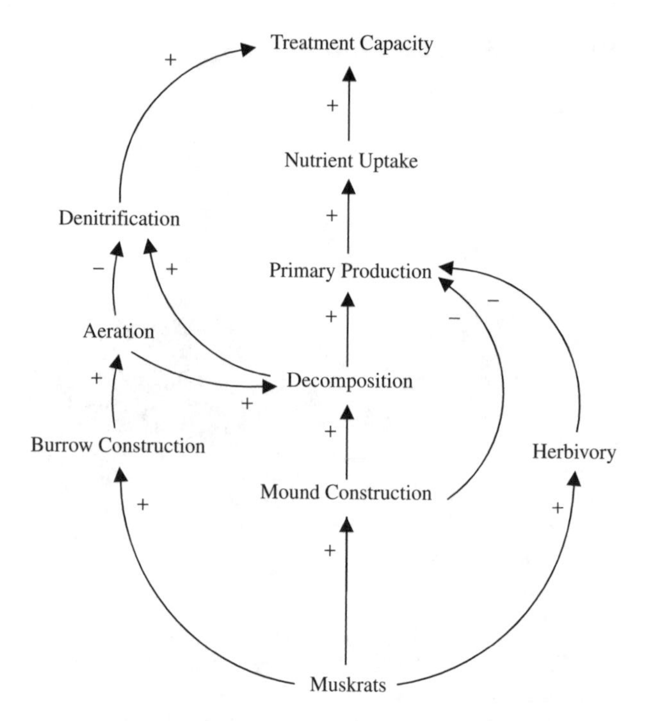

FIGURE 2.22 Causal diagram of direct and indirect effects of muskrats in treatment wetlands. (Adapted from Latchum, J. A. 1996. Ecological Engineering Factors of a Constructed Wastewater Treatment Wetland. M.S. Thesis, University of Maryland, College Park, MD.)

Drenner et al., 1997; Edwards and Densem, 1980; Gordon et al., 1982; Roels et al., 1978). An interesting example of a sequential treatment system using "controlled eutrophication" (Ryther et al., 1972) is shown in Figure 2.23. The system was a constructed marine food chain capable of producing several kinds of biomass along with clean water. An obvious risk of this kind of system is the incorporation of pathogenic microbes or chemical toxins into food products grown in wastewater. Korringa (1976) reviewed this issue and suggested that strict control over the use of these artificial food chains through monitoring systems, quarantine measures, and purification plants is possible. However, these actions are expensive and they reduce the economic viability of sewage farming. Other techniques, such as the production of species that provide nonfood products (i.e., ornamental plants or aquarium fishes) may be more viable but may lack extensive markets (see "living machines" in Chapter 8). Ecological engineering designs will certainly continue to be tested in the future to take advantage of sewage as a resource, including the production of many kinds of species that yield value to humans.

Coprophagy and Guanotrophy

One goal of ecological engineering is to design and test new treatment ecosystems that have high biodiversity and more effective treatment efficiency. Organisms that utilize feces may be good candidates for these ecosystems since they may be

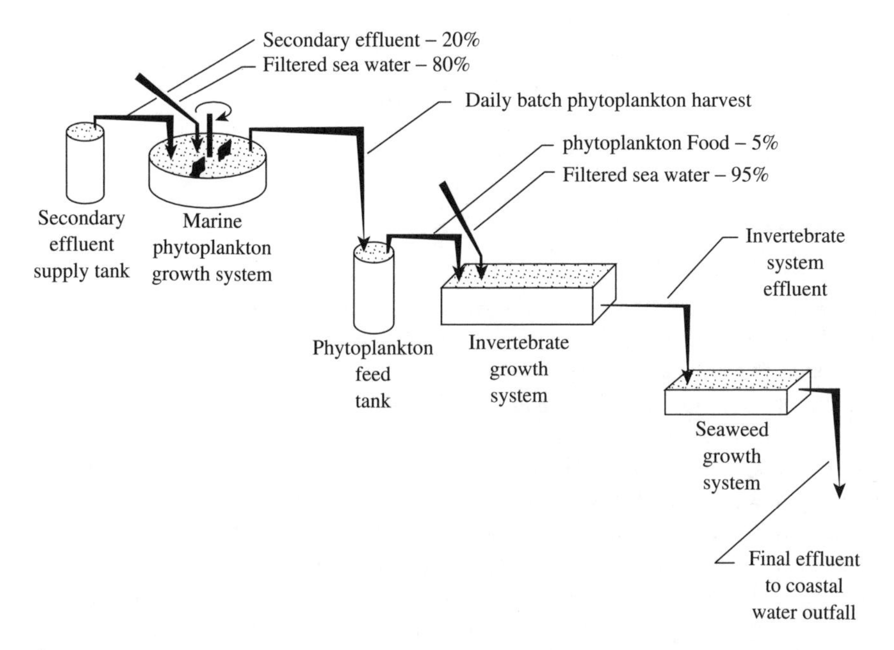

FIGURE 2.23 View of the Woods Hole wastewater treatment–aquaculture system. (From Goldman, J. C. et al., 1974a. *Water Research.* 8:45–54. With permission.)

preadapted to sewage treatment. An ecology of feces exists in the ecological literature (see, for examples, Angel and Wicklow, 1974; Booth, 1977; Wotton and Malmqvist, 2001), and the many terms used for feces to some extent indicate the broad range of this literature: guano, frass, scat, fecal pellets, dung, droppings, and coprolites (fossil feces). The consumption of feces by animals is termed coprophagy or guanotrophy. Animals consume feces because of its relatively high nutritive value (see, for examples, Hassall and Rushton, 1985; Rossi and Vitagliano-Tadini, 1978; Wotton, 1980). Mohr (1943) described a diversity of species involved in successional stages during the breakdown of cattle droppings in an Illinois pasture. This and other guano-rich environments (Leentvaar, 1967; Poulson, 1972; Ugolini, 1972) could be searched for food chains that might be transformed into treatment ecosystems through biodiversity prospecting. For example, dung beetles (*Scarabaeidae*) (Hanski and Cambefort, 1991; Waterhouse, 1974) might be ideal as the basis for a sewage sludge recycling system.

PARALLEL EVOLUTION OF DECAY EQUATIONS

The quantitative design analysis of treatment wetlands has followed, and in fact copied, the approach traditionally used in sanitary engineering for design of other types of wastewater treatment systems. In an abstract sense, a wastewater treatment system is considered to be a bioreactor, or "… a vessel in which biological reactions are carried out by microorganisms or enzymes contained within the reactor itself.

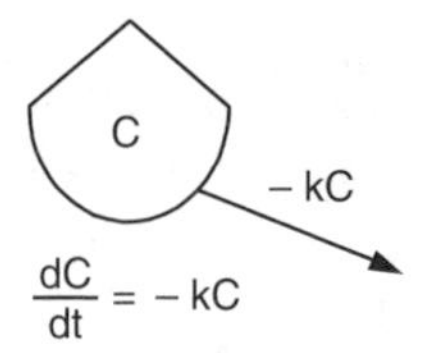

FIGURE 2.24 Energy circuit diagram of decay.

In hazardous, municipal, or industrial waste treatment, bioreactors are used primarily to reduce the concentration of contaminants in incoming wastewaters to acceptably low levels" (Armenante, 1993). Equations have been developed which describe the process of reduction of contaminants in bioreactors (see Figure 2.16) and design is based on these equations. This approach has evolved over time and can be traced by examining various editions of textbooks on sanitary engineering. For example, Leonard Metcalf and Harrison Eddy produced a standard text that has spanned the design evolution of wastewater treatment. Early versions of their text contained essentially no equations, and design was based on practical experience with the available systems, such as trickling filters or activated sludge units (Metcalf and Eddy, 1916, 1930). This text evolved with the field with revisions by George Tchobanoglous and, by the 1970s, it was filled with equations (Metcalf and Eddy, 1979). These equations are quantitative expressions of the practical experience developed over time by engineers. In many ways the equations are the heart of the engineering method because of their role in design. A simple but fundamental equation is reviewed below to demonstrate this approach to design.

Perhaps the simplest equation used to describe the biodegradation of organic materials (BOD) in domestic wastewaters is the first-order reaction:

$$Ce/Co = e^{-kt} \tag{2.1}$$

where

Ce = effluent BOD, in mg/l
Co = influent BOD, in mg/l
k = first-order reaction rate constant, in 1/day
t = time of flow through the system or hydraulic residence time, in days

This equation is the integrated form of the model shown in Figure 2.24. Basically, an initial amount of organic materials (Co) is degraded over a given amount of time (t) according to the rate constant (k) that depends on the temperature at which the reaction occurs. It has been used to describe many kinds of wastewater treatment systems, and it is used as a starting point for considering BOD removal in treatment wetlands by Reed et al. (1995) and Crites and Tchobanoglous (1998). It was first used in sanitary engineering to describe the BOD concept and to develop the test procedure (Gaudy, 1972). Early sanitary engineers struggled with standardizing BOD tests and much literature on the subject can be found in the *Sewage Works*

Journal of the 1930s (see, for example, Hoskins, 1933). Eventually the BOD definition was standardized as the amount of oxygen consumed in a 250-ml glass bottle filled with wastewater over a 5-day period, with suitable dilutions. As an aside, it is interesting that they termed this *demand*, which suggests a connection with the law of supply and demand in economics. Oxygen is "supplied" by various forms of reaeration (diffusion and primary productivity) and it is "demanded" by microorganisms that degrade the organic materials through aerobic respiration. This overall conception is described by the Streeter–Phelps equation mentioned earlier in terms of wastewater disposal in rivers (Figure 2.4). It is also interesting to note that this very basic equation had its origin in a method of measuring biodegradation, i.e., putting wastewater in a bottle and measuring the oxygen concentration decrease. The real object of concern was the organic materials in the water, but oxygen was recorded because it was relatively easy to measure and its concentration changed in direct proportion to the change in organic materials.

Engineering design equations for wastewater treatment are often stated with the form given above in order to show effectiveness of biodegradation (Ce/Co) on one side of the expression, essentially in terms of percent removal. Design criteria are often given in these terms. For example, it might be required that 90% of the influent BOD be degraded by the treatment system in order to meet a regulatory requirement. Design to meet this requirement is done by sizing the treatment system. For this step, the original equation is expanded with an expression for the hydraulic residence time:

$$t = LWD/Q \tag{2.2}$$

where

t = hydraulic residence time, in days
L = length of the treatment system, in meters
W = width of the treatment system, in meters
D = depth of the treatment system, in meters
Q = average flow rate, in m^3/day

Plugging this expression into the original equation explicitly places dimensions that can be altered by design into consideration:

$$Ce/Co = e^{(-kLWD/Q)} \tag{2.3}$$

Values of Ce and Co are given and form the design criteria. Values of k and Q are known for the particular situations being designed for. Design consists of finding combinations of L, W, and D (i.e., size of the treatment system) that match with the situation. In other words, the above equation is solved for size, knowing all of the other parameters. Much of wastewater treatment engineering involves creating and solving design equations for the size of the treatment system in a similar fashion. Many alternative treatment systems exist and many equations have been developed to describe them. Some of these equations use theoretical expressions for reaction rates while others use empirical relationships based on practical experience. Various extensions for recirculation are often required and are incorporated into equations. Kadlec and Knight's (1996) text provides the state of the art in treatment wetland design equations and all indications are that this knowledge will continue to be increased and refined, as was true for traditional sanitary engineering.

A form of parallel evolution has occurred in ecology with equations for decomposition. In this case the goal was to develop equations that describe the process of decay, rather than equations that can be manipulated to meet biodegradation criteria. The model shown in Figure 2.24 is the same model arrived at by ecologists to model biodegradation in their contexts. It was first used by Jenny et al. (1949) and is still the basic approach taken, at least as a starting point. Decomposition, as described by this equation, is as important in ecology as is primary productivity. Another parallel is that the origin of this model was largely method-based, as was true with wastewater treatment engineering in terms of BOD measurement. In this case ecologists place known amounts of organic materials (usually leaf litter) into mesh bags and use them to measure mass loss over time. These bags are placed into the environment being evaluated, and the mesh material of the bags allows access by at least those decomposer organisms that are smaller than mesh size. A set of bags is placed in the environment at the beginning of the study, and a subset is picked up at intervals and weighed throughout the study. In this way mass loss is recorded and the decay constant, k, can be found as the slope of the curve of mass loss vs. time. This is called the litter bag method of studying decomposition (Shanks and Olson, 1961), and Olson (1963) set out an early mathematical description of the modelling, which includes an analog of the Streeter–Phelps equation. Although advancements have been discussed (Boulton and Boon, 1991; Wieder and Lang, 1982), this approach is still the foundation for understanding decomposition in ecology.

In essence then, the same kind of thinking occurred in ecology to describe decomposition as occurred in sanitary engineering to describe reduction in BOD of wastewaters. Both the scientists (ecologists) and engineers (sanitary engineers) studied the decay process in their particular systems and came up with the same equation. The engineers took one further step of being able to manipulate the equation for design, but this was not required for the scientists. The parallel evolution of thinking on this single topic provides a connection for understanding the new field which combines ecology and engineering.

ECOLOGY AS THE SOURCE OF INSPIRATION IN DESIGN

The hypothesis that has emerged from the examination of the history of the use of wetlands for wastewater treatment is that the technology primarily evolved from ecologists working on applied problems, rather than from engineers discovering ecosystems as useful systems. If this hypothesis is true, it suggests that the field of ecological engineering has a unique approach compared with other forms of engineering. It is not just a variation of environmental engineering but a whole new branch of engineering, and perhaps of ecology. Detailed descriptions of the design

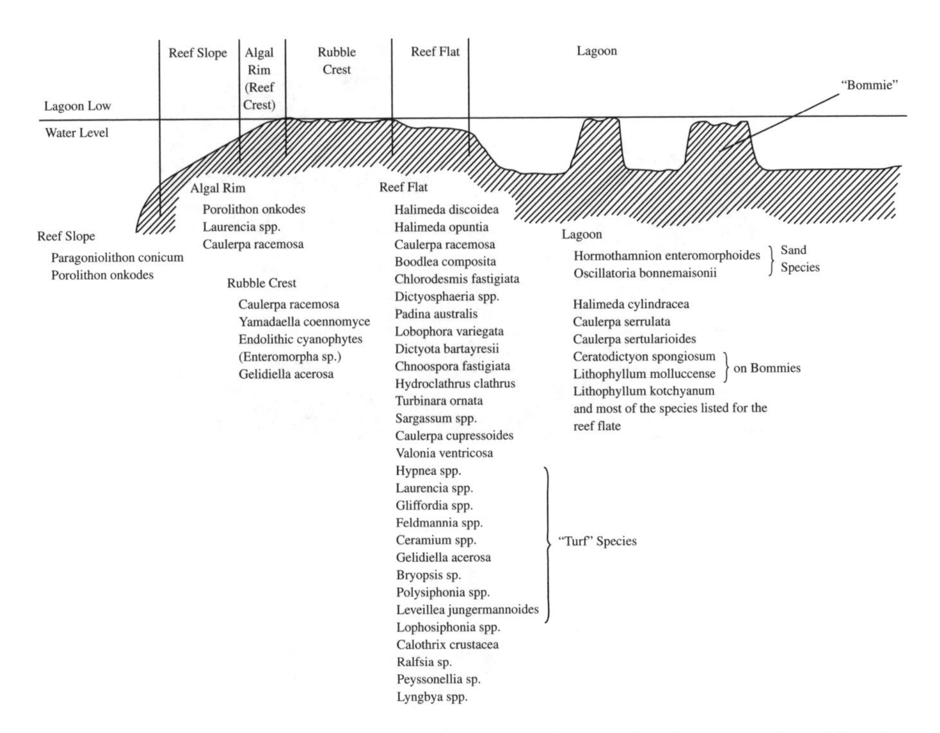

FIGURE 2.25 Cross-section through a coral reef showing the distribution of algae. Note the location of the algal rim (reef crest) where algal turfs can be found. (From Berner, T. 1990. *Coral Reefs: Ecosystems of the World.* Vol. 25. Z. Dubinsky (ed.). Elsevier, Amsterdam, the Netherlands. With permission.)

histories of two variations on the treatment wetland technology are presented to further explore this idea.

Algal Turf Scrubbers

Walter Adey has developed a unique wastewater treatment system, termed *algal turf scrubbers*, which utlize algae to strip pollutants out of water (Adey and Loveland, 1998). Although there is a long history of trials employing algae for wastewater treatment in the sanitary engineering field (Bartsch, 1961; Gotaas et al., 1954; Laliberte et al., 1994; Oswald, 1988; Oswald et al., 1957; Wong and Tam, 1998), Adey came upon his version of technology from studies of basic ecology. Adey is a coral reef ecologist who published much work, especially on algae, in the 1970s (Adey, 1973, 1978; Adey and Burke, 1976; Connor and Adey, 1977). Algae are the most important primary producers on coral reefs and they occupy many microhabitats (Figure 2.25). The algal turf scrubber technology is based on Adey's adaptation of algal turfs from coral reefs (Adey and Goertemiller, 1987; Adey and Hackney, 1989; Adey and Loveland, 1998). Algal turfs are short, moss-like mats of algal filaments covering hard surfaces found at the reef crest where wave energy is highest (Figure 2.26). Adey created artificial algal turfs by growing the algae on a screen in a shallow trough over which water was passed (Figure 2.27), with artificial lights

FIGURE 2.26 The reef crest on the barrier reef of Belize.

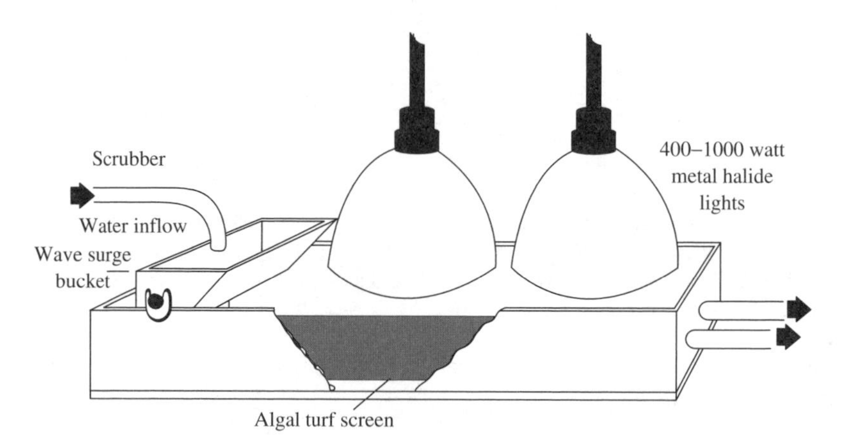

FIGURE 2.27 View of an algal turf scrubber unit. (From Adey, W. H. and K. Loveland. 1998. *Dynamic Aquaria*, 2nd ed. Academic Press, San Diego, CA. With permission.)

and wave energy generated by a surge bucket. The algae grow very quickly and strip nutrients out of the flowing water through uptake. By scraping the algae off the screens periodically, nutrients are permanently removed from the system and water quality is improved.

Adey came upon this technology while trying to design large coral reef aquaria for research and exhibit purposes (see Chapter 5). His challenge was to maintain narrow water quality conditions necessary for survival of the sensitive coral reef organisms. Adey's (1987) description of the discovery of the algal turf scrubber, after many unsuccessful trials of commercially available filter systems, reveals the basic ecological knowledge embodied in the design:

> Finally, I decided to try to remove a piece of the primary photosynthetic component, the algal turf from the reef itself, and to allow that plant community to develop and

> photosynthesize under appropriately high light intensity in a side branch of the entire system, during hours when the reef was in darkness.
>
> From our experience with both the wild and the microcosm algal turfs, we concluded that we might make this work if, along with high light intensity, we supplied wave action, water flow, a porous surface (to prevent overgrazing) and constant harvesting (to prevent community succession). Thus, we created a device called "the algal turf scrubber" and attached it to the 7 kl system late in 1979. The algal turf scrubber was extraordinary successful, in that it achieved primary production rates characteristic of a wild reef, and also simulated the effects of high-quality ocean water by adding oxygen to the system and scrubbing nutrients from it. Most important, it could be operated at night, when water quality is likely to decline, and it was controllable in many ways, since by adjusting light, wave action, water flow and harvest rates we could maintain water chemistry in the microcosm reef much as ocean flow maintains it in the wild.

An algal turf scrubber was attached to Adey's coral reef model to simulate a larger body of water that would normally surround a reef and buffer its water quality. Specifically, the scrubber is lighted in a cycle opposite to the model reef so that oxygen would be supplied during the night and so that nutrients and CO_2 released by nighttime respiration would be taken up. These critical functions allowed a high diversity of animal life to survive in the model reef. Adey patented the technology in 1982 and applied it to a number of living ecosystem models (Adey and Loveland, 1998). He later scaled the design up in size and applied algal turf scrubbers to a variety of types of wastewaters (Adey et al., 1993, 1996; Blankenship, 1997; Craggs et al., 1996). Overall, this represents an excellent example of ecological engineering by utilizing ecological knowledge and the principle of preadaptation. Specifically, Adey recognized that the natural algal turf was preadapted for wastewater treatment. The design process used by Adey is proposed in Figure 2.28. He studied natural algal turfs on coral reefs (references shown in the upper box) and then had a creative inspiration that allowed him to use the natural system in an engineered design to treat wastewater (references shown in the lower box). The creative inspiration is shown by the arrow connecting studies of the natural system with examples of engineered designs. This kind of insight is the essence of ecological engineering!

Living Machines

John Todd has developed a unique wastewater treatment system, termed *the living machine*, which is the product of a long design history (Guterstam and Todd, 1990; Todd, 1988a, 1988b, 1990, 1991; Todd and Todd, 1994). The development of the design started at the New Alchemy Institute on Cape Cod, which Todd helped create in the early 1970s. The New Alchemy Institute was an organization devoted to developing and demonstrating integrated environmental technologies involving energy systems, architecture, and sustainable agriculture (Todd and Todd, 1980, see Chapter 9). One of the principal elements in these integrated systems was aquaculture. Especially with William McLarney and Ronald Zweig, Todd tried many configurations of fish culture tanks (McLarney and Todd, 1977; Zweig, 1986; Zweig et al., 1981). He settled on a large cylindrical tank (up to 1000 gal or 3790 l capacity)

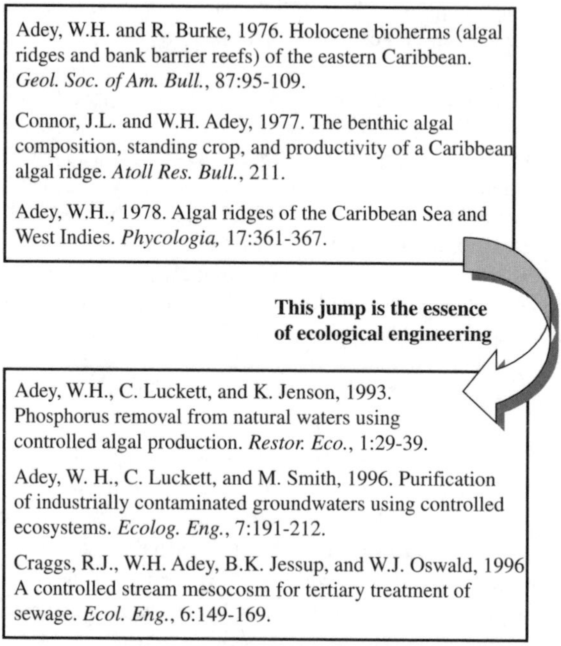

Adey, W.H. and R. Burke, 1976. Holocene bioherms (algal ridges and bank barrier reefs) of the eastern Caribbean. *Geol. Soc. of Am. Bull.*, 87:95-109.

Connor, J.L. and W.H. Adey, 1977. The benthic algal composition, standing crop, and productivity of a Caribbean algal ridge. *Atoll Res. Bull.*, 211.

Adey, W.H., 1978. Algal ridges of the Caribbean Sea and West Indies. *Phycologia,* 17:361-367.

This jump is the essence of ecological engineering

Adey, W.H., C. Luckett, and K. Jenson, 1993. Phosphorus removal from natural waters using controlled algal production. *Restor. Eco.*, 1:29-39.

Adey, W. H., C. Luckett, and M. Smith, 1996. Purification of industrially contaminated groundwaters using controlled ecosystems. *Ecolog. Eng.*, 7:191-212.

Craggs, R.J., W.H. Adey, B.K. Jessup, and W.J. Oswald, 1996 A controlled stream mesocosm for tertiary treatment of sewage. *Ecol. Eng.*, 6:149-169.

FIGURE 2.28 The intellectual leap taken by Walter H. Adey in developing the algal turf scrubber technology.

made of translucent material as a basic module (Figure 2.29). Table 2 in *Tomorrow Is Our Permanent Address* (Todd and Todd, 1980) lists much of the design knowledge embodied in the aquaculture system, labeled as "biologically designed versus engineered closed-system aquacultures." The living machine technology evolved from the basic aquaculture module, first by combining them in series and then by using the series as a sequential wastewater treatment system (Figure 2.30). This basic sequential system has evolved over time with the different modules becoming specialized to perform critical functions. Todd (1990) credits the need for maintaining water quality in the aquaculture systems as a kind of inspiration in the design evolution of living machines for wastewater treatment, as noted in the following quote:

> For over fifteen years, beginning at New Alchemy, I had raised fish and had learned innumerable tricks to purify water in order to keep the fish healthy. It seemed logical to use the same biological techniques and apply them to purifying water, sewage and other waste streams. An ecosystem approach, while dramatically different from conventional waste engineering, seemed to me to be the best long-term solution to upgrading water quality not only on Cape Cod, but throughout the country.

This design evolution is discussed further in a paper entitled "Biology as a Basis for Design" (Todd and Todd, 1991), which captures the essence of the technology.

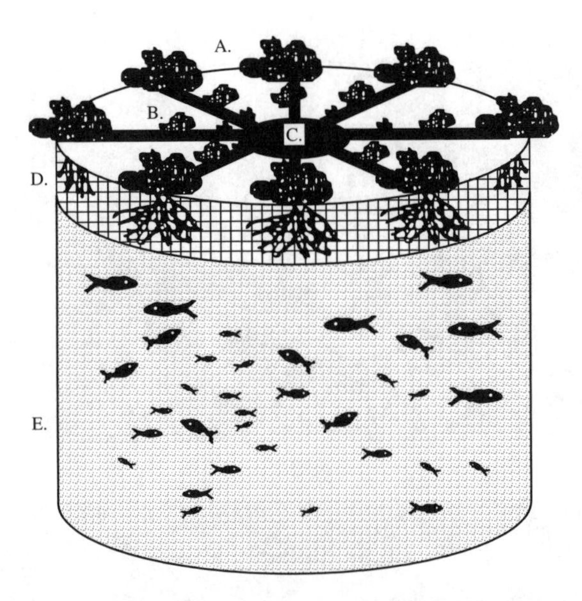

FIGURE 2.29 View of a single tank used in early work on aquaculture by John Todd. (A) Hydroponic vegetables on top of pond. (B) Styrofoam flotation and guides for plants. (C) Central core opening for fish feeding. (D) Mesh cage to prevent fish from eating plant roots. (E) Fish rearing area in pond. (From Zweig, R. D. 1986. *Aquaculture Magazine.* 12(3):34–40. With permission.)

Principal Steps of the Wastewater Treatment at the Stensund Aquaculture:

1. Raw Sewage
2. Anaerobic Treatment
3. Aerobic Treatment
4. Phytoplankton-Bacteria Basin
5. Zooplankton Basin
6. Polyculture (Fish, Crayfish, Plants)

- 2–3 are Steps of Mineralization and Detoxification
- 4–6 are Steps of Aquaculture

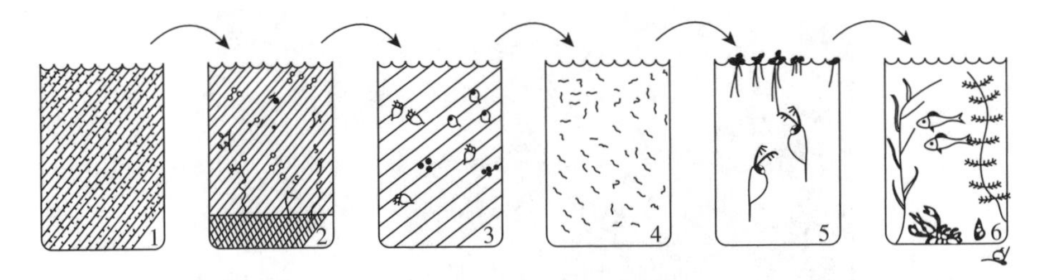

FIGURE 2.30 The idea of using a series of tanks for wastewater treatment, with aquaculture at the final stages. (From Guterstam, B. and J. Todd. 1990. *Ambio.* 19:173–175. With permission.)

A number of living machines have been built and tested (Figure 2.31) with much description presented in Todd's own journal (Josephson, 1995; Josephson et al., 1996). A significant amount of work on living machines has also been done in Sweden (Etnier and Guterstam, 1991; Guterstam, 1996; Guterstam and Todd, 1990).

FIGURE 2.31 Views of living machines. (a) Frederick, Maryland. (b) Naples, Florida. (c) Burlington, Vermont.

TABLE 2.5
Design Principles for John Todd's Living Machine Technology

Principle	Design Objective
Mineral diversity	Include igneous, sedimentary, and metamorphic rocks to provide a foundation for complex biological chemistry
Nutrient reservoirs	Provide nutrients in available forms in order to maintain balanced cycling
Steep gradients	Connect subsystems with very different physical–chemical conditions
High exchange rates	Maximize the surface area of living biomass that is exposed to wastewater
Periodic and random pulsed exchanges	Pulsing in physical–chemical conditions leads to robust adaptations
Cellular design and the structure of mesocosms	Provide some degree of small-scale autonomy within the larger scale context by using cells as units of design
Minimum number of subsystems	Incorporate at least a small number of linked subsystems to enhance stability
Microbial communities	Encourage microbial diversity because of its critical role in overall performance
Solar-based photosynthetic foundations	Utilize solar energy as an energy subsidy by incorporating photosynthesis in the design with plant populations
Animal diversity	Incorporate many animal populations, especially filter feeding invertebrates, into the design for added control functions
Biological exchanges beyond the mesocosm	Frequent seeding from different external sources adds adaptability to the design
Microcosm, mesocosms, macrocosm relationships	Include different ecological scales in the overall design concept

Source: Adapted from Todd, J. and B. Josephson. 1996. *Ecological Engineering.* 6:109.

Much of the design knowledge on living machines is recorded in the book entitled *From Eco-Cities to Living Machines* (Todd and Todd, 1994), and some of it is included in Table 2.5. Even a children's book on living machines has been produced (*Bang*, no date). Overall, this is an excellent example of an ecologically engineered technology utilizing the principle of preadaptation early in the design. The aquaculture tanks were preadapted to be organized into a wastewater treatment system with each containing a different unit process.

A critical component of the living machine is the sequential nature of the treatment process. Possibly Todd was influenced by John Ryther's project combining aquaculture and wastewater treatment that started in the early 1970s at Woods Hole

(Figure 2.23) while Todd worked there. A tie between Todd's design work and Ryther's project might be indicated by McLarney's role with Ryther in a major text on aquaculture (Bardach et al., 1972). Although Ryther's project continued through the 1970s and was well documented (Dunstan and Tenore, 1972; Goldman and Ryther, 1976; Goldman et al., 1973; 1974a, 1974b; Ryther et al., 1972; Tenore et al., 1973), it apparently led to no commercial development unlike Todd's living machine concept. Todd's work has generated a design company named "Living Technologies," another company named "Ecological Engineering Associates" (Teal and Peterson, 1991, 1993; Teal et al., 1994), and most recently, "Ocean Arks International."

3 Soil Bioengineering

INTRODUCTION

The transformation of watersheds is a characteristic of human civilization. Humans transform natural landscapes into various kinds of "land use" that provide them with habitation and resources. Altered hydrology and soil erosion occur as a consequence of these transformations, which are problems that must be addressed. The main kinds of transformations include development of agriculture, urbanization, and alterations of streams, rivers, and coastlines. In all cases natural vegetation is removed or changed and land forms are simplified (usually leveled). Society generally accepts that these direct impacts must occur to accommodate human land use, but indirect impacts such as erosion are not acceptable and require engineering solutions and/or management.

Erosion is a major environmental impact that results in loss of agricultural productivity, aquatic pollution, and property damages among other problems. Although the impact of erosion has long been recognized (Bennett and Lowdermilk, 1938; Brown, 1984; Judson, 1968), it remains a challenge to society. Costs due to urban, shoreline, and agricultural erosion are tremendous, and a major industry of businesses and technologies has arisen for erosion control.

A set of ecological engineering techniques has evolved with the industry for erosion control; that is the subject of this chapter. This subdiscipline has been referred to as *bioengineering*, and it involves a combination of conventional techniques from civil or geotechnical engineering with the use of vegetation plantings (Table 3.1). It is an interesting field that is growing rapidly as a cost-effective solution to erosion problems. Most workers in the field are not concerned about (or perhaps not even aware of) problems with overlapping meanings of the term *bioengineering,* which is often used in other contexts (Johnson and Davis, 1990). Schiechtl and Stern (1997) provide some background discussion and end up suggesting the term *water bioengineering* for some applications. Gray and Leiser's (1982) use of the phrase "biotechnical slope protection and erosion control" is perhaps more appropriate but too long and awkward as a descriptor. Here, the field is referred to as *soil bioengineering* as a compromise term that is used by many workers.

The central basis of soil bioengineering from both a philosophical and a technical perspective is an understanding of the interface between hydrology, geomorphology, and ecology. Hydrology integrates the landscape, especially by water movements, and helps create an interactive relationship between landform and ecosystem. An old subdiscipline of ecology called *physiographic ecology* in part covered this topic. Physiographic ecology was a descriptive field analysis of vegetation and topography that flourished briefly around the turn of the 20th century (Braun, 1916; Cowles, 1900, 1901; Gano, 1917). These studies are detailed descriptions that convey a rich, though static, understanding of landscape ecology. Like many kinds of purely

TABLE 3.1
Comparisons of Definitions of Soil Bioengineering

Flyer from a Rutgers University Short Course

Soil bioengineering is an emerging science that brings together ecological, biological and engineering technology to stabilize eroding sites and restore riparian corridors. Streambanks, lakeshores, tidal shorelines and eroded upland areas all may be effectively revegetated with soil bioengineering techniques if designed and implemented correctly.

Advertisement for a Commercial Company (Bestman Green Systems, Salem, Massachusetts)

Bioengineering is a low-tech approach for effective yet sensitive design and construction using natural and living materials. The practice brings together biological, ecological, and engineering concepts to vegetate and stabilize disturbed land … Once established, vegetation becomes self-maintaining.

Advertisement for a Commercial Company (Ernst Conservation Seeds, Meadville, Pennsylvania)

Bioengineering is a method of erosion control for slopes or stream banks that uses live shrubs to reduce the need for artificial structures.

Bowers, 1993

Bioengineering is the practice of combining structural components with living material (vegetation) to stabilize soils.

Schiechtl and Stern, 1997

Bioengineering: an engineering technique that applies biological knowledge when constructing earth and water constructions and when dealing with unstable slopes and riverbanks. It is a characteristic of bioengineering that plants and plant materials are used so that they act as living building materials on their own or in combination with inert building materials in order to achieve durable stable structures. Bioengineering is not a substitute; it is to be seen as a necessary and sensible supplement to the purely technical engineering construction methods.

Escheman, no date

By definition, soil bioengineering is an applied science which uses living plant materials as a main structure component … In part, soil bioengineering is the re-establishment of a balanced living, native community capable of self-repair as it adapts to the land's stresses and requirements.

descriptive sciences, physiographic ecology fell into disfavor and disappeared as experimental approaches began to dominate ecology in the mid-1900s. Few studies combining geomorphology and ecology occurred afterwards, probably due to the difficulties with conducting experiments at the appropriate scales of space and time. There was a renewal of interest in these kinds of studies in the 1970s, especially for barrier islands (Godfrey and Godfrey, 1976; Godfrey et al., 1979) where the time scales of vegetation and geomorphic change are fast and closely matched. Swanson (1979; Swanson et al., 1988) provided a modern review of the topic and synthesized his discussion with a summary diagram (Figure 3.1). This diagram traces the many interactions that occur between the realms of geomorphology and ecology that are of interest in soil bioengineering. Another view illustrating the unity of ecology and

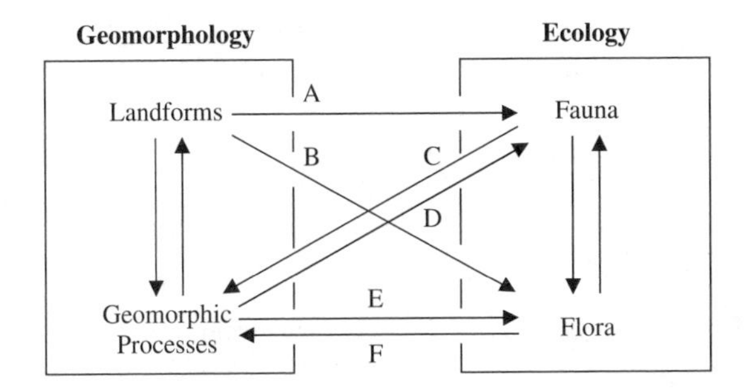

FIGURE 3.1 Relationships between geomorphology and ecology. (A) Define habitat, range. Effects through flora. (B) Define habitat. Determine disturbance potential by fire, wind. (C) Affect soil movement by surface and mass erosion. Affect fluvial processes by damming, trampling. (D) Sedimentation processes affect aquatic organisms. Effects through flora. (E) Destroy vegetation. Disrupt growth by tipping, splitting, stoning. Create new sites for establishment and distinctive habitats. Transfer nutrients. (F) Regulate soil and sediment transfer and storage. (From Swanson, F. J. 1979. *Forests: Fresh Perspectives from Ecosystem Analysis.* R. H. Waring (ed.). Oregon State University Press, Corvallis, OR. With permission.)

geomorphology is Hans Jenny's CLORPT equation. This is a conceptual model originally created for discussing soil formation (Jenny, 1941) but later generalized for ecosystems (Jenny, 1958, 1961). The basic form of the original equation is:

$$S = f(CL, O, R, P, T) \tag{3.1}$$

where

S = any soil property
CL = climate
O = organisms or, more broadly, biota
R = topography, including hydrologic factors
P = parent material, in terms of geology
T = time or age of soil

Soil is, therefore, seen as a function of environmental factors including biota of the ecosystem (O) and geomorphology (R). Jenny used the CLORPT equation for understanding pedogenesis and as a basis for his view of landscape ecology (Jenny, 1980). Updates on uses and development of this classic equation are given by Phillips (1989) and Amundson and Jenny (1997). More recently the term *biogeomophology*, and related variations, is being used for studies of ecology and geomorphology (Butler, 1995; Howard and Mitchell, 1985; Hupp et al., 1995; Madsen, 1989; Reed, 2000; Viles, 1988). This term is analogous to *biogeochemistry,* which is an important subdiscipline of ecology dealing with the cycles of chemical elements in landscapes.

The history of studies of geomorphology and ecology document that natural ecosystems control or regulate hydrology and the geomorphic processes of erosion and sedimentation. Soil bioengineering attempts to restore these functions in watersheds that have been altered by human land use. The combined use of vegetation

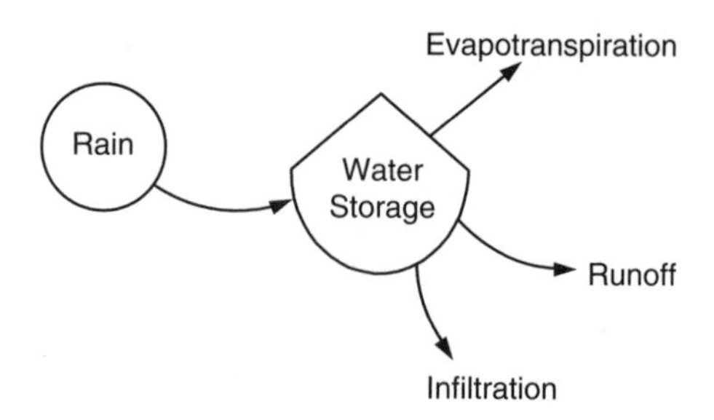

FIGURE 3.2 Energy circuit diagram of the basic hydrologic model.

plantings and conventional engineering that is involved makes this subdiscipline an important area of ecological engineering.

STRATEGY OF THE CHAPTER

Basic elements of geomorphology are covered first in the chapter to provide context for a review of soil bioengineering designs. Old and new approaches are referenced with an emphasis on a systems orientation and energy causality. Next, basic concepts of soil engineering are introduced. Like other forms of ecological engineering, this discipline represents a new way of thinking, even though some of its ideas can be traced back to Europe in the 1800s and to the Soil Conservation Service in the 1930s in the U.S. Advantages and disadvantages of soil bioengineering designs are mentioned. The philosophical implications of the field are covered, including possible connections to Eastern religions. Finally, four case studies are included which add detail to the review. The self-building behavior found in several ecosystems is highlighted as a special feature appropriate for ecological engineering designs.

THE GEOMORPHIC MACHINE

An understanding of geomorphology begins with hydrology. In very dry or very cold environments other factors are also required, but here the focus is on the more-or-less humid environments where human population density is highest. A mini-model of the hydrologic balance is shown in Figure 3.2. Precipitation is a source or input of water storage, while evapotranspiration, runoff, and infiltration are outputs. The energetics of this model are critical but straightforward. Movements of liquid water have kinetic energy in proportion to their velocity, and the storage of water has potential energy in proportion to the height above some base level. The energetics of hydrology drive geomorphic processes and create landforms.

In humid environments geomorphology involves mainly erosion, transport, and deposition of sediments. The action of these processes has been metaphorically referred to as the "geomorphic machine" in which hydrology drives the wearing down of elevated landforms (Figure 3.3). Leopold's (1994) quote for the special case of rivers given below describes this metaphor:

FIGURE 3.3 A machine metaphor for geomorphology. (From Bloom, A. L. 1969. *The Surface of the Earth.* Prentice Hall. Englewood Cliffs, NJ. With permission.)

> The operation of any machine might be explained as the transformation of potential energy into the kinetic form that accomplishes work in the process of changing that energy into heat. Locomotives, automobiles, electric motors, hydraulic pumps all fall within this categorization. So does a river. The river derives its potential energy from precipitation falling at high elevations that permits the water to run downhill. In that descent the potential energy of elevation is converted into the kinetic energy of flow motion, and the water erodes its banks or bed, transporting sediment and debris, while its kinetic energy dissipates into heat. This dissipation involves an increase in entropy.

The machine metaphor is especially appropriate in the context of ecological engineering and brings to mind John Todd's idea of the living machine (see Chapter 2). In fact, vegetation regulates hydrology and therefore controls the geomorphic machine described above. For example, the role of forests in regulating hydrology is well known (Branson, 1975; Kittredge, 1948; Langbein and Schumm, 1958). Perhaps the most extensive study of this action was at the Hubbard Brook watershed in New Hampshire. This was a benchmark in ecology which involved measurements of biogeochemistry and forest processes at the watershed scale (Bormann and Likens, 1979; Likens et al., 1977). It was an experimental study in which replicate forested watersheds were monitored. One was deforested to examine the biogeochemical consequences of loss of forest cover and to record the recovery processes as regrowth occurred. The forest was shown to regulate hydrology in various ways by comparing the deforested watershed with a control watershed that was not cut. Deforestation increased streamflow in the summer through a reduction in evapotranspiration, changed the timing of winter streamflow, reduced soil storage capacity, and increased

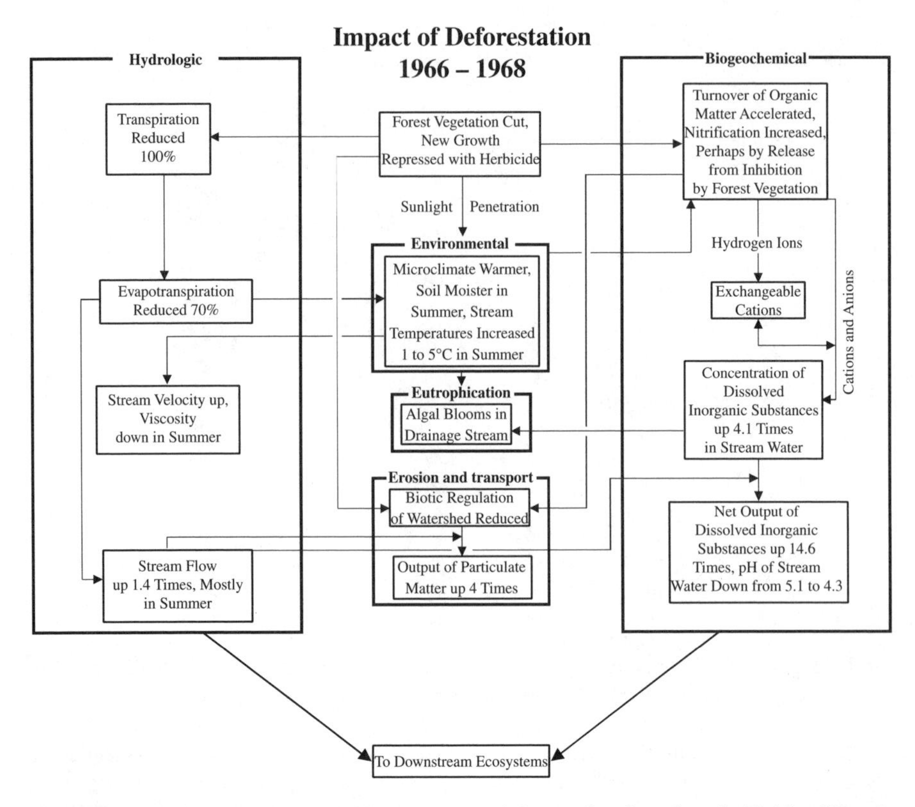

FIGURE 3.4 Sequence of watershed responses to deforestation, based on the Hubbard Brook experiment. (From Likens, G.E. and F.H. Bormann. 1972. Biogeochemical cycles. *Science Teacher*. 39(4):15–20. With permission.)

peak streamflows during storms. The summary diagram of the deforestation experiment illustrates an increased erosion rate (Figure 3.4) and thus the connection between the ecosystem and landform. Soil bioengineering systems are designed to restore at least some of this kind of control over hydrology and geomorphic processes.

To further illustrate the geomorphic machine, the three main types of erosion in humid landscapes are described below with minimodels. Emphasis is on geomorphic work, so other aspects of hydrology are left off the diagrams. In each model, erosion is shown as a work gate or multiplier that interacts an energy source with a soil storage to produce sediments.

Upland erosion is shown in Figure 3.5. Initially, precipitation interacts with soil in splash erosion. Vegetation cover absorbs the majority of the kinetic energy of rain drops, but when it is removed or reduced in agriculture, construction sites, or cleared forest land, this initial form of erosion can be significant. Sheet and rill erosion occur as the water from precipitation runs off the land. Various best management practices (BMPs) are employed to control runoff and the erosion it causes as will be discussed later.

Channel erosion is shown in Figure 3.6. Stream flow, which is runoff that collects from the watershed, is the main energy source along with the sediments it carries.

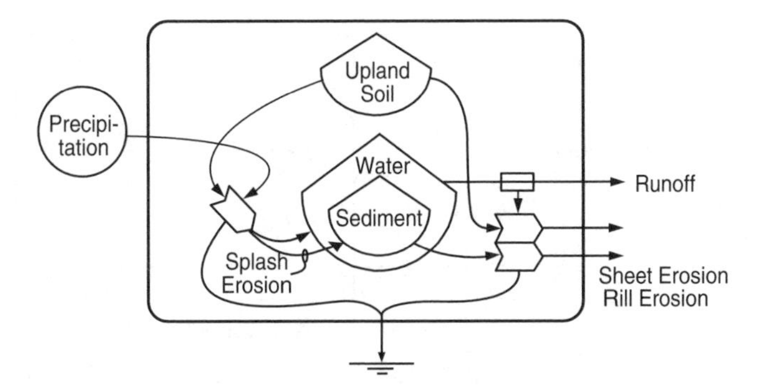

FIGURE 3.5 Energy circuit model of the types of upland erosion.

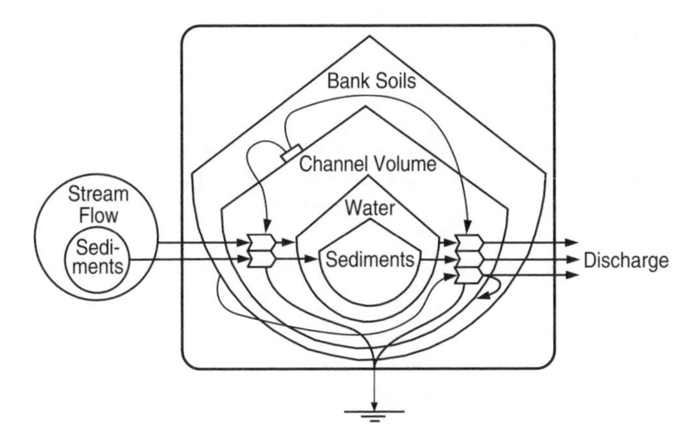

FIGURE 3.6 Energy circuit model of stream channel erosion.

The system itself is depicted as a set of concentric storages: the bank soils contain the channel volume, which contains the stream water, which contains suspended sediments. Movement of water through the system erodes bank soils and simultaneously increases channel volume. The term for output from the system is *discharge*, which includes the stream water and the sediment load that it carries through advection. The behavior of this system is covered by the subdiscipline of fluvial geomorphology. Velocity of stream water is of critical importance since it is a determinant of kinetic energy and erosive power. A typical relationship for velocity is shown below (Manning's equation; see also Figure 3.22):

$$V = 1.49(R^{2/3}S^{1/2})/n \tag{3.2}$$

where

V = mean velocity of stream water
R = mean depth of the flow
S = the stream gradient or slope
n = bottom roughness

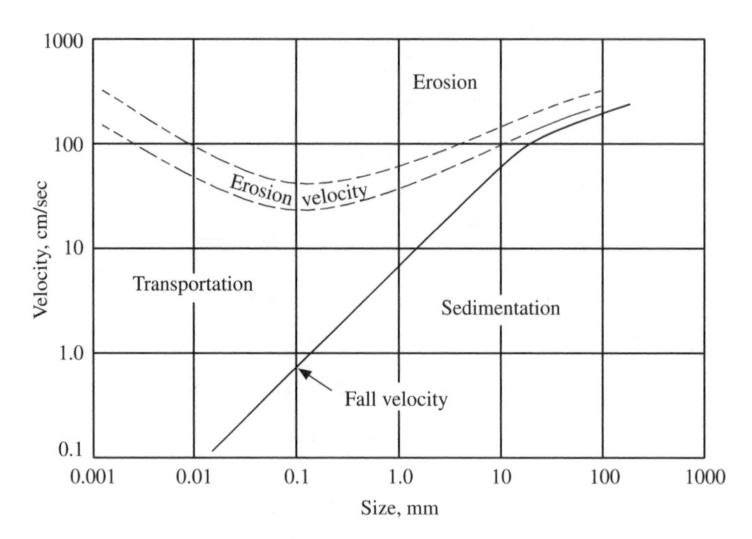

FIGURE 3.7 Complex patterns of sediment behavior relative to current velocity in a stream environment known as the Hjulstrom relationship. (Adapted from Morisawa, M. 1968. *Streams, Their Dynamics and Morphology.* McGraw-Hill, New York.)

Thus, velocity is directly proportional to depth and gradient and inversely proportional to roughness. This relationship will be explored later in terms of design of soil bioengineering systems.

The work of streams and rivers depends on velocity according to the Hjulstrom relationship, which is named for its author (Novak, 1973). This is a graph relating velocity to the three kinds of work: erosion, transportation, and sedimentation, relative to the particle size of sediments (Figure 3.7). Sedimentation dominates when particle sizes are large and velocities are slower, transport dominates at intermediate velocities and for small particle sizes, while erosion dominates at the highest velocities for all particle sizes. Based on this relationship, particle sizes of a stream deposit are a reflection of the velocity (and therefore the energy) of the stream that deposited them.

Fluvial or stream systems develop organized structures through geomorphic work including drainage networks of channels and landforms such as meanders, pools and riffle sequences, and floodplain features. Vegetation plays a role in fluvial geomorphology by stabilizing banks and increasing roughness of channels.

Coastal erosion is modelled in Figure 3.8. The principal energy sources are tide and wind, which generates waves. River discharge is locally important and, in particular, it transports sediments eroded from uplands to coastal waters. Coastlines are classified according to their energy, with erosion dominating in high energy zones and sedimentation dominating in low energy zones. Inman and Brush (1973) provide energy signatures for the coastal zone with a global perspective. Wave energy is particularly important and it is described below by Bascom (1964):

> The energy in a wave is equally divided between potential energy and kinetic energy. The potential energy, resulting from the elevation or depression of the water surface,

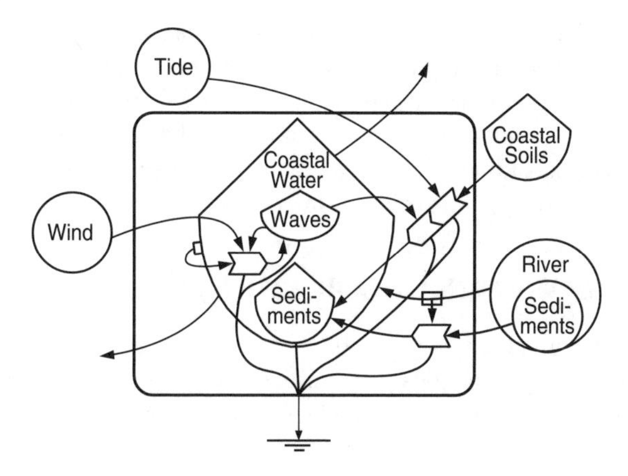

FIGURE 3.8 Energy circuit model of coastal erosion.

advances with the wave form; the kinetic energy is a summation of the motion of the particle in the wave train and advances with the group velocity (in shallow water this is equal to the wave velocity).

The amount of energy in a wave is the product of the wave length (L) and the square of the wave height (H), as follows:

$$E = (wLH^2)/8$$

where w is the weight of a cubic foot of water (64 lb).

Geomorphic work in the coastal zone builds a variety of landforms including channels and inlets, beaches, dunes, barrier islands, and mudflats. Vegetation is an important controlling factor in relatively low energy environments but with increasing energy, vegetation becomes less important, and purely physical systems such as beaches are found.

While early work in geomorphology focused on equilibrium concepts (Mackin, 1948; Strahler, 1950; Tanner, 1958), more recently nonequilibrium concepts are being explored (Phillips, 1995; Phillips and Renwick, 1992), such as Graf's (1988) application of catastrophe theory and Phillips' (1992) application of chaos. This growth of thinking mirrors the history of ecology (see Chapter 7). Drury and Nisbet (1971) provided a comparison of models between ecology and geomorphology, indicating many similarities that have developed between these fields. Like ecosystems, geomorphic systems can be characterized by energy causality, input–output mass balances, and networks of feedback pathways. They therefore can exhibit nonlinear behavior and self-organization as described by Hergarten (2002), Krantz (1990), Rodriguez-Iturbe and Rinaldo (1997), Stolum (1996), Takayasu and Inaoka (1992), and Werner and Fink (1993). Cowell and Thom's (1994) discussion of how alternations of regimes dominated by positive and negative feedback can generate complex coastal landforms is particularly instructive and may provide insight into analogous ecological dynamics. While these developments are exciting and can stimulate cross-disciplinary study, it is somewhat disappointing that geomorphologists have written little about the symbiosis between landforms and ecosystems. Knowledge of both disciplines and how they interact is needed to engineer and to manage the altered watersheds of human-dominated landscapes. Workers in soil bioengineering are developing this knowledge and probably will be leaders in articulating biogeomorphology to specialists in both ecology and geomorphology.

CONCEPTS OF SOIL BIOENGINEERING

The approach of soil bioengineering is to design and construct self-maintaining systems that dissipate the energies that cause erosion. Soil bioengineering primarily involves plant-based systems but also includes other natural materials such as stone, wood, and plant fibers. In fact, materials are very important in this field, and they are a critical component in designs. The materials, both living and nonliving, must be able to resist and absorb the impact of energies that cause erosion. Design in soil bioengineering involves both the choice of materials and their placement in relation to erosive energies. Grading — the creation of the slope of the land through earth-moving — is the first step in a soil bioengineering design. Shallower slopes are more effective than steep slopes because they increase the width of the zone of energy dissipation and therefore decrease the unit value of physical energy impact.

Soil bioengineering designs are becoming more widely implemented because (1) they can be less expensive than conventional alternatives and (2) they have many by-product values. Soil bioengineering designs have been shown to be up to four times less expensive than conventional alternatives for both stream (NRC, 1992) and coastal (Stevenson et al., 1999) environments. In addition, the by-product values of soil bioengineering designs include aesthetics, creation of wildlife habitat, and water quality improvement through nutrient uptake and filtering. The wildlife habitat values are often significant and may even dominate the design as in the restoration of streams for trout populations (Hunt, 1993; Hunter, 1991) or the reclamation of strip-mined land. Although soil bioengineering systems are multipurpose, in this chapter the focus is on erosion control. Chapter 5 covers the creation of ecosystems whose primary goal is wildlife habitat or other ecological function. As an example, Figure 3.9 depicts a possible design for stream restoration that would serve dual functions.

In some situations soil bioengineering is truly an alternative for conventional approaches to erosion control from civil or geotechnical engineering. However, other situations with very high energies require conventional approaches or hybrid solutions. Conventional approaches to erosion control involve the design and construction of fixed engineering structures. These include bulkheads, seawalls, breakwaters, and revetments which are made of concrete, stone, steel, timber, or gabions (stone-filled wire baskets). Such structures are capable of resisting higher energy intensities than vegetation. The most common and effective type of structure for bank protection along shorelines or in stream channels is a carefully placed layer of stones or boulders known as riprap (Figure 3.10). The rock provides an armor which absorbs the erosive energies and thereby reduces soil loss. Rock fragments which make up a riprap revetment must meet certain requirements of size, shape, and specific gravity. A

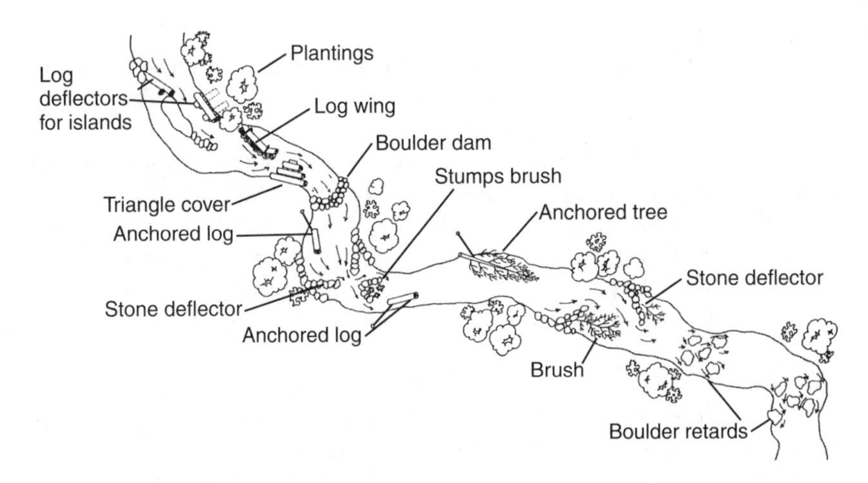

FIGURE 3.9 A typical stream restoration plan. (From Kendeigh, S. C. 1961. *Animal Ecology*. Prentice Hall, Englewood Cliffs, NJ. With permission.)

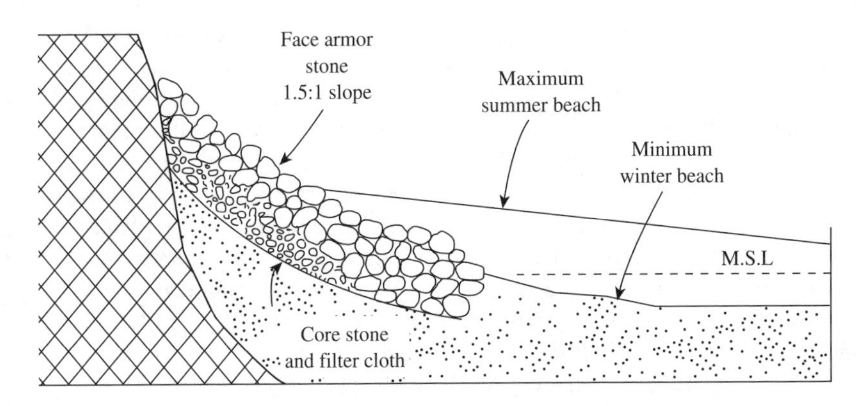

FIGURE 3.10 Use of riprap for erosion control. (From Komar, P. D. 1998. *Beach Processes and Sedimentation*, 2nd ed. Prentice Hall, Upper Saddle River, NJ. With permission.)

sample design equation for the weight of rock fragments to be used in coastline protection, known as Hudson's formula (Komar, 1998), is given below:

$$W = (dgH^3)/k(S-1)^3 \cot A \tag{3.3}$$

where

W = weight of the individual armor unit
d = density of the armor-unit material
g = acceleration of gravity
H = height of the largest wave expected to impact the structure

k = a stability coefficient
S = specific gravity of the armor material relative to water
A = angle of the structure slope measured from the horizontal

Gray and Leiser (1982) have given a related design relationship for riprap stone weight for a stream channel situation in regard to current velocity.

In addition to the structures described above, conventional approaches to erosion control employ various geosynthetics, which are engineered materials usually made of plastics. These take the form of mats used to stabilize soils, and they include geotextiles, geogrids, geomembranes, and geocomposites (Koerner, 1986).

The heart of soil bioengineering is new uses of vegetation for erosion control that can replace or augment the conventional approaches. Soil bioengineering designs are covered in several important texts (Gray and Leiser, 1982; Morgan and Rickson, 1995; Schiechtl and Stern, 1997) and in trade journals such as *Erosion Control* and *Land and Water*. A few designs are reviewed below as an introduction, but detailed case studies are covered in subsequent sections of the chapter for urban, agricultural, stream, and coastal environments. This is a very creative field with many sensitive designs that have been derived through trial and error and through observation and logical deduction about physical energetics at the landscape scale. Various kinds of vegetation are employed to control erosion, depending on the environment. Woody plants such as willows (*Saliaceae*) are used in stream environments and mangroves on tropical coastlines; herbaceous wetland plants such as cattails (*Typha* sp.) are used in freshwater and cordgrass (*Spartina* sp.) in saltwater environments. Direct mechanisms of erosion control by living plants include (1) intercepting raindrops and absorption of rainfall energy, (2) reducing water flow velocity through increased roughness, and (3) mechanical reinforcement of the soil with roots. Living plants also indirectly affect erosion through control of hydrology in terms of increased infiltraton and evapotranspiration. Plants are used in soil bioengineering designs in many ways. Individual plants are planted either as rooted stems or as dormant cuttings that later develop roots. Groups of cuttings are also planted as fascines (sausage-like bundles of long stems buried in trenches), brush-mattresses (mat-like layers of stems woven together with wire and placed on the soil surface), or wattles (groups of upright stems formed into live fences). Willows in particular are preadapted for use in soil bioengineering along streams because of their fast growth and their ability to produce a thick layer of adventitious roots (i.e., roots that develop from the trunk or from branches), and also because their stems and branches are elastic and can withstand flood events (Watson et al., 1997). Schiechtl and Stern (1997) show many line drawings of how these and other kinds of plantings are used in slope protection. Often plantings are used in hybrid designs along with conventional approaches as shown in Figure 3.11. Protection of the "toe" or lower portion of a slope with resistant materials is especially important because this location receives the highest erosive energy. Thus, a typical hybrid design would include rock armor at the toe of the slope with plantings on the upper portion of the slope.

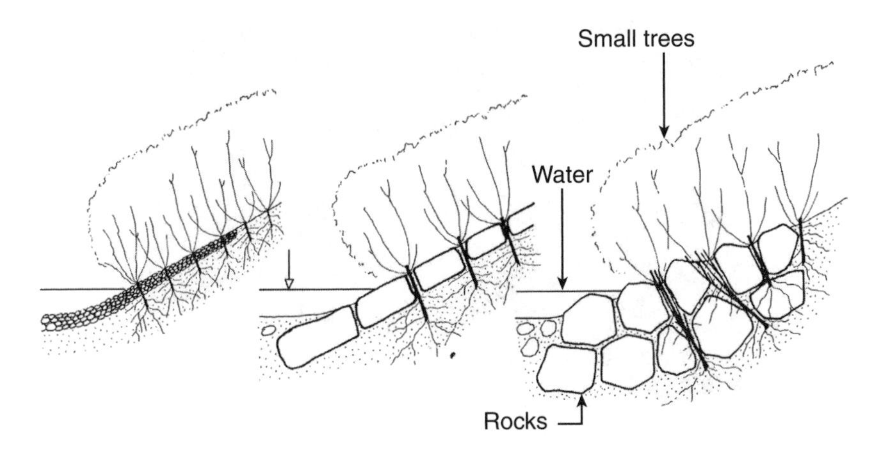

FIGURE 3.11 The combined use of riprap and vegetation plantings for a soil bioengineering design. (From Schiechtl, H. M. and R. Stern. 1997. *Water Bioengineering Techniques for Watercourse, Bank and Shoreline Protection.* Blackwell Science, Cambridge, MA. With permission.)

Other natural, nonliving materials besides stone are often included in soil bioengineering designs. For example, tree trunks are used in several ways. Log deflectors have a long history of use in streams to divert flow away from banks. Owens (1994) describes a similar though more elaborate kind of structure using trunks with branches which he terms *porcupines*. Root wads — tree trunks with their attached root masses (Figure 3.12) — also have been used as a kind of organic riprap in streams to absorb current energy (Oertel, 2001). All of these uses are made even more effective when the trees to be used are salvaged from local construction sites rather than harvested from intact forests. Other examples of natural nonliving materials used in soil bioengineering designs include hay bales, burlap, and coir, which is coconut fiber. Coir is an especially interesting natural material used as a geotextile to stabilize soil and provide a growing media for plants. Its special properties include high tensile strength, slow decomposition rate due to high concentrations of lignin and cellulose, and high moisture retention capability. Uses of coir are described by Anonymous (1995b) and by Goldsmith and Bestmarn (1992) whose company has patented several fabrication methods for coir geotextiles.

DEEP ECOLOGY AND SOFT ENGINEERING: EXPLORING THE POSSIBLE RELATIONSHIP OF SOIL BIOENGINEERING TO EASTERN RELIGIONS

Design in soil bioengineering is mostly qualitative, intuitive, and perhaps even "organic," especially in contrast to conventional approaches to erosion control. It clearly requires a sophisticated understanding of water flows and energetics that cause erosion but, as noted by Shields et al. (1995), "Despite higher levels of interest

FIGURE 3.12 View of tree trunks extending from root wads in a stream restoration project in central Maryland.

in vegetative control methods, design criteria for the methods are lacking." One interesting exception is the design analysis of root reinforcement of soil reviewed by Gray and Leiser (1982), but even this effort covers only a limited range of applications and only a few types of plant root systems. Design knowledge in soil bioengineering involves basic concepts but quantitative relationships, such as Hudson's formula described earlier for riprap rock criteria, have not been developed. Most design is based on a heuristic interpretation of the spatial patterns of erosive energies of a site, and it consists of careful choice and placement of plant species and natural materials to dissipate these energies. Because the systems are living and will self-organize, growth and development of the ecosystem over time must be integrated into the design decisions to a significant extent. Because of this nature of design knowledge and because of the qualities of materials used (i.e., live plants vs. concrete), the field has been referred to as "soft engineering" as compared with the more conventional "hard engineering" approaches from the civil and geotechnical disciplines (Gore et al., 1995; Hey, 1996; Mikkelsen, 1993).

Another dimension of design is that "plant-based systems have greater risk because we have less control" (Dickerson, 1995). The idea of control is fundamentally inherent in all kinds of engineering, where the behavior and consequences of designs must be known and understood with a high degree of assurance. However, in soil bioengineering as in all examples of ecological engineering, the designs are living ecosystems which are complex, self-organizing, and nonlinear in behavior. Design knowledge of the systems has developed sufficiently to the point that they can be used reliably but uncertainties remain because of the inherent nature of living systems.

All of the aspects of soil bioengineering design described above: qualitative, intuitive, "organic," and, to a degree, reduced human control, suggest possible connections with Eastern religions, which share these qualities. Religions are philosophies that help humans decide how to act and how to think. The discussion that follows is an attempt to show how a consideration of one particular set of religions may provide perspective and insight on design in soil bioengineering. The suggestion is that, to an extent, there is congruence between these two activities that may be profitably explored and exploited.

The Eastern religions of Hinduism and various forms of Buddhism are a related set of beliefs based on the search for enlightenment. The state of enlightenment is the goal of individuals who believe in these religions, and it represents a condition of harmony and contentment between the individual and the cosmos. Enlightenment is achieved through introspective meditation and living one's life according to certain rules and beliefs. It is a mystical state of being that is not connected to normal human reality. Thus, belief in these religions causes one to strive to lead the appropriate kind of life that results in enlightenment. These religions do not rely on supreme beings for insight and wisdom but rather on the individual's search for the right way of life.

Two books are especially relevant for relating Eastern religions to ecological engineering in general and, in particular, to soil bioengineering. Pirsig (1974) in *Zen and the Art of Motorcycle Maintenance* introduces Zen Buddhism indirectly through a story about a cross-country motorcycle trip. This is an intensive philosophical work with the subtitle, "An Inquiry into Values." The most directly relevant sections of the book involve the discussion of how the everyday maintenance of the motorcycle can provide an expression of the Zen philosophy. An analogy from this discussion can be drawn for the relationship between the ecological engineer and the ecosystem that he or she creates and maintains. Capra's (1991) book entitled *The Tao of Physics* is a more extensive treatment in that it explicitly reviews all of the Eastern religions (Hinduism, Buddhism, Chinese thought, Taoism, and Zen) while describing parallelisms with modern physics. This work discusses many direct relations between Eastern religions and physics, which are applicable to considerations of soil bioengineering, such as ideas on the importance of harmony with nature, the roles of intuitive wisdom, and the concepts of change and spontaneity. Capra provides detailed descriptions of the Eastern religions that provide quick introductions for readers from Western traditions. One passage about Taoism, which is the set of beliefs referenced in the title of the book, is given below:

> The Chinese like the Indians believed that there is an ultimate reality which underlies and unifies the multiple things and events we observe: ... They called this reality the Tao, which originally meant "the Way." It is the way, or process, of the universe, the order of nature. In later times, the Confucianists gave it a different interpretation. They talked about the Tao of man, or the Tao of human society, and understood it as the right way of life in a moral sense.
>
> In its original cosmic sense, the Tao is the ultimate, undefinable reality and as such it is the equivalent of the Hinduist Brahman and the Buddhist Dharmakaya. It differs from these Indian concepts, however, by its intrinsically dynamic quality, which, in the Chinese view, is the essence of the universe. The Tao is the cosmic process in which all things are involved; the world is seen as a continuous flow and change.

One particular example of possible application of Eastern religion to ecological engineering is the dualist notion of life situations represented by the polar opposites,

FIGURE 3.13 The diagram of the supreme ultimate in Taoism. The symmetrical pattern of yin and yang.

yin and yang. This is shown in Figure 3.13 with the "diagram of the supreme ultimate" (Capra, 1991):

> This diagram is a symmetric arrangement of the dark yin and the bright yang, but the symmetry is not static. It is a rotational symmetry suggesting, very forcefully, a continuous cyclic movement … The two dots in the diagram symbolize the idea that each time one of the two forces reaches its extreme, it contains in itself already the seed of its opposite.
>
> The pair of yin and yang is the grand leitmotiv that permeates Chinese culture and determines all features of the traditional Chinese way of life.

In the Taoist beliefs a principal characteristic of reality is the cyclic nature of continual motion and change. Yin and yang represent the limits for the cycles of change and all manifestations of the Tao are generated by the dynamic interplay between them. Thus, it is a form of organization. Although the yin and yang represent opposites, there is a harmony between them. Ecology, too, can be characterized by the interplay between polar opposites such as primary production and respiration from ecosystem energetics (see Figure 1.2) or in the growth (r) and regulation (K) terms in the classic logistic equation from population biology:

$$dN/dt = rN(K{-}N/K) \tag{3.4}$$

where

N = number of individuals in a population
t = time
r = population reproductive rate
K = number of individuals of a population that can be supported by the environment (i.e., the carrying capacity)

In this model, growth of the population over time is directly proportional to the intrinsic rate of increase, *r*, but inversely related to the population's carrying capacity, *K*. Factors related to *r* cause the population to grow while factors related to *K* cause the population to remain stable. Species also tend to adapt towards either the growth states (*r*-selected) or the stable states (*K*-selected) as discussed in Chapter 5. Thus, growth versus stability might represent polar opposites, like yin and yang. There are also examples from geomorphology such as the opposite processes of erosion and deposition, and the opposite zones found in the inner and outer banks of meanders and in pool and riffle sequences, both of which involve alterations between erosion and deposition. Obviously, design in soil bioengineering involves an understanding of these opposites and a plan for their balance on any particular site, perhaps in a fashion similar to the way a Taoist would relate yin and yang in life experiences.

Capra's work is especially relevant because he has begun to think about Eastern religions as being ecological due to their reliance on holism and the interconnectedness of all things. He has contributed to the growing philosophy called *deep ecology* (Capra, 1995; Drengson and Inoue, 1995), which attempts to articulate beliefs about sustainability for human societies. In these efforts the science of ecology is a model for developing an alternative world view or cosmology.

A few direct connections between Eastern religions and ecology and ecological engineering have been made in the literature. Cairns (1998) mentioned Zen in a paper on sustainability but did not develop the connection very much. However, Barash (1973) discussed Zen and the science of ecology in some depth. This paper, though obscure, is remarkable for having been published in a very empirically based scientific journal (*American Midland Naturalist*). One wonders how the paper survived peer review in this context. Sponsel and Natadecha (1988) make direct ties between Buddhism and conservation in Thailand, and they suggest that recent examples of environmental degradation may be the result of a decline in faith caused by westernization of the culture. More general reviews are given by Callicott and Ames (1989) and Sponsel and Natadecha-Sponsel (1993). Finally, a particularly interesting example of the connection between Eastern mysticism and ecology is found in the work of Ed Ricketts, who is best known as the model for the character "Doc" in John Steinbeck's (1937) novel entitled *Cannery Row*. Ricketts was a marine biologist who wrote an important guidebook to the intertidal ecology of the Pacific coast (Ricketts and Calvin, 1939). This book is significant as an early example of the modern approach to animal–environment relations. It is a highly refined form of descriptive ecology, especially in placing macroinvertebrates in their habitats. Ricketts also wrote philosophy, inspired by ideas of holism and interconnectedness from his ecological field work, which had similarities with Eastern religions (Burnor, 1980). In fact, Hedgpeth (1978b) described Ricketts (with additional reference to his interest in music) as a man whose driving force in life was "an urge to bring Bach and Zen together in the great tidepool." Thus, an introductory knowledge of Zen Buddhism enriches the reading of Rickett's guidebook and may lead to a deeper understanding of intertidal ecology. As an aside, Rickett's association with John Steinbeck is one of the remarkable stories in the history of ecology. Here, a marine biologist and a novelist more or less collaborated to produce a kind of mythical bond during the Depression years and into the 1940s (Astro, 1973; Finson and Taylor,

1986; Kelley, 1997). Steinbeck's (1939) *The Grapes of Wrath* which won the Pulitzer Prize for literature was published within weeks of Rickett's book, indicating that these two men reached high levels of achievement (and enlightenment?) together. Their collaboration may be best represented in the record of their scientific collection expedition to the Gulf of California, later published as *Sea of Cortez: A Leisurely Journal of Travel and Research* (Steinbeck and Ricketts, 1941). Their collaboration was cut short by Rickett's accidental death in 1948, after which it has been said that the quality of Steinbeck's writing declined.

Several workers have briefly mentioned connections between ecological engineering and Eastern religions in particular. Todd and Todd (1994) mention feng shui, which is a set of principles from Chinese philosophy for organizing landscapes and habitats. Jenkins (1994) in his review of composting systems included a chapter entitled "The Tao of Compost" which makes a case for integrating waste disposal into people's lifestyles. Finally, Wann (1996) described related thoughts as noted below:

> It's clear that we need more sophisticated, nature-oriented ways of providing services and performing functions. Many designers and engineers are taking an approach I call aikido engineering. Essentially, the Eastern martial art discipline of aikido seeks to utilize natural forces and succeed through nonresistance. Aikido never applies more force than is necessary. Its goal is resolution rather than conquest. We can and should use this approach to find solutions that avoid environmental and social problems.

Mitsch (1995a) compared ecological engineering in the U.S. and China with emphasis on technical aspects. He found some differences in approaches that are culturally related but may also reflect philosophy. The Chinese utilize ecological engineering applications widely (Yan and Zhang, 1992, plus see the many papers in Mitsch and Jørgensen, 1989, and in the special issue of *Ecological Engineering* devoted to developing countries: Vol. 11, Nos. 1–4 in 1998). They also have been practicing soil bioengineering for centuries, as illustrated by an ancient manuscript on the subject shown in the text by Beeby and Brennan (1997, see their Figure 6.14). Do Chinese philosophies of design differ from Western examples? If so, they deserve special study in order to enrich Western thinking and design.

In conclusion, the point of this section is to suggest relationships between Eastern religions and design in soil bioengineering and, to some extent, more broadly in ecological engineering. Successful soil bioengineering often depends on the ability of the designer to "read" a landscape and arrive at a design through observation, intuition, and experience. An understanding of the interconnectedness of hydrology, geomorphology, and ecology is needed along with a respect for aspects of complexity and change. Thus, it is suggested that the soil bioengineer is like the Zen master, similar to the description given by Barash (1973). David Rosgen's (1996) approach to restoring streams is a good example that is based on a deep understanding of nature. Thus, similarities between a stream restoration plan (Figure 3.9) and a Zen water garden (Figure 3.14) appear to be superficial but may be more closely related. Is a bed of riprap rocks similar to a Zen rock garden?

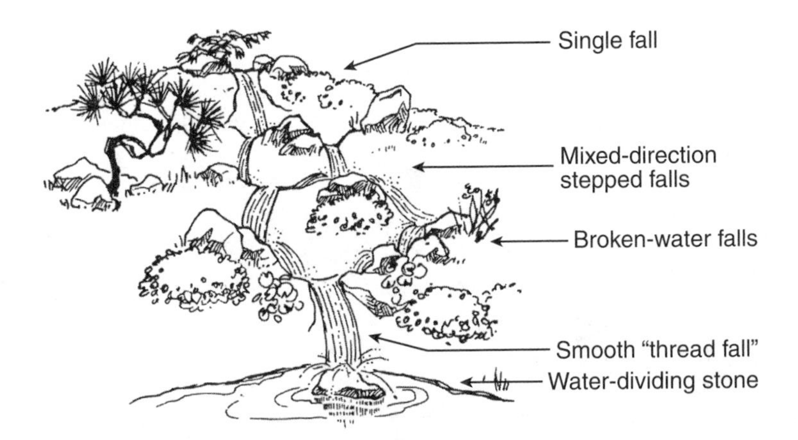

FIGURE 3.14 A typical Zen water garden. Note the similarity between the arrangement of components here as compared with the stream restoration plan shown in Figure 3.9. (From Davidson, A. K. 1983. *The Art of Zen Gardens: A Guide to Their Creation and Enjoyment.* G. P. Putnam's Sons, New York. With permission.)

CASE STUDIES

Individual case studies are presented below to review issues and designs of soil bioengineering in more depth. Four different situations are included to cover the range of applications in the field. For each case study one particular design is highlighted as an example of how ecosystems are utilized to address erosion control with engineering approaches.

Urbanization and Stormwater Management

Urbanization removes the cover of vegetation and replaces it with land use that is dominated by hard surfaces including buildings, roads, and parking lots. *Imperviousness* is the term used to describe the extent to which a watershed is made up of hard surfaces, and this parameter has been shown to influence hydrology dramatically. The most significant influence is on runoff volume. Figure 3.15 plots imperviousness vs. the runoff coefficient, which expresses the fraction of rainfall volume that is converted into surface runoff during a storm, illustrating a direct relationship between hard surfaces and runoff volume. The increased runoff in urbanized watersheds in turn creates increased flooding and increased channel erosion in streams draining the landscape. A threshold seems to exist at about 10% imperviousness, above which hydrology becomes seriously altered and thereby causes significant impacts (Schueler, 1995). Stream ecosystems in cities are degraded by these impacts, with loss of habitat and pollution by a number of contaminants (Paul and Meyer, 2001).

One way to visualize the imperviousness of watersheds is with a comparison of hydrographs. The hydrograph is a plot of discharge rate or flow of a stream as a function of time. Many different time scales are of interest to hydrologists, but here

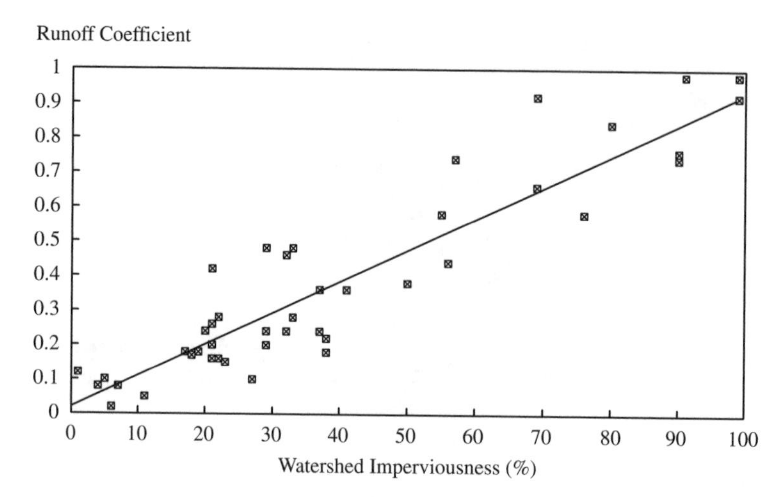

FIGURE 3.15 A relationship between runoff and impervious surfaces in a watershed. (Adapted from Schueler, T. R. 1995. *Watershed Protection Techniques*. 2:233–238.)

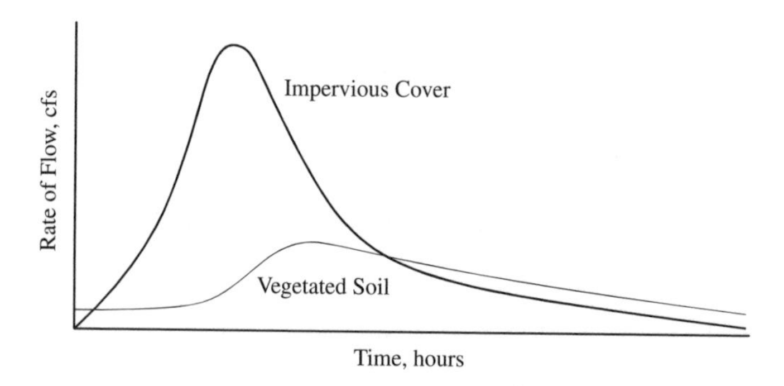

FIGURE 3.16 Comparison of hydrographs from rural (i.e., vegetated soil) and urban (i.e., impervious cover) areas. (Adapted from Ferguson, B. K. 1998. *Introduction to Stormwater: Concept, Purpose, Design*. John Wiley & Sons, New York.)

the focus is on storms so that units of hours or days are most relevant. Hydrographs provide a wealth of information as noted by Hewlett and Nutter (1969): "A hydrograph tells more about the hydrology of a drainage basin than any other single measure." A hydrograph represents a functional response of a watershed in relation to the water balance, and its shape is determined by two sets of factors: (1) characteristics of the watershed such as imperviousness, and (2) weather factors such as quantity, intensity, and duration of rainfall; distribution of rainfall over the watershed; and temperature (which is important in terms of freezing of soil or melting of snow and ice). Storms strongly influence hydrographs because they release large volumes of rainfall over short periods of time. A storm hydrograph is hump-shaped with a rise and fall of discharge as the stream drains the runoff generated by rainfall. Because urbanized watersheds have more runoff than less developed watersheds, their hydrographs differ in shape (Figure 3.16). The important features of a storm hydrograph from an urbanized watershed are the increased peak discharge (the highest point of the hump) and the shortened duration (the length of time between the rise and fall of the hump). Basically, the shape of the urban storm hydrograph shows that a large amount of water is moving quickly through the watershed over the surface, with consequent impacts of flooding and erosion. In a less developed watershed some of this water would have infiltrated into the ground and entered the stream over a longer time period as baseflow. There are also water quality impacts associated with storms since pollutants are washed into streams with runoff. This is an important type of nonpoint source pollution because the pollutants are advected by runoff moving over the watershed, as opposed to point source pollution that is generated by a discrete outfall such as from a wastewater treatment plant or a factory. Makepeace et al. (1995) provide a review of the pollutants in urban stormwater runoff, and Hopkinson and Day (1980) provide an example of a simulation model that combines urbanization and stormwater.

Stormwater management involves engineering of BMP structures that mitigate and control both the water quantity (flooding and erosion) and quality (nonpoint source pollution) impacts of storms in urban landscapes. Their role is to reduce the peak discharge of urban streams during storms. Stormwater management has a long tradition in civil engineering which has evolved into a kind of "pipe and pond" conventional approach (Urbonas and Stahre, 1993). In this approach, storm runoff is collected into centralized systems and stored temporarily in large detention ponds. Water in the ponds is released over a longer period of time, thus reducing peak discharge. While effective, this conventional approach has a number of problems associated with it, and over time new kinds of BMPs have been developed. These designs include wetlands, infiltration systems, filter strips or buffers, and porous pavement (Schueler, 1987). These designs are growing in diversity and implementation, and a whole new approach to urban stormwater management seems to be emerging. The new approach is very much a kind of ecological engineering, which is referred to by some workers as *bioretention* (Table 3.2). This is a very different approach compared with traditional stormwater management. The goal is to mimic natural hydrology through use of BMPs that emphasize vegetation. A strong effort is made to integrate BMPs into the site plans of new developments so that they become part of the landscaping rather than large, unattractive, and unsafe structures that create liabilities. Also, new ways of retrofitting stormwater management systems are being devised for sites that are already developed. This is a very creative field where workers must understand and utilize traditional engineering along with hydrology and ecology. The basic philosophy is to apply many small scale BMPs throughout the watershed, dispersing runoff rather than concentrating it. A key is to keep the drainage basin for each individual BMP small so that runoff volumes are more manageable and do not overwhelm the system's ability to function. The emphasis is on infiltration and evapotranspiration rather than drainage, and preliminary results indicate that these systems are less expensive than conventional alternatives. Bioretention is still a new approach and designs are evolving rapidly, as indicated by reports in such journals as *Watershed Protection Techniques* from the Center for Watershed Protection in Ellicot City, MD.

TABLE 3.2
Comparison of Approaches for Stormwater Management

Conventional Approach (i.e., pipes and ponds)	New Approach i.e., bioretention)
Philosophy	
Collect runoff to one point; centralize it.	Locate BMPs where runoff is produced; keep it dispersed.
Increase storage and drainage.	Increase infiltration and evapotranspiration.
A few large detention basins.	Many small retention basins.
Design	
Complex, large scale.	Simple, small scale.
Role of Vegetation	
None.	Significant, several functions.
Functionality	
One-dimensional.	Multidimensional with added benefits of aesthetics and water quality improvement.
Cost	
Relatively higher.	Relatively lower.

One example of a bioretention BMP is the rain garden, which is a modified infiltration system (Ferguson, 1994). This BMP was developed in the late 1980s by Larry Coffman in Prince George's County, MD (Bitter and Bowers, 1994; Engineering Technologies Associates and Biohabitats, 1993), and it is similar to other biofiltration systems. A rain garden is an engineered BMP designed to treat stormwater from a small drainage basin such as a parking lot or rooftop (Figure 3.17). It consists of an area with reconstructed soil stratigraphy and planted vegetation that is oriented in such a way as to receive runoff from the drainage basin. The soil of the rain garden is designed to encourage infiltration. The first layer (30 cm) is typically composed of a mixture of 50% sand, 30% top soil, and 20% mulch. This is the active zone in which most pollutant absorbtion takes place in terms of nutrients and metals. Sand or gravel are sometimes used below this layer, and the latest designs employ an under drain, as in a septic tank drain field, leading to a stormwater catchment system. The rain garden is intended to model a terrestrial system rather than a wetland in order to encourage infiltration. This objective requires design so that ponding occurs but is minimized. This is a critical element that can have long-term hydrologic implications. If ponding is too long, wetland conditions are favored which reduce infiltration capacity. The rain garden is thus designed to absorb the first flush of storm runoff and then to overflow with excess runoff leading to other

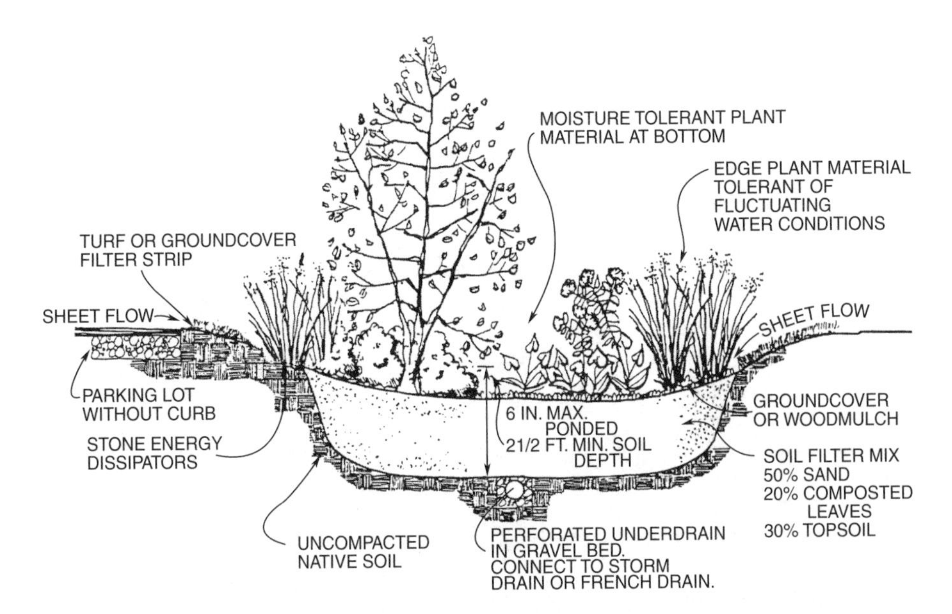

FIGURE 3.17 View of the rain garden concept. (From Coffman, L. S. and D. A. Winogradoff. 2001. *Design Manual for Use of Bioretention in Stormwater Management.* Watershed Protection Branch, Prince George's County, MD. With permission.)

devices (such as a collection system or a cascade of other BMPS). Vegetation plays several roles in rain garden function. The root systems of plants improve infiltration, and plant growth absorbs some pollutants and increases evapotranspiration. A variety of species can be planted and a landscaping approach is usually used in their design. This makes the rain garden an attractive system that improves the aesthetic values of the surrounding landscape. The rain garden system is new and long-term maintenance requirements are not completely known. They may need to be periodically excavated and rebuilt to avoid soil crusting, clogging, or sedimentation. As with any new system, design knowledge can be expected to grow as more examples are built and studied over time.

Agricultural Erosion Control

Erosion from agricultural systems is a serious problem in rural landscapes (Clark et al., 1985; Harlin and Berardi, 1987; Pimentel et al., 1987). This kind of erosion is accelerated because the natural vegetation is removed and replaced with cropping or grazing systems that provide less protective coverage of the soil. In fact, some cropping systems involve periods of time during and after tillage when the soil can be completely exposed to the driving forces of erosion (wind and rain). Agricultural erosion has been studied by applied scientists for centuries, and it is fairly well understood. The universal soil loss equation, shown in Figure 3.18 is one example of a practical model of agricultural erosion (Foster, 1977; Wischmeier, 1976). The equation is meant to be used to evaluate erosion problems for individual fields, and it is based on established, empirical relationships. Through the use of the equation,

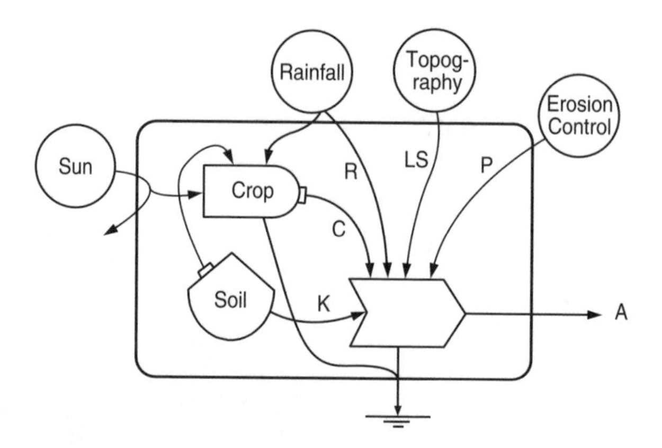

FIGURE 3.18 Energy circuit model of the universal soil loss equation, showing the erosion rate (A) as a function of a number of factors.

agricultural extension agents can advise farmers about control practices that reduce erosion.

A number of erosion control practices have evolved including techniques for controlling water flows such as contour planting and terracing and different methods of providing coverage of bare soil such as cover crops, manure from animals, plant mulches, and no-till cropping. These practices must be integrated into the overall farm system and their use is at the discretion of the individual farmer. Organic farming is a comprehensive approach of these and other techniques that has been shown to reduce erosion and improve soil fertility (Mader et al., 2002; Reganold et al., 1987).

Some of the practices listed above involve engineering approaches while others might better be thought of as management strategies. Terracing is a good example of a technique that involves some traditional engineering design in terms of spacing, grades, and cross-sections (Ayres, 1936). This technique can be traced back to prehistoric times, and it has evolved independently in many cultures (Donkin, 1979). Windbreaks are analogous systems for controlling wind erosion (Stoeckeler and Williams, 1949), but they are composed of living species (trees) rather than nonliving terraces. An example of a technique that is more management oriented is no-till cropping (Little, 1987; Phillips et al., 1980). This is a particularly interesting technique because it represents a major shift in the approach to agriculture. Traditionally, crop agriculture relied on tillage of the soil (i.e., plowing and disking) to prepare for seeding and especially to control weed growth. This practice exposes the soil to erosion but its benefits, which result in high yield, were viewed as being more significant than the costs. However, the development of selective herbicides after World War II created an alternative method of weed control. A new form of agriculture subsequently evolved substituting herbicide use for tillage, along with the creation of new seeding methods. Rachel Carson (1962) called this *chemical plowing* in her famous book on pesticide effects entitled *The Silent Spring*. This new approach has been found to have significantly less erosion than the conventional tillage approach because the soil is not disturbed and a cover of biomass is retained between crops. The litter and plant growth in no-till fields has been called a *living mulch* because of its role in nutrient conservation (Altieri, 1994). A significant number of farmers have switched to no-till agriculture, though concerns remain about possible environmental impacts of herbicides and possible buildups of insect pests.

Much work on erosion control and other aspects of agriculture is done by agricultural engineers whose special function is to apply engineering principles and approaches to farming and grazing. They design machines, study system performance, and must deal with soils, water quality and quantity, and all taxonomic levels of biodiversity, both domestic and pest. Because of these roles and because agricultural systems are really simplified ecosystems (i.e., agroecosystems, see Chapter 9), the discipline of agricultural engineering is related to ecological engineering. The main distinction is in the complexity of ecosystems that are involved. Conventional agroecosystems are clumsy and simple compared with natural ecosystems with low diversity, high runoff and erosion, and the use of manufactured chemicals as fertilizers and toxins in place of evolved ecological relationships. For example, Van Noordwijk (1999) contrasts the complex cycling of nutrients in natural ecosystems with the simple input–output flows of nutrients in agroecosystems. Agricultural systems are completely designed by humans with little positive input from nature and with few or no by-product values. These qualities make agroecosystems appear very different from other, more natural ecosystems, but some basic similarities remain. Study of agroecosystems will continue to be instructive in ecological engineering, as another context of design. Also, each farm is an experiment with a unique mix of ideas from the farmer, which offer insights into the connections that develop between human designer (i.e., farmer) and constructed ecosystem (i.e., farm). Formal relationships between the old discipline of agricultural engineering and the new discipline of ecological engineering should be encouraged to improve the design of constructed ecosystems in general.

Ecological engineering may be able to contribute to the development of alternative agricultural systems. Many problems with conventional agriculture have been described, and much work is needed to develop more sustainable alternatives (Keller and Brummer, 2002). For example, as noted by Orr (1992a):

> Since 1945 mainstream agriculture — by which I mean that espoused by agronomy departments in land-grant universities, the United States Department of Agriculture, and major farm organizations — has pursued a model of agriculture based on the industrial metaphor. Its goal has been to join land, labor, and capital in ways that maximize productivity. Farming is regarded not as a way of life but as a business. Like other businesses, it has led to highly specialized farms that grow one or two crops, or raise thousands of animals in automated confinement facilities. Like other businesses, agribusiness invested heavily in technology, became dependent on "inputs" of chemicals, fertilizer, feed, and energy, and went heavily into debt to finance it all. Farmers were advised to plow fence row to fence row, buy out their less-efficient neighbors, substitute monoculture for crop diversity, cut down windbreaks, and replace people with machinery. The results are there for all to see. The ongoing farm crisis of the 1980s suggests that it did not work economically (except for those who learned how to farm the tax code). From dying rural towns across the United States one can infer that it did not work socially. And neither does it work ecologically.

There is an interesting movement to use natural ecosystems as a model for the development of new, more sustainable agroecosystems, which is relevant to ecological engineering. For example, a National Research Council (NRC) report states that one of the goals in the development of alternative agricultural systems is a "… more thorough incorporation of natural processes such as nutrient cycles, nitrogen fixation, and pest–predator relationships into the agricultural production process" (NRC, 1989a). In this regard a number of ideas have come from the examination of rural agricultural systems in the tropics. Ewel (1986) and Hart (1980) have discussed mixed species communities that imitate plant succession sequences, and Perfecto et al. (1996) describe similarities between shade coffee plantations and natural forest habitats. In a related example, Gomez-Pompa et al. (1982) and Gliessman (1991) have studied forms of wetland agriculture that have been used since pre-Colombian times in Middle America. Finally, Altieri et al. (1983) and Gliessman (1988) have shown many parallels between traditional agricultural practices in the tropics and natural ecological systems.

Relatively less work has been done on using nature as a model for agriculture in the temperate zone. A notable exception is the research of Wes Jackson at the Land Institute in Kansas. Jackson is developing a form of agriculture for the U.S. Great Plains that he calls a *domestic prairie* (Jackson, 1980). His idea is to create a polyculture (i.e., a multispecies mix) of herbaceous perennial species in a no-till cropping system to replace the present grain-producing monocultures of herbaceous annual species (such as corn or wheat). Jackson's domestic prairie uses the natural prairie ecosystem as a model (Bender, 1995; Jackson, 1999), and he projects that it would have many advantages over the existing grain agriculture: reduced erosion, reduced fossil fuel consumption, reduced pesticide dependency, reduced dependency on commercial fertilizers, and a larger genetic reservoir that would provide benefits such as increased disease resistance. The key to realizing this new form of agriculture is breeding varieties of herbaceous perennial species that will produce high yields of seeds and fruits, as do the existing annual species. Jackson, who is trained as a geneticist, has started this breeding program with focus on promising species in genera of forage grasses (*Bromus*, *Festuca*, *Sporobolus*, *Lolium*, *Agropyron*, and *Elymus*), several of which are important in natural prairies. The existence of some perennial species, especially from the sunflower family, which naturally produce high yields of seeds, provides support for the work.

One of the obvious benefits of the "domestic prairie" agriculture would be reduced erosion. Because the crop species would be perennial in a no-till system, the soil would remain undisturbed and would be resistant to erosion. Furthermore, prairie species are known to have well-developed root systems that would hold the soil especially firmly (Stanton, 1988). Much work was done by John Weaver in the early 1900s on prairie root systems, which remains today as some of the best research available (Weaver, 1919, 1938, 1954, 1961). Weaver developed an excavation technique for studying roots which was labor intensive but which revealed great detail on the extent of prairie species root systems. His published work is characterized by wonderful drawings of the intricate root systems (Figure 3.19), and Jackson used one of these for the cover of his book which describes the domestic prairie idea (Jackson, 1980). The ecology of the belowground portion of terrestrial ecosystems

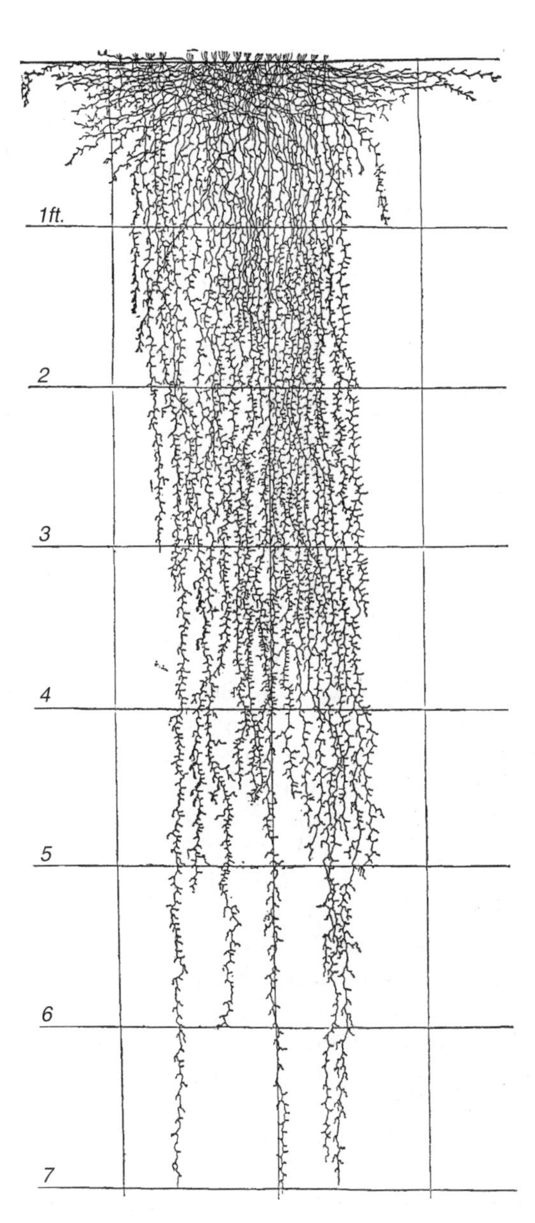

FIGURE 3.19 View of a typical grassland plant's root system. (From Weaver, J. E. and F. E. Clements. 1938. *Plant Ecology*, 2nd ed. McGraw-Hill, New York. With permission.)

is still underappreciated and poorly known (Coleman, 1985, 1996; Wardle, 2002), but the belowground system is the location from which much of the soil's resistance to erosion comes. There is probably no better engineering design for erosion control than the root system of a prairie, which is one of the most distinctive qualities of Jackson's alternative approach to agriculture.

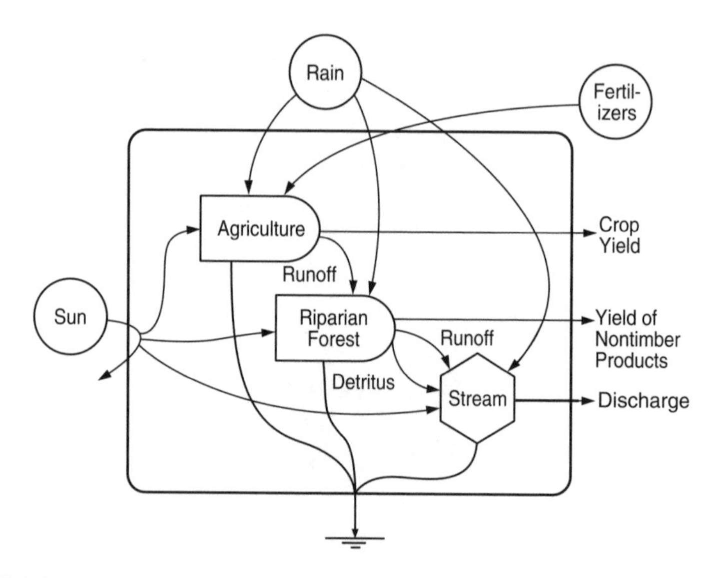

FIGURE 3.20 Energy circuit diagram of the role of the riparian forest as a best management practice in an agricultural landscape.

Another ecological system of importance in erosion control is the riparian forest. These are narrow strips of forest growth along streams that drain agricultural watersheds. They act as buffers for the surrounding fields in absorbing nutrients and sediments from runoff waters (Figure 3.20). Although relatively little engineering design is involved with the use of these forests as a BMP, they are becoming more important for controlling erosion and other forms of nonpoint source pollution (Lowrance, 1998; Lowrance and Crow, 2002; Smith and Hellmund, 1993). Riparian forests are particularly significant because of the by-product values they provide including bank stabilization, control of stream temperature through shading, and input of wood debris and leaf litter to streams and wildlife habitat (Gregory et al., 1991). Robles and Kangas (1999) also describe potential economic benefits that can be realized by farmers from the sale of nontimber products harvested from the forests. The creation of this kind of economic payback may be important in making the use of riparian forests more attractive to farmers who must take land out of production in order to implement the BMP.

Debris Dams, Beavers, and Alternative Stream Restoration

Erosion control in streams is a form of restoration that is rapidly developing for landscapes disturbed by human land use (Riley, 1998). Several examples of designs were mentioned earlier such as the use of root wads as organic riprap. These applications are advancing primarily in the creative work of consulting firms that are hired to restore streams. In this section one particular design strategy, the use of woody debris in stream channels, is highlighted because of its several interesting features.

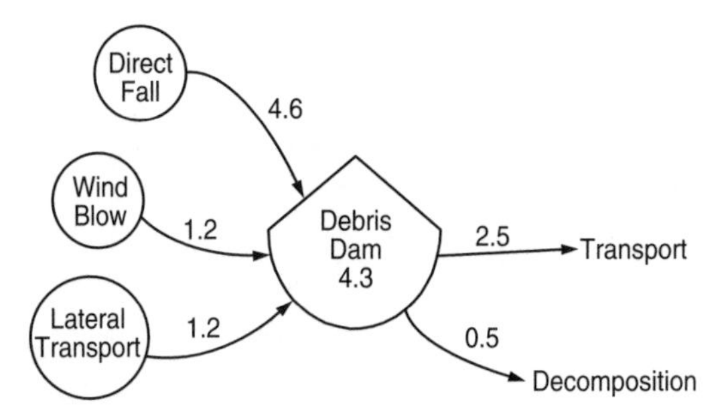

FIGURE 3.21 Energy circuit model of the inputs and outputs from a debris dam system. Data are for an estimated annual budget in tons of wood. (Adapted from Gregory, K. J. 1992. *River Conservation and Management*. P. J. Boon, P. Calow, and G. E. Petts (eds.). John Wiley & Sons, Chichester, U.K.)

Woody debris, which enters the stream as treefall and branch-fall from the riparian forest, commonly accumulates in dams and deposits by being transported in the current and by becoming lodged in the channel. These debris dams or jams are important to the ecology of stream organisms for providing habitat especially for insect larvae and fishes and as a source of food for decomposers (Bilby and Likens, 1980; Maser and Sedell, 1994; Smock et al., 1989). Perhaps even more significantly, these debris dams influence stream flow and channel form, therefore affecting the entire physical stream system. In terms of this geomorphic role, debris dams belong to the class of structures created by streams to dissipate energy as water drains through the watershed. Other examples are meanders and pool-and-riffle sequences. These structures take the form of organized patterns of materials that require the energy of the stream to be created and maintained. They can be thought of as storages of energy and materials with inputs and outputs, analogous to biological populations or trophic levels in ecology. Their presence in the channel provides feedbacks to the stream to amplify water flow and sediment transport. This conception of channel features is somewhat different from the way geomorphologists and hydrologists normally think about them (i.e., Langbein and Leopold, 1966; Yang, 1971a, 1971b), but it allows such structures to be viewed from a general systems perspective. An example of a mass balance of a debris dam is shown in Figure 3.21 in which the dam itself is depicted as a storage. Beyond the storage of wood, debris dams embody the energy of the stream that creates them. Streb (2001) studied the energetics of debris dams and their effects on channel characteristics in a flume. With physical measurements of current velocity in flumes with and without model debris dams, he was able to quantify their energy dissipation. This energy was used in making the debris dam itself and the associated pools and bars that evolved with it. Thus, the energy dissipation is a measure of the organization of the system.

Many studies have demonstrated the effects of debris dams in streams in terms of altered current flow patterns and influence on scour and deposition of sediments (Andrus et al., 1988; Heede, 1985; Robison and Beschta, 1990). One view of the role of debris dams is as an addition to the roughness of the channel. A number of

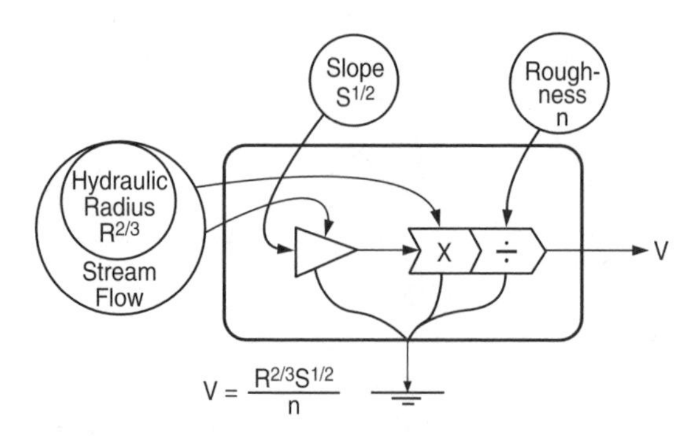

FIGURE 3.22 Energy circuit diagram translation of Manning's equation.

equations have been developed to describe the effect of channel roughness on stream flow such as the Manning's equation (Figure 3.22, see also Equation 3.2). This relationship was developed for conditions of uniform flow to design flood control channels, but it is commonly used for nonuniform flow conditions in natural channels. In this equation velocity (*V*) is directly proportional to the slope (*S*) and size of the channel (*R*) but inversely proportional to the roughness of the channel (*n*). Manning's *n* is termed the roughness coefficient, and it is an index of the friction of the channel on the stream flow. Some values of the roughness coefficient have been measured (Chow, 1964; Dunne and Leopold, 1978), and they range from 0.012 ft 1/6 for smooth concrete channels to 0.050 ft 1/6 for streams with rocky beds or dense aquatic vegetation. Manning's equation can be used for ecological engineering design as values of the roughness coefficient become better known. Debris dams have been assigned high values of Manning's *n* which is a reflection of their role in causing a localized reduction in current velocity and consequent changes in sediment transport (see the Hjulstrom relationship in Figure 3.7).

A diversity of debris dams exists in streams like different species in ecosystems. Figure 3.23 shows a gradient of debris dams ranging from megajams which span 10 times bankfull depth of the channel to microjams and individual log pieces. Table 3.3 describes three kinds of debris dams from a river in the Pacific Northwest of the U.S. Each type of dam has a unique architecture and different influences on stream flow and channel form.

An exciting strategy is the active manipulation of woody debris in streams to control erosion. This action has been done in the context of creating habitat for stream organisms, but its application for erosion control is recent (Streb, 2001). A carefully placed debris dam can divert current away from critical channel locations where erosion is to be controlled, while having several by-product values in the ecology of the stream. This kind of management can be especially important for watersheds in which riparian forests have been cut, thereby removing a natural source of wood to streams. Sedell and Beschta (1991) discuss this strategy in a paper with the compelling title of "Bringing Back the 'Bio' in Bioengineering."

Another important and interesting aspect of debris dams is the self-building behavior that they exhibit. This is the accumulation process in which wood builds up to form a dam. A key feature of the process is a positive feedback relationship

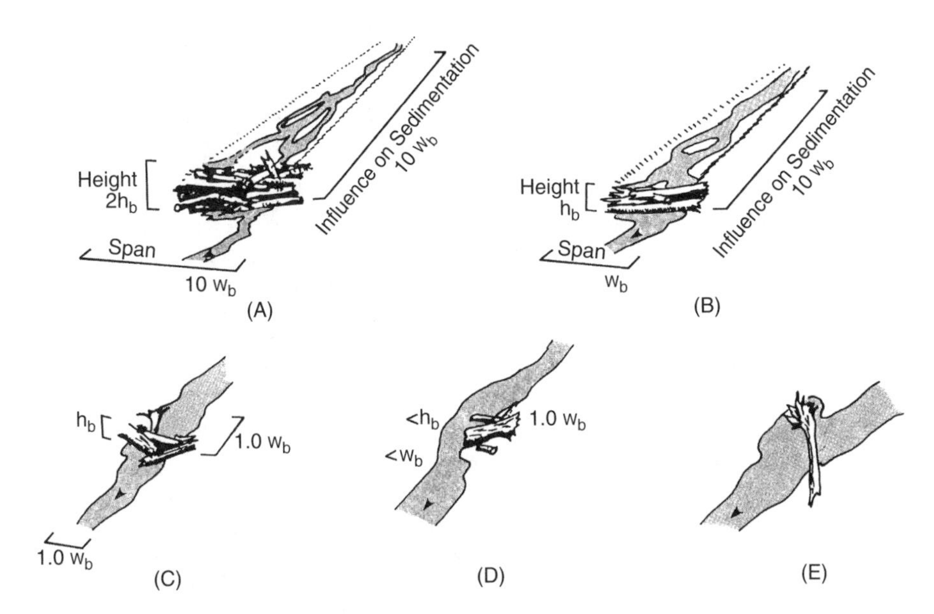

FIGURE 3.23 Examples of different types of debris dams. (A) Megajam. (B) Macrojam. (C) Mesojam. (D) Microjam. (E) Individual jog pieces; h_b = bankfull depth; W_b = width of the channel. (From Church, M. 1992. *The Rivers Handbook: Hydrological and Ecological Principles*. Vol. 1. P. Calow and G. E. Petts (eds.). Blackwell Scientific, Oxford, U.K. With permission.)

in which the initial pieces of wood in the dam catch increasingly more wood because of the action of the dam in creating a growing obstruction to flow. In this way the dam builds itself through input of wood pieces being carried in the current. The products of the self-building behavior are dams with complexity of architecture and size being determined by wood supply, channel dimensions, and current velocities. In a sense woody debris dams are fascinating structures because they represent complex systems that emerge from simple rules and elements. The self-building behavior of debris dams is a special case of self-organization, which is autocatalytic in creating structure. Other related examples are traffic jams (Edie, 1974), ice jams in rivers (Ashton, 1979; Beltaos, 1995), and even jams of signals in communication systems such as telephone networks (Alfredo Nava, personal communication). A description of autocatalysis is given by H. T. Odum (1983) below:

> Many naturally occurring units in the real world store energy and then feed it back internally to facilitate in the inflow of other energy. The feedback acts as a control, often as a multiplier, and catalyzes the inflow. Such units are sometimes termed autocatalytic. The process of storing and using the storage to pump additional energy tends to accelerate growth and maximize power. Such modules are frequent in all kinds of system.

It is suggested here that the wood of the debris dam feeds back upon the current in the stream to bring more wood into the dam and therefore it grows autocatalytically.

TABLE 3.3
Architectural Descriptions of Three Types of Debris Dams

Bar Top Jam (BTJ)

A random accumulation of logs with little vertical stacking characterize BTJs, which form a loose mat deposited on a bar top during receding flows. Although logs in a BTJ are oriented in all directions relative to the depositing flow, most are oblique … Bar top jams are relatively unstable as they are mobilized at discharges approaching bankfull. Hence they have little appreciable effect on channel morphology.

Bar Apex Jam (BAJ)

The more stable BAJ has a distinctive architecture characterized by three primary structural components: a key member nearly parallel to flow, normal members orthogonal to flow, and oblique members oriented 10–30 degrees to flow … the key members appear to be invariably a large log with an attached rootwad facing upstream. The deposition of a key member significantly reduces the effective width of flow within a channel. The LWD (large woody debris) that otherwise might be flushed through that portion of the channel is deposited, usually by racking up against the key member and contributing to a further reduction in the effective channel width. Normal members rack up against the key member rootwad orthogonal to flow, whereas oblique members deposit along the flanks of the key member. The sequential deposition of normal and oblique members commonly results in the vertical stacking of five or more interwoven layers. The formation of a jam introduces a local control on channel hydraulics that leads to distinctive changes in channel morphology and riparian forest structure. Stable LWD structures such as the BAJ provide a barrier to high velocity flows, creating sites of sediment aggradation that can lead to floodplain formation. Stable LWD structures also resist channel migration, thereby providing refugia for forest development.

Meander Jam (MJ)

Meander jams become the most common of the stable jams with increasing channel size. Unlike the BAJ, MJs have only two principal structural components: key members and racked members. An MJ has two or more key members that are initially deposited at the upstream head of a point bar and oriented nearly parallel to bankfull flows. Key members usually have rootwads facing upstream and are within approximately one rootwad diameter of one another. Racked members of various sizes accumulate normal to key member rootwads, stacking on top of one another to heights of 6 m or more … As the river migrates laterally, a stable MJ forms a revetment halting local bank erosion, often measurably compressing the river's radius of curvature and changing the orientation of the flow relative to the jam … These jams eventually armour the concave outer bank of a meander and harbour riparian forest patches proportional in size to the size of the jam.

Source: Adapted from Abbe, T. B. and D. R. Montgomery. 1996. *Regulated Rivers: Research and Management.* 12:201–221.

All living systems display autocatalytic behavior, but when it occurs in nonliving systems, it is especially significant and interesting. Ecological engineers should try to take advantage of self-building behavior because it is another way to incorporate free natural energy inputs into a design.

The discussion of self-building behavior brings to mind abstract concepts of self-reproducing machines. John von Neumann (1966) was the first to successfully explore the logic of this ambitious concept (see also Penrose, 1959). He imagined a self-reproducing robotics system and then proved with mathematical logic that it could exist in the form of a cellular automaton, which has been called the von Neumann machine (Sipper et al., 1998). The latest developments in self-reproducing machines are the computer programs of artificial life (Langton, 1989; Levy, 1992) and new directions in robotics based on distributed structures of simple units (Brooks, 1991, 2002; Lipson and Pollack, 2000; Nolfi and Floreano, 2000). Development of autonomous robots (robots that build themselves) is a goal of the latter work which is relevant to self-building behavior. Progress in this field is occurring (Webb and Consi, 2001). Examples of simple robotics that exhibit complex behavior are being built, for example, mobile, insect-like machines that can locate and detonate land mines. One robotics researcher has described these new robots as *living machines* (Trachtman, 2000) because of their lifelike behavior. This is the term used by John Todd to describe his concept of constructed ecosystems (see Chapter 2). A robot living machine is a nonliving device (technology only) that exhibits lifelike behavior, while Todd's systems are hybrid devices (ecosystems plus technology) that combine living, biological components with nonliving, engineered components. Although the ecosystems of Todd's living machines can reproduce themselves, the associated technology of tanks and pumps cannot. It is interesting to speculate that future ecological engineering development might someday merge and create new forms of living machines with more intimate linkages between the physical machine and the living system. Gastrobots (food-powered robots) have already been built (Wilkinson, 2000) that conceptually could develop mutualistic relationships with species in ecosystems. The 1970 science fiction movie *Silent Running* presaged this future development with its portrayal of robots managing greenhouse-based life-support systems in a space station. Isaac Asimov's (1950) three laws of robotics (protect humans, obey humans, protect yourself) act as guides to robot behavior in his fictional future and might be reprogrammed for the future robotic ecological engineer as protect the ecosystem, serve the ecosystem, protect yourself.

Beavers (*Castor canadensis*) relate to this discussion because they actively build dams and take advantage of self-building behavior in several ways. Moreover, they have a role in erosion control that may offer a tool for ecological engineers under certain circumstances. Beavers are remarkable animals because of their many building behaviors that create a collection of "artificial" structures and that physically change the landscape. The structures they build include bank burrows, food caches, lodges, and dams. The dams are the most amazing of these structures, as noted by Johnson (1927): "The dam is generally the most conspicuous, and impressive of the beaver's works. The total amount of labor involved is often prodigious. The size may vary from one only a few inches in length and height, damming a tiny trickle,

TABLE 3.4
Quotes on the Relation of Beaver Behavior to Engineering

Reference	Quote
Allred, 1986	Even more remarkable than the beaver's ability to build structures which yield so much control over his environment, rests his ability to employ sound engineering principles in both construction and selection of construction sites.
Cullen, 1962	In the lakes and the streams of our nation, a very special corps of engineers has been busily at work for thousands of years. sitting on their haunches, propped up by long, flat, scaly tails, these hard-working creatures double and triple in brass, not only doing the work of engineers but of contractors and construction men as well, as they go about their ordained task of building dams.
Beakley, 1984	Perhaps the animal that has been most closely associated with civil-engineering exploits is the beaver.
von Frisch and von Frisch, 1974	... beavers are experts not only in the building of dwellings but also in hydro-engineering, and have performed tremendous feats in this line long before man attempted anything of the kind.
Finley, 1937	The conservation history of America reveals many examples of killing the goose that laid the golden egg, the most striking of which is the trappers' campaign against the humble beaver to get quick profits on his hide ... Nature's engineer, the beaver, has a good warm coat, but his greatest service has been in creating our earliest industry of conserving soil and water. In the West he proved to be the most valuable wild animal in existence and one that built up a vast amount of wealth.

to vast structures several feet in height and hundreds — even thousands — of feet in length." Many authors have more or less casually made the analogy between beavers and engineers in regard to dam building behavior (Table 3.4), but the recent concept of "organisms as ecosystem engineers" brings more depth to the matter. Jones et al. (1994) introduced the concept of organisms as ecosystem engineers and defined it as follows: "Ecosystem engineers are organisms that directly or indirectly modulate the availability of resources to other species by causing physical state changes in biotic or abiotic materials. In so doing they modify, maintain and create habitats." Details on the general concept of organisms as ecosystem engineers have been described (Alper, 1998; Gurney and Lawton, 1996; Jones et al., 1997; Lawton and Jones, 1995) and some examples are listed in Table 3.5.

Beavers are one of the best examples of the concept of ecosystem engineers, especially as described by the research of Robert Naiman, Carol Johnson, and their co-workers (Johnston, 1994; Naiman et al., 1988; Pollock et al., 1995). By selectively

TABLE 3.5
References on the Concept of Organisms as Ecosystem Engineers

Species	Action	Reference
Beaver (*Castor canadensis*)	Dam building and hydrologic modification	Pollock et al., 1995; Johnston, 1994
Earthworms	Modification of soils	Lavelle, 1997; Lawton, 1994
Moss (*Sphagnum fuscum*)	Growth and nutrient accumulation	Svensson, 1995
Fish (*Prochilodus mariae*)	Sediment processing	Flecker, 1996
Mussel (*Mytilus edulis*) and tube-worm (*Lanice conchilega*)	Modification of current flows	Hild and Gunter, 1999
Midges (Chironomidae)	Improving the porosity of filter beds	Wotton and Hirabayashi, 1999
Deposit feeding invertebrates	Sediment processing (bioturbation)	Levinton, 1995
Termite (*Macrotermes michaelseni*)	Nutrient accumulation and improvement of soil physical properties	Dangerfield et al., 1998

cutting riparian forests and by altering stream hydrology and geomorphology, beavers have significant influences on biogeochemistry and biodiversity of the landscapes they inhabit. Also, because they abandon their complexes after the preferred food tree species have been depleted, they initiate succession sequences of vegetation that last on the order of tens to hundreds of years. From the human perspective, the positive roles of beavers have been long recognized. Historically, beavers were harvested for their fur almost to the point of extinction through the 1800s, but this use is now insignificant. Starting in the early 1900s the indirect positive roles of beavers due to their dam building behavior were acknowledged in terms of soil and water conservation. For example, Finley (1937) described benefits from beavers, especially in the western U.S., in conserving water and controlling erosion with the ponds created by their dams. In terms of erosion control, the ponds reduce the velocity of stream flow, causing sedimentation according to the Hjulstrom relationship. This role was particularly relevant during the human dam building era of the mid-1900s for storing sediments that otherwise would have accumulated in the man-made reservoirs of the western U.S. These reservoirs provide several water resource benefits, which become impaired as sedimentation reduces their water storage capacity. Techniques were even described (representing an early form of ecological engi-

TABLE 3.6
Listing of the Principal Artificial Structures Built by Beavers

Structure	Form	Function
Bank burrows	Excavated tunnels and chambers in stream banks	Habitation and protection from predators
Food caches	Piles of wood (tree branches and small trunks)	Source of food, especially in winter
Lodges	Dome-shaped constructed wood piles with a central chamber	Habitation and protection from predators
Dams	Constructed channel obstructions made up of wood and sediments	Creation of a pond that facilitates movements and protection from predators

neering) for reintroducing beavers into watersheds where they had been trapped-out to restore their soil and water conservation values (Couch, 1942). As the human population has grown, however, the positive roles of beavers are being counterbalanced by their negative roles in causing property damage through flooding and tree cutting. In modern times, beavers are often viewed as pests or, at best, as curious anachronisms, which is in stark contrast to their pre-Colombian role as dominant factors in landscapes throughout boreal and temperate regions.

Although the ecological roles of beavers as described above are well known, the evolution of their building behaviors has not been treated in a systematic fashion. Ideas about various behaviors either exist as scattered references in old publications or have not been explored. The evolution of the diverse behaviors of beavers is intriguing, and it is discussed here because of possible connections to debris dams and to self-building processes. The products of the principal building behaviors of beavers are listed in Table 3.6. All of these structures can be found in a single beaver pond complex, but under certain circumstances the animals will only use burrows and food caches if sufficient water is available. Beavers do create other structures such as canals for transporting wood or scent mounds for territorial marking, but the main structures are covered in Table 3.6. The behavior that results in the construction of these structures is instinctive and thus genetically based within a phylogenetic context. Beavers share with other rodents traits such as mobile hands and large, sharp teeth that grow continuouly throughout the life of the animal. An evolutionary theory of building behaviors must start with an ancestral beaver with these kind of rodent characters, and it must consist of a series of stages that are logically arranged as products of selection pressures.

A theory does exist for the evolution of lodge building that was first described by Morgan (1868) in his classic work on the beaver. This theory suggests that the lodge evolved from the living chamber of a burrow as noted below:

> The burrows of beavers inhabiting river banks are said to be occasionally detected by a small pile of beaver cuttings found heaped up in a rounded pile, a foot or more high, at the extreme end of each burrow. It is affirmed by the trappers, and with some show of probability, that this is a contrivance of the beavers to keep the snow loose over the ends of their burrows, in the winter season, for the admission of air. I have never seen these miniature lodges, and therefore can not confirm the statement, either as to their existence or use; but if, in fact, they resort to this expedient, it is another reason for inferring that the lodge was developed from the burrow with the progress of experience. It is but a step from such a surface-pile of sticks to a lodge, with its chamber above ground, with the previous burrow as its entrance from the pond. A burrow accidentally broken through at the upper end, and repaired with a covering of sticks and earth would lead to a lodge above ground, and thus inaugurate a beaver lodge out of a broken burrow.

This theory received support from Johnson (1927) and is illustrated with a figure by von Frisch and von Frisch (1974), which adds a rising water level as a driving force for lodge development. This is a logical theory which appeared early in the examination of beaver natural history. It is somewhat amazing that in more than 100 years since the publication of Morgan's book no parallel theory for the evolution of dam building has appeared.

The theory of dam building evolution presented here begins with a bank burrowing ancestral beaver, which is consistent with the paleontological evidence (Wood, 1980). This ancestor would have fed on the inner bark of certain tree species along with herbaceous aquatic plants when available. As an aside, a separate line of beaver evolution led to a non–dam-building animal the size of a bear (genus *Castoroides*), about 2 m in length, which lived like a manatee feeding on aquatic macrophytes in large river deltas (Kurten, 1968) and which became extinct in the Pleistocene age. The ancestor of the modern beaver was more versatile and inhabited large rivers or lakes with sufficient depth to support the animal's semiaquatic life style. This is important because the beaver uses standing water especially to aid the transport of wood for feeding. These animals lived in northern climates where herbaceous aquatic plants were only available during the summer months. Thus, the ability to eat the bark off tree branches and trunks gave wide access to many aquatic ecosystems with riparian forests. Competition for space, accentuated by the territorial trait of many kinds of rodents, eventually would have forced the ancestral beaver from the existing standing waters into smaller streams where dam building could have evolved.

The critical element of the theory of dam building is the food cache. This is a constructed pile of sticks stored by the beaver as a source of food for winter months. Food caching is a common behavior found in many kinds of animals (Smith and Reichman, 1984; Vander Wall, 1990). For beavers the food cache reduces their need for foraging during winter which provides security from predators and conserves metabolic energy. Wood is collected by beavers and placed near the entrance to the burrow or the lodge. In fact, Morgan (1868) mentions "false lodges" which are piles of wood at the entrance to burrows along river banks which provide protection and act as a source of food. This might represent the ancestral food cache. The more usual behavior is to submerge the food cache by sticking the wood into the sediments and creating an underwater pile by adding wood to the growing structure. The submerged food cache is easily accessed under ice cover in the winter providing further protection from predators. Because beavers eat only the bark, a great deal of wood is needed to be cached to last throughout the winter. Warren (1927) gives an idea of the size in the following quote:

> The size of some of these foodpiles is sometimes quite extraordinary. Mills gives the size of one in the Moraine colony as three feet deep and 124 feet in circumference. To make this 732 aspen saplings were gathered, also several hundred willows. Another harvest pile mentioned by him was four feet high and ninety feet in circumference. One foodpile which I saw in Gunnison County, Colorado, consisted entirely of willows, the large ends of which were stuck into or against the bank of the pond. The stuff was from three to seven or eight feet long, placed in water four feet or more deep, from the bottom up to the surface, and extending along the shore of the pond for over a hundred feet. Another brush heap which I saw not far away must have contained over eight hundred cubic feet of willow boughs … .

Thus, the ancestral beaver lived in a bank burrow with a large food cache near the entrance to the burrow. It is proposed here that the dam evolved as a modified food cache. This could have occurred in a small stream where the current compacted the food cache, perhaps during a flood event, into a structure that would be equivalent to a debris dam. This occurrence could have been facilitated by an existing debris dam that received input from a food cache. A natural debris dam creates an obstruction to flow and could easily have been enlarged by the current moving a food cache into it during a flood. This beaver protodam (natural debris dam plus food cache) would have created additional aquatic habitat that would have given the beaver selective advantage. In subsequent years the beaver could have actively added fresh wood to the protodam, still using it as a food cache. This activity would have continually created a more effective dam, therefore providing more selective advantage. Wilsson (1971) actually describes an experiment that shows how some existing beavers converted a food cache into a dam:

> [After two beavers] had raised the water level by building a dam inside their run, they began to build a winter store in the water a short distance downstream from the entrance to their lodge. Later we lowered the water level with the help of siphons so the winter store was partially exposed and the water ran audibly through the gaps in it. The animals immediately reacted by fixing peeled sticks in it and by pushing mud against the upstream side. The store was thus quickly transformed to a dam, and the water level rose again outside the lodge.
>
> When after a time the siphons became clogged, the dam at the outlet from the enclosure began to function again. The water level thus rose within the whole enclosure and the winter store was again submerged. The animals then removed the peeled sticks using them for building on the lodge and the dam at the outlet, and again began to fix branches with edible bark at the store.

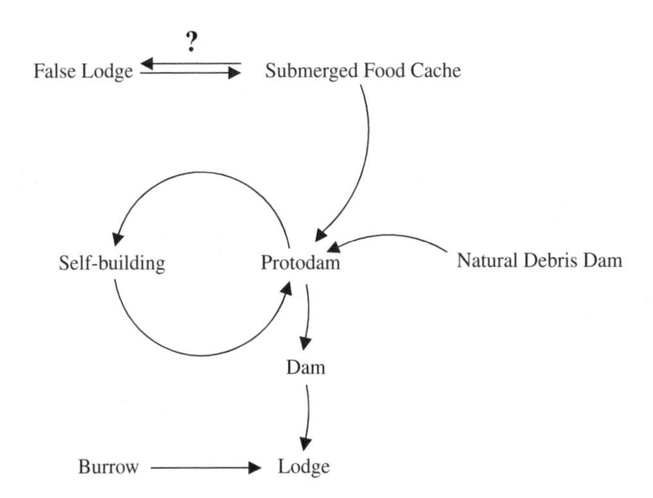

FIGURE 3.24 A hypothetical sequence of steps in the evolution of building behaviors by beavers.

Natural obstructions including debris dams have been mentioned as locations where the modern beaver initiates dam building (Johnson, 1927; Morgan, 1868; Warren, 1927), which adds support to the possibility that an ancestral beaver may have used them to anchor a food cache.

The development of the beaver protodam would have been assisted by two forms of self-building processes. First, the protodam would have continued to act like a natural debris dam in accumulating wood and sediments carried by the current. This kind of self-building also has been described for existing beaver dams by Johnson (1927) and Warren (1927). A second and perhaps more interesting process involves the wood that the beaver uses as a source of food. Once the bark is eaten off the wood, it is discarded into the water where it can add to the dam. Wilsson (1971) provides a description of this process:

> Peeled sticks and other waste is always taken out of the lodge every morning and, if no building activity is in process, is thrown into the water where it sinks to the bottom. Material deposited in this way during the course of the year constitutes the most important source of building material in the autumn. In places where there are no natural obstacles interfering with the flow of the water the pile of accumulated waste can itself sometimes serve as an obstacle, forming the base on which the beavers begin to build their dam. Such behaviour was observed in animals 4A–5A which, during their second autumn in the enclosure, began to build a dam on the pile of sticks that had accumulated on the bottom during the previous summer.

Thus, a logical set of steps can be theorized for the evolution of dam building from an ancestral non–dam-building beaver (Figure 3.24). Dam building is suggested to have arisen from food caching whereby a submerged food cache becomes a protodam, possibly with combination of a natural debris dam. The protodam then enlarges with the aid of two forms of self-building processes. Selective advantage for the beaver comes from the increase in aquatic habitat area formed by the protodam,

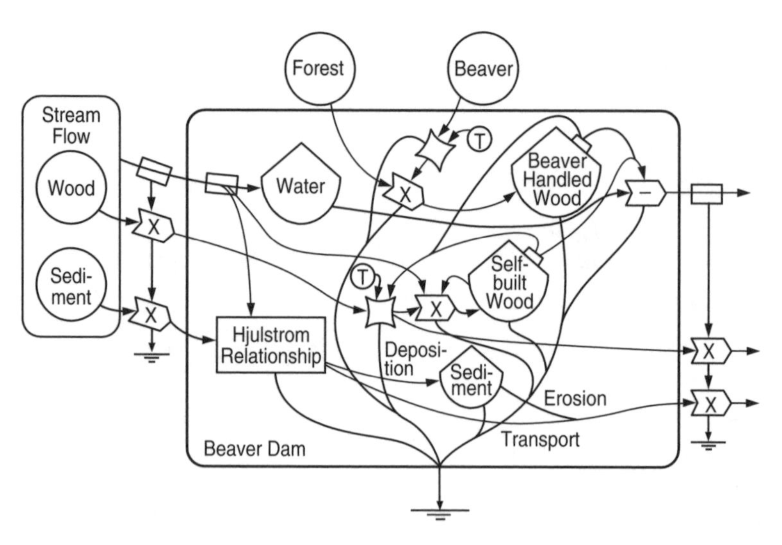

FIGURE 3.25 Energy circuit diagram of a beaver dam system including wood from debris dam processes and wood deposited by beavers.

which reduces mortality from terrestrial predators, aids in wood transport, and increases access to riparian food resources. The true dam emerges when beavers start to actively add wood to the protodam. In a sense, then, beavers were preadapted for dam building by their behavior of cutting trees and caching wood underwater. This theory could be partially tested by attempting to build a beaver protodam through the addition of wood to a natural debris dam. If this experiment could increase aquatic surface area (i.e., cause ponding), then the theory would be supported. Furthermore, the size distribution of wood that makes up modern dams may reveal clues of the origin of dam building, if there is a relationship with natural debris dams. Alternative theories for the origin of dam building, such as through the modification of lodge building, can be imagined but they seem less likely. In fact, it seems more likely that increased water levels from the action of the beaver protodam may have triggered the development of lodges as the bank burrows became flooded.

Figure 3.25 summarizes the structure of the beaver dam–debris dam system. In a debris dam, the storage of self-built wood accumulates naturally by self-organization from wood being carried by stream flow. This wood storage interacts with water flow to reduce velocity, which in turn causes sediments and more wood to deposit. Beavers amplify this natural structure by directly adding wood from the forest to the storage of beaver-handled wood through several behaviors (caching, feeding, and dam building). Overall, the beaver dam consists of two storages of wood plus some of the storage of sediments. The storage of water behind the dam provides services to the beaver, providing an evolutionary selective force reinforcing various behaviors. Thus, the beaver acts like a true ecological engineer in taking advantage of the free energies and the self-design properties of the stream to create a system useful to itself!

A major challenge for ecological engineering is to find ways to use the beaver for positive roles for humans, such as erosion control. A similar problem was described in Chapter 2 for muskrats in treatment wetlands. These animals are essentially programmed by nature to perform certain kinds of hydrologic manipulations. In the wrong locations these are the actions of a pest species, but if the right locations can be found, then these animals can provide ecosystem services for free. There are limits to beavers' dam-building capabilities in terms of the size of stream and the need for adequate food supply of riparian trees, but within these limits beavers can be a tool for ecological engineering design.

The Role of Beaches and Mangroves in Coastal Erosion Control

Strategies for controlling coastal erosion depend on the energy level of the coast. High-energy coastlines with strong waves and/or tidal flows are usually protected by hard engineering alternatives while vegetation-based systems (i.e., soft bioengineering) can be used in low-energy settings. Protection is required for average conditions and for storm events along all coasts.

The hard engineering alternatives for coastal erosion control are well known and include breakwaters, groins, seawalls, and revetments (Bascom, 1964). These are structures made of rock or concrete that dissipate the kinetic energy of waves and tide before erosion can occur. This is an expensive approach for erosion control, but it is often necessary to protect shorelines with important human land use. The discipline of coastal engineering covers the design, construction, and operation of these structures. Ecological engineering has little to offer to this topic because the structures are nonliving. However, the hard engineering structures do provide habitat for organisms and, thus, they have by-product values to the surrounding ecosystems. A contribution from the ecological perspective to coastal engineering would be to add habitat values into the design process for hard structures. An example would be certain artificial reefs described in Chapter 5. Montague (1993) also describes a number of ways that coastal engineering can create habitat for sea turtle nesting.

The optimal design from nature for high energy coasts is the beach. This is a fascinating geomorphic structure that dissipates wave energy and naturally protects the shoreline. As noted by Sensabaugh (1975),

> There are two principal features of the beach which make it particularly effective in protecting the upland. First, it has a sloping surface that gradually dissipates the energy of a wave as the wave flows up the slope. Second, since it is made of sand, the beach is flexible and the slope can change as the waves change.

Of course, beaches are highly valued by humans, especially for recreational uses. Beaches can erode both from natural changes and from changes caused by humans. One approach to controlling this erosion process is through beach nourishment (Bird, 1996), which involves the artificial addition of dredged sand to compensate for losses due to erosion. This strategy is an example of "soft engineering" that can be less expensive than the use of hard structures.

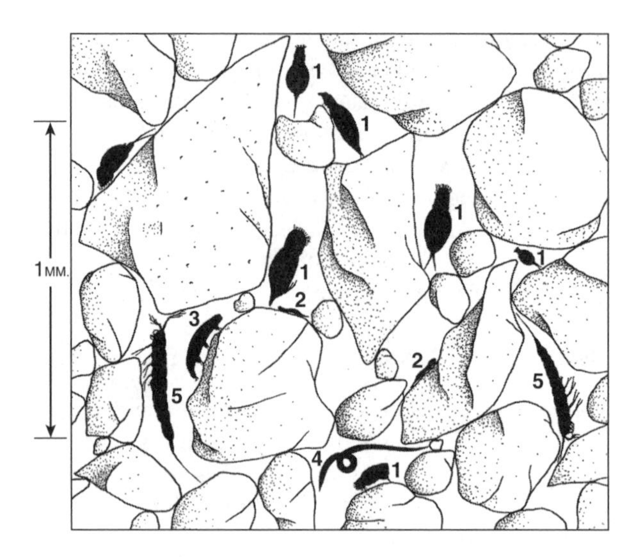

FIGURE 3.26 Interstital meiofauna in beach sediments. (1) Rotifers. (2) Gastrotrichs. (3) Tardigrade. (4) Nematode. (5) Harpacticoid copepods. (From Pennak, R. W. 1989. *Fresh-Water Invertebrates of the United States, Protozoa to Mollusca,* 3rd ed. John Wiley & Sons, New York. With permission.)

While beaches may appear to have little biota, in fact a rich and complex ecosystem exists within the intertidal sands (McLachlan, 1980; McLachlan and Erasmus, 1983; McLachlan et al., 1981; Pearse et al., 1942). This system is dominated by interstital meiofauna (Giere, 1993) such as nematodes and turbellarians, along with burrowing macrofauna such as mole crabs and surf clams. Algae and protozoans also can be important as in the "living sands" of coral reef sediments described by Lee (1995). An interesting challenge for ecological engineering would be to attempt to add some of this kind of biota (Figure 3.26) to sand filters used in wastewater treatment (Anderson et al., 1985; Crites and Tchobanoglous, 1998). Sand filters are beds of medium to coarse sands, usually on the order of 1 to 2 m in depth, that are underlain with gravel containing collection drains. Effluent is applied intermittently to the surface and percolates through the sand to the bottom of the filter. The under drain collects the filtrate which is either recirculated back to the bed or discharged. Sand filters are designed to be aerobic and are very effective at removing BOD and TSS and in nitrification. Like most wastewater treatment systems, only the microbial organisms are considered to be relevant to the operation of the system (Calaway, 1957). However, because sand filters are somewhat analogous to beaches, the ecological engineer could try to design a more complex food web for the sand filter based on knowledge of beach ecosystems and the extensive literature on animal–sediment relationships (Aller et al., 2001; Gray, 1974; Rhoads and Young, 1970). A more complex food web might upgrade the performance of sand filters for wastewater treatment by improving porosity or reducing clogging (see Wotton and Hirabayashi, 1999). These actions could reduce maintenance costs and lead to a more effective sand filter for commercial use. Another ecological analog for sand filters is the hyporheic zone in streams (Findlay, 1995; Stanford and Ward, 1988). This is the upper portion of the sediment layer which is fed by water from the channel rather than from the groundwater. Like beaches, the hyporheic zone is dominated by meiofauna, whose metabolism can match or exceed that of the ecosystem within the stream channel.

Vegetation-based systems can be used for erosion control along low-energy coastlines. These are wetland ecosystems dominated by plant species with special adaptations for flooding and salt tolerance. Along the Earth's coastlines, vegetation type generally is determined by the presence or absence of frost. In temperate and arctic regions, marshes with perennial herbaceous vegetation are found which die back aboveground each winter due to frost stress (i.e., saltmarsh). In the lowland tropics where temperatures are never below freezing, woody tree vegetation (i.e., mangroves) is found which is evergreen and has no adaptation to frost. This section focuses on mangroves because of the interesting literature on their self-building behavior, which relates to the earlier discussions on debris dams and beaver dams. saltmarshes are discussed in Chapter 5 in terms of restoration ecology.

Mangroves include a number of plant families with representative tree life forms that grow in the coastal zone. In the neotropics and Africa these are low diversity swamp forests with only a few tree species, while in Asia and Australia many species of mangroves are found in the coastal swamps. Mangroves as a group exhibit a number of special adaptations including physiological salinity control, vivipary (i.e., seeds germinate while they are still attached to the parent tree), and modified roots. Of these, the root systems are most relevant to a consideration of erosion control. The special lateral root systems of mangroves provide support in soft sediments and expose surface area to facilitate aeration for living tissues in the anoxic muds (Figure 3.27). These fall into two main groups: prop root systems in which the lateral root is aerial (Figure 3.27a) and cable root systems in which the lateral root is belowground, usually with aerial extensions called *pneumatophores* (Greek for "breathing roots") (Figure 3.27b–e).

Prop roots are found on species of the genus *Rhizophora* (Gill and Tomlinson, 1971, 1977) which is usually found at the edge of the coastline where erosion processes are dominant. These aerial roots create a "dense baffle which is highly effective in reducing current strength" (Scoffin, 1970). Scoffin measured tidal flows in the field and found that currents of 40 cm/sec velocity at the edge of the prop root zone are reduced to zero only 1 m inside the forest. Because erosion is directly related to flow rate, the *Rhizophora* prop roots have an important role in erosion control by reducing current velocity. These props roots are massive biogenic structures (Figure 3.28), and Golley et al. (1975) found that they make up 25% of the total aboveground biomass in a mangrove swamp in Panama.

The ability of mangroves to reduce erosion, along with the dispersal adaptation of vivipary, led a number of early workers to conclude that mangroves actually build land in the coastal zone by their growth and by the sediment accumulation they cause. Davis (1940) gave the most extensive treatment of this role, which created a

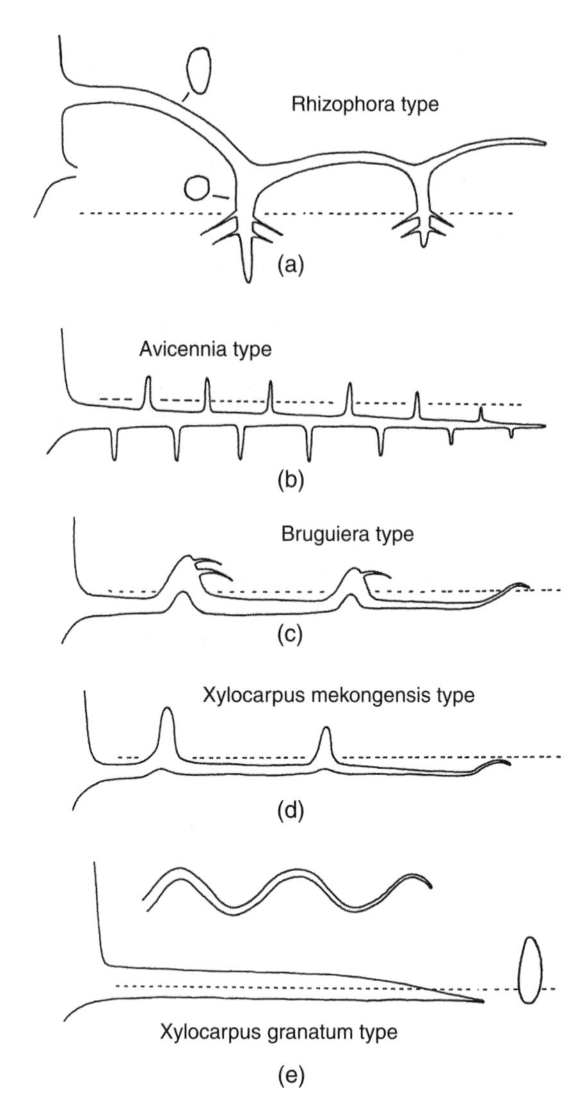

FIGURE 3.27 Views of lateral root systems of mangrove tree species. (a) Pop root system. (b–e) Cable root systems. (From Tomlinson, P. B. 1986. *The Botany of Mangroves*. Cambridge University Press, Cambridge, U.K. With permission.)

controversy in the literature. Many workers, starting in the 1950s, studied particular sites and reported in opposition to Davis that mangroves follow silting rather than cause it. This opinion is summarized by Frank Egler (1952a) in his unique style as noted below:

> As for the Rhizophora community actively "walking out to sea" (an action suggested by the way the prop roots extend out from the front of the community), and as for the fruits plunking into the mud below and planting themselves (an action suggested by

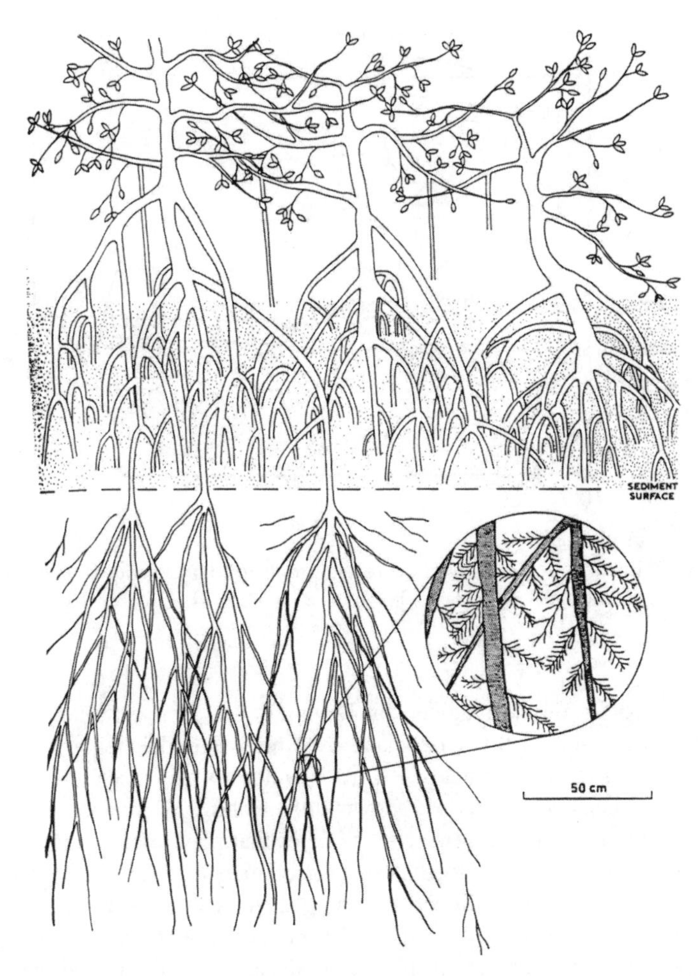

FIGURE 3.28 View of the aerial root system of the red mangrove, *Rhizophora mangle.* (From Scoffin, T. P. 1970. *J. Sed. Petrol.* 40:249–273. With permission.)

> their plummet-shape and green upper parts), these stories appear to be part of the arm-chair musings of air-crammed minds of a century ago. I have seen sporadic instances of such phenomena: they are not impossible as accidental events. On the other hand, for its being a normal community activity accounting for the bulk of the Rhizophora belt, I place its probability — for the regions I know — on par with that of a chimpanzee bowed before a typewriter and batting out the Sermon at Benares.

Thom's work (Thom, 1967, 1984; Thom et al., 1975) is most often cited in opposition to the theory that mangroves build land (see also the review by Carlton, 1974). The modern view is, then, that "Mangroves establish once sediments have built up sufficiently by physical processes but once established, mangroves contribute to the process of accretion by accelerating the rate of land elevation, preventing erosion and by stabilizing the accretion-erosion cycle" (Lugo, 1980). This action,

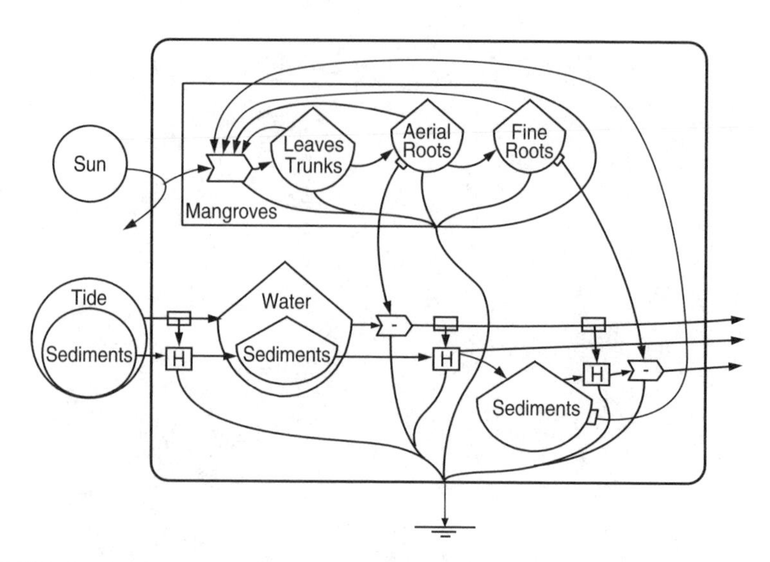

FIGURE 3.29 Energy circuit model of the roles of roots in controlling erosion in a mangrove swamp. Sediment transport is shown with the box symbol, H, representing the Hjulstrom relationship.

though not as dramatic as imagined by Davis (1940), represents another kind of self-building behavior. In this case mangroves help to build their own forests by stabilizing and accreting sediments. Figure 3.29 depicts this mechanism for a tidal flow situation. Mangrove biomass is shown with an aboveground storage that drives primary production along with two root storages that play roles in erosion control. Aerial roots reduce current velocity causing sediments to deposit, and fine roots reduce the loss rate of the sediments once deposited. The self-building feedback is shown in the pathway from sediments to the production work gate. This represents the necessary role of sediments in providing substrate for the mangroves to grow on. The early view (i.e., Davis, 1940) and the modern view (i.e., Thom, 1984) of mangrove sedimentation agree that this self-building process occurs but they differ in the extent of new land that has been created. This disagreement is partly due to the complication of the rising sea level that has been occurring globally since the end of the last ice age. Humans are apparently accelerating sea level rise through the greenhouse effect, and an interesting literature is developing on impacts to existing coastal wetlands (see Chapter 5).

The goal of ecological engineering is to actively utilize the erosion control potential of mangroves and other types of coastal vegetation. Carlton (1974) provides a view of this use (Figure 3.30), which combines hard engineering (bulkhead seawall) with soft bioengineering (mangroves) for the best coastal development. This kind of action would provide by-product value through the detritus food webs and habitat

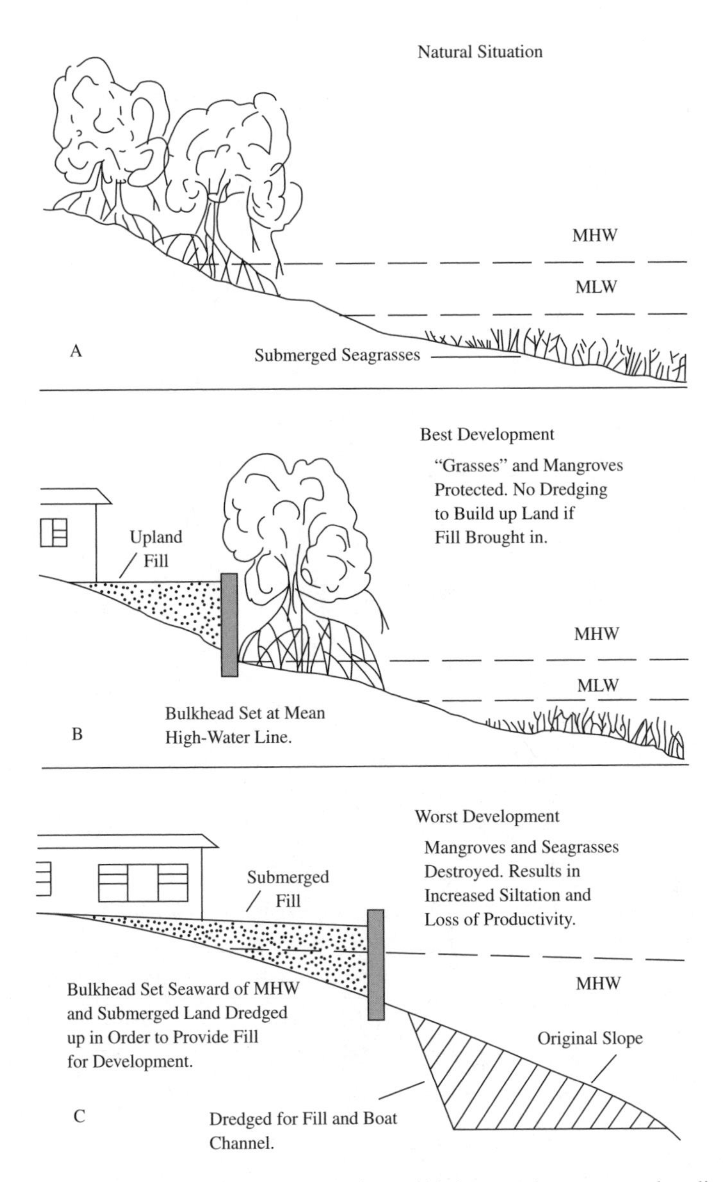

FIGURE 3.30 Comparisons of shoreline designs. (A) A natural mangrove shoreline. (B) A developed shoreline with bulkhead and mangroves. (C) A developed shoreline with dredging and bulkhead protection. (From Carlton, J. M. 1974. *Environmental Conservation* 1:285–294. With permission.)

values that mangroves provide to local fisheries, while controlling erosion and loss of human property.

4 Microcosmology

INTRODUCTION

Microecosystems or microcosms are relatively small, closed or semi-closed ecosystems used primarily for experimental purposes. As such, they are living tools used by scientists to understand nature. *Microcosms* literally means "small world," and it is their small size and isolation which make them useful tools for studying larger systems or issues. However, although they are small, as noted by Lawler (1998),

> microcosms should share enough features with larger, more natural systems so that studying them can provide insight into processes acting at larger scales, or better yet, into general processes acting at most scales. Of course, some processes may operate only at large scales, and big, long-lived organisms may possess qualities that are distinct from those of small organisms (and vice versa). Because large and small organisms differ biologically, it will not be feasible to study some questions using microcosms. However, to the extent that some ecological principles transcend scale, microcosms can be a valuable investigative tool.

Microcosm, as a term, was originally used in ecology as a metaphor to imagine a systems concept (Forbes, 1887; see also Hutchinson, 1964). More recently, Ewel and Hogberg (1995) and Roughgarden (1995) used *microcosm* as a metaphor for islands, which have been used profitably as experimental units in ecology (Klopfer, 1981; MacArthur and Wilson, 1967). Microcosms are a part of ecological engineering because (1) technical aspects of their creation and operation (often referred to as boundary conditions) require traditional engineering and design, and (2) they are new ecological systems developed for the service of humans.

A large literature exists on the uses of microcosms primarily to develop ecological theory and to test effects of stresses, such as toxic chemicals, on ecosystem structure and function. This literature demonstrates a high degree of creativity in design of experimental systems as surveyed in the book length reviews by Adey and Loveland (1998), Beyers and H. T. Odum (1993), and Giesy (1980). Adey (1995) graphs microcosm-based publications/year from 1950 through 1990, showing a steady increase in literature production over time. Lawler (1998) suggests that production is about 80 microcosm-based publications/year, while Fraser and Keddy (1997) find more than 100 per year for the mid-1990s. Microcosm research covers a tremendous range from gnotobiotic systems composed of a few known species carefully added together (Nixon, 1969; Taub, 1969b) to large mesocosms composed of thousands of species seeded from natural systems, such as Biosphere 2 [see Table 1 in Pilson and Nixon (1980) for an example of the variety of microcosms used in ecological research]. Some are artificially constructed systems kept under controlled environmental conditions while others are simple field enclosures exposed to the natural environment. Philosophies of microcosm use vary across these kinds of

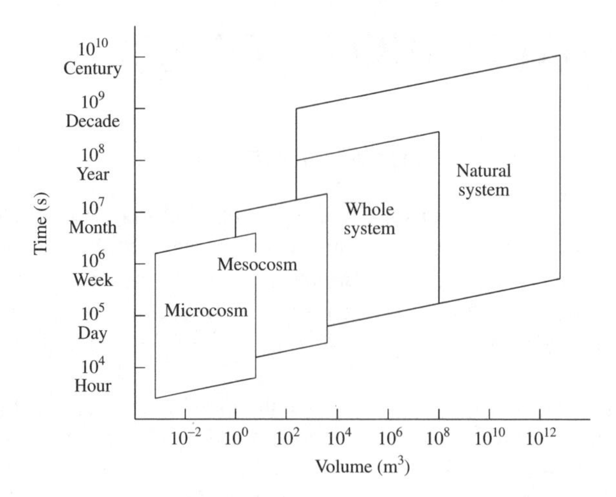

FIGURE 4.1 Comparisons of time and space scales showing the appropriate dimensions for use of microcosms and mesocosms. (From Cooper, S. D. and L. A. Barmuta. 1993. *Freshwater Biomonitoring and Benthic Macroinvertebrates*. D. M. Rosenberg and V. H. Resh (eds.). Chapman & Hall, New York. With permission.)

experimental gradients, which makes this a rich and interesting subdiscipline of ecological engineering.

One useful size distinction occurs between microcosms and mesocosms, with microcosms being smaller and mesocosms being larger experimental systems. Although there is no consensus on the size break between microcosms and mesocosms, several ideas have been published. Lasserre (1990) suggests a practical though arbitrary limit of 1 m^3 (264 gal) volume to distinguish laboratory-scale microcosms from larger-scale mesocosms used outside the laboratory. Lawler (1998) prefers to base the distinction on the scale of the system being modelled:

> Whether the term "microcosm" or "mesocosm" applies should depend on how much the experimental unit is reduced in scale from the system(s) or processes it is meant to represent. A microcosm represents a scale reduction of several orders of magnitude, while a mesocosm represents a reduction of about two orders of magnitude or less … The distinction between terms is admittedly rough, but I hope it is preferable to an anthropocentric view where a microcosm is anything small on a human scale (smaller than a breadbox?) and mesocosms are somewhat larger.

Cooper and Barmuta (1993) combine time and space scales in a diagrammatic view that portrays overall experimental systems used in ecology (Figure 4.1). Taub (1984) suggests that microcosms and mesocosms serve different purposes and answer different questions in ecology (Table 4.1). Clearly, by their relatively larger size, mesocosms contain greater complexity and exist at different scales of space and time compared with typical laboratory-scale microcosms (Kangas and Adey, 1996; E. P. Odum, 1984). However, both microcosms and mesocosms share the aspects of ecological engineering noted earlier and are treated together in this chapter.

TABLE 4.1
Comparisons between Microcosms and Mesocosms

Microcosms	Smaller, with more replicates
	Usually used in the laboratory with greater environmental control
	More easily analyzed for test purposes
	Often focus on certain components or processes
Mesocosms	Larger, with fewer replicates
	Often used outdoors with ambient temperature and light conditions
	Realistic scaling of environmental factors
	Give maximum confidence in extrapolating back to large-scale systems
	Provide greater realism by incorporating more large-scale processes

Source: Adapted from Taub, F. B. 1984. *Concepts in Marine Pollution Measurements*. H. H. White (ed.). Sea Grant Publ., University of Maryland, College Park, MD.

Several authors have almost playfully referred to the use of microcosms in ecology as *microcosmology*, implying a special world view (Beyers and H. T. Odum, 1993; Giesy and E. P. Odum, 1980; Leffler, 1980). Adey (1995) has also hinted at this kind of extensive view by suggesting the term *synthetic ecology* for the use of microcosms. The issue is one of epistemology, or how we come to gain knowledge, and the suggestion seems to be that microcosms provide a unique, holistic view of nature perhaps by reducing the scale difference between the experimental ecosystem and the human observer. In this way a special insight is conferred on the scientist from use of microcosms or at least it is easier to achieve than when dealing with ecosystems of much greater scale than the human observer.

Perhaps the most important philosophical aspect of the use of microcosms is their relationship to real ecosystems. Are they only models of analogous real systems or are they real systems themselves? Leffler (1980) provided a Venn diagram which shows that microcosms overlap with real systems but also have unique properties (Figure 4.2). Likewise, the real-world systems have unique properties such as disturbance regimes and top predators that are too large to include in even the largest mesocosm. Clearly, there are situations when a microcosm is primarily used as a model of a real system. For example, it is obviously advantageous to test the effect of a potentially toxic chemical on a microcosm and be able to extrapolate to a real ecosystem rather than to test the effect on the real system itself and risk actual environmental impact. When a microcosm is meant to be a model of a particular ecosystem, the design challenge is to create engineered boundary conditions that allow for the microcosm biota to match the analogous real system with some significant degree of overlap in ecological structure and function. While this use may be the most important role of microcosms, there are situations when the microcosm need not model any particular real system, such as their use for studying general ecological phenomena (i.e., succession) or their direct functional use as in wastewater treatment or in life support for remote living conditions. Natural micro-

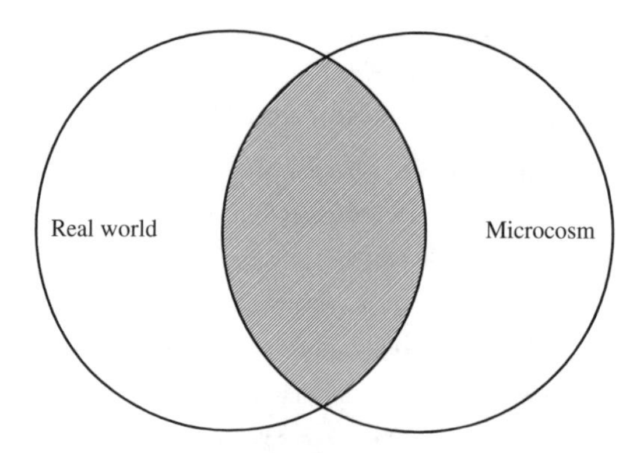

FIGURE 4.2 Venn diagram of the philosophical bases of microcosmology. (Adapted from Leffler, J. W. 1980. *Microcosms in Ecological Research.* J. P. Giesy, Jr. (ed.). U.S. Dept. of Energy, Washington, DC.)

cosms, such as phytotelmata (Kitching, 2000; Maguire, 1971), depressions in rock outcrops (Platt and McCormick, 1964), and tide pools (Bovbjerg and Glynn, 1960), demonstrate that systems on the scale of even the smallest microcosm are real systems whose study can yield insights as valid as from any other real-world system. In fact, there may be value in purposefully creating microcosm designs that do not match with any existing real ecosystem in order to study the ability of systems to adapt to new conditions that have never existed previously. In this case the portion of the microcosm set outside the zone of overlap with the real world in Figure 4.2 is of great interest. This sense is somewhat analogous to the use of islands in ecology mentioned earlier. In classic island biogeography, the islands are not necessarily meant to be models of continents but rather natural experiments of different ages, sizes, and distances from continents. Therefore, the position taken in this chapter is that microcosms are real systems themselves, but they may or may not be models of larger ecosystems depending on the nature of the experiment being undertaken. See Shugart (1984) for a similar discussion about the relationship of ecological computer simulation models and real ecosystems, which includes a Venn diagram similar to Figure 4.2.

STRATEGY OF THE CHAPTER

This chapter reviews the uses of microcosms and mesocosms as experimental ecosystems. Numerous excellent reviews have been published on this topic, and many are cited for further reading throughout the text. An effort is made to focus on elements of relevance to both the engineering side and the ecological side of applications. In relation to engineering, design aspects of microcosms are covered, including scaling, energy signatures, and complexity. The controversy between ecologists and engineers over the role of microcosms in research on space travel life support systems is given special attention as a case study in ecological engineering. In relation to ecology, aspects of the new systems that have emerged from microcosm research are highlighted. The new qualities show up in (1) examples of micrcocosm replication, and (2) when microcosms are compared with real analog ecosystems.

MICROCOSMS FOR DEVELOPING ECOLOGICAL THEORY

Microcosms have a long tradition of use for developing theories about most of the hierarchical levels covered by ecology: organism, population, community, and ecosystem. While some of this work has been descriptive, most has relied on experiments. In the experimental approach, replicate microcosms are developed and partitioned into groups with some being held as controls and others being treated in some fashion. The experiment is analyzed by statistically comparing the control group with the treated group(s) after a given period of time. Such an experiment can be a challenge to carry out in nature due to the difficulty in establishing replicates and the difficulty in changing only one factor per treatment group. On the other hand, it is easy to carry out this kind of controlled experiment with microcosms, which allows them to be used as valuable tools in ecology.

The earliest microcosm work was done on species change during succession of microbial communities (Eddy, 1928; Woodruff, 1912), but most research using microcosms dates after the 1950s. Uses of microcosms for developing ecological theory generally fall into two groups: one in which the ecosystem itself is of interest (ecosystem scale) and the other in which the ecosystem provides a background context and population dynamics or interactions between species are of interest (community or population scale). In both cases, microcosms often are used in a complementary fashion with basic field studies and mathematical models as part of an overall research strategy.

Many of the important figures in modern ecology used microcosms in early studies of ecosystems including Margalef (1967), Whittaker (1961), and H. T. Odum (Armstrong and H. T. Odum, 1964; H. T. Odum and Hoskin, 1957; H. T. Odum et al., 1963a). Robert Beyers, H. T. Odum's first doctoral student, also was an early proponent of microcosms (1963a, 1963b, 1964) and, together with H. T. Odum, coauthored probably the most comprehensive text on the subject (Beyers and H. T. Odum, 1993). The early studies outlined the basic processes of energy flow (primary production and community respiration) and biogeochemistry (nutrient cycling), which are the foundations of ecosystem science today. One example of the contribution of microcosms to ecosystem science can be seen in papers by E. P. Odum and his associates on succession (Cooke, 1967, 1968; Gordon et al., 1969). These papers described ecosystem development under both autotrophic (initial conditions of high nutrients and low biomass) and heterotrophic (initial conditions of low nutrients and high biomass) pathways in laboratory microcosms. These studies directly contributed to E. P. Odum's development of a tabular model of ecological succession (see Chapter 5) as can be seen by comparing their summary tables [Table 2 in Cooke (1967) and Table 12 in Gordon et al. (1967)] to E. P. Odum's tabular model [Table 1 in E. P. Odum (1969) and Table 9.1 in E. P. Odum (1971)]. E. P. Odum's model compares trends expected through succession for 24 ecosystem

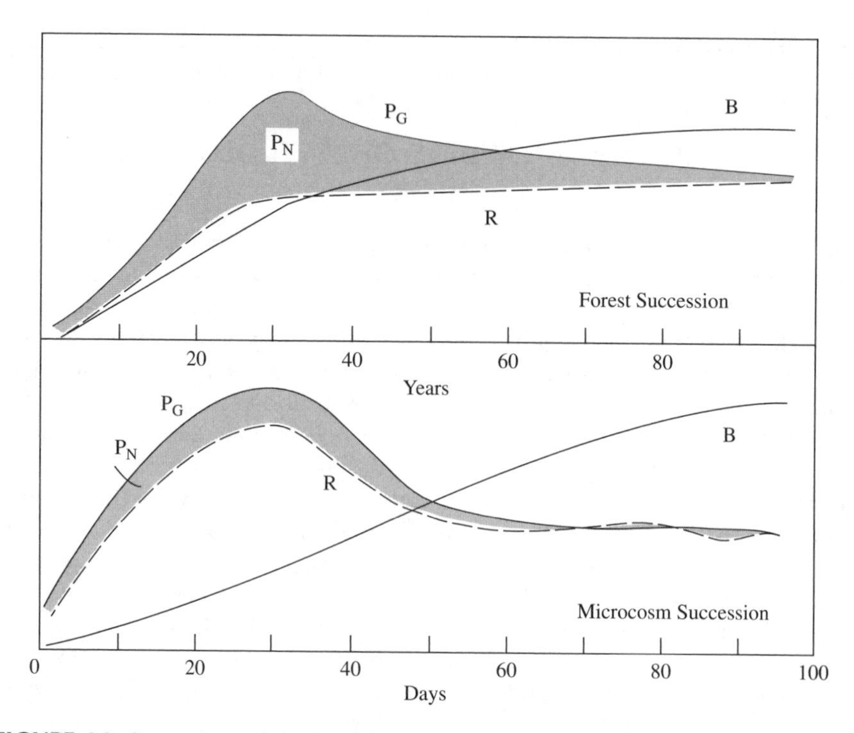

FIGURE 4.3 Comparison of the development of a forest ecosystem with a microcosm. The time patterns are similar but the time scaling is different. P_G = gross production; P_N = net production; R = total community respiration; B = total biomass. (From Odum, E. P. 1971. *Fundamentals of Ecology*, 3rd ed. W. B. Saunders, Philadelphia, PA. With permission.)

attributes and is an intellectual benchmark in the synthesis of ecosystem science. E. P. Odum (1971) also used data from Cooke's (1967) work to illustrate the generality of certain metabolic patterns of succession by comparing small-scale microcosm results with field-scale results (Figure 4.3). This figure is particularly interesting in showing a kind of self-similarity or scaling coefficient on the order of days for the microcosm and years for the forest. Although many other examples could be cited, Hurlbert's studies of pond microcosms (Hurlbert and Mulla, 1981; Hurlbert et al., 1972a, b) are especially detailed examples of ecosystems comparing effects of fish predation and insecticides on ecosystem structure and function.

For another line of research, the microcosm provides only a context for studies of population dynamics or species interactions. Recent reviews of this work are given by Drake et al. (1996), Lawler (1998), and Lawton (1995). Included here are some of the fundamental studies of ecology such as those by Gause (1934) and Park (1948). G. F. Gause was a Russian scientist who studied interactions among protozoan populations in glass vials. He is credited with the first expression of the competitive exclusion principle which states that when two species use similar resources (or occupy the same niche), one species will inevitably be more efficient and will drive the other extinct under limiting conditions (see Chapter 1). He also conducted laboratory experiments on predator–prey relations such as shown in Figure 4.4. *Paramecium caudatum* was the prey population in these laboratory

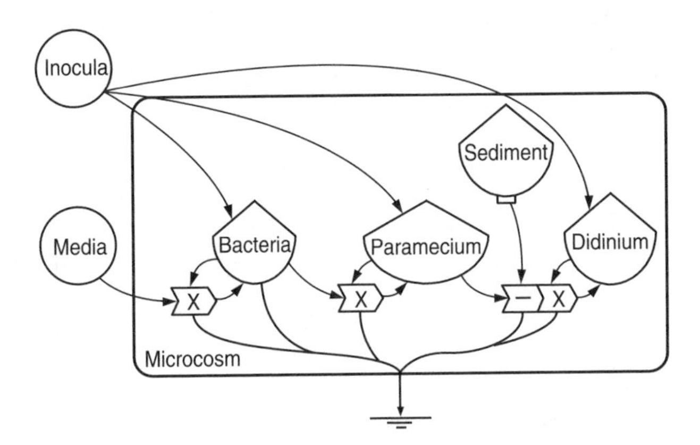

FIGURE 4.4 Energy circuit diagram of Gause's classic microcosm. Note the series connections characteristic of predator–prey relations.

cultures, which was supported on an undefined set of bacteria at the base of the food chain, and *Didinium nasutum* was the predator population. Much work was required to design an effective growth media for all of the species (Gause, 1934). Three conditions were demonstrated by the experiments. With no special additions, the predator consumed all of the prey and they both went extinct (Figure 4.5 A). When sediment was placed in the bottom of the vials, it acted as a refuge for the prey to escape the predator. In this case the predator eventually went extinct and the prey population grew after being released from predation pressure (Figure 4.5B). Finally, when periodic additions of both prey and predator were used to simulate immigration, the oscillations characteristic of simple mathematical equations were found (Figure 4.5C).

Thomas Park also studied basic population dynamics and competition with laboratory cultures of flour beetles (Figure 4.6). More than 100 papers were produced by Park and his students over a 30-year period on this extremely simple ecological system, which laid the foundation for important population theory. The microcosm consisted of small glass vials filled with a medium of 95% sifted whole-wheat flour and 5% Brewers' yeast. A known number of adult beetles of one or two species (depending on the experiment) in equal sex ratios were added to the media and were incubated in a growth chamber for 30 days. At that time the media were replaced and the beetles were censused and returned to the vials. This procedure was followed for up to 48 censuses (1,440 days), which was "roughly the equivalent of 1,200 years in terms of human population history" when scaled to human dimensions (Park, 1954)! Obviously, the engineering involved in these microcosms was minimal but elegant in providing such a powerful experimental tool for the time period. Also, the flour beetles themselves were preadapted for use in the microcosms because they spend their entire life cycle in flour. The focus of Park's work was on the population rather than the ecosystem, though it did simulate a natural analog of food storage and pests (Sinha, 1991). Park (1962) described the experimental system with a machine analogy as follows:

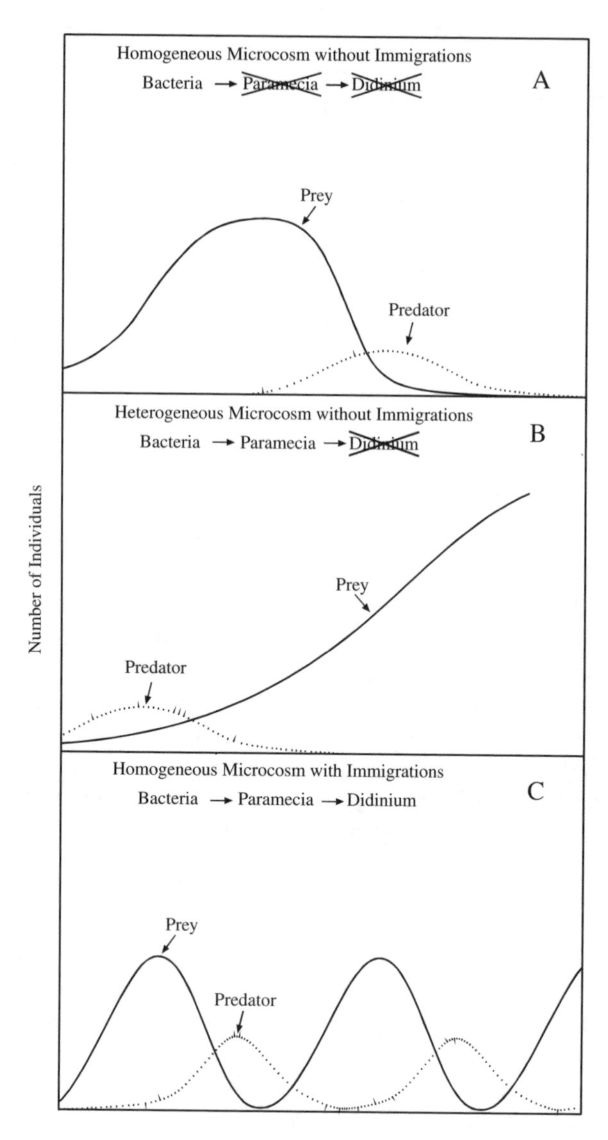

FIGURE 4.5 Outcomes of Gause's experiments on the role of predation. (A) Result of experiment with no sediment or species additions. (B) Result of experiment with the addition of sediment which acts as a refuge for the prey *Paramecium*. (C) Result of experiment with periodic additions of both the predator *Didinium* and the prey *Paramecium* resulting in oscillations of population sizes. (Adapted from Gause, G. F. 1934. *The Struggle for Existence*. Williams & Wilkens, Baltimore, MD.)

Let us begin with two seemingly unrelated words: beetles and competition. We identify competition as a widespread biological phenomenon and assume (for present purposes at least) that it interests us. We view the beetles as an instrument: an organic machine which, at our bidding, can be set in motion and instructed to yield relevant information. If the machine can be properly managed and if it is one appropriate to the problem,

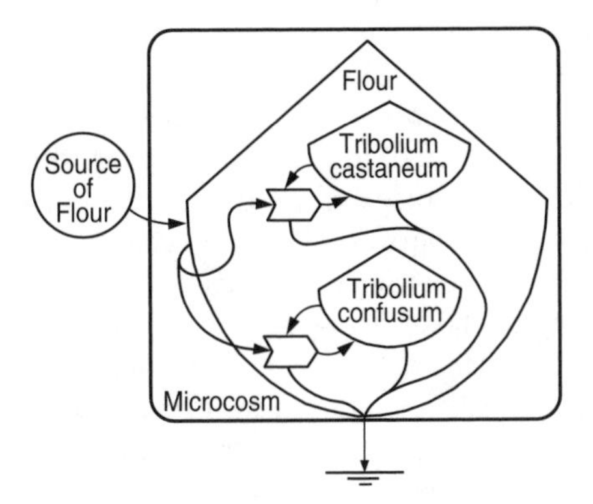

FIGURE 4.6 Energy circuit diagram of Park's classic microcosm. Note the parallel connections of competition between the two *Tribolium* species.

> we are able to increase our knowledge of the phenomenon. … Obviously, there exists an intimate marriage between machine, its operator, and the phenomenon. Ideally, this marriage is practical, intellectual, and esthetic: practical in that it often, though not immediately, contributes to human welfare; intellectual in that it involves abstract reasoning and empirical observation; esthetic in that it has, of itself, an intrinsic beauty. Perhaps these rather pretentious reflections seem far removed from the original words — beetles and competition. But I do not think this is the case.

Basic scientific research on populations and communities at the mesocosm scale began with the work of Hall et al. (1970) on freshwater pond systems. Historically, most mesocosm studies have been directed at applied studies of ecotoxicology but, as noted by Steele (1979), this work almost always also yields insights on general ecological principles. One of the best examples of basic mesocosm research may be the work of Wilbur (1987, 1997) and his students on interactions among amphibians in temporary pond mesocosms. These studies of life history dynamics, competition, and predation have led to a detailed understanding of the community structure of this special biota. The mesocosms consist of simple metal tanks, and an interesting dialogue on Wilbur's experimental approach is given in a set of papers in the journal *Herpetologia* (Jaeger and Walls, 1989; Hairston, 1989; Wilbur, 1989; and Morin, 1989). Much discussion has been recorded on the trade-offs between realism and precision in this type of research (see, for example, Diamond, 1986), and Morin (1998) describes mesocosms as hybrid experiments at a scale between the laboratory and the field with an optimal balance between the two extremes of experimental design.

MICROCOSMS IN ECOTOXICOLOGY

Microcosms are important as research tools in ecotoxicology for understanding the effect of pollutants on ecosystems. Experiments in which treatments are various concentrations of pollutant chemicals can be conducted in microcosms with replication and with containment of environmental impacts due to isolation from the environment. Although this role for microcosms in ecotoxicological research is well established, their potential role within formal regulatory testing or screening protocols in risk assessment is controversial. Challenges for ecological engineering include the design and operation of microcosms that are effective for both research and risk assessment in ecotoxicology. Uses for risk assessment will be emphasized in this section owing to the controversial debate about the role of microcosms and the wide potential applications of microcosm technology that are involved.

Testing or screening of chemicals is regulated by the Environmental Protection Agency (EPA) in the U.S. This regulation is necessary because of the tremendous number of new chemicals that are produced each year for industrial and commercial purposes. Many of these chemicals are xenobiotic or man-made, whose potential environmental effects are unknown. Thus, uncertainty arises because natural ecosystems have never been exposed to them and species have not adapted to them. Special concern is needed for pesticides because they are intentionally released into the environment and are intended to be toxic, at least to target organisms. The primary examples of legislation covering regulatory testing and screening of chemicals are the Toxic Substances Control Act and The Federal Insecticide, Fungicide, and Rodenticide Act, along with several others (Harwell, 1989). An interaction has developed among the EPA, the chemical industry, environmental consulting firms, and academic researchers in relation to risk assessment of new chemicals, which has in turn created opportunities for applications of ecologically engineered microcosm technology.

EPA's risk assessment approach for chemicals (Norton et al., 1995) has evolved over time since early work in the 1940s on methods for measuring the effects of pollutants. The purpose of risk assessment is to evaluate potential hazards in order to prevent damage to the environment and human health. The basis for testing or screening is a hierarchical (tiered) protocol of sequential tests. Physical and chemical properties are tested at the lowest tier, and acute and chronic toxicity data along with estimated exposure data are gathered for several aquatic species at intermediate tiers, followed at least in principle by simulated field testing at the highest tier (Hushon et al., 1979). The intention is to minimize the number of tests required to assess a chemical's hazard and at the same time to include a comprehensive range of tests. Each tier level can trigger testing at higher levels by comparison of test results to established end points which determine whether or not the chemical is considered to be toxic or hazardous. Choice of end points is important because they are the criteria for determining regulatory action. Concern exists at all levels about tests that result in false negatives (results which indicate that a chemical is toxic when it is in fact not toxic) and false positives (results which indicate that a chemical is not toxic when it is in fact toxic). Cairns and Orvos (1989) suggest that

> the sequential arrangement of tests that were used from simple to the more complex possibly reflects, in a broad, general way, the historical development of the field. As a consequence, tests with which there is a long familiarity are placed early in the sequence and more recent and more sophisticated tests that are still in the experimental stage or development are placed last.

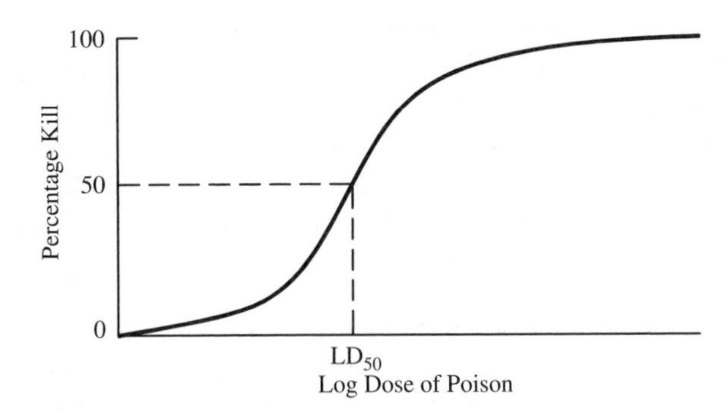

FIGURE 4.7 A typical dose–response curve from ecotoxicology.

Microcosms and/or mesocosms occupy the highest tier in this type of protocol, but they are seldom used by regulators because they can be expensive, time consuming, variable, and difficult to evaluate in terms of end points.

Most regulatory decisions are made based on the intermediate tier from single-species tests in which data from toxicity experiments are compared to estimated environmental exposure data. Thus, test populations of certain species are grown in the laboratory and tested for short-term (acute) vs. long-term (chronic), and lethal (causing mortality) vs. sublethal (causing stress but not mortality) dose experiments. The organisms most often used are the green alga *Selenastrum capricornutum*, the microcrustacean *Daphnia magna* (water flea), and the fish *Pimephales promelas* (flathead minnow). This selection of species provides a broad range of organismal responses to the chemicals being tested rather than focusing on a single taxonomic group. Typical acute tests would last 48 to 96 h and would test for end points in terms of survival of *Daphnia* and the flathead minnow or photosynthesis of the alga. Typical chronic tests would last up to a month and would test for end points in terms of reproduction of *Daphnia* and growth of the flathead minnow. Such tests are illustrated in Figure 4.7 with a dose–response curve. Thus, test populations are raised in a series of containers with increasing doses of the chemical that is being assessed (plotted along the x-axis of the figure) and their mortalities are recorded (plotted along the y-axis of the figure). The dosage of the end point (LD50 or lethal dose for 50% of the initial test population) is found by interpolation on the curve. This dosage is compared with the estimated environmental exposure dosage to complete the test. Note that the end point, death, is simple, definite, and easy to evaluate. The classic shape of the dose–response curve is sigmoid, though a u-shaped curve is also important for certain cases (Calabrese and Baldwin, 1999).

A controversy has arisen about the kinds of tests required in risk assessment of chemicals. A number of ecologists have insisted that single-species tests are inadequate for a full evaluation of ecosystem level impacts and that multispecies toxicity tests should be required. The principle issue is whether results from the single-species tests can be extrapolated to higher levels of ecological organization (Levin, 1998). Arguments against the ability to extrapolate have been provided by the Cornell University Ecosystems Research Center (Levin and Kimball, 1984; Kimball and

Levin, 1985; Levin et al., 1989), by Taub in relation to her work on the standardized aquatic microcosm (Taub, 1997), and most strenuously, by John Cairns over three decades of writing (Cairns, 1974, 1983, 1985, 1986a, 1995a, 2000). The main argument against reliance on single-species tests in risk assessment is that they provide no information on indirect and higher order effects in multispecies systems, which many ecologists believe are important. Taub (1997) has summarized the situation as follows:

> Single-species toxicity tests are inadequate to predict the effects of chemicals in ecological communities although they provide data on the relative toxicity of different chemicals, and on the relative sensitivity of different organisms. Only multispecies studies can provide demonstrations of: (1) indirect trophic-level effects, including increased abundances of species via increased food supply through reduced competition or reduced predation; (2) compensatory shifts within a trophic level; (3) responses to chemicals within the context of seasonal patterns that modify water chemistry and birth and death rates of populations; (4) chemical transformations by some organisms having effects on other organisms; and (5) persistence of parent and transformation products.

Thus, two categories of the effect of a pollutant are included in ecotoxicology: (1) direct impact on a species, derivable from single-species toxicity tests, and (2) indirect impacts due to interactions between species, best derivable from multi-species toxicity tests. The study of indirect effects is an important topic in ecology (Abrams et al., 1996; Carpenter et al., 1985; Miller and Kerfoot, 1987; Strauss, 1991 and; Wootton, 1994), and some researchers believe that the indirect effects are quantitatively more significant than the direct effects. For example, Patten's theoretical work (Higashi and Patten, 1989; Patten, 1983) indicates a dominance of indirect effects in ecosystems. Based on matrix mathematics and information on direct trophic linkages, Patten and his co-workers have developed a number of concepts and indices of network structure and function that quantify indirect effects and that challenge conventional thinking about ecological energetics (Fath and Patten, 2000; Higashi et al., 1993; Patten, 1985, 1991; Patten et al., 1976). This is a unique theory, termed network environs analysis, that represents a fascinating, though controversial, view of ecology (Loehle, 1990; Pilette, 1989; Weigert and Kozlowski, 1984). An example of an indirect effect caused by trophic interactions would be the increase in a prey population, which occurs when a predator population is eliminated by a toxin. In this case the direct effect is the impact of the toxin on the predator, which in turn causes the indirect effect of the release of the prey from control by the predator. Nontrophic interactions such as facilitation may also be involved in indirect effects (Stachowicz, 2001).

Ecologists, as indicated above, have criticized regulators for relying on single-species tests. Cairns and Orvos (1989) were particularly outspoken. They said "The development of predictive tests has been driven more by regulatory convenience than by sound ecological principles." And, "In an era where systems management is a sine qua non in every industrial society on earth, it is curious that the archaic fragmented approach of quality control is still in practice for the environment. Probably the reason for this is that the heads of most regulatory agencies are lawyers and sanitary engineers rather than scientists accustomed to ecosystem studies."

Regulators, on the other hand, find that multispecies toxicity tests (microcosms and mesocosms) have problems that limit their utility in risk assessment, including issues of standardization, replication, cost, and clarity of end points. Furthermore, regulators point to the existence of at least some comparisons between single-species tests and tests with microcosms and mesocosms which suggest that results from single-species tests can be extrapolated to higher levels of organization (Giddings and Franco, 1985; Larson et al., 1986). An example of the interplay between ecologists and regulators is provided in a special issue of the journal *Ecological Applications* (Vol. 7, pp. 1083–1132) which provides discussion about EPA's decision to formally drop the use of mesocosms as the high tier in testing of pesticides. Apparently, there is a fundamental lack of agreement between ecologists and regulators about the need for multispecies toxicity tests (Dickson et al., 1985).

This situation presents an ecological engineering design challenge to create multispecies toxicity tests in the form of microcosms and mesocosms that will satisfy both ecologists and regulators. A large volume of literature has developed on various systems design and testing protocols (Hammons, 1981; Hill et al., 1994; Kennedy et al., 1995a; Pritchard and Bourquin, 1984; Sheppard, 1997; Voshell, 1989). Much of this work is funded by the EPA and the chemical production industries. For example, starting in the 1980s, the EPA funded center-scale research first at Cornell University, then at the Microcosm Estuarine Research Laboratory (MERL) facility on Narragansett Bay, RI, and presently at the Multiscale Experimental Ecosystem Research Center (MEERC) at the University of Maryland. Earlier work by the University of Georgia scientists on end points for microcosm testing of chemicals is a good example of efforts by ecologists to develop simple designs and appropriate end points (Hendrix et al., 1982; Leffler, 1978, 1980, 1984). They used small aquatic microcosms and tested for the influence of chemical inputs on a variety of system parameters listed below:

Biomass
Chlorophyll a
Net daytime production
Nighttime respiration
Gross production
Net community production

From this work Leffler (1978) derived a formal definition of stress with several metrics that could be useful as end points (Figure 4.8). Stress is evident and quantified by the difference between the experiment and control microcosms in Leffler's definition. Unfortunately, this approach is relatively complicated compared with the simple LD50 toxicity test on single species, which regulators prefer. However, the University of Georgia research described above represents the kind of efforts ecologists are taking to meet the needs of regulators for multispecies toxicity tests.

Some of the most valuable progress at bridging the gap between regulators and ecologists has been in the development of standardized microcosms. Regulators value precision (low variance) and reproducibility (Soares and Calow, 1993), and these preferences have led some ecologists to design, build, and operate small, simple

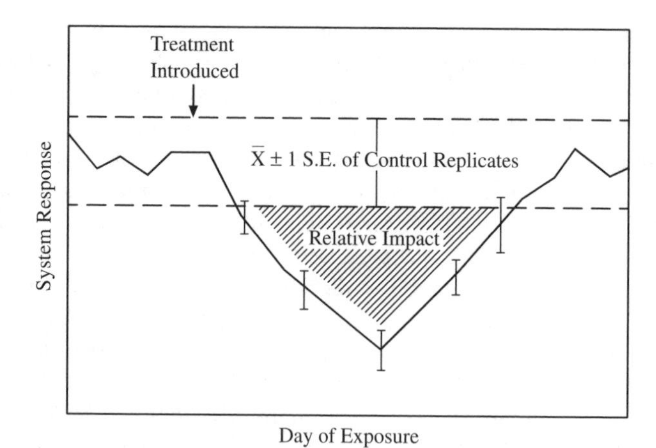

FIGURE 4.8 Definition of stress as a deviation in system response in a microcosm experiment. (From Leffler, J. W. 1978. *Energy and Environmental Stress in Aquatic Systems*. J. H. Thorp and J. W. Gibbons (eds.). U.S. Dept. of Energy, Washington, DC. With permission.)

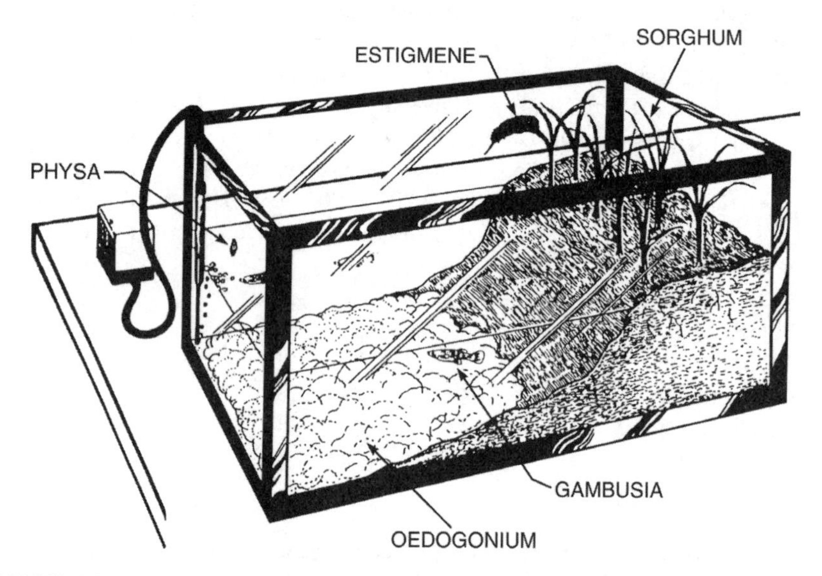

FIGURE 4.9 Metcalf's microcosm which simulated a farm pond and an adjacent field. (From Anonymous. 1975. The Illinois Natural History Survey Reports 152. With permission.)

microcosms as test systems. Precision and reproducibility in a test system provide the confidence in results that regulators appreciate for decision making. Beyers and H. T. Odum (1993) called these "white mouse" microcosms, drawing on the analogy of standard experimental animals used in medical research. The first example of a standardized microcosm in ecotoxicology was developed by Robert Metcalf (Metcalf, 1977a, b; Metcalf et al., 1971), who was an entomologist with an interest in

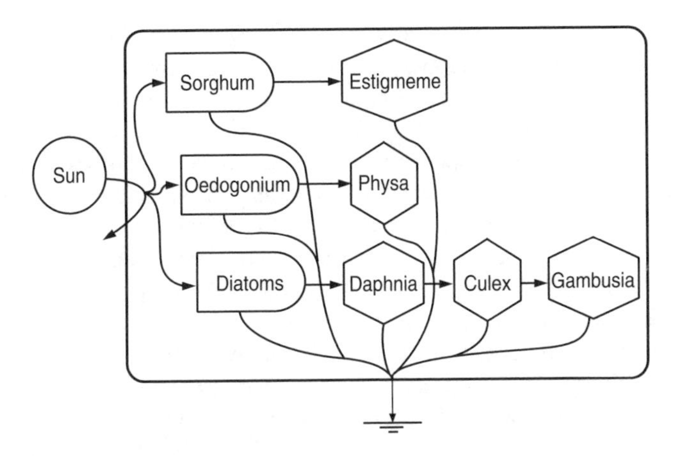

FIGURE 4.10 Energy circuit diagram of the food chains in Metcalf's microcosm.

the environmental effects of pesticides. Metcalf tried several different designs, but most of his work was done with glass aquariums containing an aquatic–terrestrial interface representative of an agricultural field and a farm pond (Figure 4.9). The aquarium was seeded in a standardized schedule with the following organisms which formed three food chains (Figure 4.10):

Aquatic Habitat

200 *Culex pipiens quinquefasciatus* (mosquito larvae)
3 *Gambusia affinis* (mosquito fish)
10 *Physa* sp. (snails)
30 *Daphnia magna* (water fleas)
A few strands of *Oedogonium cardiacum* (a green alga)
A few milliliters of plankton culture

Terrestrial Habitat

50 *Sorghum halpense* seeds (a flowering plant)
10 larvae of *Estigmeme acrea* (caterpillar)

Radio-labelled test chemicals were added to the system and their biomagnification and biodegradation were studied routinely. Experiments were run for a standard 33 days and the timing of additions of different organisms was designed for the sorghum, *Daphnia*, and mosquito larvae to be completely eaten by the end of the experiment! Thus, Metcalf's microcosm was not intended to be self-sustaining, but rather it was designed to collapse ecologically and be a short-term model, especially of food chain biomagnification. Metcalf and his students studied more than 100 pesticides and other chemicals with this system mostly in the 1970s, and the microcosm was modified and used by other researchers (Gillett and Gile, 1976).

Frieda Taub developed a standardized aquatic microcosm (SAM) which continues to be used (Taub, 1989, 1993). This system was reviewed by Beyers and H. T. Odum (1993), including an energy circuit diagram of the system. Taub's microcosm consists of a nearly gnotobiotic, 3-l flask culture with 10 algal species (blue-greens, greens, diatoms), five animal species (protozoa, Daphnia, amphipods, ostracods, and rotifers), and a mix of bacteria which cover a range of biogeochemical niches and feeding types. The system is run with a standard protocol for 63 days, and has been studied and verified to such a degree that it has been registered with the American Society for Testing and Materials as a standard method (ASTM E1366-90). The system is especially significant in ecological engineering because it represents the culmination of several decades of research design by Taub and her co-workers. The system is widely known and the chemically defined media and the microcosm itself are named after Taub, which is a reflection of her long record of work on its development and use. The development of the SAM can be traced back to the 1960s with early work on gnotobiotic microcosms (Taub, 1969a, 1969b, 1969c; Taub and Dollar, 1964, 1968).

The design research required to develop the SAM is an example of the kind of trial-and-error study required in ecological engineering to create ecosystems which perform specific functions, in this case to serve as a model test system for ecotoxicology. Here the engineering is in the design/choice of growth chamber, container, media, and organisms that make up the ecosystem, rather than in the "pumps and pipes" type design characteristic of conventional engineering. Living organisms are not completely understood and are not easy to combine into working systems, unlike the case for well understood engineering systems such as hydraulics or electronics. Thus, ecological engineering design differs from conventional engineering design because of the unknown factors associated with biological species. If organisms were completely understood, as perhaps approximated with Thomas Park's flour beetles, then the ecological system becomes a "machine" with a level of design equivalent to conventional engineering. Perhaps Park's flour beetle microcosm, in its elegant simplicity, is like the pencil or the screw, both equally elegant and simple machines whose engineering histories are described in book length treatments by Petroski (1989) and Ryeczynski (2000), respectively.

DESIGN OF MICROCOSMS AND MESOCOSMS

Design of microcosms depends on the nature of the experiment to be conducted and requires a number of straightforward decisions about materials, size and shape of container, energy inputs, and biota. The combination of these elements into a useable configuration is the design challenge. Although there are good reasons to standardize design for some purposes, the literature is filled with unique and ingenious microcosms that demonstrate a wide creativity for this subdiscipline of ecological engineering. General design principles for microcosms are covered by Adey and Loveland (1998) and Beyers (1964). Design of aquatic microcosms historically derives in part from the commercial aquarium hobby trade (Rehbock, 1980) and aquarium magazines can be a source of inspiration about possible microcosm designs. Terrestrial microcosms, on the other hand, seem less related to terrariums in terms of design. As with all constructed systems, cost is an important constraint on microcosm design. Cost is often proportional to size and number of replicates, and must include both construction (capital) and operation figures.

Physical Scale

The primary challenge of microcosm design is physical scaling, in terms of both time and space (Adey and Loveland, 1998; Dudzik et al., 1979; Perez, 1995; Petersen et al., 1999). Scaling of hydraulic models in civil engineering is well developed (Hughes, 1993) and may be a guide to designers of ecological microcosms. The appreciation of scale as a fundamental consideration in ecology has been recognized only in the past 20 years (Gardner et al., 2001; Levin, 1992; H.T. Odum, 1996; O'Neill and King, 1998; Peterson and Parker, 1998; Schneider, 2001), though Hutchinson (1971) mentioned the subject much earlier. The basic way to portray scale is with a "Stommel diagram" where different systems are plotted on a graph with axes of time and space (Stommel, 1963). Figure 4.1 is this type of diagram, showing the relative scale of microcosms and mesocosms in relation to natural ecosystems. Figure 2.14 is another variation of a scale diagram, in this case for biota (see also the related early graph given by Smith, 1954). Scale is a somewhat abstract concept that is still being explored theoretically and empirically. As noted by O'Neill (1989):

> Scale refers to physical dimensions of observed entities and phenomena. Scale is recorded as a quantity and involves (or at least implies) measurement and measurement units. Things, objects, processes, and events can be characterized and distinguished from others by their scale, such as the size of an object or the frequency of a process … Scale is not a thing. Scale is the physical dimensions of a thing.
>
> Scale also refers to the scale of observation, the temporal and spatial dimensions at which and over which phenomena are observed … The scale of observation is a fundamental determinant of our descriptions and explanations of the natural world.

Scale is an important concept because ecosystems contain components and processes that exist at different scales and because the ability to understand and predict environmental systems depends on recognizing the appropriate scalar context. For example, a forest may adapt to disturbances such as fire or hurricane winds, and to understand the ecosystem it must be recognized that the fire or hurricane is as much a part of the system as are the trees or the soil, even though the disturbance may occur only briefly once every quarter century. Obviously, microcosms often (though not always) are smaller scale than real ecosystems. This is an intentional sacrifice to provide for the benefits or conveniences of experimentation: ease of manipulation, control over variables, replication, etc. However, the reduction in scale affects the kind of ecosystem that develops in the microcosm and, according to some, limits the ability to extrapolate results (Carpenter, 1996).

Microcosm scaling issues fall into two broad categories that can be difficult to separate: fundamental scaling effects and artifacts of enclosure (Petersen et al., 1997, 1999). Fundamental scaling effects are those that apply in natural ecosystems as well as microcosms. These are primarily issues of sizing and temporal detail. In terms of sizing, perhaps the most often cited example is the work of Perez et al. (1977) in designing small-scale microcosms to model the open water ecosystem of Narragansett Bay, RI. Their design consisted of replicate plastic containers with 150 l of seawater from the bay. Paddles driven by an electric motor provided turbulence and fluorescent lamps provided light, timed to a diurnal cycle. A plastic box of bottom sediment from the bay was suspended in the containers to represent the benthic component of the system. Scaling was done to match Nararagansett Bay for surface-to-volume ratio and water volume to sediment surface area, along with underwater light profiles and turbulent mixing. Comparisons were made for plankton systems between the bay and microcosm. Microcosm zooplankton densities matched the bay, but phytoplankton densities were higher, perhaps due to the absence of large grazing macrofauna (fish, large bivalves, and ctenophores). The authors maintained that detailed attention to scaling was necessary for the microcosm to simulate conditions in the bay, and Perez (1995) has elaborated on this philosophy for ecotoxicology applications.

Other examples of scaling tests have compared different sizes of the same microcosm type (Ahn and Mitsch, 2002; Flemer et al., 1993; Giddings and Eddlemon, 1977; Heimbach et al., 1994; Johnson et al., 1994; Perez et al., 1991; Ruth et al., 1994; Solomon et al., 1989; Stephenson et al., 1984). There seems to be a tendency in these studies for plankton-based microcosms to have gradients with size, but benthic-based systems seem less affected by changes in size alone. These studies have the practical application of identifying the smallest sizes of microcosms that can be extrapolated to natural systems while minimizing cost. The most elaborate scaling test of this sort was done at the MEERC project of the University of Maryland's Horn Point Laboratory. This study examined plankton-based systems from the Choptank River estuary for three sizes of microcosms along both constant depth and constant shape (as expressed by constant radius divided by depth of tanks) gradients (Figure 4.11). Petersen et al. (1997) found that gross primary productivity scaled proportional to surface area under light-limited conditions and to volume under nutrient-limited conditions. These results represent a first step towards developing a set of "scaling rules that can be used to quantitatively compare the behavior of different natural ecosystems as well as to relate results from small-scale experimental ecosystems to nature" (Petersen et al., 1997).

Time scaling has received much less attention than spatial scaling of microcosms though both time and space are coupled. A sensitivity to time is often demonstrated in microcosm work in such aspects as diurnal lighting regimes and by the need to conduct experiments during different seasons. However, the central issue of time scaling is the duration of experiments. Most microcosm experiments are run only on the order of weeks or months in order to focus on special treatments such as the effect of a nutrient pulse or a toxin. Longer durations result in successional changes that can complicate the interpretation of these experiments. While the need for short-term studies is necessary for certain types of experiments, there does seem to be a bias in the literature against long-term studies of microcosms. This situation is unfortunate because long-term studies are necessary in ecology to understand many kinds of phenomena (Callahan, 1984; Likens, 1989). In fact, as a rule of thumb, most field ecological studies should be conducted for a minimum of 3 years so that

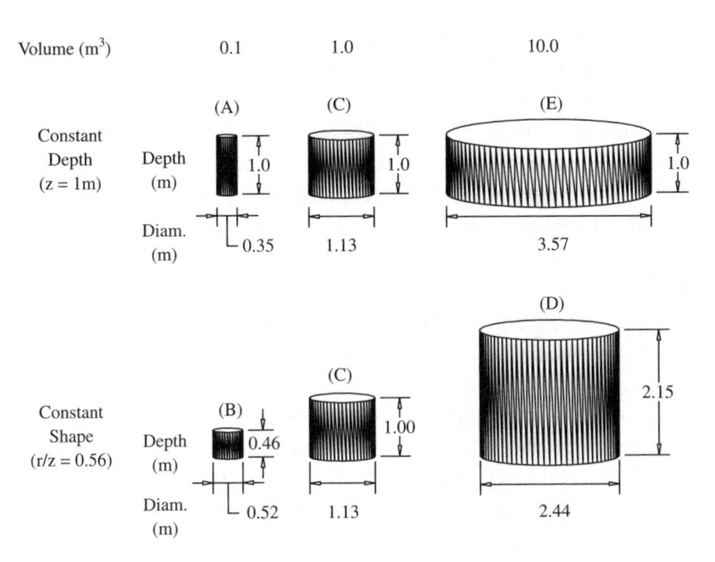

FIGURE 4.11 Scales of experimental units from the pelagic–benthic research at the Multiscale Experimental Ecosystem Research Center (MEERC) at the University of Maryland's Center for Environmental Science. (Adapted from Petersen, J. E. 1998. Scale and Energy Input in the Dynamics of Experimental Estuarine Ecosystems. Ph.D. dissertation, University of Maryland, College Park, MD.)

inter-year variability can be examined. One approach to accommodate this issue of time scaling is to study communities of protozoans and other microorganisms whose generation times are short. These kinds of microcosms have been called *biological accelerators* (Lawton, 1995) because they allow the examination of long-term ecological phenomena, such as predator–prey cycles and succession, with short real-time durations. These kinds of microcosms are essentially scaled on a one-to-one basis with their real-world analogs and thus they have been commonly used for ecological experimentation. A major challenge of microcosm work is to design and operate experimental systems that allow for reproduction of larger animals, such as fish, and for completion of complex life cycles, as exhibited by organisms that have planktonic larvae and sedentary adults (e.g., oysters and corals). In some cases this may require simply enlarging the size of the experimental unit (from flasks or tanks to ponds), but there is also a need for pumps and water circulation systems that do not destroy larvae. As demonstration of this need, for the short time that the EPA required aquatic mesocosm screening of pesticides, they mandated that mesocosms be large enough to include a reproducing population of bluegill sunfish (*Lepomis macrochirus*) (Kennedy et al., 1995).

The other category of scaling concern has been termed *artifacts of enclosure* (Petersen et al., 1997, 1999), which includes wall effects and missing components. The first aspect of wall effects is the composition of the walls of the container themselves. A wide variety of wall materials has been used in microcosms. Most are rigid (such as fiberglass), but flexible walls (such as plastic) are used for limnocorrals or other large in-situ enclosures. Schelske (1984) has covered possible chemical effects of walls that must be considered in design decisions: (1) walls should be nontoxic, (2) nutrients should not leach out of the walls, and (3) walls should not sorb substances added in experiments. An example of the latter issue of sorbtion was discussed by Saward (1975) who found that copper absorbsion was very low for fiberglass walls of an aquatic microcosm whereas absorbsion of oils and organochlorine was high. The other aspect of wall effect is that walls act as substrate for a biofilm of attached microorganisms (bacteria, algae, fungi, and protozoans). This biofilm, which begins to develop within hours to days, can have dramatic and undesirable effects on an experiment, especially if it is designed to study a plankton system suspended in a water column (Dudzik et al., 1979; Pritchard and Bourquin, 1984). As noted by Margalef (1967):

> When experiments are performed with a wide assemblage of species taken from natural populations, the systems develop a flaw — a fortunate flaw, because it throws light on the dynamics of populations in estuaries and in other natural environments. Species able to attach themselves to the walls of the culture vessels become more successful in competition. …
>
> The adherence of organisms to the walls is a most serious inconvenience in the use of chemostats as analogues of plankton systems. Species that are used often as models of planktonic algae, as *Nitzchia closterium*, and even some small species of *Chaetoceros*, are found attached in some way. Propensity to attachment seems to be different according to conditions of nutrition, to accompanying bacterial flora, and to the time elapsed from the start of the experiment. The role of possible mutants cannot be excluded. Stirring does not check attachment of algae to the walls. The design of a reliable chemostat for experimenting with complex planktonic populations awaits the improbable discovery of a bottle without walls. Ice walls do not help.

Can ecological engineers design a microcosm without walls, as mentioned by Margalef? Remarkably, he seems to have tried. Although he doesn't elaborate, Margalef's ice walls presumably were intended to reduce biofilm growth and thereby eliminate the wall effect. The biomass and metabolism of the biofilm on walls can quantitatively dominate a microcosm, thereby significantly influencing normal biogeochemical and toxin cycling. In general this kind of wall effect is proportional to wall surface area and inversely proportional to container volume. To the extent that artificial surface area in a microcosm exceeds that area found in the intended natural analogs, the microcosm represents a new system and may not be appropriate for extrapolation of experimental results. Many workers have recognized this problem and devised methods of removing the biofilm from the walls during experiments. The study by Chen et al. (1997) in the MEERC tanks (Figure 4.11) may be the most detailed study of wall effects. They found a number of relationships between biofilm growth and design factors of estuarine plankton tanks, along with quantifying the dominance of biofilm metabolism over plankton metabolism. Figure 4.12 is an energy circuit diagram of their system showing the dimensional effects of microcosm wall area (A) and volume (V) on biofilm and plankton components, respectively. Also shown is a new pathway that emerged with zooplankton, which are normally pelagic, feeding on the wall growth of the system. These kinds of wall effects are

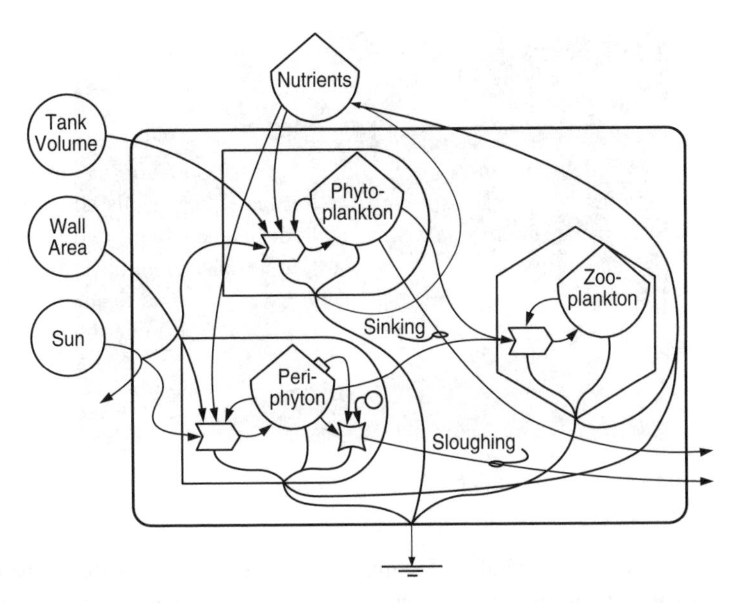

FIGURE 4.12 Energy circuit diagram of the influences of wall area and tank volume on the MEERC microcosms.

reminiscent of the classic concept of edge effects in natural ecosystems. Edge effect is the "tendency for increased variety and density at community junctions" (E. P. Odum, 1971). Community junctions are also known as *ecotones* (Risser, 1995a). The edge effect concept was coined by Aldo Leopold (1933) in relation to wildlife species that take advantage of qualities in communities along both sides of the ecotone; for example, foraging in one community and nesting or roosting in the other. Studies of species distributions along community transitions have identified some as "edge species" and others as "interior species," especially in terms of birds (Beecher, 1942; Kendeigh, 1944). Because some of the edge species are game animals, such as deer, wildlife managers have historically tried to maximize the amount of edge in landscapes. However, this wisdom is being questioned, especially for plants and nongame wildlife that seem to be negatively affected by edge (Harris, 1988). The classic concept of edge effect is related to wall effects in microcosms in the way the walls represent a discontinuity. A true edge effect occurs when two communities or habitats are in juxtaposition. Few microcosm studies have tried to model this situation of a true ecotone, which seems to represent a significant design challenge (John Petersen, personal communication). Metcalf's microcosm (Figure 4.9) was intended to include ecotones of an agricultural landscape (cropland and farm pond), but it was too simple to represent the concept.

The other aspect of artifacts of enclosure is that certain characteristic species or phenomena are left out of microcosms due to closure. Walls of a microcosm act as a barrier to movements of organisms and thus they limit genetic diversity inside the system. In some cases characteristic organisms are just too large or difficult to maintain within the confines of a microcosm. For example, sharks simply won't fit inside small marine microcosms even if they are the characteristic top predators in

FIGURE 4.13 Experimental burning of the marsh mesocosms at the MEERC facility, in Cambridge, MD.

the pelagic system of the natural analog marine ecosystems. Some species are always left out of experimental microcosms, and their absence can cause artifacts to arise, such as larger than normal prey populations in the absence of predators. Human actions are sometimes required to simulate top predators by removing prey individuals from a microcosm in order to maintain specified conditions (Adey and Loveland, 1998). Another important class of missing features in microcosms is the large-scale disturbances that influence ecosystems. Some workers have simulated disturbances such as fire (Figure 4.13; Schmitz, 2000; see also Richey, 1970) and storm events (Oviatt et al., 1981), but more research is required to test microcosm responses. Disturbances are large-scale phenomena in that they occur infrequently and act over large areas. They may be appropriately left out of short-term experiments, but their inclusion in micrcocosms can add to the accuracy of modelling of real ecosystems.

The Energy Signature Approach to Design

The use of energy signatures is one approach for the physical scaling of microcosms. The concept can be used to design microcosms by matching, as closely as possible, the energy signature of the natural analog system with the energy signature of the microcosm. The most straightforward approach to this matching of energies is to construct the microcosm in the field where it is physically exposed to the same energies as natural ecosystems. Examples are the pond ecosystems commonly used in ecotoxicology and *in situ* plastic bags floated in pelagic systems (called limnocorrals when used in lakes). In the lab the challenge of matching energies is greater. Significant effort is usually taken to match sunlight with artificial lighting whose intensity, spectral distribution, and timing can be controlled. Perhaps the most abstract examples of laboratory scaling are the origin-of-life microcosms (Figure 4.14). Here the challenge is to bring together the prebiotic physical–chemical conditions on the earth in a bench-scale recirculating systems in order to examine the chemical reactions that may have led to the origin of life. As an example of this

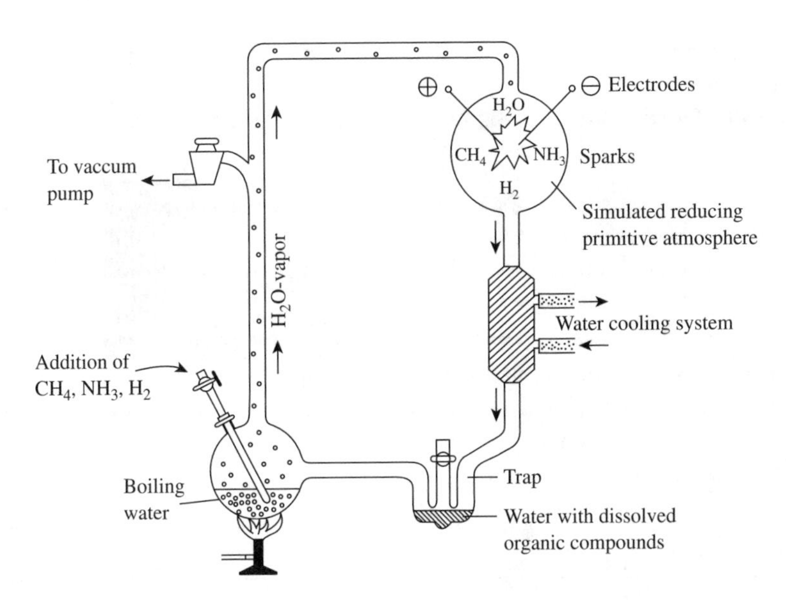

FIGURE 4.14 Miller's origin of life microcosm. (From Schwemmler, W. 1984. *Reconstruction of Cell Evolution: A Periodic System.* CRC Press, Boca Raton, FL. With permission.)

kind of study, Miller (1953; 1955; Miller and Urey, 1959; Bada and Lazcano, 2003) used an energy signature of the earth as a guide for designing their microcosm. In Miller's experiments, electrical discharge into a simulated prebiotic atmosphere produced a number of organic molecules including amino acids. This was a significant breakthrough, but there was still nothing alive in the microcosm after the experiments. Obviously creating a microcosm that generates life from nonliving components is the greatest design challenge!

A more modest but still difficult design challenge is providing turbulent mixing in pelagic microcosms. Turbulence is important in pelagic systems in providing physical–chemical mixing and reducing losses from sinking for phytoplankton and, to a lesser extent, for zooplankton. Turbulent mixing is reduced or eliminated when enclosing a water column with a microcosm because it is driven by larger-scale processes of water circulation and wind that are excluded. These larger-scale processes that generate turbulence represent auxiliary energy inputs to the plankton system. Early studies of pelagic microcosms, especially the floating bags in lakes and marine waters, completely excluded mixing energies, and artificial successions of phytoplankton occurred with dominance of motile species and losses of heavier, nonmotile species such as diatoms (Bloesch et al., 1988; Davies and Gamble, 1979; Takashashi and Whitney, 1977). This led to criticism of these studies; for example Verduin (1969) stated, "… before a lot of people buy a lot of polyethylene, I suggest that such companion experiments be performed and their validity versus the big bag be assessed and reported." Recognition of the problem also led to designs that generated turbulence in pelagic microcosms, including bubbling the water column with compressed air within floating bags (Sonntag and Parsons, 1979) and mechanical mixing with plungers or propellers in fixed tanks (Estrada et al., 1987; Nixon

TABLE 4.2
Steps in Developing a Living Model of an Ecosystem

1. Set up physical environmental parameters which provide the framework for the model.
2. Account for chemical and biological effects of adjacent ecosystems as imports and exports with either attached functioning models or simulations.
3. Add first biological elements which provide structure to the model. Typically these are plants or animals in reef structures (oysters or corals).
4. Begin biological additions in community blocks which are manageable units of soil or mud.
5. Repeat biological "injections" to enhance species diversity.
6. Add the larger, more mobile animals, particularly predators or large herbivores last, after plant production and food chains have developed.
7. The human operator takes over functions left out of the model, such as cropping top predators.

Source: Adapted from Adey, W. H. and K. Loveland. 1998. *Dynamic Aquaria*, 2nd ed. Academic Press, San Diego, CA.

et al., 1980; Petersen et al., 1998). The study by Nixon et al. (1980) is particularly interesting in describing the incorporation of turbulent mixing in the MERL tanks as a design challenge with many comparisons of measurements of turbulence both within the microcosms and in Narragansett Bay. Their plunger rotated in an elliptical fashion with a variable number of revolutions per minute. Thus, there was considerable engineering required to design, manufacture, operate, and maintain the plunger apparatus. Finally, Sanford (1997) provides a complete review of the issue with great attention to physical processes and assessments of alternative design options. He notes that no existing designs match microcosm turbulence within the real world but some options are better than others.

Walter Adey has developed an approach to building aquatic microcosms that includes matching forcing functions between a model (i.e., the microcosm) and the natural analog. His approach probably derives from his field work, especially on coral reef ecology, where he has shown the importance of "synergistic effects" of different external influences on ecosystems (Adey and Steneck, 1985). This attention to matching forcing functions is included in Adey's stepwise instructions for building effective model ecosystems, as shown in Table 4.2. An example of this approach is the Everglades mesocosm built in Washington, DC near the Smithsonian Institution's National Museum of Natural History where Adey works. This was a greenhouse scale model that was built as a prototype for one of the ecosystems in Biosphere 2. Like the real Everglades it included a gradient of subsystem habitats ranging from freshwater to full seawater (Figure 4.15). The model was successfully operated for more than a decade (Adey et al., 1996), which is a major accomplishment for a system of this size and complexity. The success of the mesocosm was partly due to a matching of forcing functions between the Washington, DC, greenhouse and the Florida Everglades. Figure 4.16 shows an example of this matching for annual temperature patterns. Temperature inside the greenhouse matched closely with data from southwest Florida while temperatures outside the greenhouse in Washington,

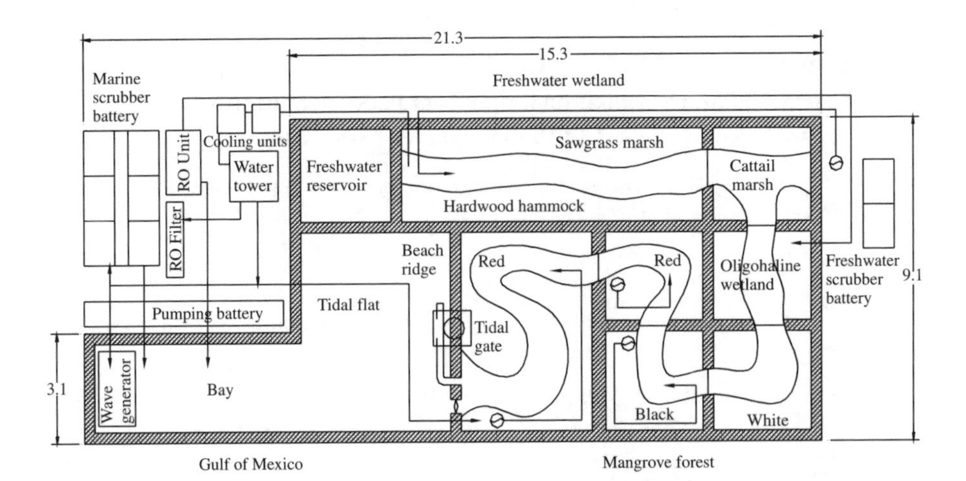

FIGURE 4.15 Floor plan of the Smithsonian Institution's Everglades mesocosm in Washington, DC. Note: Lengths are in meters. (Adapted from Adey, W. H. and K. Loveland. 1998. *Dynamic Aquaria*, 2nd ed. Academic Press, San Diego, CA.)

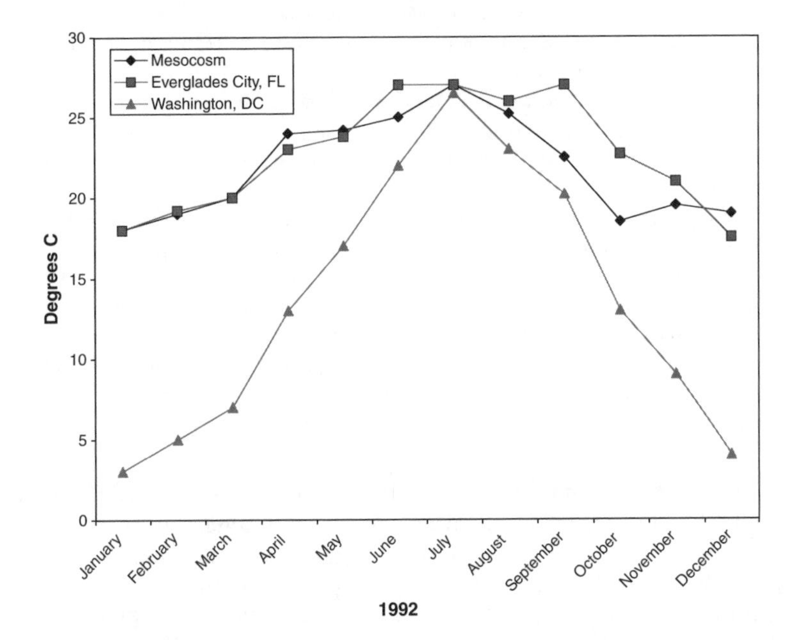

FIGURE 4.16 Comparison of temperature regimes for the Everglades mesocosm. (Adapted from Lange, L., P. Kangas, G. Robbins, and W. Adey. 1994. *Proceedings of the 21st Annual Conference on Wetlands Restoration and Creation*. F. J. Webb, Jr. (ed.). Hillsborough Community College, Tampa, FL.)

DC, were very different. Streb et al. (in press) analyzed the energy signature of the Everglades mesocosm by using the emergy analysis method (H. T. Odum, 1996). The method involves quantitative derivation of energy inputs to a system in standard

TABLE 4.3A
Energy Signature Evaluation for the Everglades in Southwest Florida for an Area Equivalent to the Everglades Mesocosm

Energy	Actual Energy (J/year)	Transformity (seJ/J)	Emergy (seJ/year)
Sun[a]	2.60×10^{12}	1	2.60×10^{12}
Wind[b]	3.55×10^{8}	1496	5.31×10^{11}
Tide[c]	2.08×10^{8}	16,842	3.50×10^{12}
Rain[d]	6.09×10^{8}	18,199	1.11×10^{13}
Waves[e]	3.75×10^{9}	30,550	1.15×10^{14}
Total	2.60×10^{12}	—	1.32×10^{14}

[a] Average solar insolation for Southwest Florida is approximately 7.00×10^9 J/m^2/year (E. P. Odum, 1971). Total solar energy is (7.00×10^9 J/m^2/year) (372.1 m^2).

[b] Wind energy = (0.5)(density of air) (wind velocity^2)(eddy diffusion coefficient) (height of boundary layer). Density = 1.2×10^3 g/cm^3. Wind velocity = 378.3 cm/sec (Ruttenber, 1979). Eddy diffusion coefficient = 1×10^4 cm^2/sec (Kemp, 1977). Height of boundary layer = 1×10^4 cm. Area affected = 130.5 m^2.

[c] Tidal energy = (0.5)(area elevated) (tides/year) (tidal height^2)(density of water) (gravitational acceleration). Area = 60 m^2. Tides/year = 706 (H. T. Odum, 1996). Tidal height = 100 cm (Carter et al., 1973). Density = 1.025 g/cm^3. Gravitational acceleration = 980 cm^2/sec.

[d] Chemical potential of rain = (area)(rainfall) (Gibbs free energy of water). Area = 90 m^2. Rainfall at Ft. Myers, FL = 1.37 m/year (Drew and Schomer, 1984). Gibbs free energy = 4.94 J/g (H. T. Odum, 1996).

[e] Wave energy = (shore length) (1/8)(density of water) (gravitational acceleration) (wave height^2) (velocity) (from H. T. Odum, 1996). Shore length = 3.1 m. Density = 1000 kg/m^3. Gravitational acceleration = 9.8 m/sec^2. Wave height = 0.1 m (assumed). Velocity = (gravity × depth)$^{1/2}$, where depth = 1 m (assumed).

units of joules per unit time. These standard unit flows are then scaled with transformation ratios to account for the degree to which different energies are concentrated (i.e., embodied energy or emergy), measured in equivalent units of solar joules per unit time. This scaling accounts for differing concentrations (in terms of ability to do work) of energies. Table 4.3A shows energies for Southwest Florida, which is the natural analog for the Everglades mesocosm, whose energy signature is given in Table 4.3B. Wave energy dominates the emergy budget in the real Everglades of Southwest Florida, while natural gas and electricity, which are needed for heating, cooling, and running machinery (such as the wave generator) have the highest emergies in the mesocosm. Such analysis illustrates how much more total energy is required to operate a mesocosm compared with the actual system it models. In this case the mesocosm required two orders of magnitude more emergy to operate than the analog ecosystem. Beyers and H. T. Odum (1993) include a similar analysis of

TABLE 4.3B
Energy Signature Evaluation for the Everglades Mesocosm in Washington, DC

Energy	Actual energy (J/year)	Transformity (seJ/J)	Energy (seJ/year)
Sun[a]	2.05×10^{12}	1	2.05×10^{12}
Tap water[b]	2.47×10^{8}	18,199	4.50×10^{12}
Labor[c]	—	—	4.05×10^{15}
Electricity[d]	3.35×10^{11}	1	$74,0005.83 \times 10^{16}$
Gas[e]	1.60×10^{12}	48,000	7.68×10^{16}
Total	3.99×10^{12}	—	1.39×10^{17}

[a] Average insolation for Washington, DC, is approximately 5.50×10^9 J/m²/year (E. P. Odum, 1971). Total solar energy is (5.50×10^9 J/m²/year)(372.1 m²).
[b] Chemical potential energy of water added to the mesocosm = (volume) (density) (Gibbs free energy). Volume used was 53,400 l/year (Lange, 1998). Density = 1000 g/l. Gibbs free energy = 4.62 J/g (H. T. Odum, 1996).
[c] Labor requirements for the mesocosm were 20 h/week or 43.33 d/year multiplied by the Energy use/person of 9.35×10^{13} seJ/day (H. T. Odum, 1996).
[d] Based on power consumption and operational times of all pumps, heaters, fans, etc., the total electrical use was 3.35×10^{11} J/year in the mesocosm.
[e] Based on power consumption and operational time of gas heaters, the total gas use was 1.60×10^{12} J/year in the mesocosm.

the MERL mesocosms, which showed that turbulence had the highest emergy input, which is perhaps appropriate for a pelagic system.

Seeding of Biota

Seeding a microcosm with biota is usually done after physical scaling considerations. For example, Adey and Loveland (1998) included several procedures for introducing biota in their stepwise instructions for microcosm set up (Table 4.2). This can be an intricate task, for example, in setting up a coral reef system, but in other cases the task is simpler, as for an algal mat system described by H. T. Odum (1967):

> At times one can strip the mat from the bay bottom and roll it up like a carpet. When the water is completely blown off of a section by the wind so the mat dries, it is not immediately killed and can be reactivated in a day by putting it back into water. It is a transferable package that Dr. Robert Beyers called an "instant ecosystem."

Microcosm design can even be like a cooking recipe as noted by Darnell (1971) for a protozoan culture in a teaching laboratory workbook:

Preparation of Broth: Add 150 gm of dried grass to 2000 ml of distilled water, and boil for 15 minutes. Cool and strain quickly, once through a double thickness of cheesecloth and once through a thick nest of glass wool in a large funnel. Dilute with distilled water to a volume of 3000 ml, and store in a refrigerator until needed.

Preparation of Cultures: To a one-gallon battery jar add 200 ml of broth (shake well before pouring) and 1800 ml of distilled water. Innoculate with 5 gm of pond mud, and stir vigorously. Partly cover with a glass plate (leaving enough of an opening for gas diffusion), and place in a dark incubator set at 30 degrees C. Bring out for class use on the appropriate day, but keep dark at all times. In such cultures succession proceeds rapidly during the first week and more slowly thereafter, and most major changes will have taken place by the end of three weeks.

In general, there are two basic approaches to seeding a microcosm with biota: (1) use of natural assemblages of organisms obtained from local sources, and (2) the gnotobiotic approach of a synthesized system using standard species (as in Taub's SAM). These are fundamentally different approaches that require different degrees of design by the ecological engineer. In essence, when using natural assemblages as a seed source, the ecological engineer relies completely on self-organization to develop the food web and nutrient cycles within the microcosm. However, when using the gnotobiotic approach, the ecological engineer takes on a significant role as designer of the ecological organization within the microcosm. As an example of this design effort, Taub has noted that much trial and error was required to develop her SAM as a useful tool in ecotoxicology.

In addition to these two fundamental seeding approaches, cross-seeding and reinoculation are often-used techniques in setting up microcosms. Cross-seeding involves mixing innocula between replicate microcosms to reduce variability. This is usually done in early stages of the experiment. Reinoculation is done for specific species which do not develop sustainable populations from the initial seeding. This is usually necessary to maintain desirable species, such as target organisms in ecotoxicology work or species characteristic of the natural analog ecosystem in academic modelling experiments.

H. T. Odum has advocated an approach termed *multiple seeding* for developing a microcosm. In this approach innocula from several natural assemblages are mixed together to provide a species pool which subsequently becomes self-organized into stable, sustainable ecological circuits. This approach speeds up the self-organization process by providing an excess number of species for internal selection of viable circuits of energy flow and nutrient cycling. All of these approaches and techniques for biotic seeding represent an input of genetic information to the microcosm. In that sense they can be considered as part of the energy signature of the system, though, because the actual energies involved with genetic information are so small, this is seldom done. Perhaps techniques of biotic seeding are best thought of as modelling the processes of colonization or immigration that occur in natural ecosystems.

Island biogeography theory is relevant for explaining some of the features associated with the seeding and development of microcosm biota. As noted at the beginning of this chapter, islands have been important experimental units in ecology,

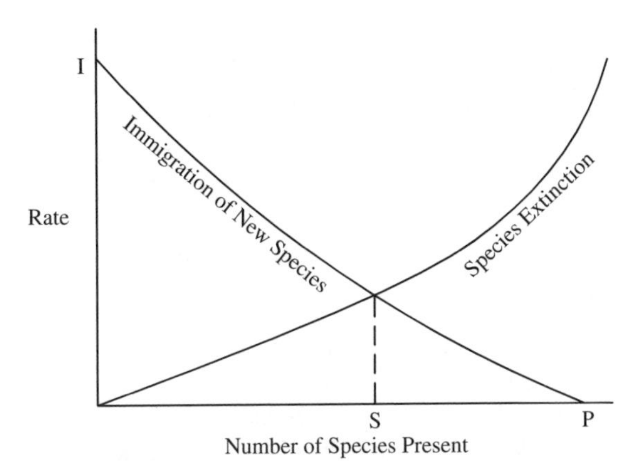

FIGURE 4.17 The species equilibrium concept from the theory of island biogeography. I = immigration rate at the beginning of colonization; S = number of species; P = number of species in the species pool available for colonization. (Adapted from MacArthur, R. H. and E. O. Wilson. 1967. *The Theory of Island Biogeography.* Princeton University Press, Princeton, NJ.)

and they have been used as metaphors with microcosms. Natural islands are isolated systems, and when small enough, such as the mangrove islands in Florida Bay studied by Simberloff and Wilson (1969, 1970), they are able to be manipulated in experiments. Islands also have been artificially constructed or intentionally fragmented from existing, especially larger systems for experimental purposes. Thus, there are natural similarities between islands and microcosms, and microcosms have been used to test island biogeography theory, as reviewed by Dickerson and Robinson (1985). As an aside, the unintentional fragmentation of forests and other ecosystems into habitat fragments is a major environmental problem affecting landscapes (Haila, 1999; Harris, 1984; Saunders et al., 1991). Impacts occur because the fragments are isolated and surrounded by a nonforest environment and because of the reduction in area of the habitat fragment. In a sense all microcosms suffer from these same impacts.

The basic tenet of island biogeography theory is that the number of species found on an island is determined by the balance between the immigration rate of species reaching the island from an outside species pool and the extinction rate of species on the island (MacArthur and Wilson, 1967). It has been termed the equilibrium model because the number of species on the island is actually a dynamic steady state (or equilibrium) in which the composition is changing but the number is constant. Thus, the intersection of the immigration rate curve and the extinction rate curve represents the equilibrium number of species to be expected (Figure 4.17). This is a simple, elegant model and, as represented in the energy circuit language in Figure 4.18, immigration is seen as an input or energy source to the system.

A more detailed view of the theory covers islands that are small vs. large and close vs. distant from the species pool, which is usually a continent in some sense. It is controversial but still a useful paradigm for understanding ecological organization (see Chapter 5). This theory is applicable to microcosms in terms of the number of species that can be supported by a closed system. In most cases, seeding of a

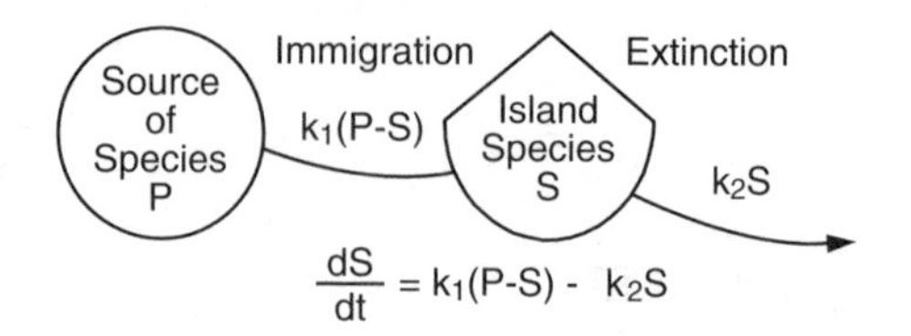

FIGURE 4.18 Energy circuit diagram of the species equilibrium concept from the island biogeography.

microcosm is done initially in an experiment over a period of time and then it is stopped. This is analogous to cutting off the immigration rate to an island at a certain point in time. When this situation occurs, a reduction in species results because only extinction takes place. While in a simple mathematical model the number of species goes to zero, in a real microcosm some species maintain themselves in a sustainable organization of energy flow and nutrient cycling. The reduction in species upon removing immigration has been termed *relaxation*, and is a common phenomenon in microcosm development. It represents a self-organization process whereby those species which fail to find roles within the networks of energy and nutrient flows go extinct. The resulting community after extinctions represents a stable set of species that in a sense are preadapted to the microcosm environment, based on their evolutionary histories. Two examples from the Everglades mesocosm illustrate this phenomenon. First, Lange's (1998) study of fish populations within the mesocosm shows a steady decline in species across the different habitats included in the mesocosm from the initial seeding which started in 1987 through an 8-year period (Figure 4.19). The maximum relaxation of species occurred in the marine tanks of the mesocosm where fish richness diversity declined from 25 to 8. A second example is Swartwood's (in preparation) study of a single species, the mangrove tree snail (*Littorina angulifera*), within the mesocosm. This species was originally seeded into the mesocosm but it went extinct, perhaps due to insufficient humidity or other habitat factors. The species was reintroduced in 1996 along with an attempt at modification of the microclimate. Populations declined after the reintroduction, perhaps converging on a lower density similar to that found in the natural analog. In both of these cases, preadaptation may explain the survivors, in terms of species with the fish community studied by Lange and in terms of individuals with the snail population studied by Swartwood.

The number of species or species diversity supported by a microcosm can be important to the designer for various reasons. Diversity is an important parameter in ecosystems as mentioned earlier in the book. It has been used as an index of ecosystem complexity and has been linked with stability in the most controversial relationship in ecology. The diversity–stability relationship was formally introduced in the 1950s from empirical observations (Elton, 1958) and theoretical explorations (MacArthur, 1955). Basically, the relationship suggests that more species provide more opportunities for the system to adapt to environmental changes and thus diversity promotes stability. In part because it has a strong commonsense appeal, the diversity–stability relationship has long intrigued ecologists, even though the evidence has not been found to be consistent (Goodman, 1975; Johnson et al., 1996;

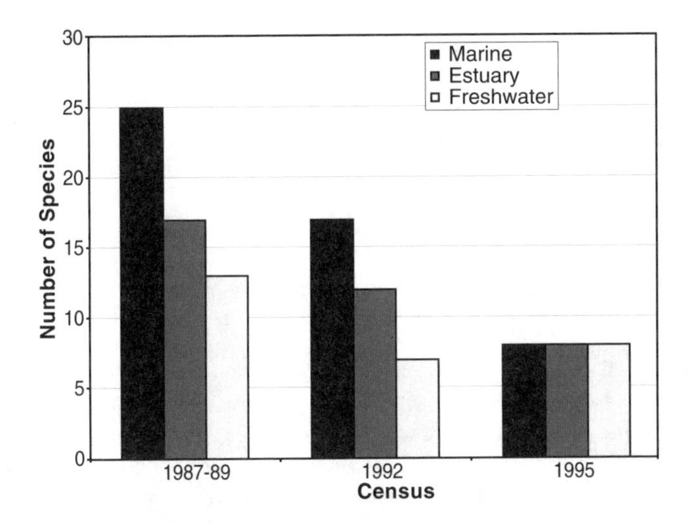

FIGURE 4.19 Declines in the number of fish species found in the Everglades mesocosm from different salinity zones. (Adapted from Lange, L. E. 1998. An Analysis of the Hydrology and Fish Community Structure of the Florida Everglades Mesocosm. M.S. thesis, University of Maryland, College Park, MD.)

May, 1973). In reference to the problems with this relationship, Paine (1971) spoke of it as "the ecologist's Oedipus complex" in a book review of the first symposium on the topic. Semantics is a factor in developing generalities about diversity and stability, especially in relation to the latter concept. For example, Grimm and Wissel (1997) review the use of stability in ecology and find 163 definitions of 70 different concepts (see also Figure 4.4 in Peters, 1991). Holling's (1973, 1996) classification of two main types of stability seems to be accepted by most ecologists. Resilience is the extent to which a system returns to a previous state after perturbation, while resistance is the extent to which a system maintains itself without change during perturbation. These two concepts cover the ability of an ecosystem to withstand (resistance) and to recover from (resilience) from a perturbation. Other related stability concepts, using analogies from engineering, have been introduced, including strain (Deevey, 1984; Kersting, 1984) and elasticity (Cairns, 1976). Recently, emphasis seems to have shifted away from the diversity–stability relationship toward more general relationships between biodiversity and ecosystem function (Hart et al., 2001; Naeem et al., 1999; Risser, 1995b; Tilman et al., 1997).

Microcosm work has played a role in the quest for a valid relationship between diversity and stability (Hairston et al., 1968; Van Voris et al., 1980 as examples), but the evidence remains inconclusive. This is not to deny the importance of diversity in its own right as an ecological characteristic for describing microcosm designs. While most microcosms support low diversities due to various factors, some designs support higher levels. For example, Small et al. (1998) found 534 species supported by a 5 m^2 microcosm of a Caribbean coral reef, with another 30% suspected to be present. The diversity of this microcosm was used as a basis of estimating global diversity of coral reefs, and according to their analysis, the authors suggested that existing global estimates are three times too low!

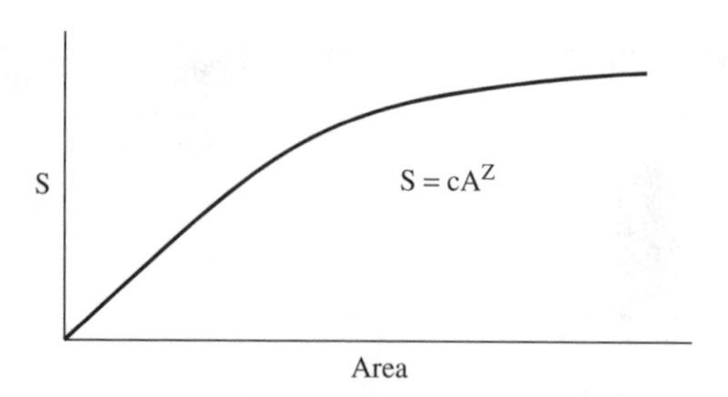

FIGURE 4.20 The species–area relationship from ecology.

In general a designer has three strategies available to increase diversity in microcosms: (1) continually add species to balance extinctions, (2) add refuges (physical complexity), and (3) increase the size. The first strategy is based on the species equilibrium model of island biogeography mentioned earlier. Some microcosms which rely on flow-through water systems have the potential to continually introduce planktonic species with the turnover of water. However, most microcosm experiments are more static because the seeding of species is stopped at some early point in time. Adding refuges can also increase diversity by providing spaces within the microcosm where inferior competitors or vulnerable prey may escape and be sustained. Early work on microcosms (Gause, 1934; Huffaker, 1958) demonstrated this role for refuges and it contributed to the increase in awareness by ecologists of the importance of spatial heterogeneity in ecosystems (Hastings, 1977, 1978; Roff, 1974; Wiens, 1976). Finally, increasing the size of a microcosm increases the number of species that can be supported according to the species–area relationship (Figure 4.20). This is probably the most robust relationship in ecology in that it has been found to apply to all examples that have been studied. Early work on species–area curves was done for the practical purpose to study the optimal sample areas needed to describe a plant community (Arrhenius, 1921; Gleason, 1922). Later, the relationship was used to describe diversity on islands of different sizes in studies of island biogeography (MacArthur and Wilson, 1967). The relationship is actually a general expression of the organization of ecological communities by combining aspects of the species–abundance relation and the spatial pattern of habitats (H. T. Odum, 1983; Pielou, 1969). McGuinness (1984) reviews the history and alternative forms of the species–area curve. Size or area is always a constraint in microcosm design, but as a rule of thumb, bigger is better in terms of the amount of diversity that can be supported. The largest "microcosm" ever built is Biosphere 2, which covers 1.25 ha (3.0 acres), and is described in the next section.

Closed Microcosms

One of the most interesting microcosm experiments has been the construction of closed systems. These systems are actually open to energy exchange in terms of sunlight as an input and heat as an output but closure applies to all other materials. Thus, a system is seeded with biota, an atmosphere, media, and any other structure and then sealed shut. The microecosystem then self-organizes into a stable system

FIGURE 4.21 Some of Folsome's original microcosms on display at Biosphere 2 in Arizona.

of biotic and abiotic components. These closed systems provide models of the whole biosphere, which makes them of special interest (Jones 1996).

The first modern closed microcosms were constructed and studied by Clair Folsome and his students and collaborators (Folsome and Hansen, 1986). Folsome had the title of Director of the Exobiology Laboratory at the University of Hawaii. He was a microbiologist with an interest in the origin of life (Folsome, 1979) and a commitment to microcosm research. His systems consisted of vials or flasks which were filled with defined media and seeded with either gnotobiotic assemblages or mixed cultures from the environment, including bacteria, algae, other microbes, and at least one metazoan or nonmicrobe, the crustacean *Halocaridina rubra.* Folsome started enclosing microcosms in 1967 and most of the research on them was published in the journal *BioSystems* (Kearns and Folsome, 1981; Obenhuber and Folsome, 1984, 1988; Takano et al., 1983; Wright et al., 1985). Some of Folsome's oldest systems have maintained microbial activity for more than 30 years, demonstrating that self-organization can result in stable systems (Figure 4.21).

More recently, interest in global change as a result of the buildup of CO_2 and other atmospheric changes due to human activity has led to renewed research on closed systems. Most of this work consists of short-term studies of the effects of elevated CO_2 levels. For example, Korner and Arnone (1992) built small (17 m^3 or 600 ft^3) closed greenhouses with tropical plant communities to study the effects of CO_2 on various measures of system structure and function. Their microcosms were run for 3 months after an initial stabilization period of 1 month, and they developed many patterns similar to tropical forest communities, such as representative light extinction curves. Much more of this line of research can be expected with several national initiatives focused on understanding expected global changes.

The great challenge is the design of a closed system that contains a human. Some details have been worked out for living underwater (Miller and Koblick, 1995), but the challenge remains for manned space flight and the long-term occupation of extraterrestrial environments such as space stations. For this purpose a life support

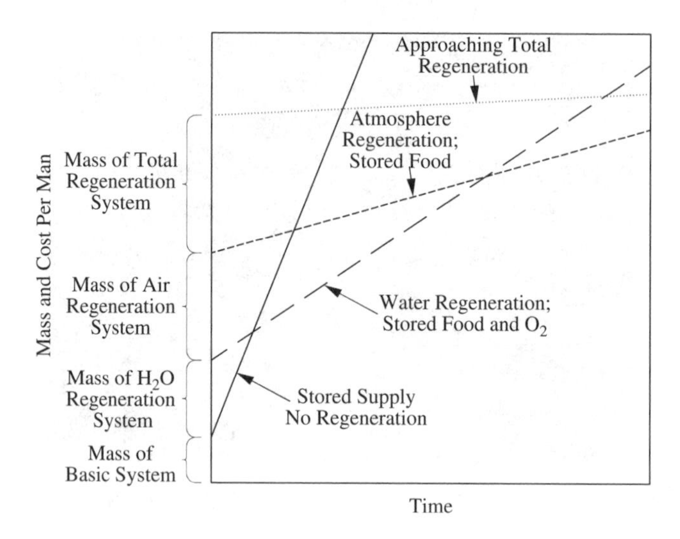

FIGURE 4.22 Scaling relationships of life support alternatives. Nonregenerative storage is best when mission time is short and total regenerative systems are best when mission time is long. (Adapted from Myers, J. 1963. *American Biology Teacher* 26:409–411.)

system is required that will supply at least the minimum human needs (oxygen, water, and food) and eliminate human wastes (carbon dioxide, urine, feces, and heat). Size and weight are obvious considerations of such a life support system as are many other concerns. Figure 4.22 illustrates different life support alternatives and how the choice between them changes with the mission duration of a space flight. This curve was introduced in the 1960s (Myers, 1963) but remains unquantified (Eckart, 1994). Stored supply is the best choice for short durations followed by systems which recycle water at intermediate durations. Water recycling is a technical problem whose engineering was worked out long ago. Only the longest durations will require a totally regenerative system, and this is the design challenge that has not been solved after nearly 50 years of intensive research funded by government agencies interested in space travel. The totally regenerative system must maintain an atmosphere, provide food, and recycle wastes for the human occupants of the life support system. Most agree that this kind of life support system must include biological species that maintain biogeochemical cycles, and therefore these systems have been referred to as bioregenerative.

The history of research on bioregenerative life support systems is fascinating and includes an element of great relevance to ecological engineering, that is, the design of a multispecies microcosm in which a human is a component (and of course the most important) species. This history is outlined here especially because it contains a dialogue between a small group of ecologists who advocated the multispecies approach and the majority of researchers from engineering and physiology who advocated an approach that emphasized a mechanical system with one or a few species. Taub (1974) provides an excellent review of the first two decades of research on bioregenerative life support systems. This early period was a time of creative searching of many lines of design. Table 4.4, derived from Taub's (1974) review,

TABLE 4.4
Comparisons of Early Life Support System Experiments

System Name	Biota	Function Tested
Recyclostat	Algae	Gas balance
Bioregenerative unit	Fungi, algae, rat	Gas balance, waste recycle
Algatron	Bacteria, algae	Gas balance, waste recycle
Microterella	Algae, bacteria, mouse	Gas balance, waste recycle, water recycle
Mecca	Algae, bacteria, man	Gas balance, waste cycle, water recycle

Source: Adapted from Taub, F. B. 1974. *Annual Review of Ecology and Systematics,* 5:139.

lists a few of the systems tested. Most of these early systems relied on the green algae *Chlorella pyrenoidosa*, whose laboratory culture was well known. *Chlorella* cultures could absorb CO_2 while producing oxygen and food under artificial lighting conditions. Much research was done sealing in various animals, such as mice, rats, and monkeys, with one or a few microbial cultures, such as *Chlorella*, to test for self-sufficiency, with the test lasting several days or weeks. This work was funded by the U.S. Air Force and the National Aeronautics and Space Administration (NASA) with millions of dollars spent. Eventually, these algal-based systems were abandoned due to unknowns of reliability, high weight of water required for culturing algae, and difficulties converting algae into an acceptable human food (Taub, 1974). Emphasis in the 1970s shifted to higher plants as the basis for bioregenerative life support systems and work in this direction continued both with NASA in the form of their Controlled Ecological Life Support System (CELSS), which was formally started in 1978 (Galston, 1992), and by the Soviet Union's similar approach, termed BIOS (Salisbury et al., 1997).

A small group of ecologists, including primarily H. T. and E. P. Odum, R. Beyers, G. D. Cooke, and F. Taub, became interested in the challenge of life support system design in the 1960s. The ecologists suggested that a multispecies ecosystem was required to support a human during space flight in order to ensure reliability or stability. Their basic argument relied on the diversity–stability relationship discussed earlier and on experience with microcosm experiments. In particular, microcosm experiments of ecological succession demonstrated that self-organization results in a stable ecosystem with balanced primary production and community respiration, which implies balanced gas exchange between oxygen and carbon dioxide. H. T. Odum (1963) described the process needed to design such a life support system for humans with the same approach of multiple seeding used for microcosm setup. He calculated that the multispecies life support system would require about 2 acres per human (0.8 ha) based on considerations of the energy transformation of sunlight through primary production. Such a relatively large area was required due to the low efficiency of photosynthesis in converting light energy to chemical energy in organic matter. The large area with many species was also needed to provide the "homeostatic mechanisms" (Cooke et al., 1968) required to maintain system stability. Multiple seeding and self-organization at the scale of 2 acres per astronaut did not meet with the approval of the engineers and physiologists working with single-species cultures. Such a system appeared to be too large and too heavy to be practical, which of course is a valid criticism.

The fundamental issue is long-term reliability. Once the spaceship leaves earth the astronauts must be able to rely on the life support system or risk death from lack of an atmosphere or food or buildup of waste products. The question is which approach best meets the requirement of reliability. The conventional engineering approach tried to develop and optimize subsystems for different functions, such as oxygen production, waste recycle, and food production, and then to connect the subsystems back together. The subsystems consisted of single-species cultures and mechanical components in this case. The overall design concept was to keep the system simple and well understood as noted by Brown (1966):

> What we are trying to do is ... to make this closed system as simple as we can. The virtue of simplicity is that you can understand all of the regulatory factors that you have to worry about; you can put in the proper manual or automatic controls; and the fewer the components, the better.

This is essentially the well-tested and established engineering method of design. In comparison, the ecologists' approach was quite the opposite. It involved mixing together hundreds or thousands of species, most of which are little known, and through self-design a stable system would emerge. This approach is based on a faith in the ecosystem and its long evolutionary history rather than on the conventional engineering method. Several quotes from ecologists on this dichotomy reveal the extent of the disagreement:

> When I read of schemes to create living spaces from scratch upon which human lives will be dependent for the air they breathe, for extrinsic protection from pathogens and for biopurification of wastes and food culture, I begin to visualize a titantic-like folly born of an engineering world view. At this point we don't know enough, being totally reliant on knowledge as well as physical subsidies from nature to survive on earth. In space there are no doors to open or neighbouring ecosystems to help correct our mistakes. (Todd, 1977)

> … these simple systems are inherently unstable and depend upon a very large investment in power for control. The probability of failure of the two-species (man and microorganism) linkage during a long space voyage due to successional processes or to stress is high. It would seem obvious that these simple ecosystems pose a serious risk to the astronaut, and further work on them, as the basis for a life-support system, should be abandoned. Above all, the narrow engineering approach cannot be applied to bioregeneration. Organisms cannot be "designed" and "tested" like transistors or batteries to perform "one function" or to solve "one problem"; they have evolved with other organisms as a unit and they carry out a variety of functions which must dovetail with other activities of the ecosystem. The "minimum ecosystem" for man must clearly be a multispecies one. (Cooke, 1971)

> In appraising the potential costs of closed system designs one has the alternative of paying for a complex ecosystem with self maintenance, respiration, and controls in the form of multiple species as ecological engineering, or in restricting the production to some reduced system like an artificial algal turbidistat and supplying the structure, maintenance, controls, and the rest of the functions as metallic-hardware engineering. Where the natural combinations of circuits and "biohardware" have already been selected for power and miniaturization for million of years probably at thermodynamic limits, it is exceeding questionable that better utilization of energy can be arranged for maintenance and control purposes with bulky, nonreproducing, nonself maintaining robot engineering. (H. T. Odum, 1963)

There is a kind of frustration in these quotes in opposition to conventional engineering. This sense can also be felt by reading the discussions that occurred at the symposia on bioregenerative systems where a single ecologist tried to defend the multispecies approach to a group of engineers and physiologists (Cooke in Cooke et al., 1968 and E. P. Odum in Brown, 1966).

This dialogue provides perspective on the gulf between ecologists and engineers not just in terms of design of life support systems but in more general terms. There are differences between the ecological engineering method and the conventional engineering method, and this text is an effort at bringing the differences to the forefront for consideration. In terms of life support systems design, the engineers and physiologists have temporarily won the funding battle over the ecologists, as can be seen by the heritage of NASA's CELSS program which involves essentially little or no ecology (see, for example, Brechignac and MacElroy, 1997). Before the advent of CELSS, H. T. Odum (1971) called this situation "something of a national fiasco" because NASA had "refused to recognize that multispecies designs are required for stability and that this energy cost is unavoidable." No successful long-term bioregenerative life support system has yet been designed and only time will tell if NASA made the correct choice between alternative life support system designs.

As a side note, NASA has supported a group of ecologists who published a set of papers on the rhetoric of closed systems (Botkin et al., 1979; Maguire et al., 1980; Slobodkin et al., 1980). They suggest that closed systems can be a valuable tool for developing ecological theory. Curiously, however, this group does not cite any of the work mentioned above and does not offer any discussion of the life support system design question. NASA also provided some support to Folsome's microcosm work mentioned earlier. Neither of these token efforts of support for ecological research seems to have had an influence on NASA's approach to life support.

As another side note, it is interesting that the case for the multispecies microcosm was again made here in terms of life support system design as it was made earlier in terms of ecotoxicology testing. In both cases ecologists were pitted against a rival group, and in both cases they lost the argument for funding support.

Against this backdrop of government funded research on life support systems, the Biosphere 2 project started in the desert north of Tucson, AZ. Biosphere 2 is an aircraft carrier-sized (3.0 acres/1.25 ha) mesocosm originally designed and constructed to test human closure in a bioregenerative life support system (Figure 4.23). It was an impressive project from several perspectives. First, it was a privately funded project that was run by a corporation (Space Biosphere Ventures) as a for-profit

FIGURE 4.23 Biosphere 2 in Tucson, AZ.

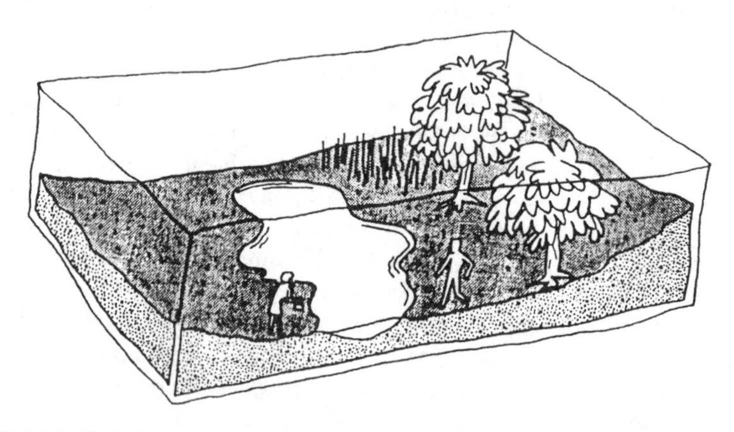

FIGURE 4.24 H. T. Odum's early view of a closed experimental system with humans, which was realized by Biosphere 2. Odum suggested the closed system could be "armory-sized." (From Odum, H. T. 1971. *Environment, Power, and Society*. John Wiley & Sons, New York. With permission.)

enterprise. Second, it was much bigger than any other closed system ever attempted. Third, it utilized to a large extent the ecological engineering approach to life support system design, previously ignored by government sponsored research. The latter point makes Biosphere 2 an important case study for ecological engineering.

H. T. Odum (1971) had anticipated the Biosphere 2 project in discussing a large life support system (Figure 4.24), and he had written several grants in the 1960s for a similar though smaller-scale project that were not funded. Biosphere 2 realized this concept but with huge technological couplings. The project was started in 1984 and it followed a tumultuous series of events (Table 4.5). It still exists today but with a very different focus. The original purposes of the project were "... to develop bioregenerative and ecologically-upgrading technologies; to conduct basic scientific and ecological research; and to educate the public in ecosystem and biospheric

TABLE 4.5
Milestones in the History of Biosphere 2 near Tucson, AZ

1984	Project begins; $150 million private enterprise venture
1988	Test module experiments begin
September 26, 1991	Closure of first crew inside Biosphere 2
October 12, 1991	One of the crew leaves for finger surgery, then immediately returns.
July 21, 1992	Science advisory committee report criticizes Biosphere 2 science program
February 1993	Science advisory committee resigns
August 1993	Oxygen is injected into Biosphere 2
September 26, 1993	Biosphere 2 opens; end of first experiment
March 1994	Closure of second crew inside Biosphere 2
April 2, 1994	Takeover of old management by owner
April 6, 1994	Two of the original biosphereans break the seal letting oxygen in
September 1994	End of second experiment
November 1995	Decision is made to transfer Biosphere 2 to Columbia University for management
1996	The movie *Biodome* opens

issues. Biosphere 2 is privately funded and is designed to be a for-profit venture, combining eco-tourism with the excitement of 'real-time' science and experimentation" (Nelson, 1992). A smaller pilot project, called the test module, was initially studied (Ailling et al., 1993) in the conventional engineering tradition. Based on this pilot project and an extensive review of the field of life support systems, the designers built the mesocosm with several ecosystem types including soils to provide the bioregenerative support. The design philosophy was given by Nelson et al. (1993):

> Biosphere 2 … has incorporated an approach that seeks to promote and ensure the self-organizing capabilities of living systems by deliberately replicating a typical range of tropical and subtropical environments with their associated diversity of life forms and metabolic pathways … To foster its diversity, Biosphere 2 includes many micro-habitats within each ecosystem type. It was deliberately over-packed with species to provide maximal diversity for self-organization and to compensate for unknown and potentially large initial species losses. Environmental technologies provide thermal control, water and wind flows, and substitutes for natural functions like waves and rain. But the facility would not function unless the biota fulfills its essential role of using energy flow for biomass production (including food from agricultural crops) and ensuring closure of essential biogeochemical cycles through diverse metabolisms. The design of Biosphere 2, which attempts to harmonize living systems with supporting environmental technologies, unites two historical approaches to life-support systems: the engineered and the ecological.

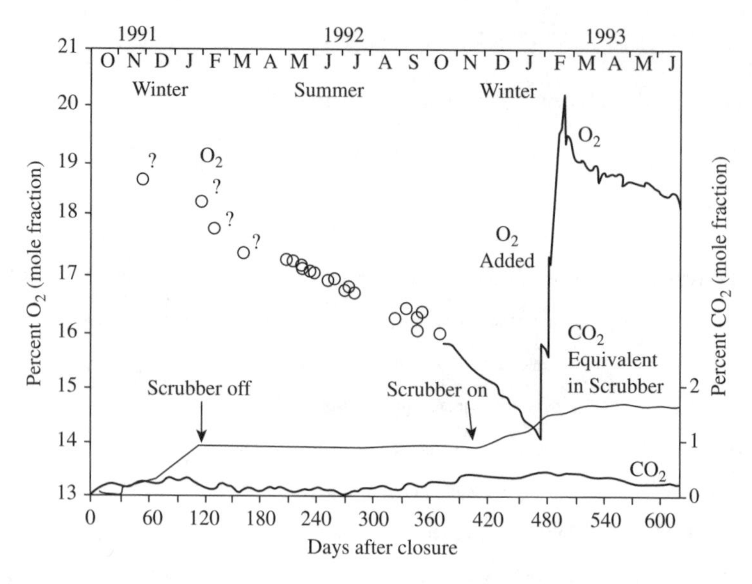

FIGURE 4.25 Gas dynamics for the first enclosure experiment at Biosphere 2. Note the decline in oxygen before artificial additions were required. (From Cohen, J. E. and D. Tilman. 1996. *Science* 274:1150–1151. With permission.)

It is amazing how fast the project developed with less than a decade between initial plans and final completion. This fast development stands in contrast to NASA's research efforts on bioregenerative life support systems, which are still incomplete after more than 50 years of effort.

The first manned experiment consisted of enclosure of a crew of eight people for a 2-year period from September 1991 to September 1993. During the experiment the project became engulfed in controversy, from both the outside scientific community and the general public. An emphasis on popularization for education and media purposes dominated the early stages of the project (Allen, 1991; Alling and Nelson, 1993). This was a positive initiative but was not immediately backed up with hard scientific documentation and full discussion of problems inherent in any project of the scale of Biosphere 2. Pivotal in the controversy was the report of an outside scientific advisory panel, which was critical of project management in some respects but complimentary in others. Ultimately, relationships between the scientific advisory committee and the management team broke down and the committee resigned. Meanwhile, oxygen levels inside Biosphere 2 dropped (Figure 4.25), threatening the health of the original crew and requiring additions from outside. Some viewed this bailout as failure of the project, and miscommunications about this and other problems occurred between the management team and outsiders. Eventually the original management team was disposed of by the owner and Biosphere 2 was transferred to Columbia University to be used for global change research with no emphasis on bioregenerative technologies. While no one can know all of the details of the events that took place in the project's history, much can be learned from the updates published in the News & Comments and other sections of

FIGURE 4.26 View of the Mars Society's Desert Research Station including the habitat and an early design for the greenhouse.

Science magazine (Appenzeller, 1994; Kaiser, 1994; Stone, 1993; Watson, 1993; Wolfgang, 1995). Was the original Biosphere 2 project judged unfairly by the scientific community and the general public? E. P. Odum (1993), a member of the original scientific advisory committee, wrote that Biosphere 2 represented a new kind of science (as did Nelson and Dempster, 1993) and that it provided knowledge that could not have been acquired by other methods (as did Avise, 1994). Surprises occurred inside Biosphere 2 that challenged conventional thinking. For example, it took a graduate student to interpret the missing sink in the oxygen and carbon budgets as absorption by the concrete (Severinghaus et al., 1994). These kinds of surprises are learning opportunities and more can be expected as results from the original Biosphere 2 project are published (see, for example, the 1999 special issue of *Ecological Engineering* devoted to Biosphere 2: Vol. 13, Nos. 1–4; Cohn, 2002; Cohn and Tilman, 1996; Walford, 2002).

A related new development concerning life support systems and space travel is the work being done by the Mars Society. This is an international organization of people interested in exploration and, ultimately, colonization of the planet Mars. To some extent the society has been initiated and inspired by Robert Zubrin's (1996) plan which includes use of physical–chemical resources on Mars to make return fuel. Thus, a Mars mission would have to carry only enough fuel to get to Mars. Once there, the explorers could make the fuel for the return trip with indigenous resources. This approach reduces the required payload and makes the entire mission concept much more feasible than alternative scenarios. Research to support the Mars Society vision is being conducted with a different approach than has been used by NASA or Biosphere 2. Table 4.6 compares some differences with emphasis on life support initiatives. All three organizations envision greenhouse-based systems, but the work of the Mars Society is recent and closure of the systems has not yet been attempted. Current emphasis is on wastewater recycling in the Mars Society's greenhouse (Figure 4.26).

TABLE 4.6
Comparisons of Different Programs of Developing Life Support Systems for Space Activities

	NASA CELSS	Biosphere 2	Mars Society
Time line	1980 to present	1984–1995	1999 to present
Scale	Small	Large	Small
Relative funding level	Moderate	Large	Very small
Source of funding	Federal Government	Private	Private
Pattern of project administration	Vertical	Vertical	Horizontal
Degree of ecological engineering	Low	High	High
Level of closure	High	High	None yet

MICROCOSM REPLICATION

Replication is an issue of experimental design rather than technical design of a microcosm. It is included in this discussion because of its critical nature to the "microcosm method" (Beyers, 1964) and because its consideration leads to fundamental questions of variability (possibly chaos) of ecosystems. In fact, replication is one of the "minimal requirements of experimental design in ecology" (Hairston, 1989) as in any application of the scientific method. As noted by Sheehan (1989) "the main purpose of replication is the supply an estimate of variability (error) by which significance of treatment and control comparisons can be judged." Thus, replication is needed to distinguish between natural variation and variation due to a treatment (such as introduction of a toxin into a microcosm) in an experiment. The number of replicates required for an experiment depends on the variability of the data being collected. As has been noted, cross seeding is used in microcosm research to reduce this variability and thus "enhance" replication (Beyers and H. T. Odum, 1993).

Problems do arise when considering replication of microcosms. In a review of 360 microcosm experiments, Petersen et al. (1999) found that only 65% of the studies reported the number of replicate systems per treatment, which implies that many researchers take replication for granted. While most studies and reviews focusing on replicability have found that it can be satisfactorily achieved in microcosms (Conquest and Taub, 1989; Giesy and Allred, 1985; Isensee, 1976; Levy et al., 1985; Takahashi et al., 1975), there is disagreement. For example, Abbott (1966) studied replicability of 18 5-gal (19 l) glass carboy microcosms of an estuarine bay and found that, based on coefficient of variation calculations, "... under proper conditions aquatic microcosms show replicability comparable to that found among other types of statistical trials. This means that groups of parallel systems can be established and studied in rigorously defined experiments, but only on a statistical basis." This was the first study devoted to the question of microcosm replicability, and it has been often quoted in more recent studies as justification for the experimental use of microcosms. However, Abbott's study was criticized on the very same grounds of coefficient of variation by Hurlbert (1984) and Pilson and Nixon (1980). Furthermore, Whittaker (1961) in one of the first microcosm studies introduced the term *aquarium individuality* for "the marked differences between aquaria with similar conditions which result from minor, uncontrolled factors and are a major limitation on reproducibility and statistical adequacy of the data." Ironically though, Abbott (1966) quotes another pasage from Whittaker's paper as justification for the microcosm approach. Clearly, there is some subjectivity here, and there are no clear cut criteria on how similar replicate microcosms must be for use in experiments. Hurlbert's (1984) discussion is valuable in this regard as are papers reviewing the statistical treatment of microcosm research (Chapman and Maund, 1996; Gamble, 1990; several papers in Graney et al., 1994; Sheehan, 1989; Smith et al., 1982). Pilson and Nixon (1980) provide many practical insights on the issue. For example, they state: "The existence of variability is a severe problem, becoming even philosophically difficult to deal with. Nature herself does not replicate well. Every patch of water follows its own course to some extent, exchanging all the while with its surroundings. Every bay and cove is to some measurable extent unique." They discuss the dilemma of trying to develop a set of microcosms that replicate well but still exhibit the variability inherent in nature. Curiously, they conclude that "the better the microcosm the worse it replicates and, therefore, the worse it is as an experimental tool." They suggest long time period observations as the "way out of this morass," which is the opposite of most microcosm research that relies on short-term studies to avoid the almost inevitable divergence of replicate microcosm over time.

General results of microcosm research seem to indicate that replication becomes more difficult to achieve as the microcosm becomes larger and as the experiment lasts longer, but there is clear scientific value in large and long-term microcosms. Probably the best strategy for ecological engineering as a whole is research at small and short time scales balanced with large and long time scales.

However, it may be instructive to embrace microcosm variability and to try to learn from it (Figure 4.27). In almost all cases (though see Sommer, 1991), microcosms diverge to greater or lesser extent in characteristics over time. H. T. Odum and Lugo's (1970) study of forest floor microcosms is one example (Figure 4.28). They constructed microcosms in small plastic desiccators that were incubated on the floor of a rain forest in Puerto Rico. Each system contained clay from the forest soil and leaf litter with seeding of one each of bromeliad, fern, lichen, moss, and a clump of algae. After months of adaptation, "Each system developed in a special way; thus, by the time measurements were made, the herbs in each were different. Some chambers were well filled with leafy proliferation of one or more species, but in each system a different species was dominant, and there were differences in the relative quantity of photosynthetic tissue. Animal components of the soil were different also." The microcosms were then closed and their metabolism was studied. They found the characteristic diurnal pattern of CO_2 change with a rise in the dark period due to respiration and a decline during the light period due to photosynthesis. While the rates of metabolism were relatively similar between microcosms, the

FIGURE 4.27 Cartoon view of the problem of replication of microcosms. (From Sidney Harris, New Haven, CT. With permission.)

concentrations of CO_2 in the atmospheres were very different, ranging from 415 to 3,400 ppm and higher, compared with the ambient 300 ppm. They concluded by relating their findings to the idea that biological evolution might control atmospheric composition because different species compositions generated different atmospheric concentrations of CO_2 in microcosms.

The divergence that normally occurs in replicate microcosms may be random or perhaps an example of chaotic behavior. Chaos refers to complex pseudorandom behavior arising in nonlinear deterministic systems. Characteristics of chaos are a sensitive dependence on initial conditions and an increasing divergence over time in studies of variability. The earliest studies in ecology identified chaotic behavior in simple population equations (May, 1974b, 1976; May and Oster, 1976). This was an important result because it demonstrated that complexity (in terms of dynamics) could arise even from simple ecological systems. Reviews of chaos theory in ecology are given by Cushing et al. (2003), Hastings et al. (1993), O'Neill et al. (1982), and Shaffer (1985, 1988; Shaffer and Kot, 1985). It is very difficult to distinguish true randomness from chaotic behavior, which is actually a form of orderly dynamics. There is much interest in searching data sets for chaos (Godfray and Grenfell, 1993), and perhaps the microcosm literature may be another place to look for it, especially in terms of the importance of initial conditions in determining species compositions or ecosystem function. Most studies of chaos in ecological systems have been

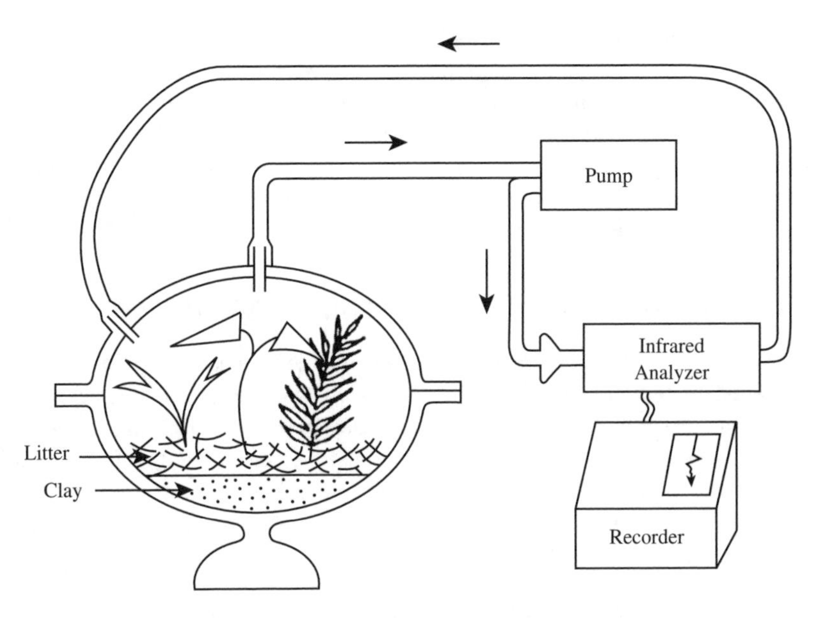

FIGURE 4.28 Closed terrestrial microcosm from the forest floor of a Puerto Rican rainforest. (From Odum, H. T. and A. E. Lugo. 1970. *A Tropical Rain Forest*. H. T. Odum and R. Pigeon (eds.). U.S. Atomic Energy Commission, Oak Ridge, TN. With permission.)

theoretical analyses of mathematical equations; it remains to be seen how much explanatory power chaos theory will have in studies of actual ecosystems.

Several other concepts exist in ecology for understanding the kind of divergence that occurs in "replicate" microcosms. Assembly theory (Drake, 1990; Weiher and Keddy, 1999) is one paradigm that seeks rules for understanding how different communities can arise from different sequences of seeding. Lawton (1995) has reviewed this literature, and interesting microcosm examples have been described for amphibian communities in temporary ponds (Alford and Wilbur, 1985; Wilbur and Alford, 1985), microbial lab cultures (Drake, 1990a,b, 1991; Drake et al., 1993), fruit flies (Gilpin et al., 1986), and wetland plants (Weiher and Keddy, 1995). The lottery model of community structure is somewhat similar in relying on historical explanation but is based on randomness. In this approach community composition depends on who colonizes a site first. Sale (1977, 1989) first proposed this nonequilibrium concept to explain the local variation of fish diversity on coral reefs. The concept of priority effects, in which the presence of one species decreases the probability of colonization by another, has been introduced to account for some aspects of lottery-type behavior in ecological communities (Shulman et al., 1983). Finally, alternative stable states may explain the divergence of microcosms (see Chapter 7 for more details on the theory of alternative stable states).

In conclusion, variability is an important consideration in microcosm research. Replication is necessary for testing hypotheses in the scientific method and the quality of replicates should not be ignored. However, the divergence of microcosms is interesting in its own right, and it might be capitalized on to understand certain aspects of ecosystem structure and function. Simberloff (1984) likened the "apparent indeterminancy in ecological systems," which is similar to the issue of microcosm replicability, to Heisenberg's Uncertainity Principle from physics, and he suggested its importance thus: "What physicists view as noise is music to the ecologist; the individuality of populations and communities is their most striking, intrinsic, and inspiring characteristic … " In dealing with microcosms, ecological engineers must learn to appreciate the individuality of each replicate while using replication as a necessary feature of experimental design.

COMPARISONS WITH NATURAL ECOSYSTEMS

In many if not most cases, microcosms are intended to be models of some real ecosystems. In this context the model is usually compared with the real analog ecosystem in order to establish success of the design. If the model matches with the analog, then experiments done with the model can be extrapolated to the real world. Therefore, comparisons between microcosms and real ecosystems are important in the microcosm method; again, Pilson and Nixon (1980) provide useful insights on the issue. They state:

> … the issue of correspondence between the living model and nature has emerged as a major question for those doing microcosm research and for those who hope to use the results from microcosms to develop management polices.
>
> This issue of correspondence is a very difficult problem … . there are no generally agreed upon qualitative or quantitative criteria for success. At this time it is not possible to give a general objective description of a successful microcosm. We do not know which parameters are most important to have in agreement or how similar they must be to be considered in agreement.

And later, they continue:

> Although we admit that no microcosm can ever be an exact replica of nature, we still want them to be "not too abnormal."

Many kinds of measurements, such as nutrient concentrations, population densities, species diversity, biomass, or metabolism, can be compared between a microcosm and its analog ecosystem, and the choice of measurements has been subjective in practice. As an example, Figure 4.29 compares litterfall, which is a measure of aboveground net productivity, for mangroves inside Biosphere 2 and in Southwest Florida over the same annual cycle (Finn, 1996). In this particular case litterfall in the microcosm (Biosphere 2) follows a seasonal pattern similar to the real analog (Southwest Florida) though values are sometimes higher in the mesocosm. Gearing (1989) provides many other examples of this kind of exercise in a review of aquatic microcosm research.

The process of developing a realistic microcosm is itself a test of ecological knowledge as noted in the following quotes: "… the task of assembling, maintaining, and predicting the behavior of even moderately complex ecosystems in the laboratory

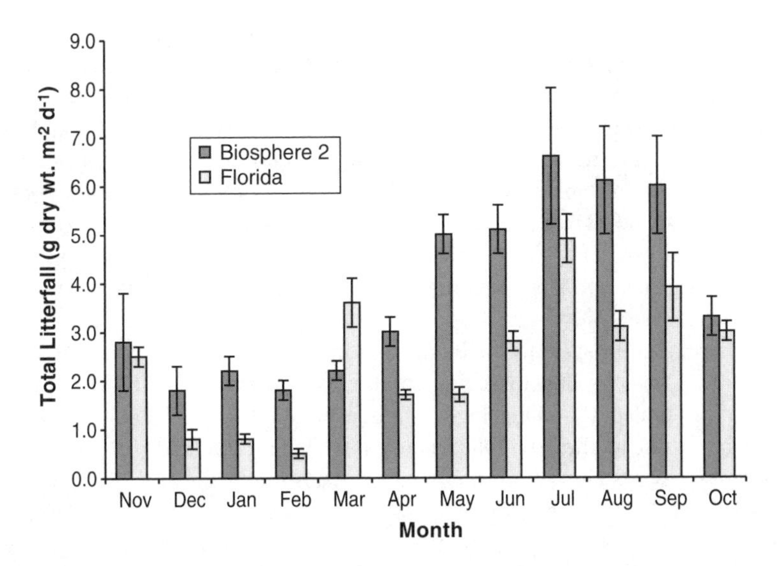

FIGURE 4.29 Comparison of litterfall from the mangrove forests in Biosphere 2 and Southwest Florida. (Adapted from Finn, M. 1996. Comparison of Mangrove Forest Structure and Function in a Mesocosm and Florida. Ph.D. dissertation, Georgetown University, Washington, DC.)

tests our understanding to the limit." (Lawton, 1995), and "If ecologists can learn to construct mesocosms that replicate important characteristic of natural ecosystems, this will provide unmistakable evidence that they really understand how these ecosystems function" (Cairns, 1988). Thus, the ability to create microcosms that adequately model real ecosystems not only allows experiments to be conducted and extrapolated, but this ability also provides a practical measure of how well real ecosystems are understood. This is the notion that Nixon (2001) was referring to when he stated that "every mesocosm is a living hypothesis."

It also may be possible to learn from microcosms that fail or from those that do not match with natural analogs. Most researchers have not followed this line of thinking and examples are not easy to find. One example of a system that "failed" was embedded in the Everglades mesocosm of the Smithsonian Institution. The problem here can be seen by comparing maps of the system soon after construction (see Figure 7 in Adey and Loveland, 1991, p. 580) and after 7 years of self-organization (see Figure 7 in Adey and Loveland, 1998, p. 417). This mesocosm was intended to be an abstract model of the gradient of systems that make up a transect across the Southwest Florida Everglades from freshwater (tank 7) through estuarine (tanks 2 to 6) to marine (tank 1) habitats (Figure 4.15). If success is judged by the stability of the ecological components within the tanks, then the overall system was a success because there was little change between the initial map and the map after 7 years of change. In fact, it was a remarkable system in containing so much biodiversity characteristic of the Everglades in engineered gradients of tide and salinity. However, tank 5 did change. This tank was originally intended to model a saltmarsh system with dominance of grasses and succulent herbs, as can be seen in the map from the first edition of Adey and Loveland's book. After 7 years, tank 5 changed to a mangrove-dominated system with white and red mangroves and mangrove ferns, as can be seen in the map from the second edition of Adey and Loveland's book (Figure 4.15). The change was due to natural succession in which mangroves were able to outcompete saltmarsh (Kangas and Lugo, 1990). In essence then, the "failure" of tank 5 is actually a verification of a successional hypothesis. Saltmarsh exists in South Forida either temporarily or in locations that exclude easy mangrove establishment. However, in the close confines of the greenhouse mesocosm environment, saltmarsh was not able to find a refuge from the competitively superior mangrove system and it was out-competed.

In other cases ecosystems have emerged in microcosm experiments that do not match natural analogs due to scaling problems. Oviatt et al. (1979) found super blooms of phytoplankton in a pelagic microcosm which were difficult to explain but were probably due to altered light climates in the lab vs. the field. The best examples of emerging new systems are probably those attached to walls of microcosms. These can sometimes become interesting in themselves even though they confound the intended systems. For example, Twinch and Breen (1978) found an attached system develop on their limnocorrals that included about 90 snails/m^2 supported by the growth of algae on the walls. Margalef (1967) recognized the significance of these types of wall growths in plankton experiments, even though they are unintended:

> The attached species progressively invade all the flasks as time advances. The concentration of organic matter on the walls and the absorption of light are new factors and the whole pattern becomes blurred. The elegant simplicity of the experiments with free-floating algae is lost. The brutal competition for dominance based on the rates of increase has given way to more subtle and interminable processes and the chemostat is prevented from attaining a stationary state. The situation is interesting as an example of development of more organization than the experimenter desires.

His last sentence in this quote is particularly relevant in suggesting how ecosystems, which surprise the experimenter by generating more organization than was intended, can emerge. Perhaps the most unusual microcosms that indicate emergence of new ecosystems are systems that are given control over their own inputs. The first example of this type was an unpublished experiment by Beyers (1974; Kania and Beyers, no date) that was diagrammed by H. T. Odum (1983) in Figure 4.30. Here Beyers' flask microcosms were interfaced through a pH meter to the timer that controlled the lights in the growth chamber where the experiment took place. Changes in pH occur diurnally with uptake and release of CO_2 in photosynthesis and respiration, respectively. Thus, the interfaced microcosm could control the duration of lighting through its own metabolism. Figure 4.31 shows light–dark patterns of the three replicate systems. Periods of lights "on" and lights "off" are plotted. Two of the replicates had longer "on" periods than "off" periods and they both eventually evolved to turn the lights on continuously. The third system had longer "off" periods than "on" periods, and it never turned on the lights continuously. This is a remarkable experiment that never got published for one reason or another. This story becomes even more interesting because Petersen (1998) performed an

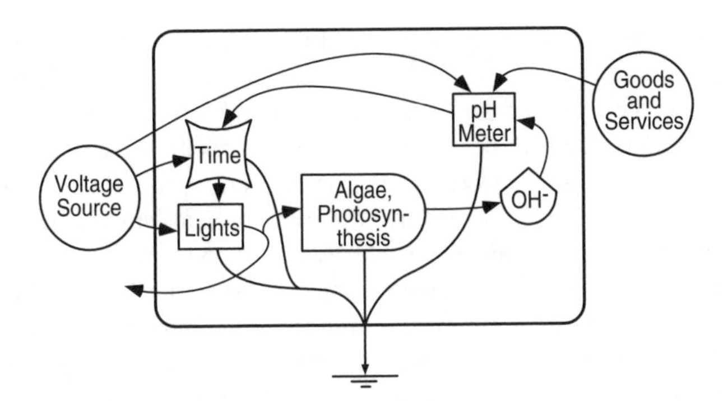

FIGURE 4.30 Energy circuit diagram of Beyers' interfaced microcosm, where the ecosystem controlled its light source. (From Odum, H. T. 1983. *Systems Ecology: An Introduction*. John Wiley & Sons, New York. With permission.)

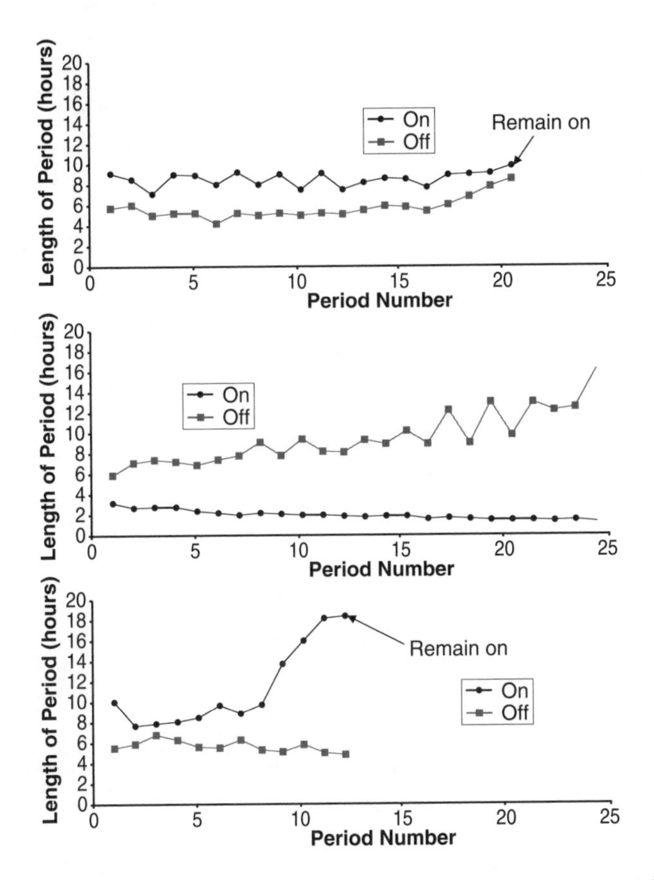

FIGURE 4.31 Sample data sets from Beyers' interfaced microcosms. (Adapted from Kania, H. J. and R. J. Beyers. No date. Feedback control of light input to a microecosystem by the system. Unpublished report. Savannah River Ecology Laboratory, Aiken, SC.)

independent experiment as part of his dissertation research that was very similar to Beyers' original work, without knowledge of it. He interfaced pelagic microcosms with an oxygen electrode and set up a routine whereby the systems could control the length of the light and dark periods. In general, his results were similar to Beyers' results, with the microcosms generating alternating periods of light and dark through their metabolism (Petersen, 2001). Both of these experiments are remarkable because they create an ecosystem that never existed previously: one that can control its energy source. In fact, many kinds of emerging new ecosystems can be developed in microcosms through creative ecological engineering. Every microcosm is a new ecosystem that never existed previously, even if it is intended to model a natural analog. Some of the most interesting microcosms may be those with strange, artificial scaling because they show what is possible in ecosystem development. In this sense, the differences between microcosms and their natural analogs are opportunities to learn, perhaps about some fundamental property of a simple food chain or of a succession sequence, or perhaps some great new truth of ecology.

5 Restoration Ecology

INTRODUCTION

Restoration ecology is a subdiscipline of ecological engineering that has been growing out of the need and desire to add ecological value to ecosystems that have been degraded by human impacts. Projects range in size from less than one hectare for an individual prairie or wetland to the entire Everglades of South Florida. It is a very general field in that any kind of ecosystem can be restored but different actions are required for each ecosystem type. An extensive literature, which is a useful guide to future restorations, is developing out of the experience of practitioners. Much work is generated by legal requirements such as the Surface Mining Control and Reclamation Act of 1977 and the "No Net Loss" policy for wetlands, both from the U.S. Another antecedent to modern restoration ecology was the early efforts to improve industrial landscapes, especially in Europe (Chadwick and Goodman, 1975; Gemmell, 1977; Johnson and Bradshaw, 1979; Knabe, 1965). Although the field can be viewed as being a recent development, as early as 1976 an annotated bibliography of restoration ecology included nearly 600 citations (Czapowskyj, 1976). Storm (2002) considers restoration in the U.S. to be the basis for a growth economy because it is attracting investment from businesses, communities, and government.

A relatively large literature involves definitions of restoration and related terms (Bradshaw, 1997a; Higgs, 1997; Jackson et al., 1995; Lewis, 1990; National Research Council [NRC], 1994; Pratt, 1994). In general, *restoration* is the term used when a degraded ecosystem is returned to a condition similar to the one that existed before it was altered. However, many other related terms are used as is indicated by the titles to books on the subject: *Recovery* (Cairns, 1980; Cairns et al., 1977), *Rehabilitation* (Cairns, 1995b; Wali, 1992), *Repair* (Gilbert and Anderson, 1998; Whisenant, 1999), *Reconstruction* (Buckley, 1989) and *Reclamation* (Harris et al., 1996). To some extent, the differences in terms relate to differences in end points expected from the respective processes (Zedler, 1999). These end points may be very different as indicated in Figure 5.1. Sometimes ecosystems are created on a site which did not exist previously, as in wetland mitigation, and in other cases entirely new systems are constructed such as the "designer ecosystems" mentioned by MacMahon (1998) or the "invented ecosystems" mentioned by Turner (1994).

Some authors such as William Jordan III focus on conceptual approaches (Jordan, 1994, 1995; Jordan and Packard, 1989; Jordan et al., 1987), while others such as Anthony Bradshaw focus on more concrete principles (Bradshaw, 1983, 1987a, 1997a). There is a continual search for deep meaning by some workers in restoration ecology, which has resulted in an unusually broad field. For example, Brown (1994) uses the "prime directive" metaphor from the science fiction series *Star Trek* to suggest ways of dealing with restoration actions, and Baldwin et al. (1994) provide a book-length review of opinions from workers in art, literature, philosophy, and

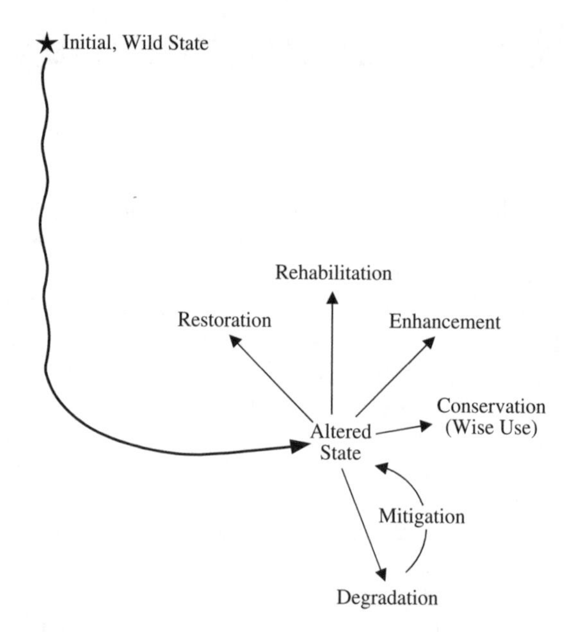

FIGURE 5.1 Different end points in various restoration processes. (Adapted from Francis, G. R., J. J. Magnuson, H. A. Regier, and D. R. Talhelm. 1979. Technical Report No. 37. Great Lakes Fishery Commission, Ann Arbor, MI.)

ecology about restoration. However, most workers share a sense of urgency about the need for restoration, as noted below in the quote by Packard and Mutel (1997a):

> Restoration today is similar to battlefield medicine. We learn, by necessity, from attempts to revive torn and insulted ecosystems. The discipline profits much from watching the results of extreme measures taken in these emergency situations. As a result, practical knowledge is far ahead of hard science. We need as much scientific knowledge as we can get to inform restoration decisions, but restorationists must often act with imperfect knowledge if they are to act at all before the biodiversity they seek to preserve disappears. Thus, restoration relies on art and intuition as well as on objective knowledge.

Restoration ecology is an important subdiscipline of ecological engineering because it involves the design, construction, and operation of new ecosystems. The use of conventional engineering varies considerably across the spectrum of restoration projects. Some restorations rely almost completely on the passive ecosystem self-organization of natural succession while others are much more active, involving costly planting programs and landscape modification with changes in geomorphology and hydrology. The relationship between ecology and engineering has not always been positive in this subdiscipline, as indicated by Clark (1997):

> We see at present an uneasy relationship between ecology and technology, with uncertainty about the proper role for each. At one extreme there is "restoration" which is

> virtually a branch of engineering. Adherents to this approach reflect the engineer's concern to build structures according to fixed plans and to a high precision, but not necessarily in sympathy with natural environmental processes. Indeed, the discipline of environmental engineering has developed in parallel to restoration ecology, and the practical objectives are often similar. For example, environmental engineers have made great progress in construction of wetlands for the purpose of water treatment. The difference from ecological restoration is that these are essentially engineered structures, perhaps requiring the building of new levees or excavating of the land in areas which could not otherwise support wetland communities; such structures often require virtually constant aftercare. At the other extreme are the wildlife conservation organizations which attempt to restore ecosystems with only hand tools and willing volunteers. The problems with this approach are that it can be very slow, can only be performed at a small scale, and the results obtained are unpredictable.
>
> One of the reasons for this uncomfortable relationship is certainly a distaste amongst some ecologists for the tools that technology provides. Bulldozers, herbicides, pesticides, chainsaws, and high explosives are, for many conservation-minded ecologists, the instruments of the Devil. It is using precisely these means that the damage that they wish to put right was created. This is an attitude, which, while perhaps understandable, is none the less a barrier to progress. No tool in itself is bad or good; what matters is how it is used.
>
> Restoration ecology must improve its use of technology, and find a middle course between these two extremes.

A goal of ecological engineering is to break down the dichotomy described above and help create the "middle course" where both ecology and engineering are used in a collaborative rather than an antagonistic way.

STRATEGY OF THE CHAPTER

In this chapter *restoration* is used as a general term to broadly cover the field. Both policy and technical aspects are included in an effort to provide an overview. The relationship of restoration to environmentalism is discussed first. As with other disciplines which utilize ecology, ecological engineering is related to society's perception of the need to care for the environment. One particular aspect is presented in this section due to its similarity to the engineering approach to design. A more radical form of environmentalism that relates to engineering is also mentioned.

Most of the chapter focuses on restoration practice. The energy signature approach is suggested as a general, guiding principle with special attention given to genetic inputs in restoration. Succession may be the most important tool in this regard and is emphasized. Bioremediation is introduced as a special type of restoration process. Procedural or policy aspects, including indicators of success and reference sites, are discussed as being important issues of the field. Finally, three case studies are described to illustrate topics covered throughout the chapter.

RESTORATION AND ENVIRONMENTALISM

The goal of restoration ecology is the restoration of a degraded ecosystem or the creation of a new ecosystem to replace one that was lost. The primary purpose of these actions is to add ecological value for its own sake, rather that to provide some useful function for society. In this sense, restoration ecology differs in emphasis from other subdisciplines of ecological engineering such as treatment wetlands or soil bioengineering where ecosystems are constructed to provide a useful function first (i.e., wastewater treatment or erosion control) and to add ecological value as a secondary objective. In fact, restoration ecology sometimes attempts to restore or replace ecosystems to a natural state that existed before human presence was dominant (except for aboriginal peoples). Thus, there is a direct and logical connection between restoration ecology and environmentalism because of the primary focus of restoring systems to their natural condition.

In general terms, environmentalism is a popular movement that arises from the social desire to maintain natural ecosystems within landscapes that are dominated by humans. In the past, when human population densities were relatively low, this movement was motivated by idealism. However, as population densities have increased, there is now a growing awareness that natural ecosystems provide real life support functions for humanity as a by-product of their natural existence, which makes the past idealism become pragmatic and adaptive. Environmentalism takes many forms, ranging from the establishment of parkland in urban environments through the protection of wilderness and endangered species, to the rise of political parties based on this theme. Here, two dimensions are explored that can be tied to engineering.

At one end of the environmentalism continuum is the application of scientific approaches to conserving biodiversity, which is the concern of the field of conservation biology. This important field combines elements of ecology and genetics along with public policy analysis for maintaining as much of the Earth's biodiversity as possible. Restoration ecology and conservation biology are related because the restored ecosystems provide habitat for species threatened by human impacts (Dobson et al., 1997; Jackson, 1992; Jordan et al., 1988). One activity in conservation biology that has some similarities with engineering practice is the design of preserves based on island biogeography (see Chapter 4). The similarities involve the use of theoretical equations for design, which justifies reviewing the topic here.

The theory of island biogeography was outlined in the 1960s by Robert MacArthur and E. O. Wilson (MacArthur and Wilson, 1963, 1967). It basically described the origin and maintenance of ecological species diversity on oceanic islands with extrapolations to habitat islands, such as patches of forest in an agricultural landscape. The theory was explosively popular among academic scientists who applied it to a tremendous number of situations in the 1970s and 1980s, such as caves (Culver, 1970), mountaintops (Brown, 1971), reefs (Molles, 1978; Smith, 1979), lakes (Keddy, 1976; Lassen, 1975), rivers and streams (Minshall et al., 1983; Sepkoski and Rex, 1974), plant leaves (Kinkel et al., 1987), host-parasite systems (Tallamy, 1983), and artificially constructed habitat islands (Cairns and Ruthven, 1970; Dickerson and Robinson, 1985; Schoener, 1974; Wallace, 1974). The simplest expression of the theory explained the number of species that could be supported on an island as a function of the area of the island and its proximity to other islands which act as sources of species that might immigrate. The equilibrium number of species that could be supported is a function of the number of species available to immigrate and the balance between immigration and extinction rates, as given by the following equation

$$dS/dt = k_1(ST - S) - k_2 S \tag{5.1}$$

where

$k_1(ST - S)$ = immigration rate
k_2S = extinction rate
S = the number of species on the island
ST = the total number of species on nearby islands that can immigrate to the island
k_1 and k_2 = proportionality constants
t = time

Thus, when an island is first exposed to colonization, as might occur after a hurricane removes the biota, the number of species increases due to an excess of immigration over extinction until a dynamic equilibrium between the two processes is reached. The number of species could decrease (or "relax") if the area of the island declines, as occurs when sea level rises forming land bridge islands. In this case extinction exceeds immigration until a new equilibrium is established. The theory also drew on the species–area curve. Area figures into the equation indirectly with the values of the proportionality constants. In general, the extinction rate decreases as island area increases, while immigration rate increases as island area increases. The proximity to source islands also leads to increased immigration rate.

Together, these expressions formed the quantitative foundation for the island biogeographic theory of MacArthur and Wilson. They were tested in many settings, and generally they were found to provide explanations for species diversity patterns. Not unexpectedly, the theory was also quickly applied to the problem of reserve design in conservation biology, which was just emerging in the 1970s. This was an obvious application because a reserve is like an island of natural species within a surrounding landscape of agricultural, urban, or other human-dominated land use. In the mid-1970s a number of papers were presented that applied island biogeography theory to reserve design (Diamond, 1975; Diamond and May, 1976; May, 1975; Sullivan and Shaffer, 1975; Terborgh, 1975; Wilson and Willis, 1975). Rules of reserve design evolved from the theory of island biogeography in a systematic fashion. Of course, the species–area equation indicated that reserves with larger areas would support greater numbers of species, which was a desirable objective. The species equilibrium equation also indicated that the number of species supported in a reserve could be increased by increasing the immigration rate. This could be achieved by placing the reserve near other reserves that provide a source of species

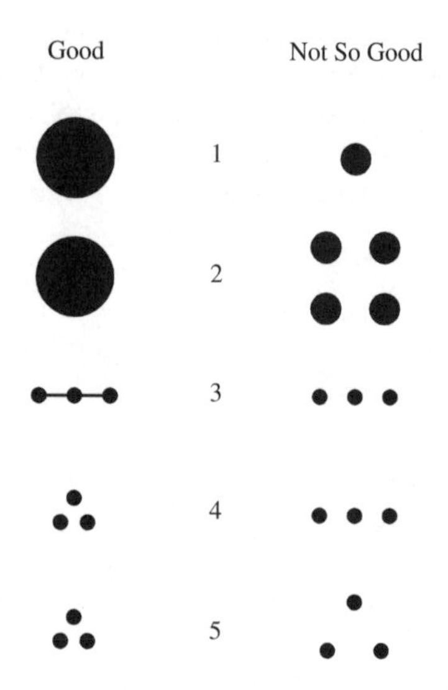

FIGURE 5.2 Extremes of reserve design based on the theory of island biogeography. (Adapted from Diamond, J.M., 1975. *Biological Conservation*. 7:129–146.)

for immigration or by the use of a corridor configuration to connect reserves and facilitate migration by species. Diamond (1975) summarized reserve design principles as shown in Figure 5.2. This use of theoretical equations for the purpose of design is reminiscent of engineering applications, such as the sizing equation given in Chapter 2 in regard to treatment wetlands. As a first approximation, the equations from island biogeography provide a quantitative basis for design decisions to be made about reserves. The equations provide predictions that can be used to make choices between alternatives and to explore the implications of possible solutions, as in engineering. However, this application was quickly and repeatedly criticized, especially by Daniel Simberloff and his associates (Simberloff and Abele, 1976; see the review in Shafer, 1990), bringing up many exceptions and controversies about the complexity of reserve design. For example, there may be situations where more diversity is maintained in a landscape with a number of small reserves that protect local patches of high species diversity rather than in one large reserve that is not able to protect all of the diversity from the scattered patches. Thus, in the present state of the art, the theory of island biogeography does not provide much valuable insight in conservation biology (Hanski and Simberloff, 1997; Simberloff, 1997; Williamson, 1989), but it does represent a historical example of design practice relevant for perspective on ecological engineering.

At another extreme, environmentalism takes on passionate, emotional displays and actions for the protection of natural ecosystems (Zakin, 1993). Perhaps the most extreme form of such passion is ecoterrorism. "Monkey-wrenching," for example, involves the destruction of equipment and impairment of work of developers and polluters who cause environmental impacts. The novelist Edward Abbey coined the term in 1975 when he described the fictional actions (some of which are listed in Table 5.1) of the "Monkey Wrench Gang" (George Washington Hayduke, Seldom

TABLE 5.1
Monkey Wrenching Activities Carried Out by a Fictional Gang in Arizona

Pushing a bulldozer into a reservoir
Setting a bulldozer on fire
Destruction of an oil drill-rig tower
Removal of geophones used for seismic oil exploration
Draining the oil from diesel engines, then starting them up and letting them run without oil
Cutting barbed wire fences on ranches
Blowing up a railroad bridge used for coal transport from a strip mine
Defacing a Smokey the Bear sign put up by the U.S. Forest Service
Cutting power lines to a coal strip mine
Pouring sand and Karo syrup into fuel tanks of bulldozers
Pulling up developers' survey stakes
Cutting up the wiring, fuel lines, control link rods, and hydraulic hoses of earth moving machines
Knocking over commercial billboards along highways

Source: Adapted from Abbey, E. 1975. *The Monkey Wrench Gang*. Avon Books, New York.

Seen Smith, Bonnie Abbzug, and Dr. Alexander Sarvis). These actions ranged from "subtle, sophisticated harassment techniques" to "blatant and outrageous industrial sabotage," but there was never any intention to threaten human life (Abbey, 1975). This kind of ecoradical activity is actually being carried out, in one form or another, by certain extreme environmental organizations. For example, it appears that one extreme environmental group may have been responsible for destruction of structures at a lab conducting research on genetic engineering of trees (Service, 2001). The subject relates to the present book because well-trained ecological engineers probably would make excellent monkey wrenchers based on their balance of knowledge between ecology and traditional engineering and their facility with destructive technology.

As an aside, one objective of Abbey's Monkey Wrench Gang was to blow up Glen Canyon Dam on the Colorado River near the Arizona–Utah border in order to return the river to its natural condition. Although the Glen Canyon Dam still stands, the gang members would be pleased to learn that dam removal is becoming a socially accepted form of river restoration across the U.S. (Grossman, 2002; Hart and Poff, 2002).

HOW TO RESTORE AN ECOSYSTEM

Restoration is a broad subject because any kind of ecosystem can potentially be restored or created. Some general technical principles are covered in the next sections, while procedures and policies are covered in the following sections.

The Energy Signature Approach

One of the fundamental principles in ecology is that each ecosystem type has a unique energy signature of sources, stresses, and other forcing functions. Thus, the first step in restoration or creation is to ensure that the appropriate energy signature is present on the site where restoration is to occur. Without this step, success of the restoration project is unlikely to occur. There are obvious examples of this approach, such as ensuring a source of water when attempting to create a wetland, but in other cases, detailed knowledge may be needed about the ecosystem. Brinson and Lee (1989) emphasized this approach for wetland restoration in stating "duplication of the energy signature of the replaced wetland is the most critical design consideration." The requirement of the appropriate energy signature is also fundamental when creating a microcosm model of an ecosystem as discussed in Chapter 4.

There are cases in which the whole restoration project revolves around restoring the energy signature itself. At least in a general sense this is true for the multibillion dollar effort to restore the Everglades in South Florida. Here the goal is to restore water flows through the subtropical savanna by reengineering roads, canals, and levees to allow water to pass more freely from Lake Okeechobee to Florida Bay and the Gulf of Mexico. While this single action will not completely restore this highly impacted landscape, it is the most critical aspect of the plan. Another classic case is the restoration of Lake Washington in the Puget Sound region of Washington State (Edmondson, 1991). This lake had been stressed by nutrient additions in secondarily treated sewage from the city of Seattle. These discharges took place through the 1940s and 1950s, until the sewage flows were diverted from the lake. Cultural eutrophication occurred due to the nutrient additions, turning the lake from an oligotrophic state with good water quality conditions to a eutrophic state with poor water quality conditions. Characteristics of the eutrophication process were reduced water clarity and blooms of the blue-green alga (*Oscillatoria rubescens*), which were stimulated by the nutrients. After diversion of the nutrients, the lake restored itself through self-organization, such that blooms disappeared and water clarity increased. Thus, the lake was restored simply by removing a source (i.e., nutrients in treated sewage) from the energy signature that was not characteristic of the natural lake conditions. Much of lake restoration involves this kind of approach as surveyed by Cooke et al. (1993). A final example of restoration through manipulation of the energy signature occurs with controlled flooding of Grand Canyon in Arizona. This is a case where restoration required the recreation of a disturbance (i.e., flooding) that was characteristic of the natural river ecosystem. The flood-pulse concept (Johnson et al., 1995; Junk et al., 1989) of rivers emphasizes the importance of annual flooding in affecting many physical–biological aspects of the river–floodplain system (see also Middleton, 2002). Flooding in Grand Canyon has been eliminated by the reservoir storage in Lake Powell (behind Glen Canyon Dam), which is located upstream from the canyon. Hydrology in the river is regulated by water storage in the reservoir and by steady low-flow releases through the dam for hydroelectric power generation. Lack of flooding has stressed the Colorado River in Grand Canyon National Park, especially by altering fluvial geomorphology and encouraging exotic plant species. An experimental flood was tested in 1996 and

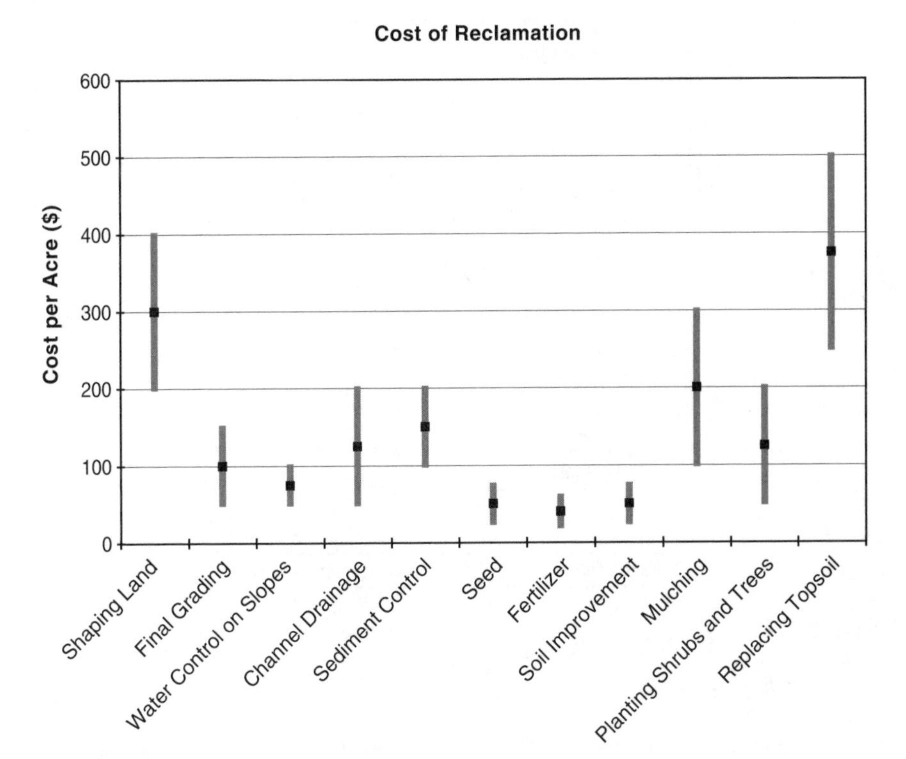

FIGURE 5.3 Costs of different aspects of strip mine reclamation. (Adapted from Atwood, G. 1975. *Scientific American.* 233(6):23–29.)

seems to have acted to restore certain natural conditions of the river ecosystem (Webb et al., 1999). Pulsing of energy sources is characteristic of many — perhaps all — ecosystems and was articulated in overview sense first by E. P. Odum (1971) in his pulse-stability concept (see also H. T. Odum, 1982; W. E. Odum et al., 1995; Richardson and H. T. Odum, 1981). Thus, full restoration may require pulsing disturbances that provide for periodic system rejuvenation as part of the energy signature. Middleton's (1999) excellent text on wetland restoration and disturbance dynamics supports this contention.

Although the examples described above focus on a single forcing function within an energy signature, most restoration involves multiple sources, stresses, etc. Figure 5.3 illustrates the many inputs to strip mine reclamation with cost data for different actions. Eleven costs are listed, ranging over an order of magnitude in cost per acre. This complex case is probably more typical of a restoration project with a diverse set of inputs required. In this particular case, it is interesting to note that restoration of soils and landforms has the highest costs, while inputs from seed and fertilizer are the lowest. This difference is indirect evidence of nature's scaling of values in a typical landscape. Soils and landforms represent storages that have developed over much longer time scales than the vegetation, which is restored with seed and fertilizers. Cost of restoration is thus directly proportional to the scale of the storage being restored.

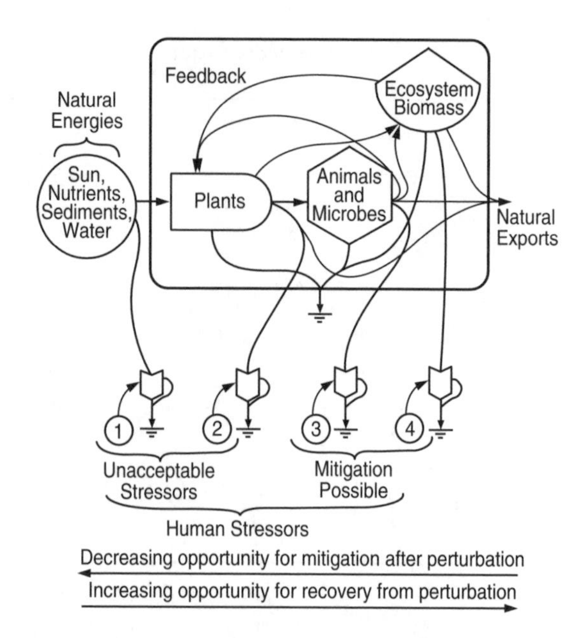

FIGURE 5.4 Energy circuit diagram depicting the role of different kinds of stress on ecosystems. (Adapted from Brown, S., M. M. Brinson, and A. E. Lugo. 1979. Gen. Tech. Rept. WO-12, USDA. Forest Service, Washington, DC.)

The energy signature approach also has the potential to clarify semantic problems between the different concepts in the field of restoration ecology, noted in the introduction to this chapter (restoration vs. recovery vs. reclamation vs. rehabilitation, etc.). Diagramming the energy signature and system structure in a restoration project provides clear notions of stressors and actions needed for mitigation. In this regard, the energy signature diagrams prepared by A. Lugo and his associates are especially instructive. Figure 5.4 from Brown et al. (1979) is an example showing a spectrum of different stressors and the relative difficulty involved in appropriate restoration actions. According to the hypothesis shown in the diagram, impacts directly involving or close to the primary energy sources are difficult to mitigate, while impacts far up the chain of energy flow have greater opportunity for recovery. Lugo and others produced a number of energy circuit diagrams illustrating this concept and a complete review of them is useful, especially for those learning this symbolic modelling language (Lugo, 1978, Figures 5 and 8; Lugo, 1982, Figure 3; Lugo and Snedaker, 1974, Figure 1; Lugo et al., 1990, Figure 4.9).

The energy signature approach emphasizes a systems perspective, but a somewhat similar approach has evolved which portrays inputs or factors necessary to support a particular species. This species-oriented approach attempts to quantify the quality of a site for a particular species based on assessments of key elements. It involves the calculation of a habitat suitability index (HSI) in a way that is reminiscent of an engineering design equation. Habitat is a critical concept in ecology and refers to a place that provides the life needs (food, cover, water, space, mates, etc.) of a species (Hall et al., 1997; Harris and Kangas, 1989). In this sense, there is one optimal habitat for each species, just as there is one energy signature for each ecosystem. Historically, habitat was a qualitative concept that was best understood after long natural history study. The HSI model approach was developed by the U.S. Fish and Wildlife Service in order to formalize the habitat concept and to create a quantitative tool for field personnel to evaluate the conservation value of sites. An HSI model is an algorithm that is solved to calculate a numerical index that ranges from 0.0 to 1.0, with 0.0 representing unsuitable habitat and 1.0 representing optimal habitat, always relative to a particular species. More than 160 HSI models were developed, mostly in the 1980s, each of which is a very interesting synthesis of scattered information about a species. The algorithm of an HSI model consists of a series of graphical assessments of individual environmental factors that are combined in an equation to calculate the index value. For example, the HSI for muskrats (Allen and Hoffman, 1984) includes nine separate relationships dealing with hydrology and marsh vegetation that are used for calculating the quality of the habitat. In one sense, this is another expression of the niche of the muskrat. Reviews of the HSI model concept and other habitat evaluation approaches are given by Garshelis (2000) and Morrison et al. (1992), and a problem solving exercise on the HSI is given by Gibbs et al. (1998). In conclusion, the HSI model is useful in species-specific restoration ecology because it indicates the key factors that must be restored or created for a particular species. It also is useful in the larger context of ecological engineering because it represents an approach that can be used, as could be a design equation in traditional engineering, when restoring a habitat.

Biotic Inputs

Although there are many inputs to restoration projects, the genetic inputs in the biota that are planted or introduced are usually the primary emphasis (even though they are not always the most costly). These inputs actually are part of the energy signature of the project but they are practically never considered in energy units. Within the biological realm, focus is most often on higher plants. This is appropriate because plants almost always provide the three-dimensional structure of an ecosystem and are necessary for full ecological development of a site.

Active planting is not a particularly complex task per se, but a great many options and considerations are involved (Table 5.2). The basic decisions are (1) what species to plant, and (2) what structure or life form to plant. A broad knowledge of natural history and ecology is useful for making these decisions. For example, there are advantages and disadvantages to seeds vs. transplanting juvenile or adult plants, depending on the species and the site conditions. Experience is the best guide to successful planting programs (Erickson, 1964), and a number of useful texts have been published as aids, such as given by Galatowitsch and van der Valk (1994), Kurtz (2001), and Packard and Mutel (1997b). Practical experience on successful planting approaches is accumulating because of the large number of restoration projects that are taking place in all kinds of ecosystems. Some projects are conducted by the large industry of environmental consultants who work mostly on legally mandated programs (such as strip mine reclamation or wetland mitigation) while others are volunteer efforts which are often local or community-based. A side result

TABLE 5.2
Planting Approaches and Considerations

Direct Seeding
Seed preparation
Breaking seed dormancy
Planting time
Seeding rate
Seeding rates and competitive interference
Planting very low seeding rates
Seeding depth
Drill seeding
Interseeding
Broadcast seeding
Seed bed requirements
Aerial seeding
Hay mulch seeding
Cultipacker-type seeding
Hydroseeding
Transplanting
Planting densities for trees and shrubs
Wildings (plants from natural settings)
Sod
Bare-root stock
Container-grown stock
Cuttings
Sprigs

Source: Adapted from Whisenant, S. G. 1999. *Repairing Damaged Wildlands*. Cambridge University Press, Cambridge, U.K.

of this surge in restoration ecology has been the development of commercial nurseries that provide both plants and information on how to do restoration. An excellent example that cuts across several of these areas is Environmental Concern, Inc. of St. Michaels, MD, which is run by Edward Garbisch. Environmental Concern includes a commercial nursery, a consulting firm, and a nonprofit educational component. Garbisch himself is one of the pioneers in wetland restoration and ecological engineering (see Chapter 4 in Berger, 1985), and his company has published a variety of useful materials on wetlands including a planting guide (Thunhorst, 1993), a curriculum plan for teachers (Slattery, 1991), and a scientific journal.

Despite the experience that is accumulating, planting programs often fail when the species that are planted die or do not contribute significantly to the restored ecosystem in the long run. Failures range across the gradient from large to small projects. An example at the large scale was the U.S. Army Corps of Engineers planting project at Kenilworth Marsh in Washington, DC. Here approximately 30 acres (12 ha) of tidal freshwater marsh was planted at a cost on the order of hundreds of thousands of dollars. A well-developed marsh ultimately self-organized on the site but the intentionally planted species made up a relatively minor part of the plant community after 5 years (Hammerschlag, personal communication). On a small scale, for example, Shenot (1993; Shenot and Kangas, 1993) described the results of plantings at three stormwater wetland sites in central Maryland. Eight species were intentionally planted but they made insignificant contributions (less than 12% of the total density and less than 1% of the total diversity at each site) to the plant communities 3 to 5 years after planting. Lockwood and Pimm (1999) reviewed 87 published studies of restoration projects (mostly wetlands or prairies) for success or failure. They found 17 failures, 53 partial successes, and 17 successes. However, their review is biased because it considered only published studies. Many failures probably go unpublished because they would have to report negative results. Of course, failures are important opportunities to learn (see Chapter 9), and the publication of negative results should be especially encouraged in the field of restoration ecology.

One cause of failure in plantings is predation by species of herbivores that are attracted to the restoration sites. Plants in natural ecosystems have a number of defenses against herbivores, such as spines or chemical deterrents, which limit herbivory to on the order of 10% of net primary productivity. Exceptions occur, such as muskrat eat-outs in marshes (see Chapter 2) and insect outbreaks, but these cases are relatively rare. Restored sites represent new ecosystems which must self-organize to conditions different from those experienced by natural ecosystems. One expression of this self-organization is the emergence of new food chains, which may be undesirable to the restoration ecologist. Some of the best examples are herbivory of wetland plants by Canada geese (*Branta canadensis*) and of terrestrial plantings by white-tail deer (*Odocoileus virginianus*). The magnitude of herbivore impact was demonstrated by May (in preparation) in his study of freshwater tidal marsh restoration at Kenilworth Marsh mentioned earlier and at other sites along the Anacostia River. He enclosed some plots with fence to keep herbivores away from marsh plants (exclosures) and left other plots with no fencing as controls. In certain areas of the marsh all vegetation was eaten by herbivores (primarily Canada geese), except those plants protected within the exclosures (Figure 5.5). This kind of study demonstrates the power of herbivory to determine success or failure in restoration plantings. Whisenant (1999) describes techniques for protecting plants such as chemical repellents and protective tubes. Extra cost is required to protect plantings, but it is sometimes necessary as a safeguard against project failure.

Failures in planting projects sometimes are due to lack of accountability. Enough projects have been conducted that common causes of failure (such as from herbivory) should be able to be avoided. Some consulting firms who contract for restoration work now guarantee plantings against failure, which is an encouraging indication of the evolution of the field. However, large sums of money are still being wasted in planting programs destined to fail. This money could surely be better invested for conservation purposes, and restoration ecologists must always include this kind of economic perspective in their work.

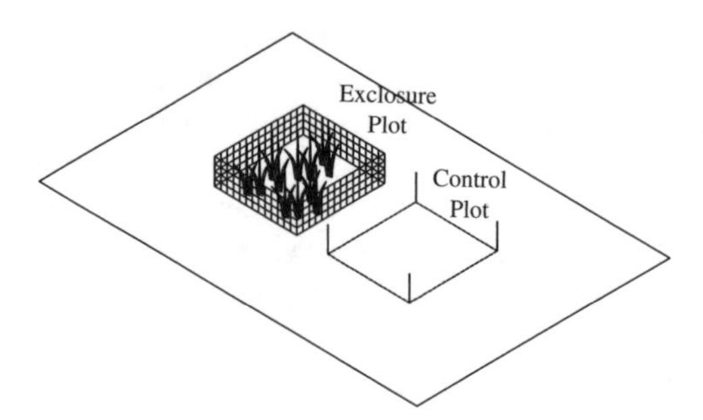

FIGURE 5.5 Experimental design to test for the effects of herbivores on marsh vegetation restoration on a mud flat (Adapted from May, P.I., in preparation.)

Often ignored in restoration projects are the free biotic inputs from nearby ecosystems. These are usually seeds that disperse into the site, germinate, and become established. A common problem in restoration ecology is to focus solely on the intentional plantings and to overlook the "volunteer" species that emigrate from the surrounding landscape. These volunteers have also been called *spontaneous species* (Fraisse et al., 1997; Prach and Pysek, 2001) because they spontaneously appear at a site even though they were not intentionally planted. In many cases these kinds of species come to dominate the site. MacLean (1996; MacLean and Kangas, 1997) was able to split a wetland mitigation site in central Maryland into four experimental cells in which three strategies of plantings were tested: low diversity intentional planting (11 species) of native wetland species typical of local mitigation projects; high diversity intentional planting (132 species) of native wetland species and others; and natural colonization without any intentional planting. The high diversity case emphasized the multiple seeding approach in an attempt to remove seed source as a possible limiting factor to plant community development. The observed plant species richness after two growing seasons is shown in Table 5.3. Some of the intentionally planted species were observed but volunteer species dominated the diversity in all of the experimental cells. This result was even more pronounced in terms of stem density counts from permanent plots at the site (Table 5.4). Facultative (FAC and FACW) and obligate (OBL) wetland species dominated all of the cells in terms of observed species and in three of the four cells in terms of numbers of individuals. Since the presence of these species is an indicator of success for wetland creation, it is interesting to note that the cell which received no intentional plantings had the highest number of wetland species (Table 5.3) and the highest number of wetland individuals (Table 5.4) of all of the experimental cells. In this case, as in many others, the surrounding landscape provided a subsidy to the restoration project through dispersal of a high diversity of species at no cost to the humans conducting the restoration. This kind of result suggests restoration ecologists are either arrogant or naive in thinking that the set of species they have chosen for intentional plantings is the most appropriate for a site. Natural selection often demonstrates that the intentional plantings are incorrect and that volunteer species from seed sources in the surrounding landscape are competitively superior. Unfortunately, knowledge of natural recruitment is not well enough developed to reliably

TABLE 5.3
Comparison of Vegetation Development under Different Restoration Planting Treatments: Species Richness

Seeding	Cell 1 Low Diversity Seeding	Cell 2 Low Diversity Seeding	Cell 3 Natural Colonization	Cell 4 High Diversity
Number of introduced species	11	11	0	132
Number of introduced species observed after 2 years	4	8	—	38
Total number of species observed after 2 years	80	83	43	99
Percent of the observed community that is made up of facultative or obligate wetland species	75	72	86	72

Note: Data are for the number of species that were observed in the mitigation wetland cells.

Source: Adapted from MacLean, D. and P. Kangas. 1997. *Proceedings of the 24th Annual Conference on Ecosystems Restoration and Creation*. Hillsborough Community College. Plant City, FL.

TABLE 5.4
Comparison of Vegetation Development under Different Restoration Planting Treatments: Stem Density

Seeding	Cell 1 Low Diversity Seeding	Cell 2 Low Diversity Seeding	Cell 3 Natural Colonization	Cell 4 High Diversity
Percent of the sampled community after 2 years that was originally introduced	0.8	4.5	—	35.6
Percent of the sampled community after 2 years that colonized naturally	99.2	95.5	100	64.4
Percent of the sampled community after 2 years that was made up of facultative or obligate wetland species	71.9	75.3	95.8	39.0

Note: Data are based on the number of plants that were sampled in 11 quarter meter square quadrants in each of the mitigation wetland cells.

Source: Adapted from MacLean, D. and P. Kangas. 1997. *Proceedings of the 24th Annual Conference on Ecosystems Restoration and Creation*. Hillsborough Community College, Plant City, FL.

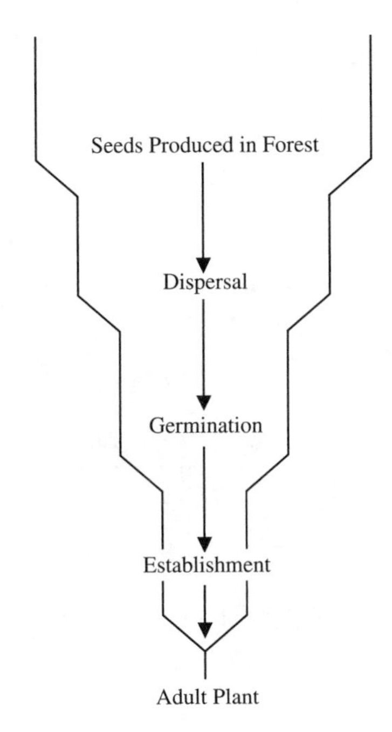

FIGURE 5.6 Sequential losses of individuals during the recruitment process for plants. (Adapted from Uhl, C. 1988. In E. O. Wilson (ed.). *Biodiversity.* National Academy of Sciences, Washington, DC.)

predict the quantity or quality of biotic inputs of volunteer species, and this probably explains the continued inefficient reliance on intentional plantings in restoration projects.

Recruitment of species through natural dispersal involves several processes. This is sometimes termed *supply side* ecology, using an economic metaphor because it involves the rate of production of individuals (i.e., the supply) at the site (Fairweather, 1991; Roughgarden et al., 1986; Underwood and Fairweather, 1989; Young, 1987). Figure 5.6 illustrates the processes involved for a plant species, showing the sequential reduction in numbers of initially available individuals as seeds relative to the number that ultimately become established as adult plants. Of course, the seed life stage is initially critical. Seed ecology of a site involves a number of aspects including seed budgets (see, for example, Kellman, 1974) and seed banks (Leck et al., 1989; Roberts, 1981). Understanding flows of seeds in dispersal is important when considering free inputs to a restoration site (Chambers and MacMahon, 1994), but storages in seed banks are also being actively manipulated in restoration projects (Brock and Britten, 1995; Maas and Schopp-Guth, 1995; van der Valk et al., 1992). Access to naturally occurring seed sources is an important design issue in any restoration project and inputs of volunteer species can be a significant free subsidy to a project.

There are then two sources of biotic inputs in any restoration project: intentional, artificial plantings (either seeds, juveniles, or adults) and natural colonization through

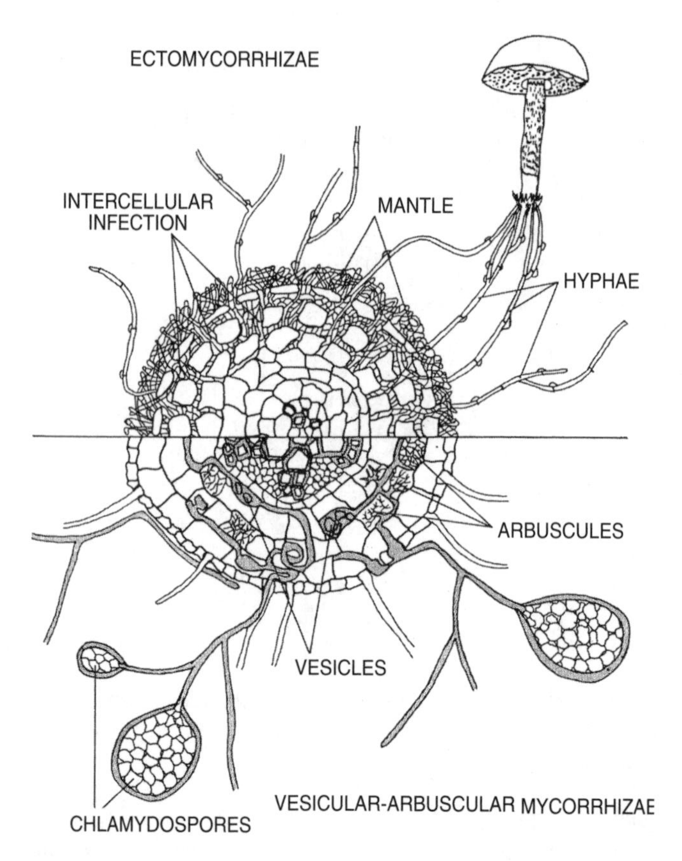

FIGURE 5.7 Cross section of an idealized mycorrhizal fungus showing both VA and ectomycorrhizal feature. (From Whitford, W. G. and N. Z. Elkins. 1986. *Principles and Methods of Reclamation Science. With Case Studies from the Arid Southwest.* C. C. Reith and L. D. Potter (eds.). University of New Mexico Press, Albuquerque, NM. With permission.)

seed dispersal. Either the intentional plantings or the natural colonization may dominate the final plant community, but a generalization seems to be emerging that species arriving through natural colonization are more successful when seed sources are near than those intentionally planted by humans. Success of any planting program is determined by natural selection operating on the total biotic input to a site. Thus, the biota self-organizes into a community that exists until conditions change. This process is often called *self-design* in restoration ecology in recognition of the fact that nature ultimately determines the composition of restored or created communities. Because nature rather than humans selects successful species and because intentional plantings are often expensive, the rationality of planting programs is being examined with greater scrutiny. The question of whether "to plant or not to plant" is being asked (Harmer and Kerr, 1995; Kentula et al., 1992) and self-design is being evaluated as a viable restoration strategy more widely (Middleton, 1999; Whisenant, 1999). William Mitsch is a leader in this effort for wetlands (Metzker and Mitsch, 1997; Mitsch, 1995b, 1998a, 2000; Mitsch and Cronk, 1992; Mitsch and Wilson, 1996; Mitsch et al., 1998) and his long-term, system-wide studies may be the most effective way to determine the optimal planting strategy.

A final consideration concerning biotic inputs is mutualism or symbiotic relationships between organisms. Mutualisms can be critical to the successful establishment of certain species. Animals in particular may play roles in this context when their actions are necessary for plant survival (Handel, 1997; Majer, 1997). Enhancing bird use of a site by providing perches is one example that increases dispersal of certain plant species (McClanahan and Wolfe, 1993; Robinson and Handel, 1993). Perhaps the most important mutualism in regard to restoration is the relationship between certain plants and mycorrhizal fungi. This mutualism occurs in the roots (Figure 5.7), and *mycorrhizae* literally means "fungus root." There are two economically important types of these fungi: ectotrophic and endotrophic (vesicular–arbuscular), which differ in their morphology. The fungi acquire all of their carbon for nutrition from the plant and, in return, they aid in nutrient uptake. For both kinds of mycorrhizae, the thallus is located within the cortex of the root, but most of the fungal biomass is in hyphal threads that grow into the surrounding soil. Ectotrophic mycorrhizae directly contribute to the breakdown of soil organic matter, while endotrophic mycorrhizae are especially efficient at nutrient uptake. It is well known that mycorrhizae stimulate host plant growth, and they have even been considered to be keystone species because of this role (Lodge et al., 1996). Their function in restoration ecology is reviewed by Haselwandter (1997), Miller (1987), and Miller and Jastrow (1992). Although strong mutualistic relationships between species such as mycorrhizae are relatively uncommon in nature, E. P. Odum (1969) suggests that they are characteristic of mature ecosystems. Thus, mutualisms should be encouraged in restorations, and their presence is an index of a successful project, according to E. P. Odum's criteria.

Succession as a Tool

Succession is the process through which ecosystems develop over time (see Figure 4.3 in Chapter 4). As such it is one of the fundamental concepts in ecology (Golley, 1977; McIntosh, 1981). Disturbance is the normal trigger for succession to begin, and different kinds of succession are recognized (primary vs. secondary), depending on the degree to which the ecosystem is set back in the development process. Species abundances change sequentially as succession proceeds because no species is adapted to the full range of environmental conditions that occur at a site from the early pioneer stages through the later, mature stages. Classifications of species strategies in relation to succession have been proposed such as r- vs. K-selection (MacArthur and Wilson, 1967; Pianka, 1970), where the letters refer to coefficients in the logistic population growth equation (see Eq. 3.4). The r-selected and K-selected species form ends of a gradient of adaptation in this theory. The r-selected species have short life-expectancy, large reproductive effort, and low competitive ability, while K-selected species have the opposite: long life-expectancy, small reproductive effort, and high competitive ability. Thus, in relation to the logistic equation, r-selected species emphasize high reproductive rates and are likely to occur in early succession when resources are not limiting. K-selected species emphasize high competitive ability, which is important when resources become limiting, as occurs in later successional stages. Applications of this theory have been criticized, but it is still elegant and useful as a generalization. Grime (1974, 1979) offered a slightly more complicated classification for understanding species strategies: competitive (similar to *K*-selected), ruderal (similar to *r*-selected), and stress-tolerant. His classification is especially significant because of the distinctions drawn between the concepts of stress and disturbance. According to Grime, stress is a forcing function that effects production, while disturbance is a forcing function that effects biomass. Dominance of either stress or disturbance leads to different life history patterns in a predictable fashion. MacMahon (1979) provides a model for different plant life forms in relation to Grime's classification.

A rich variety of life history classifications exists in the literature, sometimes with quite evocative names attached to different strategies: "spenders" vs. "savers" (During et al., 1985), "fugitives" (Hutchinson, 1951; Horn and MacArthur, 1972), "gamblers" vs. "strugglers" (Oldeman and van Dijk, 1991), "bet-hedgers" (Stearns, 1976), and "supertramps" (Diamond, 1974). Van der Valk's (1981) classification is particularly detailed for freshwater wetland plants. Twelve basic life history types are recognized based on three key traits (life span, propagule longevity, and propagule establishment requirements). This classification was developed during long-term studies of succession in prairie wetlands and has been advocated for use as a basis for wetland restoration (Galatowitsch and van der Valk, 1994; van der Valk, 1988, 1998). Whigham (1985) also has successfully applied van der Valk's approach to understanding vegetation in treatment wetlands. Clearly, knowledge of life history patterns can significantly improve restoration plans by aiding in making appropriate choices of species for intentional plantings. Other important references on life history and succession are given by Huston and Smith (1987), Noble and Slatyer (1980), and Whittaker and Goodman (1979).

Succession can be considered both at the population scale, as noted above in terms of species strategies, and also at the ecosystem scale where patterns of change in nutrient cycling and energy flow take place over time. E. P. Odum's (1969) summary is a good introduction to ecological change at both scales.

There is a direct connection between succession and restoration because both concern ecosystem development over time. Some restoration ecologists, especially those who work in terrestrial systems, hold the view that the goal of restoration is to accelerate succession (Bradshaw, 1987) or to otherwise shorten it (MacMahon, 1998). In this sense, succession is used as tool for restoration efforts. Kangas (1983a,b) examined this idea with a simulation model of succession as applied to strip mine reclamation for phosphate mines in central Florida. The model included three stages of succession characteristic of the southeastern U.S. (Figure 5.8) with grass as the pioneer stage, pine trees as the intermediate stage, and hardwoods as the mature or climax stage. Transitions between stages were controlled by shading and the development of a litter layer that regulated seed germination. Figure 5.9 compares the standard run of the model without manipulation to a simulated run with high amounts of seeding and litter addition, as might occur in restoration efforts. In this case, the time to the mature, climax stage of succession was reduced by one half, from 60 years in the standard run to 30 years in the simulation. This type of

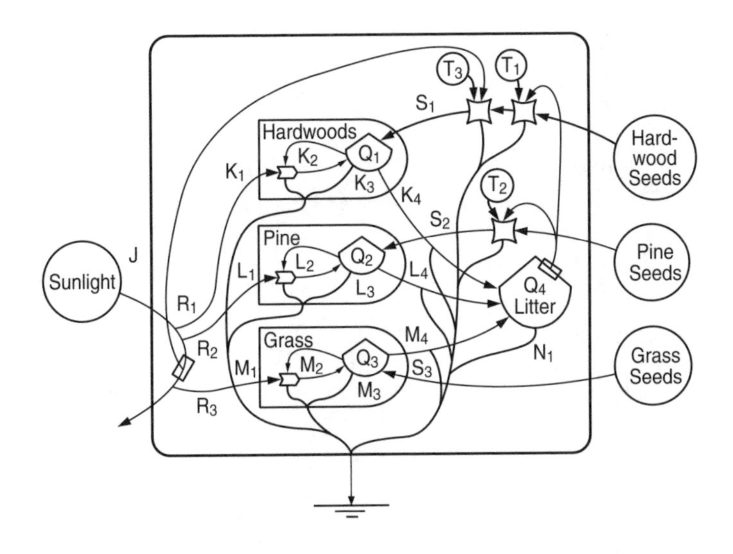

Equations for the storages are given below:

$$\dot{Q}_1 = K_2Q_1R_1 - K_3Q_1 - K_4Q_1(+S_1 \text{ IF } R_2 < T_3 \text{ AND } Q_4 < T_1) \quad (1)$$

$$\dot{Q}_2 = L_2Q_2R_2 - L_3Q_2 - L_4Q_2(+S_2 \text{ IF } Q_4 > T_2) \quad (2)$$

$$\dot{Q}_3 = M_2Q_3R_3 - M_3Q_3 - M_4Q_2 + S_3 \quad (3)$$

$$\dot{Q}_4 = K_4Q_1 + L_4Q_2 + M_4Q_3 - N_1Q_4 \quad (4)$$

$$R_1 = J/(1 + K_1Q_1) \quad (5)$$

$$R_2 = R_1/(1 + L_1Q_2) \quad (6)$$

$$R_3 = R_2/(1 + M_1Q_3) \quad (7)$$

FIGURE 5.8 Energy circuit model of succession on abandoned phosphate mines in Florida. (Adapted from Kangas, P. 1983b. *Analysis of Ecological Systems: State-of-the-Art in Ecological Modelling*. W. K. Lauenroth, G. V. Skogerboe, and M. Flug. (eds.). Elsevier, Amsterdam, the Netherlands.)

work suggests that succession can be managed to reduce cost of restoration projects and to increase the ecological value of the resulting systems. The idea of using succession as a tool is to take a systems perspective to restoration. Thus, the goal is "to plant a forest, not trees." In other words, a mature, complex ecosystem is the result of multiple successional stages at a site over time, and it is difficult and costly to skip these stages in restoration. Knowledge of successional history is fundamentally important for understanding and restoring complex ecosystems. Luken (1996) provides a summary of the use of succession as a tool with many examples of strategies related to ecosystem restoration and creation.

While knowledge of succession is clearly useful in restoration ecology, it may have another, more abstract use that is related to engineering. This is the idea of succession as a form of computation and therefore as an abstract tool for problem solving. Several concepts of biology have acted as guides or models for computa-

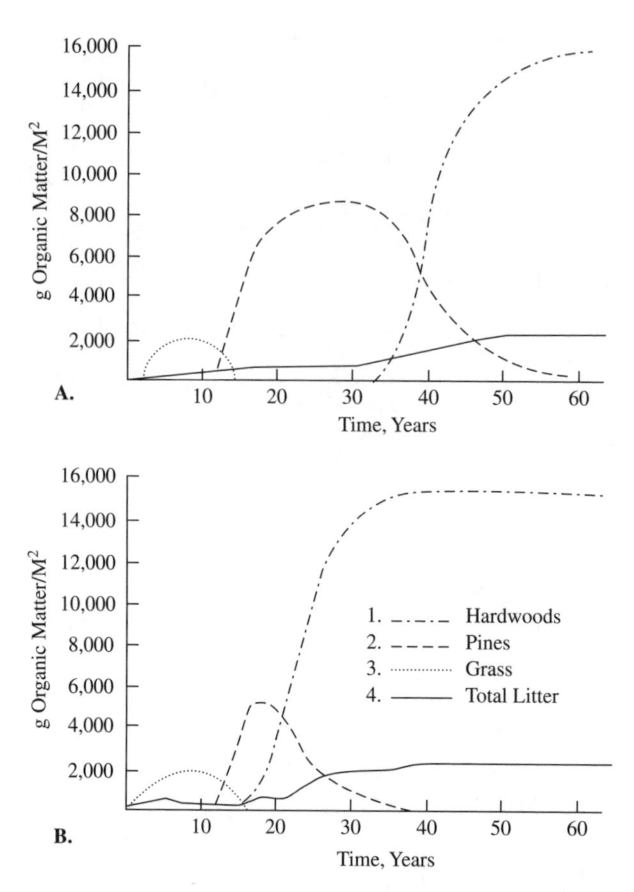

FIGURE 5.9 Comparison of simulation runs of the phosphate mine simulation model from Figure 5.8. (A) Standard run. (B) Result of increasing the seeding rate 1000 times and adding litter. (Adapted from Kangas, P. 1983b. *Analysis of Ecological Systems: State-of-the-Art in Ecological Modelling*. W. K. Lauenroth, G. V. Skogerboe, and M. Flug. (eds.). Elsevier, Amsterdam, the Netherlands.)

tional development, and succession is likewise a possible candidate (Table 5.5). The algorithmic or recursive nature of succession suggests this use. Succession is often portrayed with flowchart diagrams (Figure 5.10) that perhaps could be the basis for computational development through some form of translation. The key to this use is to understand what kinds of problems that the succession algorithms might solve. Evolution has proven to be a very robust model which has been used as a basis for several kinds of evolutionary computation, especially based on optimization (Fogel, 1995, 1999). However, evolution solves different problems than succession. Perhaps the traveling salesman problem is a model for the type of problem that succession solves. This is a kind of minimum-distance problem where the salesman in the metaphor has to find the shortest possible path between a number of towns, each of

TABLE 5.5
Areas of Computational Biology

Biological Analog	Computational Expressions	References
Evolution	Genetic algorithms	Goldberg, 1989; Mitchell, 1996
	Artificial life	Langton, 1989; Levy, 1992
	DNA computers	Lipton and Baum, 1996
Intelligence	Artificial intelligence	Feigenbaum and Feldman, 1963
	Computational neuroscience	Schwartz, 1990; Von Neumann, 1958
Social insect behavior	Distributed programming	Bonabeau et al., 1999
Immunology	Computer security programs	Dasgupta, 1999
Succession	Successional computation	This text

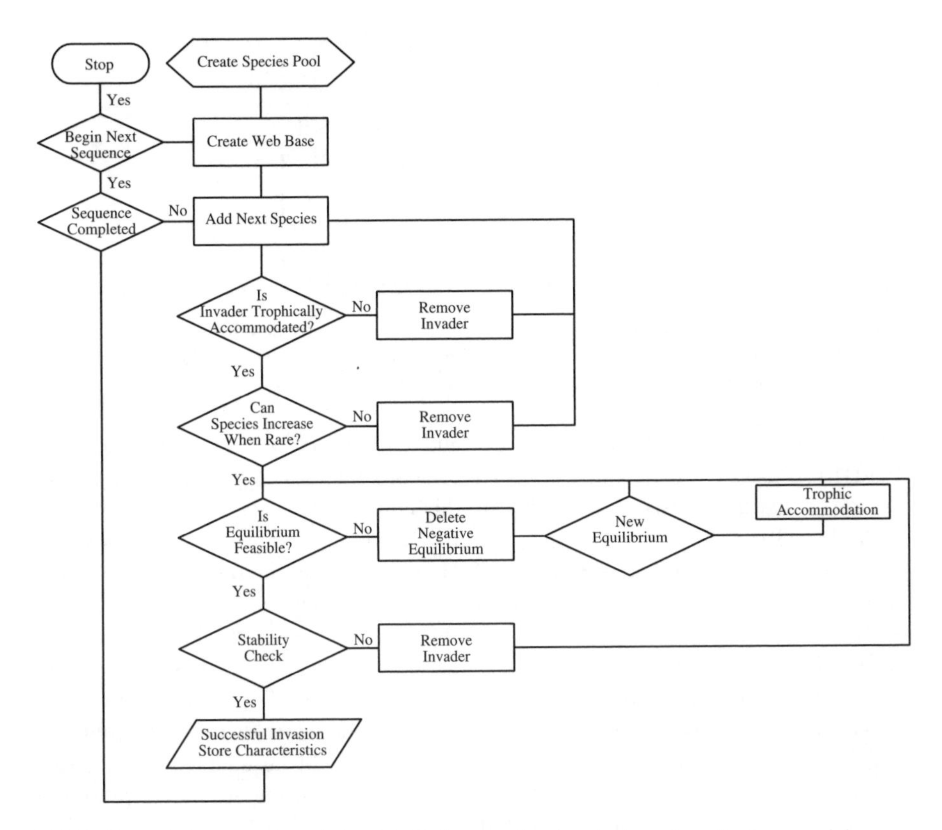

FIGURE 5.10 A successional algorithm diagram for developing diversity in a model community. (Adapted from Drake, J. A. 1990a,b. *TREE* 5:159–164.)

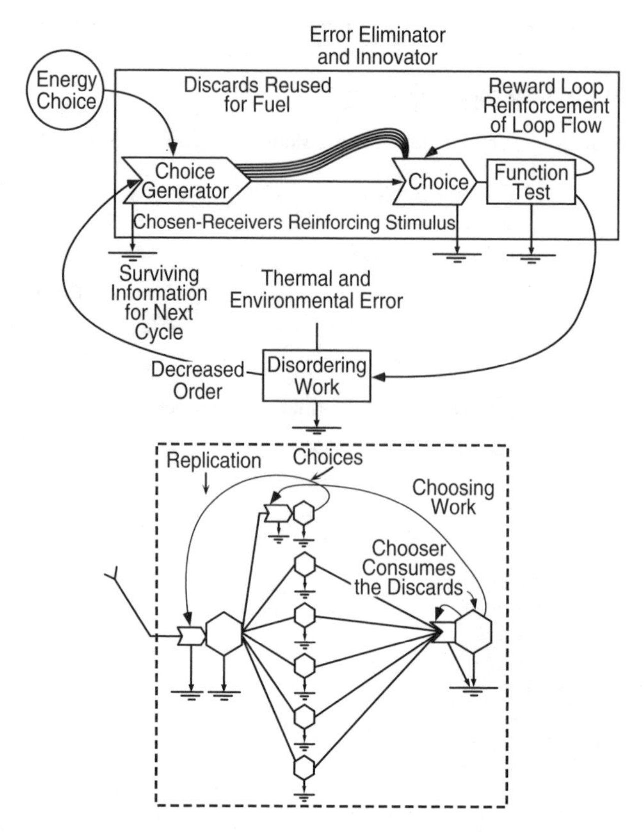

FIGURE 5.11 Two different views of the loop reinforcement model. This model represents the self-organization process in succession. (From Odum, H. T. 1971. *Environment, Power, and Society*. John Wiley & Sons, New York. With permission.)

which must be visited only once (Haggett and Chorley, 1969; Lowe and Moryadas, 1975). The different stages that succession passes through might be analogous to the towns that the salesman must visit. In this context the succession diagram with multiple pathways in the old ecological literature might provide a library of possible solutions to minimum-distance problems. However, these diagrams only show the successful links, and it may be necessary to have knowledge about links that have been selected against (i.e., towns not visited by the salesman or possible successional stages that don't occur). The travelling salesman problem is addressed with ant colony behavior by Dorigo and Gambardella (1997).

What is being suggested here is not a simulation model, such as shown in Figure 5.8, but rather a more generalized algorithm that could be adapted for abstract problem solving. H. T. Odum's (1971) loop reinforcement model may represent a possible starting point because it includes both a feedback phase and a selection phase, like evolution or learning (Figure 5.11). The quote listed below provides a summary of H. T. Odum's (1971) concept:

> Consider the central principle of self-design, which is often misunderstood and opposed as nonmechanical teleology by those who do not understand the network nature of the environment. Systems readily develop towards their successful purpose by a process which essentially may be the same essence as thinking of the brain. Systems have purpose just as people do, for both are highly mechanical and readily understood as causal processes. The self-organizing process by which a system develops a network of insulated mineral and food pathways is a special case of a process that may be termed in circuit nomenclature as "loop reinforcement" … .
>
> If the various possible pathways which are first attempted by organisms invading or evolving in a place are greater in number and variety than those which can emerge finally on the available energy budget, the ones which will prevail will be those that have a positive feedback loop since these are reinforced by resources which are drained away from those circuits not receiving loop reinforcement. In other words, the processes believed to occur in learning within an organism and the process of organizing an ecosystem are essentially the same … An ecosystem is learning when it is under successional development.

Information about succession is stored in the collective trophic and life history strategies of species that exist in the seed sources and seed banks of the landscape. This information is transmitted through time as succession proceeds and is a template for future successions. Margalef (1968) outlined similar mechanisms in his discussion of succession.

The goal in the computational effort proposed above is to develop the concept of the ecosystem as a computer. H. T. Odum (1971) briefly outlined this perception when he wrote a short section entitled *An Ecosystem as Its Own Computer*. His main thrust was to develop simulation models, but a new kind of network epistemology can be seen to emerge from his work (Kangas, 1995). Michael Conrad (1995; Conrad and Pattee, 1970) also has contributed to this work and suggests alternative approaches. The notion of the ecosystem as a computer is the ultimate in the machine analogy (see Chapter 7).

If succession can be harnessed as a form of computation, it might open a whole new area of computational biology. Perhaps the next generation of ecological engineers who learn enough about both engineering and ecology can bridge the present gaps in knowledge and will be able to develop this possibility.

Bioremediation

In some cases restoration may take the form of bioremediation. This approach covers any system that utilizes natural, enhanced, or genetically engineered biological processes to alleviate a pollution problem (Cookson, 1995). In practice, bioremediation usually refers to microbial systems (primarily bacteria and/or fungi) that degrade the pollutant through biological metabolism (i.e., biodegradation). Thus, the pollutant becomes part of the energy signature for these systems. The microbiologist Martin Alexander (1973, 1981) was the first to outline the use of microbes for bioremediation of pollution sources. He put forward the principle of microbial infallibility which states that no natural organic compound is totally resistant to biodegradation, given the appropriate environmental conditions (Alexander, 1965). This is an important idea that is fundamental to all natural ecosystems in relating to biogeochemical cycling. The principle might be restated by saying that nature always recycles. Alexander recognized that some man-made compounds (sometimes termed *xenobiotics*) resist biodegradation and consequently persist and accumulate in the environment. These are considered to be recalcitrant due to their chemical structure in terms of molecular form and bond sequences. Recalcitrant compounds are of special interest to biochemists because their chemical structure is so exotic that microbes lack enzymes to break them down. Overcoming the barriers to biodegradation of recalcitrant compounds is a primary goal of bioremediation, and Alexander (1994) believes microbial metabolism ultimately can be managed or engineered for this purpose.

The first application of bioremediation that is widely recognized as being successful occurred in 1989 at the Exxon Valdez oil spill in Alaska [Office of Technology Assessment (OTA), 1991]. Oil from the tanker contaminated more than 100 miles (160 km) of beaches along Prince William Sound. Bioremediation was tested by adding fertilizer to the contaminated beaches in order to stimulate natural microbes. Biodegradation was accelerated as much as fourfold over control beaches that did not receive fertilizers. The results of this experiment encouraged much work on bioremediation throughout the 1990s. The two basic methods of bioremediation involve *in situ* (on site) and *ex situ* (in bioreactors) applications. *In situ* methods work for low levels of contaminants and include two kinds of additions to the environment. Biostimulation involves the addition of nutrients which otherwise limit biodegradation by indigenous microbes. Bioaugmentation involves the addition of microbes to the site for cases where the local microflora lacks appropriate species to carry out biodegradation. *Ex situ* methods work for high levels of contaminants where more control over environmental factors such as temperature and pH is necessary. A final method that can be carried out either *in situ* or ex situ is the use of microbes that have been genetically engineered for enhanced biodegradation. This approach is highly regulated because of risks associated with introducing these exotic species to the environment (see Chapter 7). One idea that is being studied as a countermeasure is to engineer self-destruct genes into the genetically engineered microbes so that after they break down a pollutant, they will die off and not become invasive. However, use of genetically engineered species is still experimental and not yet a major factor in commercial applications of bioremediation.

Phytoremediation relies on plants for pollutant cleanup (Brown, 1995; Susarla et al., 2002). The primary mechanism is uptake by roots and incorporation into biomass, though other techniques involving oxygenation of the rhizosphere also are used. Some examples of species used in phytoremediation were mentioned in Chapter 2 in relation to sewage treatment: *Lemna* (duckweed), *Typha* (cattail), and *Eichhornia* (water hyacinth). This approach is best developed for a special class of plants called *hyperaccumulators* (Brooks, 1998). These plants take up and store much higher concentrations of heavy metals compared with normal plants. Examples include mustard plants (*Brassica* sp.) and sunflowers (*Helianthus* sp.). The method is to grow plants in contaminated soil or water and to harvest their biomass as a way to concentrate and remove the pollutants. This is the same approach used in the algal turf scrubber technology described in Chapter 2. The harvested biomass which is now contaminated must be disposed of either by landfilling or by incineration. Genetic engineering is being studied for enhancing phytoremediation potentials but existing applications rely on plants that are naturally preadapted for high uptake rates. Phytoremediation is a relatively new approach for pollutant cleanup and new candidate species are being sought. The spring ephemerals of temperate zone deciduous forests are a group that might lend themselves to phytoremediation. These plants grow quickly and complete their life cycles during the spring time, for the most part before the overhead canopy of trees leafs out. Familiar species of the eastern U.S. include spring beauties (*Claytonia* sp.), mayapple (*Podophyllum peltatum*) and jack-in-the-pulpits (*Arisaema* sp.). These wildflowers are a diverse group with significant nutrient uptake capacity due to their fast growth (Blank et al., 1980). In fact, Muller and Bormann (1976) suggested that spring ephemerals act like a "vernal dam" in absorbing nutrients that might otherwise be lost to the forest nutrient cycle due to leaching by snow melt and spring showers. Could the spring ephemerals be used in some kind of horticultural design for phytoremediation? Another candidate system might be the tropical rain forest subsystem of tree roots and fungal mycorrhizae that carry out direct recycling in the litter layer and upper soil layers. The direct-recycling hypothesis was put forth by Went and Stark (1968a, 1968b) from observations made in an Amazonian rain forest. The idea is that nutrients tend to cycle directly in the tropical trees from decomposing litter back into roots without passing through the mineral soil. The mycorrhizae act as "nutrient traps" by contributing to both decomposition and nutrient uptake (see Figure 5.7). This adaptation is important in tropical rain forests because leaching due to high rainfall can cause rapid removal of nutrients from the rooting zone of the soil. The Went and Stark hypothesis has been supported by experimental work (Herrera et al., 1978; St. John, 1983; Stark and Jordan, 1978) and indications of direct recycling have been found more widely in forests outside of the tropics (Fogel, 1980).

Other, less well-known systems also fall under the heading of bioremediation. For example, John Todd adapted his living machine concept (see Chapter 2) to create a lake restorer system (Todd, 1996a). This is a living machine that floats on a raft on a water body and acts to improve water quality by recirculating water through the system. Water is pumped on to the raft, where it flows through the living machine and then it is discharged back to the water body. Treatment takes place by the same kinds of processes that occur in a treatment wetland (see Table 2.3). An interesting feature of lake restorers is their autonomy. Todd has used a windmill to provide power to run the pump on his systems and a group of University of Maryland students (Yaron et al., 2000) has used solar power (Figure 5.12). The autonomy of these systems means that once created, they theoretically can act independently and with little maintenance. Development of lake restorers is still in early stages but they represent a very interesting state-of-the-art design in ecological engineering, especially because of their potential for autonomous behavior and self-organization. The largest lake restorer built by Todd's group is located in Berlin, MD, on the eastern shore of the state (Shaw, 2001). This system provides final treatment of wastewater from a poultry processing plant. The restorer is located in a lagoon that has been formed into a meandering channel by the installation of curtains of artificial textile

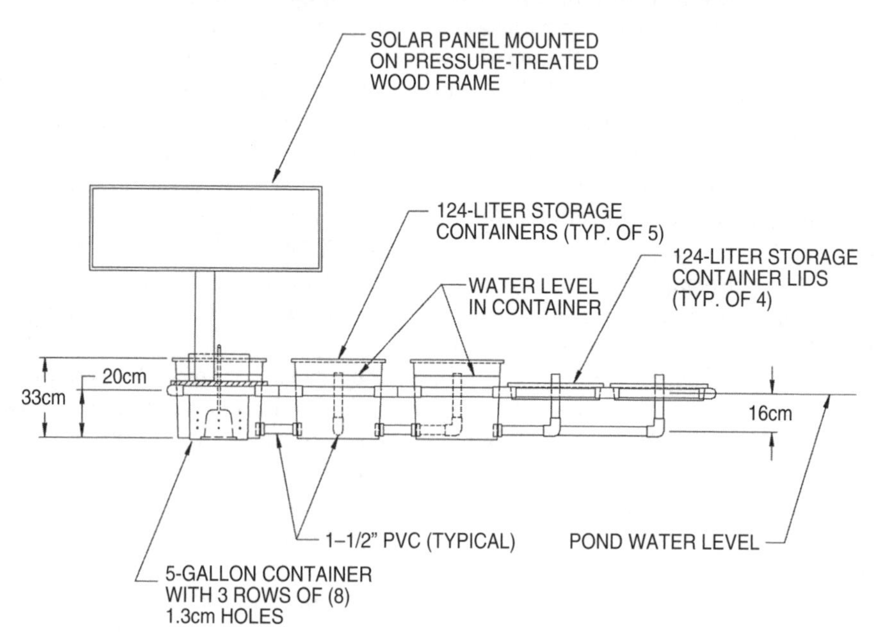

FIGURE 5.12 Views of the University of Maryland lake restorer ecosystem. (Adapted from Yaron, P., M. Walsh, C. Sazama, R. Bozek, C. Burdette, A. Farrand, C. King, J. Vignola, and P. Kangas. 2000. *Proceedings of the 27th Annual Conference on Ecosystems Restoration and Creation*. P. J. Cannizzaro (ed.). Hillsborough Community College, Plant City, FL.)

that extend from the surface down to the sediments. The restorer system consists of three components: floating piers with racks of aquatic plants that extend out into the channel, the plankton in the channel, and the curtains which are covered with attached macroinvertebrates. Unlike the smaller, autonomous lake restorers, this large system treats the wastewater as it flows through the lagoon. Treatment occurs by spiraling between the three component subsystems as the wastewater moves along the channel (Figure 5.13).

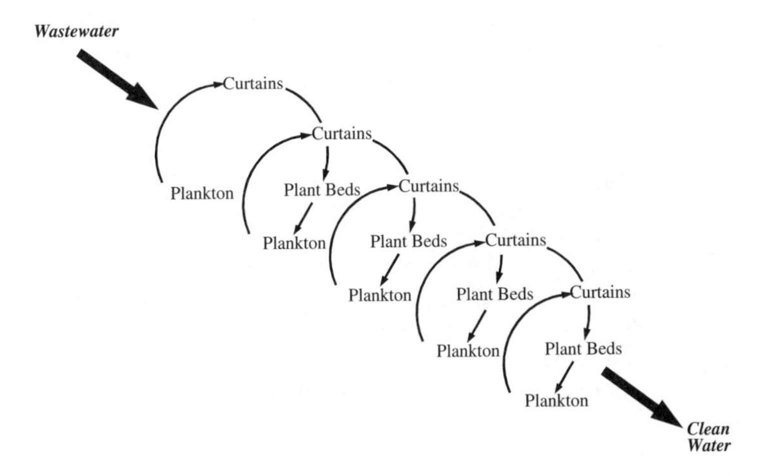

FIGURE 5.13 Spiraling wastewater treatment in the Ocean Arks' lake restorer at the Tyson Food's poultry processing plant in Berlin, MD.

Suspension feeding bivalves also have been used as a form of bioremediation to control phytoplankton and, therefore, eutrophication of aquatic ecosystems. The filtration of the water column by bivalves during feeding removes phytoplankton and reduces turbidity. Nutrients are transferred from the pelagic zone to the benthos either by biodeposition in feces or psuedofeces (materials which are ingested but quickly rejected) or by incorporation into bivalve biomass. This approach has been shown to be effective for natural reefs of oysters (especially *Crassostrea virginica*) and beds of mussels. Officer et al. (1982) provide quantitative relationships showing criteria under which bivalves can control phytoplankton, based on prey–predator equations. Reviews are given by Dame (1996, 2001), Levinton et al. (2001), and Strayer et al. (1999). It has been further suggested that bivalves can be used, through a form of biomanipulation (see Chapter 7), to actively control eutrophication. Thus, Ulanowicz and Tuttle (1992) showed with a simulation model that oyster reef restoration and raft culture could significantly impact eutrophication in the Chesapeake Bay, and Wisniewski (1990) experimentally demonstrated techniques for enhancing zebra mussel (*Dreissena polymorpha*) filtration with artificial substrates in Poland. Raft culture in particular has been shown to have very high production rates of shellfish for food (Ryther, 1969), and therefore, it has potential for use in eutrophication control. Other suspension feeders, such as sponges on coral reefs (Diaz and Rutzler, 2001), along with polychaetes (families *Sabellariidae* and *Serpulidae*) and gastropods (family *Vermetidae*) that form reefs in tropical estuaries (Mohan and Aruna, 1994; Pandolfi et al., 1998; Schiaparelli and Cattaneo-Vietti, 1999), have high filtration rates and may be candidates for future ecological engineering design for bioremediation.

PROCEDURES AND POLICIES

Procedures and policies have evolved along with the technical knowledge about restoration ecology. Procedures involve methods for organizing and understanding restoration projects so that knowledge developed from them can be effectively utilized in decision making. In turn, because of the success of restoration ecology and its methodologies, public policies have emerged that mandate restoration of certain environmental impacts.

Measuring Success in Restoration

A number of procedures have been developed to ensure success of restoration projects. These procedures address methodological challenges that are inherent in restoration. Clewell and Rieger (1997) provide a list of 15 of these issues with discussion of each problem and of possible solutions. Several of the most important issues are discussed below.

Perhaps the most critical procedure to be followed in restoration ecology is to have some kind of monitoring program for a restoration project that will provide data on the development of the restored ecosystem. Monitoring data is the basis for adaptive management. In this approach data on the ecosystem are compared with a target or set of target goals. If the monitoring data match with the target, then the restoration is making successful progress and no action is required. However, if the monitoring data do not match with the target, then some remedial measures should be taken. There are many ideas on how to conduct monitoring programs and on how to establish appropriate target goals, but as noted by Cairns (1986) below, there is little agreement on standards to follow:

> The probability of achieving anything approaching a professional consensus on the relative importance, reliability, replicability, measurement, interpretation, and a variety of other issues regarding end points at the community and ecosystem level is small.

Thus, restoration procedures are essentially subjective and will always be open to debate.

One of the most fundamental issues with restoration projects is what to measure. There are a large number of possible parameters that could be measured in an ecosystem (see for example E. P. Odum's [1969] list of 24 attributes as a starting point), but it is not possible on practical grounds, alone, to measure everything. Some single measure or set of measures must be choosen for tracking progress and establishing end points for restoration. In many cases there are obvious choices. For example, when restoration involves an active planting program, the survival and reproduction of the intentionally planted species must be a consideration. These species are judged to be desirable and their presence indicates success of the project. Conversely, invasion by undesirable species (see Chapter 7) indicates failure of the project and triggers remedial action. This particular attitude about invasive, exotic species is deeply engrained in restoration science to the degree that the Society of Ecological Restoration has established a formal bylaw against the use of exotic, nonnative species in restoration projects.

A number of measures of ecosystem structure and function have been used for evaluating restoration projects depending on ecosystem type and preference of the researcher. In some cases individual parameters are used as indicators of the overall

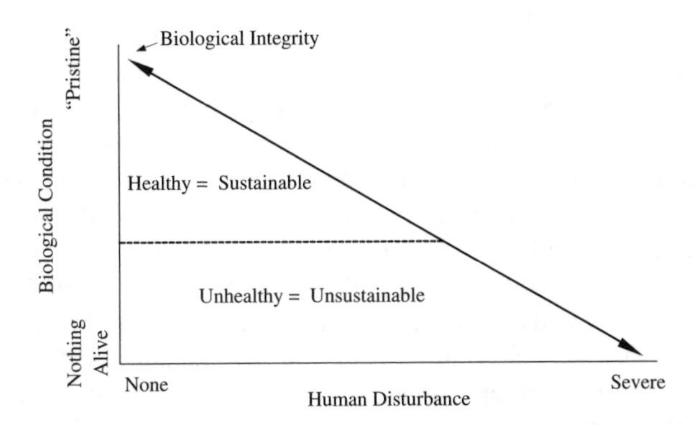

FIGURE 5.14 Matching of a disturbance gradient with an index gradient for biotic integrity. (From Karr, J. R. 2000. *Ecological Integrity, Integrating Environment, Conservation, and Health.* D. Pimental, L. Westra, and R. F. Noss (eds.). Island Press, Washington, DC. With permission.)

ecosystem, such as ant abundance (Andersen and Sparling, 1997; Majer, 1983) or decomposition rate (Durall et al., 1985; Lawrey, 1977) for terrestrial systems and phosphorus concentration (Dillon and Rigler, 1975) for lakes. In these cases the individual parameters used as indicators are judged to have special importance in the particular ecosystem context under consideration. Different individual parameters of the ecosystem develop at different rates. Thus, as mentioned earlier, marsh vegetation was restored quickly at Kenilworth Marsh in Washington, DC, but the development of organic matter content in the soil has lagged far behind (Kassner, 2001).

In other cases, sets of parameters are combined together into composite indices, such as with the wetland evaluation technique (i.e., WET, see Adamus, 1988) or the lake trophic state index (Carlson, 1977; Lambou et al., 1983). One of the best known indices is the *Index of Biotic Integrity* (*IBI*) developed by James Karr (1981, 1991; Karr and Chu, 1999) which was initially used for stream fish communities but has been applied more widely over time. The *IBI* consists of a series of attributes, termed *metrics*, that reflect both structural and functional characteristics of an ecosystem. In general, metrics are summed and an overall index is calculated that has meaning relative to reference conditions. Karr's index is especially interesting because he developed a new concept with it, termed *biotic integrity*, which is meant to synthesize the qualities of an ecosystem. Karr and Dudley (1981) define biotic integrity as "a balanced, integrated, adaptive community of organisms having a species composition, diversity, and functional organization comparable to that of natural habitat of the region." The approach is to devise an index scale that matches a general scale of ecosystem structure and function (Figure 5.14). This and other related concepts, such as ecosystem health, represent new approaches for relating human impacts to the condition of the environment. However, it must be remembered that much subjectivity is involved in choosing metrics for the *IBI*. For example, exotic species are usually left out of the index because they are judged not to contribute to the biotic integrity of the system. But exotic species do add diversity and function to the actual ecosystem where they are found, which is thus unaccounted for by the convention of the *IBI*. Semantic issues with applying the idea of health to an ecosystem also are problematic (Slobodkin and Dykhuizen, 1991). For example, a polluted stream dominated by tolerant species (Figure 2.3) might be considered unhealthy by human standards but it is perfectly adapted to the conditions it is exposed to. The community of species in the polluted stream is appropriate (and healthy) for an energy signature that includes organic wastes. In fact, the clean water or intolerant species would be indicators of unhealthy conditions in a polluted stream because they are not adapted to the stream's energy signature! Efforts at assessing ecosystem function for wetlands with an indicator approach are perhaps the most developed in the field of restoration ecology, to the point of having an established American Society for Testing and Materials standard ASTM, 1998), but even here subjectivity still exists. This sense of subjectivity in restoration ecology sets it apart from other areas of ecological engineering and engineering in general where there is little subjectivity in what to measure in order to determine the success of a design. For example, standards in terms of BOD, total nitrogen, and suspended solids are well accepted as measures of success for treatment wetlands, and there is little disagreement between workers on the subject.

An important procedure that is often used in restoration ecology is to compare measures of a restored ecosystem to a reference system in order to evaluate success. In this situation the reference ecosystem is judged to represent ideal or at least appropriate conditions, which usually means as close as possible to the natural or undisturbed state. Thus, success is gauged by the degree to which the restored ecosystem matches with the reference ecosystem. The logic of this procedure is obvious, yet difficulties arise in establishing reference sites. First, a decision must be made on the type of ideal reference and then a search must be made to discover if any of these actually exist in the landscape. Here again, subjectivity is involved, which leaves open the possibility for critical debate. A good deal of literature concerns the reference ecosystem issue (Aronson et al., 1994; Brinson and Rheinhardt, 1996; Egan et al., 2001; Findlay et al., 2002; Hughes et al., 1986; White and Walker, 1997), and Hughes (1994) lists six approaches for establishing a reference. The most objective approach is to use historical data on the original ecosystem that was impacted and is now to be restored. Unfortunately, this kind of data is seldom available. The more typical approach is to match the ecosystem to be restored with nearby, similar ecosystems that have not been impacted (see, for examples, Confer and Niering, 1992; Galatowitsch and van der Valk, 1996). Because of the inherent variability of ecosystems, it is seldom possible to locate a single individual reference ecosystem, and typically, multiple reference ecosystems are used to account for natural variation. Inevitably, these studies deal almost as much with interpreting reference conditions as they do with comparing reference sites to restored sites. However, when high quality reference sites can be established, they become very valuable as examples of natural conditions, and they deserve study in their own right and preservation for their special environmental value.

Public Policies

A number of public policies relate to restoration ecology. These are parts of the complex hierarchies of laws that regulate environmental impacts and mandate mitigation. At the federal level there are examples such as the National Environmental Protection Act and the Clean Water Act, along with others that call for restoration after strip mining, oil spills, and other impacts. States also become involved in local regulation with various legislation. All of this makes understanding the "regulatory" environment a formidable task and one beyond the scope of this text. However, the important case of wetland regulation is discussed below to illustrate some facets of public policy in regards to restoration.

Wetlands regulations have evolved in the U.S. as society has become aware of the values of these ecosystems. At one time wetlands were viewed as wastelands with no value, and they were actively filled or drained. However, in the 1960s and 1970s their natural values in hydrology, water quality, wildlife habitat, and education began to be recognized, and by the 1980s and 1990s wetland values began to be quantified (see Chapter 8). One consequence of the recognition of these values was that laws were enacted to protect wetlands. A national policy emerged called "No Net Loss," in order to reverse the destruction of wetlands and to restore both their area and function (Davis, 1989; Deland, 1992). A mitigation process has been created as part of this overall policy to deal with cases where a land owner wishes to develop a wetland for commercial or residential purposes (Beck, 1994; Berry and Dennison, 1993). Mitigation refers to a set of actions or rules that seek to preserve and even increase wetland values while accommodating economic development. Three main categories of action are meant to be applied in a sequential fashion for each case of proposed wetland impact: first, attempt to avoid the impact; second, attempt to minimize the impact; and third, provide compensation where the impact is inevitable. Compensation usually requires the creation of new wetlands or the enhancement of existing wetlands, both of which require the technology of restoration ecology. In general, wetland regulations and restoration techniques have developed concurrently (Kruczynski, 1990; Wolf et al., 1986) and both are still evolving. A multimillion dollar industry of environmental consultants also has developed for the evaluation of natural wetlands and for the design and construction of mitigation wetlands. Thus, large areas of new wetlands are being created across the country to compensate for natural wetlands that are being destroyed by economic development. Usually, a larger amount of creation is required relative to wetland destruction, such as a 2:1 ratio of created vs. destroyed acreage. This is done in an effort to ensure that wetland functions are not lost and, hopefully, that their values actually will be increased. The entire topic has become controversial, primarily because evidence is scattered and incomplete on the question of whether restoration technology can create new wetlands that are equivalent to the natural ones that are lost to development (Harvey and Josselyn, 1986; Malakoff, 1998; Race, 1985, 1986; Race and Christie, 1982; Savage, 1986; Young, 1996).

Restoration clearly can create ecosystems, but do these new ecosystems provide the same services as the natural ones that are lost? A great deal of ecological research is being conducted on this topic (Kusler and Kentula, 1990; Zedler, 1996a,b, 2000), but the mitigation question remains unresolved and contentious. Essentially the situation is the ecological equivalent of the Turing test for determining artificial intelligence in computers. The mathematican Alan Turing (1950) invented this imitation game just as digital computer technology was being developed. In the most general form of the test a human is seated at a teletype console by which he or she can communicate with a teletype in another room. The second teletype is controlled either by another human or by a computer. The programmer at the first teletype asks questions through the console to determine whether he or she is in contact with a human or a computer in the other room. If the programmer cannot distinguish between responses of a human and a computer at the second teletype, then the computer is said to have passed the test and is considered to be intelligent. In the ecological equivalent of the Turing test, ecologists sample created and reference wetlands, like the programmer asking questions of the human and the computer (for examples, see Wilson and Mitsch, 1996; Zedler et al., 1997). The created wetland passes the test if the ecologist cannot distinguish it from the reference wetland. Unfortunately, at the current state of the art, created wetlands do not seem to be passing the ecological Turing test very often (Kaiser, 2001a; Turner et al., 2001). Despite this situation, though, the "No Net Loss" policy and the mitigation process are achieving at least some kind of balance between economic development and environmentalism. At the same time, these policies are creating a major source of employment for ecological engineers whose growing experience should lead to technologies for achieving functional equivalency between created and natural wetlands (Zentner, 1999).

CASE STUDIES

Three case studies are presented in conclusion to provide perspective on some of the approaches to restoration ecology described above. These were chosen to illustrate the range of the ecosystems that have been involved in this subfield of ecological engineering. These case studies also include several examples of importance in the history of restoration ecology.

SALTMARSHES

Saltmarshes are the dominant vegetation along low energy coastlines in the temperate zones of the world. However, because human development also is focused along these coastlines, saltmarshes have been converted to commercial and residential land uses through dredging and filling in many areas. The concern about losses of saltmarshes became even more important as their value to society began to be recognized through ecological research in the 1960s. saltmarshes are important as a source of, and nursery zone for, fish and shellfish species that are harvested for seafood, and due to their role in shoreline protection. In fact, the first major ecological valuation study was done for saltmarshes (see Chapter 8) and is a benchmark in ecological economics. Thus, because of their losses due to human development and because of the recognition of their values to society, saltmarshes became a focus of conservation and restoration along the U.S. east coast in the 1970s. saltmarshes became the first ecosystems to be restored on a large scale and the technology is now well developed (Zedler, 2001).

The history of saltmarsh restoration is particularly interesting because it involves a coevolution between dredge disposal activities conducted by the U.S. Army Corps of Engineers and planting research by ecological scientists. Perhaps without contributions from both the Corps and the scientists, the development of saltmarsh restoration might have been inhibited or might have taken a different course. This coevolution is even more remarkable because dredging and filling activities conducted by the Corps were, in part, the cause of saltmarsh losses before the coevolution began!

An introduction to the Corps of Engineers is useful before describing the development of saltmarsh restoration. The Corps is the largest engineering organization in the world and has been important in several aspects of ecological engineering. While the environmental record of the Corps has not been flawless, as noted in the introductory chapter of this book, major changes are under way, and in the future the Corps may become a leader in ecological engineering and in areas of environmental management. The Corps has always had a role in water management as noted by Hackney and Adams (1992) below:

> It is difficult to find anyone or a single publication that presents an unbiased view of the U. S. Army Corps of Engineers and their activities in U.S. Waters. In the beginning, 16 March 1802, the Corps was devoted exclusively to military operations. As the one organized group of engineers "on call" for the U.S. government, they quickly became associated with the construction and maintenance of waterways and harbors through which the U. S. military could rapidly move ships, troops, and supplies. The lack of a national policy related to transportation and defense became obvious to many American leaders after the War of 1812 with Britain. In 1824 the U.S. Army Corps of Engineers was officially given legislative authority to participate in civil engineering projects. … Of perhaps greatest importance was the fact that the Corps of Engineers not only undertook projects directed by the military, but planned and directed projects that were primarily related to civilian commerce. Clearing rivers of snags, building canals and roads, erecting piers and breakwaters all became part of the role of the Corps of Engineers before the Civil War.
>
> After the Civil War both the limited accepted role of the Corps in civilian projects and the annual appropriations from Congress expanded dramatically. The Rivers and Harbors Act of 1899 further expanded the Corps of Engineers' authority by granting them regulatory authority of all construction activities in navigable waters. This not only gave them authority over individual projects, but also gave them preeminence over all other agencies and boards when it came to potentially navigable waters.
>
> All U.S. Army Corps of Engineers activities are mandated by Congress. Although the Corps may recommend certain activities (usually after a directive from Congress for study), their activities are mostly driven by various individuals and agencies through their elected official … Almost from the beginning civilians have had an influence in initiating what later became Corps of Engineers projects. While some projects were suggested by community-spirited individuals, many had the potential to bring large profits to individuals or certain industries. Congress, however, ultimately directs all

such projects through annual appropriations. Corps of Engineer project were often used to bring jobs to an area and became pork barrel projects for elected officials.

According to some, the Corps developed a questionable record of concern about the environment starting in the 1930s through their flood control efforts along inland rivers and through their dredging and filling activities, especially along the coasts. To some extent this reputation is unfair because society as a whole in the U.S. did not generally recognize the importance of environmental values until after the first Earth Day in 1970. However, the Corps' reputation developed because they were directly responsible for destroying large areas of natural ecosystems and broadly impairing ecosystem services due to their initiatives and mandates from Congress.

The Corps' environmental record is changing and the case study of saltmarsh restoration is one example. The contribution of the Corps to saltmarsh restoration has been catalyzed by its mandate for dredge and fill activities (Murden, 1984). This is a major function as described by Hales (1995):

> The U.S. Army Corps of Engineers (USACE) is involved in virtually every navigation dredging operation performed in the United States. The Corps' navigation mission entails maintenance and improvement of about 40,000 km of navigable channels serving about 400 ports, including 130 of the nation's 150 largest cities. Dredging is a significant method for achieving the Corps' navigation mission. The Corps dredges an average annual 230 million cu m of sedimentary material at an annual cost of about $400 million (US).

The Corps must dispose of the dredge materials, which is a major challenge. Dredge material is a waste product of dredging and disposal takes place both on land and in waterways. Disposal can cause environmental impacts if a natural ecosystem is filled, making this activity a significant concern. One major solution has been the idea to use dredge material as a substrate for building new saltmarshes in restoration. In this way a waste by-product is used as a resource, which is a key principle in ecological engineering. Moreover, because disposal itself is costly, use of dredge material in saltmarsh restoration can result in money savings for the overall project.

The idea to use dredge material as a planting substrate seems to have come from a group of scientists interested in saltmarshes at North Carolina State University (Seneca et al., 1976):

> In 1969, we approached the U.S. Army's Coastal Engineering Research Center, Fort Belvoir, Virginia, with the proposition that stabilization of intertidal dredged material might reduce channel maintenance costs by preventing such material from being washed back into the same channels from which it had been dredged. Further, stabilization of the material with *S. alterniflora* would result in salt-marsh being established and thus replace some of the surface that had been lost through dredging operations. The Coastal Engineering Research Center was receptive to our ideas and supported our efforts to explore the possibility of stabilizing dredged material in the intertidal zone and the concomitant initiation of salt-marsh.

The Corps thus supported the first research on ecological restoration of saltmarshes by the NCSU group and soon afterwards by other researchers (Johnson and McGuinness, 1975; Kadlec and Wentz, 1974), including Edward Garbisch as noted earlier. It is significant for understanding the nature of ecological engineering that the idea came from outside the Corps rather than from inside. The Corps might have been pioneers in this subfield of ecological engineering, but they followed the stimulus from ecologists rather than being leaders. The explanation may be that the pre-1970s Corps was made up mostly of civil engineer types with little ecological training. Ecological engineering activities such as restoration require an interdisciplinary perspective that was lacking in the pre-1970s Corps, but it emerged as a coevolution when stimulated by ecologists.

The use of dredge materials for restoration was quickly taken up by the Corps and incorporated into their operations (Kirby et al., 1975; Landin, 1986) after the coevolution began. The North Carolina State group became leaders in saltmarsh restoration research with support from the Corps, resulting in the development of a large literature and a sound technology (Broome, 1990; Broome et al., 1986, 1988; Seneca, 1974; Seneca and Broome, 1992; Seneca et al., 1975, 1976; Woodhouse and Knutson, 1982). Most of this work involved horticulture of *Spartina alterniflora* or, in other words, basic planting techniques for dredge materials. saltmarsh restoration evolved from this early work as a two-step process. First, an appropriate site is chosen that is protected from waves, wind, and boat wakes, and dredge materials is deposited. This step takes into account the energy signature of the site in order to avoid high-energy sites where erosion will occur. The second step is planting saltmarsh species, which is essentially horticulture with considerations of soils, nutrient levels, and plant materials. In general, this two-step process has been successful in developing saltmarsh vegetation in many locations. The technology has developed since the 1970s and now includes alternative methods of dredge disposal such as spraying (Ford et al., 1999) and use of bioengineering materials (Allen and Webb, 1993). There also has been a broadening of interest to additional aspects of ecosystem structure and function, beyond plant survival and growth, when considering the success of saltmarsh restorations (Haven et al., 1995; Moy and Levin, 1991; Niering, 1997; Zedler, 1988, 1995, 2001). Much of this work is summarized by Matthews and Minello (1994) and in the proceedings of the Hillsborough County Community College Annual Conference on coastal restoration ecology that dates to the early 1970s (see also the interesting independent research being carried out in China as described by Chung, 1989).

One of the complexities with the use of dredge materials for restoration involves the system that becomes filled to create the marsh. The ecological values of these systems are lost when they are converted to marshes. Thus, there is an environmental impact when a marsh is created with dredge material. This is usually ignored because marshes have high value and are endangered. However, problems can arise with the assumption that marshes are more valuable than the systems they replace. For example, in the Anacostia River in Washington, DC, tidal freshwater marshes are being created by the Corps dredge disposal program. Existing mud flat ecosystems are filled with dredge materials to raise the surface to an appropriate level for marsh plant growth. May (2000) showed the value of the mud flats as shorebird habitat

(Table 5.6). When the mud flats are filled, the shorebird habitat is lost because these

TABLE 5.6
Bird Survey Results at a Mudflat in the Kenilworth Marsh in Washington, DC

Bird Species	Numbers of Birds	
	1997	1998
Killdeer	197	231
Canada Goose	140	233
Mallard	46	112
Ring-billed Gull	54	91
Great Blue Heron	71	42
Great Egret	61	15
Greater Yellowlegs	27	32
American Crow	16	31
Herring Gull	21	1
Belted Kingfisher	13	10
Double Crested Cormorant	12	4
Black Duck	3	16
Bufflehead	—	14
Wood Duck	—	8
Osprey	3	4
Solitary Sandpiper	—	7
Spotted Sandpiper	—	6
Hooded Merganser	5	—
Pintail	4	—
Red Tail Hawk	3	—
Bald Eagle	2	1
Green Heron	2	1
Forster's Tern	—	3
Lesser Yellowlegs	1	1
Bonaparte's Gull	—	2
Common Merganser	1	—
Least Sandpiper	1	—
Semipalmated Plover	—	1
Total	689	866

Note: The numbers are totals for 36 observations per year.

Source: Adapted from May, P. I. 2000. *Proceedings of the Annual Ecosystems Restoration and Creation Conference*. Hillsborough Community College, Plant City, FL.

species need open, exposed sediments rather than vegetated marshes to meet their life needs. Because shorebirds may be more endangered than the marshes, the wetland creation project may be generating less environmental value than if no action was taken and the mud flats were preserved. Perhaps a more in-depth analysis is needed in cases such as this one concerning marsh restoration through dredge disposal.

The Corps of Engineers has upgraded its ecological capabilities over time in response to critics and due to the need for a broader environmental awareness. One example is the multimillion dollar Corps wetland research program which started in 1990. However, there is still a civil engineering emphasis (see, for example, Palermo, 1992) which, in part, is appropriate and important in restoration work. It will be interesting to observe if and how the Corps responds to the growing paradigm of ecological engineering. Although Corps projects in saltmarsh restoration generally have been successful, this kind of ecosystem naturally has a low complexity relative to other ecosystems. Corps efforts at restoring the more complex tidal freshwater marshes have not been as successful (see the discussion of Kenilworth Marsh earlier in this chapter). Questions remain about the ability of the Corps to combine ecology and engineering. Does the military administration of the Corps inhibit interdisciplinary thinking needed for ecological engineering? Is the ecosystem too complex for the traditional civil engineering approaches of the Corps? The hope is that both the field of ecological engineering and the U.S. Army Corps of Engineers will benefit from future collaborations such as those between the dredge disposal program and saltmarsh ecologists in the early 1970s.

A final consideration about saltmarsh restoration involves the secular sea level rise that is presently occurring along global coastlines. Sea level rise causes an encroachment of the flooded tidal lands on the adjacent uplands and submergence of existing coastal ecosystems. If coastal wetlands can grow both upward and in an inland direction, they may be able to avoid submergence. However, if coastal wetlands are restricted in area and/or cannot match sea level rise by vertical accretion, then a loss of these ecosystems will occur. This situation has been discussed for mangrove ecosystems (Ellison and Stoddart, 1991; Field, 1995; Woodroffe, 1990), and Rabenhorst (1997) has called for a new approach to understanding coastal marshes in relation to sea level rise, which he terms the *chrono-continuum.* saltmarsh restorations are also susceptible to this problem. Thus, sea level rise may submerge and therefore destroy restored saltmarshes as quickly as they are created in some areas (J. Court Stevenson, personal communication; see also Stevenson et al., 2000). This issue will complicate the restoration of saltmarshes in the future (Christian et al., 2000).

Artificial Reefs

Artificial reefs are structures of human origin used in aquatic ecosystems to increase fish production. Informed design is employed in the construction and placement of these devices, relying on both conventional and ecological engineering. These artificial reefs come to resemble natural reefs in both ecological structure and function, and can even generate more fish production than their natural analogs under certain circumstances.

A great number of different designs have been tried in both marine and freshwaters. Fish aggregating devices are usually included under the topic of artificial reefs, although they either are suspended in the water column or floated at the surface to attract pelagic fishes. More commonly, artificial reefs refer to structures that rest on the bottom substrate and attract benthic fishes, similar to natural oyster or coral reefs.

TABLE 5.7
Sequence of Marine Fouling Organisms Found in Succession on Submerged Hard Surfaces

Slime forming organisms	Bacteria, diatoms, microalgae, protozoa
Primary fouling organisms	Barnacles, hydroids, serpulids, polyzoa
Secondary fouling organisms	Mussels, ascidians, sponges, anemones
Adventitious organisms	Polynoids, sabeliids, cirratulids, nudibranchs, ostracods, amphipods

Source: Adapted from Crisp, D. J. 1965. *Ecology and the Industrial Society.* John Wiley & Sons, New York.

Though the intended focus of artificial reefs is the fishes that are attracted to them, they are constructed ecosystems. Aquatic organisms colonize the surface of the artificial reefs. These organisms attach to the surfaces with various adaptations, and are sometimes referred to as "fouling" organisms. "Fouling" is a successional process in which attached organisms colonize a submerged hard surface (Crisp, 1965). The name itself is anthropocentric; the verb "to foul" has negative connotations because the organisms that attach to certain human-produced surfaces, such as pipe outfalls or ship bottoms, can cause significant problems. Table 5.7 provides a sequential listing of typical marine fouling organisms that might colonize an artificial reef in temperate marine waters. Colonization is by natural dispersal of life stages carried by currents, and artificial seeding is practically never necessary. Principles of island biogeography (see Chapter 4 and the discussion earlier in this chapter) have been useful in understanding the development of artificial reef communities because of the role of natural colonization and the insular qualities of reefs themselves (Bohnsack et al., 1991; Molles, 1978; Walsh, 1985). Fishes are attracted to artificial reefs because they provide food, shelter from predators, and sites for orientation and reproduction, i.e., habitat (Bohnsack, 1991). The use of artificial substrates for the scientific monitoring of benthic ecosystems (Cairns, 1982) is related to the topic of artificial reefs because both kinds of structures have similar design considerations. In particular, the materials used for both artificial reefs and scientific monitoring substrates must be similar to natural materials so that attachment by organisms is not inhibited.

The leaders in the use of artificial reefs have been the Americans and the Japanese, but they have taken very different pathways (Stone et al., 1991). In Japan artificial reefs are a highly developed technology that supports commercial fishing (Grove et al., 1994; Mottet, 1985; Yamane, 1989). Records of Japanese artificial reefs date to the 1600s when rock formations were constructed as reefs in shallow waters along the coast. In the present day tens of millions of dollars are spent annually in government supported reef programs on a national scale. Japanese artificial reefs are characterized by sophisticated prefabricated designs. For example, Grove and Sonu (1985) describe 68 different kinds of reef structures and report that more than 100 are in use. Knowledge of fish ecology, life history patterns, and behavior is well developed. In some cases, reefs are designed, constructed, and sited to support particular species, based on this well-developed knowledge base (Nakamura, 1985).

Artificial reef use in the United States differs markedly from the Japanese approach. In the U.S. most artificial reefs are constructed for recreational fishing. They are smaller scale projects supported by local governments or private interest groups such as fishing clubs. Scrap materials are often utilized in designs which are sometimes quite ingenious but still unsophisticated compared with the Japanese models. Artificial reefs were first employed in the U.S. in the 1800s, but usage increased greatly after World War II (Stone, 1985). There are many more freshwater examples in the U.S. than in Japan. Methods for these systems were described as early as Hubbs and Eschmeyer's (1938) important work on fish management in lakes. An interesting development in the U.S. is the use of artificial reefs for mitigation of habitat damage (Foster et al., 1994), as was described earlier in this chapter in relation to wetland restoration. This usage emerged as studies have demonstrated the development of comparable ecosystem structure and function between artificial and natural reef systems.

As an aside, restoration of natural reefs is also an important topic in the U.S. In particular, efforts are under way to restore oyster reefs in many coastal areas such as Chesapeake Bay (Leffler, undated). Oyster populations collapsed in the late 1800s and early 1900s due to cumulative impacts including overfishing, disease, and water quality decline. All of these impacts must be dealt with before full recovery is possible, but restoration efforts are being initiated. Techniques for growing oyster reefs are similar to those used for artificial reefs and they have long been known (Brooks 1891). In areas with sufficient current velocities to carry their food source (particulate organic matter), oysters will attach to hard surfaces and grow into self-sustaining reef structures. Old oyster shells are often used as substrate to start new reefs, mimicking the positive feedback that took place on natural oyster reefs. An interesting example of coral reef restoration is the work of Todd Barber of Reef Balls, Inc. (Menduno, 1998). He has developed his own design for artificial substrates which are made of concrete (Figure 5.15). These are called reef balls and they are being used around the world in restoration projects. The hydrodynamic shape of the reef balls facilitates colonization by pelagic larvae of fouling organisms, including corals. General aspects of coral reef restoration with artificial substrates are described by Spieler et al. (2001).

Unlike many other examples of restoration ecology, creation of artificial reef systems requires a significant amount of conventional engineering, including aspects of materials and structural stability along with siting criteria, which is perhaps more closely related to ecological engineering (Sheehy and Vik, 1992). A variety of materials have been used to construct artificial reefs including natural materials (such as brush, quarry rock, and logs), manufactured products (such as poured concrete, fiberglass, and plastic) and scrap (automobile tires and bodies, rubble from construction sites, and scuttled vessels). Considerations in choice of materials include availability, cost, durability, and complexity of surfaces. Because reef materials are submerged and exposed to a number of destructive processes, durability is a critical quality that often determines the life expectancy of the reef structure. Most conventional engineering knowledge used in artificial reef design involves analyses of

FIGURE 5.15 A small reef ball made by a group of University of Maryland undergraduate students.

stability. For example, Mottet (1985) applied Hudson's formula (see Chapter 3) to evaluate reef stability in relation to wave energy and offered suggestions for similar stability equations in relation to current velocity. Other examples such as calculation of reef block strength are given by Grove et al. (1991) and Sheng (2000).

Siting is a particularly important step in artificial reef development that involves a number of considerations. This requires a knowledge of the energy signature of the site including substrate type, bottom topography, relations to adjacent reefs, and especially current and wave energy. The reef must be exposed to appropriate levels of current energy to advect fouling organism life stages to the reef for colonization, to advect food for fouling organisms, and to attract fishes. There are also features which must be avoided such as interference with navigation, areas used for commercial fishing with nets which might snag on the reef, and sites with very strong tidal currents. Overall, the ideal site would be one with a depth of 30 to 40 m in order to attract large benthic fish species and only a few kilometers offshore in order to facilitate access by fishermen.

Scrap tires are used to construct one of the most common kinds of artificial reef in the U.S. In this type of reef, tires are combined together in various ways to create complex structures that support fouling communities and attract fishes (Figure 5.16). As noted by Candle (1985), "The same tire qualities that are advantageous to motorists, strength, durability and long life are the keys to the advantage of tires as reef-building materials." They are also plentiful, cheap, and easy to handle, process,

FIGURE 5.16 Some different configurations of artificial reefs made from scrap tires. (From Grove, R. S. and C. J. Sonu. 1985. *Artificial Reefs: Marine and Freshwater Applications*. F. M. D'Itri (ed.). Lewis Publishers, Chelsea, MI. With permission.)

and transport to the reef site. However, in order to use scrap tires, they must be purged of air. This is usually accomplished by punching holes in them or splitting them in half. Ballast is also necessary to add stability against wave surge or bottom currents. Tire reefs have been shown to provide effective substrates for aquatic ecosystems (Campos and Gamboa, 1989; Reimers and Brandon, 1994) in both marine and freshwaters. Use of scrap tires for artificial reefs is a good ecological engineering example of turning a waste by-product into a valuable product. Hundreds of millions of scrap tires are produced annually worldwide, creating a disposal problem. This problem is turned into an advantage when tires are used as reefs. Although on a net basis artificial reefs made of scrap tires do require input of money for labor, ballast material, and ship time required in reef placement, a savings is integrated into the project in terms of the disposal fee for landfilling that would otherwise be required. Hushak et al. (1999) provide an analysis of one artificial reef that documents a net surplus income for the overall system, including the local economy.

Exhibit Ecosystems

Exhibit ecosystems are those designed, built, and operated primarily as exhibits for educational purposes. The best examples may be large public aquaria and botanical gardens that represent specific ecosystem types. Exhibit ecosystems require human maintenance but range across a gradient of relative contributions from humans vs.

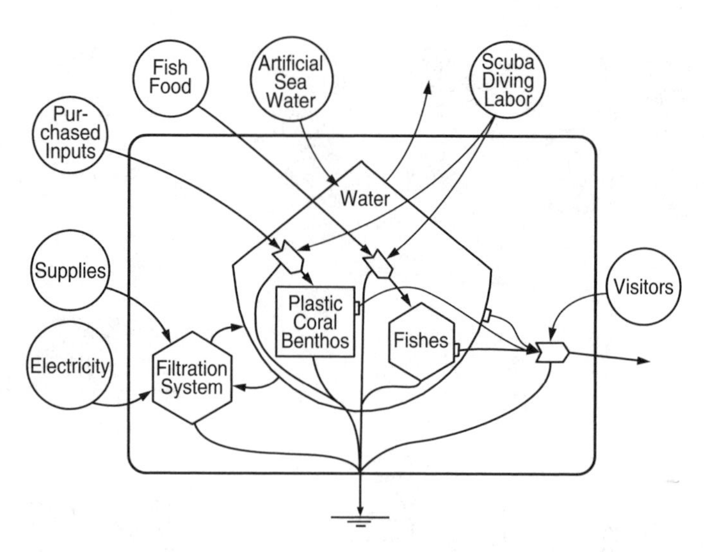

FIGURE 5.17 Energy circuit diagram of the coral reef exhibit ecosystem at the National Aquarium in Baltimore, MD.

natural self-sustainability. Although these systems are more or less artificial, they have special value for teaching aspects of ecology to students and to the general public. Significant design ingenuity is often required to help make them appear natural, which is necessary for an optimal education experience. Several examples of exhibit ecosystems are described below.

Perhaps the most complex ecosystem on the biosphere is the coral reef. These tropical ecosystems occur in shallow, clean, high-energy waters and have high biodiversity. Two basic approaches have been employed to create exhibits of coral reefs in public aquaria. On the one hand Walter Adey has developed a holistic ecological design method that emphasizes mimicking the energy signature of aquatic ecosystems as a form of modelling. He and his co-workers at the Smithsonian Institution have developed coral reef systems that have been displayed in a number of settings (Luckett et al., 1996). His systems represent a major design advancement because they support representative samples of the high diversity of a coral reef in a sustainable fashion. His first major coral reef exhibit was displayed at the National Museum of Natural History in Washington, DC, starting in 1980 (Miller, 1980; Walton, 1980). This was a 13,000 l (3,430 gal) tank system with more than 200 tropical marine species. One of the most important aspects of Adey's designs is the simple algal turf scrubber system attached to the coral reef aquaria which provides water filtration and oxygenation needed to support the biota (see also Chapter 2). Adey has continued to develop his design approach and the principles are described in his text entitled *Dynamic Aquaria* (Adey and Loveland, 1998). The largest coral reef models developed with this approach are the 2.5 million l (0.7 million gal) Great Barrier Reef Aquarium in Townsville, Australia, and the 3.4 million liter (0.9 million gal) ocean tank at Biosphere 2 near Tucson, AZ.

At the other extreme are typical coral reef exhibits such as at the National Aquarium in Baltimore, MD (Figure 5.17). This system contains a live fish community characteristic of a coral reef, but it is completely artificial otherwise. Thus, a complex filter system is employed with physical–chemical–biological components to maintain clean water, and fishes are supported by artificial feeding. Most remarkably, the tank is lined with nonliving, plastic corals that provide a quite realistic appearance but no feedback to the reef system. The result is an energy intensive, highly designed, artificial ecosystem which serves the purpose of providing an educational setting for aquarium visitors to learn about coral reefs, but it is mostly nonliving. While both of these extremes are equally valid approaches to the development of exhibit coral reefs, clearly Adey's method involves much more ecological engineering design.

The artificial approach also has been taken in developing tropical rain forest exhibits across the U.S. and in other countries. Rain forests are as complex as coral reefs and, thus, represent similar challenges in terms of exhibit ecosystem design. Most examples are highly artificial, often with plastic plants and rocks along with a few living species. They are, however, interesting systems that attract a great deal of attention from the general public (see, for example, the description of the National Zoo's Amazonia exhibit by Park, 1993). An interesting study would be to survey many of these exhibit rain forests and compare living vs. nonliving components. How much actual ecology is involved in these ecosystems? A similar survey could be made for engineering aspects, which would probably reveal some interesting features that are unique to exhibit ecosystems relative to other ecologically engineered systems.

Another example of these artificial systems is the case of environmental enrichment of zoo exhibits (Ben-Ari, 2001; Markowitz, 1982; Shepherdson et al., 1998). This situation was defined by Shepherdson (1998) as follows:

> Environmental enrichment is an animal husbandry principle that seeks to enhance the quality of captive animal care by identifying and providing the environmental stimuli necessary for optimal psychological and physiological well-being. In practice, this covers a multitude of innovative, imaginative, and ingenious techniques, devices, and practices aimed at keeping captive animals occupied, increasing the range and diversity of behavioral opportunities, and providing more stimulating and responsive environments … On a larger scale, environmental enrichment includes the renovation of an old and sterile concrete exhibit to provide a greater variety of natural substrates and vegetation, or the design of a new exhibit that maximizes behavioral opportunities. The training of animals can also be viewed as an enrichment activity because it engages the animals on a cognative level, allows positive interaction with caretakers, and facilitates routine husbandry activities. Indeed, with correct knowledge, resources, and imagination, caretakers can enrich almost any part of the environment that the captive animal can perceive.

Environmental enrichment attempts to increase the amount of stimulation and complexity of the environment, to reduce stressful stimuli, and to provide for species-appropriate behaviors in captive animals. It is an interesting topic that has engineering dimensions (Forthman-Quick, 1984), but is focused primarily at the species level rather than the ecosystem, unlike most of ecological engineering.

At a much larger scale are the restored tall grass prairies of the midwestern U.S. Although it is somehow unfair to call these systems exhibits since they range in size from less than one to hundreds of hectares, the restored prairies are still a small part of the landscape and their primary function is in education. They are not artificial in the same way as exhibit rain forests but they require controlled burns by humans for their maintenance. Most restored prairies are park-like with interpretative trails and associated displays.

In the pre-Colombian vegetation of the U.S., the tall grass prairie (5 to 8 ft or 1.5 to 2.4 m in height) bordered the temperate forests to the east. It occupied a zone stretching from Illinois and Minnesota in the north to Texas in the south. In this zone a dynamic relationship occurred between forests and grasslands mediated by shade competition which favored trees and fire resistance which favored grasses and forbs. To the west, zones of midgrass (2 to 4 ft or 0.6 to 1.2 m in height) and short grass prairie (0.5 to 1.5 ft or 0.2 to 0.5 m in height) extended across the Great Plains to the Rocky Mountains, completing the vast grassland biome or biotic region. All of these natural grasslands were eventually replaced by crop agriculture and rangeland as human development proceeded through the 1800s, leaving only scattered prairie remnants in small plots of land such as along railroad and highway rights-of-way and in unmaintained cemetaries. A movement to restore prairies began slowly in the 1930s and continues to the present time throughout the grassland biome. The prairie remnants were the seed sources for these original restorations but now nurseries have taken over this role. The oldest and best-known restored prairies are in the tall grass region, especially in southern Wisconsin and in northern Illinois. The first prairie restoration occurred at the University of Wisconsin Arboretum in Madison, WI, and was conducted by the famous conservationist Aldo Leopold, starting in the 1930s (Meine, 1999). This prairie was subsequently named after the Wisconsin plant ecologist John T. Curtis who applied a scientific approach to developing restoration techniques there. In fact, Curtis seems to have been able to develop the first scientific evidence for the importance of fire in maintaining prairie ecosystems through his research on restoration methods (Curtis and Partch, 1948). The Curtis Prairie is a lowland system with deep organic soils and a diversity of over 300 native prairie plant species (Cottam, 1987). The Greene Prairie, which is an upland system, was later added to the Wisconsin Arboretum. Restoration of this prairie was carried out by H. C. Greene, starting in the 1940s with collaboration from Curtis (Greene and Curtis, 1953). Long-term studies of both of these prairies have been made by several academic generations of Wisconsin ecologists and these studies have provided a simple, reliable technology for restoration. The basic procedure is to (1) clear and plow the soil of the site which is to be restored, (2) plant a mix of grass and forb seed, and (3) keep the area free of woody and non-native weeds with periodic, controlled burns. This is, of course, a rather simple procedure, but it requires attention to scheduling of planting and burning, and to matching seed mixes to soil types. Of particular interest is the need for fire, which represents a

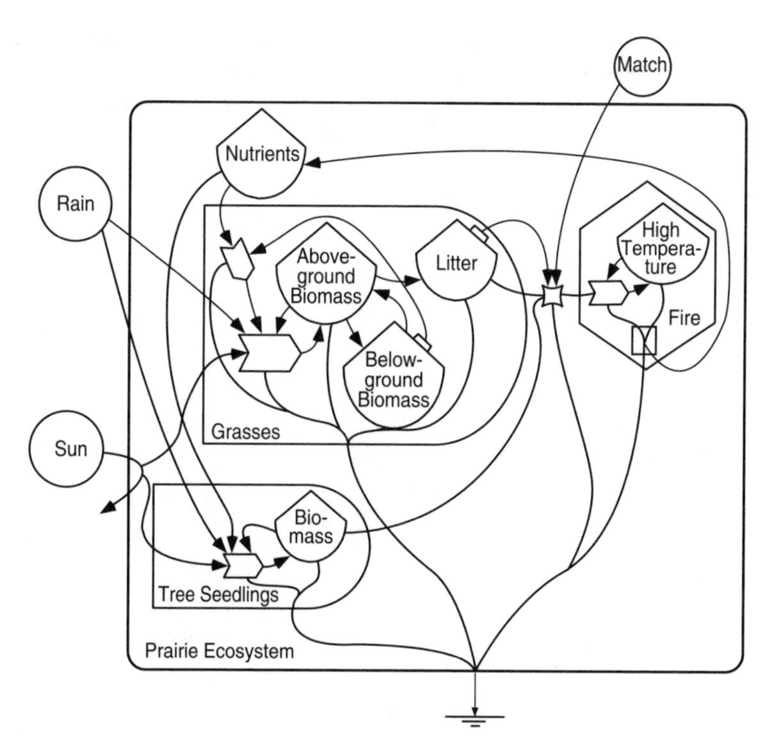

FIGURE 5.18 Energy circuit diagram of a prairie ecosystem. Note that the fire disturbance is shown as a consumer in combusting litter and recycling nutrients.

disturbance input in the restoration's energy signature. Figure 5.18 is an overview model of a prairie ecosystem. Fire is depicted with a consumer group symbol since it actually consumes biomass, similar to a herbivore. The storage of fire is composed of the concentration of high temperatures from combustion, which exists only for a short time period. Fire was initiated in the natural prairie by lightning, but controlled burns by humans are a form of technology in which fire is used as a tool. Controlled burns are usually implemented in the spring or fall to clear away dead vegetation and to kill plant species lacking fire adaptation. Native prairie species survive fires by having living portions below ground whose growth can actually be stimulated by burning, though details of fire adaptation are still not completely worked out (Anderson, 1982). Dominant grass species in most tall grass prairie restorations are little bluestem (*Schizachyrium scoparius*), big bluestem (*Andropogon gerardi*), switch grass (*Panicum virgatum*), and indiangrass (*Sorghastrum nutans*), along with a variety of non-grass, forb species such as asters (*Aster* sp.) and sunflowers (*Helianthus* sp.). Other historically important tall grass prairie restorations are the Schulenberg prairie at the Morton Arboretum (Schulenberg, 1969) and the Fermi Laboratory prairies which even have a small herd of buffalo (Thomsen, 1982). Both of these restorations are located in the Chicago region of northern Illinois. A popular account of tall grass prairie restoration is given by Berger (1985) in his Chapter 8, and technical references are given by Kurtz (2001), Packard and Mutel (1997b), and Shirley (1994).

6 Ecological Engineering for Solid Waste Management

INTRODUCTION

Humans generate solid wastes as by-products from all of their activities. Disposing of these solid wastes has become a challenge, especially as population densities have grown. Waste deposits (i.e., garbage dumps) have been associated with human habitation since prehistoric times. For example, the ancient cultures of the East and Gulf coasts of the U.S. left huge piles of shells from marine molluscs they had eaten. Called *middens*, these piles indicate past settlement patterns. In this regard, one of the most interesting studies on modern solid wastes has been conducted from the perspective of the archaeologist (Rathje and Murphy, 1992a, 1992b; Rathje and Psihoyos, 1991). This was the "Garbage Project" which spanned more than two decades at the University of Arizona. Its approach was to view solid waste as a reflection of the material culture of modern society, in the same way that archeologists have studied past civilizations.

Solid waste consists of a diversity of objects from a variety of sources. In most cases, materials from different sources are collected and mixed together to form municipal solid waste. Approximate contributions to municipal solid waste in the U.S. are as follows: 50% residential, 25% commercial, 12.5% industrial, and 12.5% institutional (Hickman, 1999). The composition of this waste, from the Garbage Project, is shown in Figure 6.1. This data came from actual excavations of modern landfills, conducted like "archaeological digs." Paper, including packaging, newspapers, telephone books, glossy magazines, mail-order catalogs, etc., dominates all other waste categories in this survey. On a more personal basis, Table 6.1 shows production from the author's household for the year 2000 when the first draft of this book was being written. On a seasonal basis, peaks in total trash production occurred in March with spring cleaning and in December with extra holiday trash. The average solid waste generation was 2.7 lbs/person/day (1.2 kg/person/day) with a ratio of nearly 2:1 of waste that was recycled vs. waste that went to the local sanitary landfill. Paper, including the categories for newspaper, glossy paper, and part of the unsorted trash, again dominated other categories of waste. Because all kinds of paper carry information, especially the newspapers, these analyses of waste composition may be the best indication that modern society has passed from the industrial age to the age of information.

Routine disposal of solid wastes requires a highly organized solid waste management system (Hickman, 1999; Tammemagi, 1999). A number of actions are

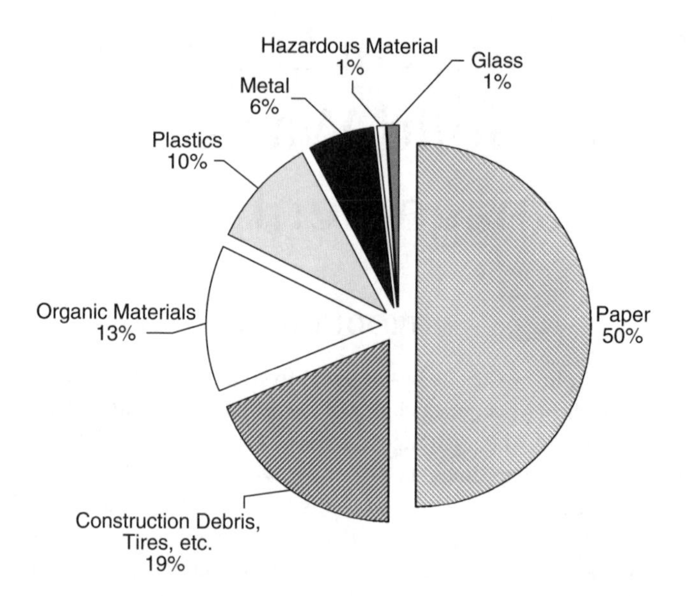

FIGURE 6.1 Composition of solid waste from analyses by the Garbage Project of the University of Arizona. (Adapted from Tammemagi, H. 1999. *The Waste Crisis: Landfills, Incinerators, and the Search for a Sustainable Future*. Oxford University Press, New York.)

involved including collection, transport, separation, storage, and treatment. This is a major commercial enterprise in the U.S. with associated government involvement and regulation. Ultimately, the methods of managing solid waste are as follows:

1. Source Reduction — prevention of solid waste generation
2. Recycling — diversion of specific items from the solid waste stream for other uses (such as composting)
3. Combustion — combustion of solid waste to reduce volume and in some cases to generate energy
4. Landfilling — disposal of solid waste by burial

Several of these management methods involve constructed ecosystems that can be considered as forms of ecological engineering.

STRATEGY OF THE CHAPTER

Ecological engineering approaches are appropriate for the organic component of municipal solid waste, especially for categories such as food wastes and yard wastes. In these cases, the organic materials provide an energy source for detritus food webs. Traditionally, only microbes have been considered important decomposers of organic solid wastes, as is characteristic of sanitary engineering in general. However, other forms of biota can be engineered into more complex ecosystems for solid waste

TABLE 6.1
Household Solid Waste Generation from the Kangas Residence during 2000

Waste Category	Waste Mass Generated Weekly (lb/week)												
	Months												
	Jan.	Feb.	Mar.	Apr.	May	June	July	Aug.	Sept.	Oct.	Nov.	Dec.	Average
Unsorted trash	13	8	39	7	6	6	11	6	19	7	8	29	13.2(34.7%)
Newspaper	11	12	15	14	15	14	16	14	12	14	17	13	13.9(36.6%)
Glossy paper	5	2	2	4	10	4	1	5	3	1	6	9	4.3(11.3%)
Plastic & tin	2	3	2	2	4	4	3	1	1	2	3	3	2.5(6.6%)
Food waste	3	2	2	2	1	3	2	1	4	1	2	3	2.3 (6.1%)
Cardboard	1	2	3	1	1	3	1	2	1	1	2	3	1.8(4.7%)
Total	35	29	63	30	37	34	34	29	40	26	38	60	38.0(100%)

Note: Data are one-week samples per month from a two-person household. All waste categories were recycled except for unsorted trash which went to the local landfill in the weekly pickup. Average solid waste generation was 2.7 lb/person/day with a ratio of nearly 2:1 of recycled vs. landfill waste.

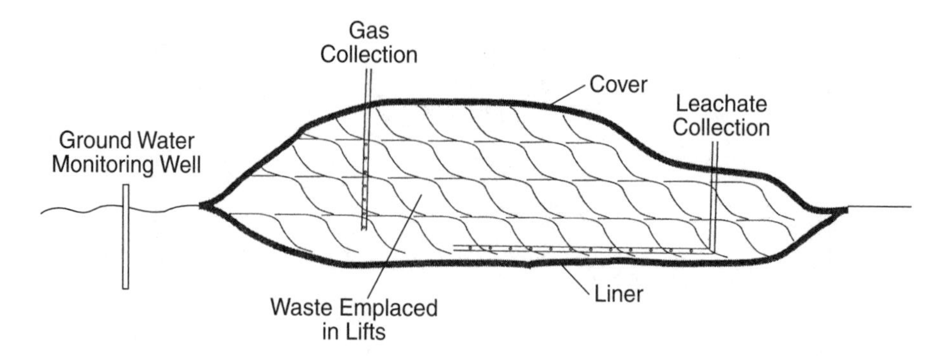

FIGURE 6.2 Cross-section view of a typical sanitary landfill. (From Tammemagi, H. 1999. *The Waste Crisis: Landfills, Incinerators, and the Search for a Sustainable Future*. Oxford University Press, New York. With permission.)

management. The two main kinds of ecosystems associated with solid wastes are the sanitary landfill and composting systems. The sanitary landfill has very slow decomposition rates because (1) only a small part of the wastes are high quality, organic materials, (2) some toxic, hazardous wastes may be present, and (3) little oxygen and water are available to support decomposition. Conversely, composting systems have relatively fast decomposition rates because all factors controlling the process are engineered to be optimal. These decomposer-based systems are described in this chapter with emphasis on ecological dimensions. The new field of industrial ecology is also introduced as a possible example of reverse engineering of ecosystems for solid waste management. Finally, comments are made on economics, especially in terms of the by-product values of solid waste materials.

THE SANITARY LANDFILL AS AN ECOSYSTEM

The most common method of solid waste management is burial in a sanitary landfill (Figure 6.2). In this way, solid waste is stored in a manner that isolates it from human exposure. The nonorganic fraction of waste in a landfill is essentially stored permanently but the organic fraction can decompose. Thus, the landfill is an ecosystem because it includes living processes. Although landfills are often viewed negatively by the general public, they do provide a necessary function and are the least expensive option available. There are also examples of storage systems in nature that have some analogies with the landfill and that provide useful functions (Table 6.2).

The conventional engineering aspects of landfills are well developed (Bagchi, 1990) and are the result of a long design history. Landfills evolved from dumps where solid waste is left in the open on the surface of the ground. Solid waste is buried in trenches or depressions of the landfill and covered every day with at least 15 cm (6 in.) of clean dirt. The daily covering is done to exclude pests and to prevent the outbreak of fires. The practice of landfilling began in the early 1900s, but it became commonplace after World War II. From the beginning, landfills were

TABLE 6.2
Examples of Storage Components from Natural Ecosystems That Have Some Analogies with Landfills

Example	Composition	Useful Function
Peat	Deposits of partially decomposed plant materials in certain wetlands	Substrate for living plants, low-grade energy source for humans
Hypolimnion of a eutrophic lake	The bottom of a lake that doesn't mix with the surface due to density stratification	Anaerobic nutrient transformations
Snags	Standing dead trees in a forest	Habitat for wildlife, source of firewood for humans
Oyster shell	Accumulations of shells from dead oysters that once made up living reefs	Substrate for live oysters, source of construction materials for humans
Guano	Deposits of feces from seabird colonies on oceanic islands	Source of fertilizers for humans

designed to be covered over and landscaped after they were filled with solid wastes, in order to provide some useful end function. Thus, one common practice was to site a landfill on wetlands in order to reclaim the land for human land use. This practice was recognized as faulty starting in the 1950s and 1960s after hazards of liquids draining from the landfills (i.e., leachates) were identified and after the values of wetlands began to be understood. Thus, early landfills often were sited in the worst possible locations! An example is the Fresh Kills landfill outside of New York City, which is the largest landfill in the world. Fresh Kills was located on tidal wetlands in 1948. It has become a major source of pollution to the local coastal watersheds and is scheduled to be closed and redeveloped (Fulfer, 2002). Modern sanitary landfills are now sited to avoid groundwater drainage networks and are lined with up to a meter of dense clay and sheets of plastic in order to collect leachate and prevent it from reaching the groundwater.

Landfills produce gases from the decomposition processes that occur inside them. The sequence of gas production from a landfill is a reflection of the succession of microbial communities involved in decomposition (Figure 6.3). Landfill gas is composed mostly of methane and carbon dioxide along with a number of components in much smaller quantities. These gases are evidence of the ecological processes occurring in the landfill and can be collected, purified, and used as an energy source by humans. El-Fadel et al. (1997) provide simulation models for these gas dynamics. Figure 6.4 illustrates the landfill system including gas release and leachate collection.

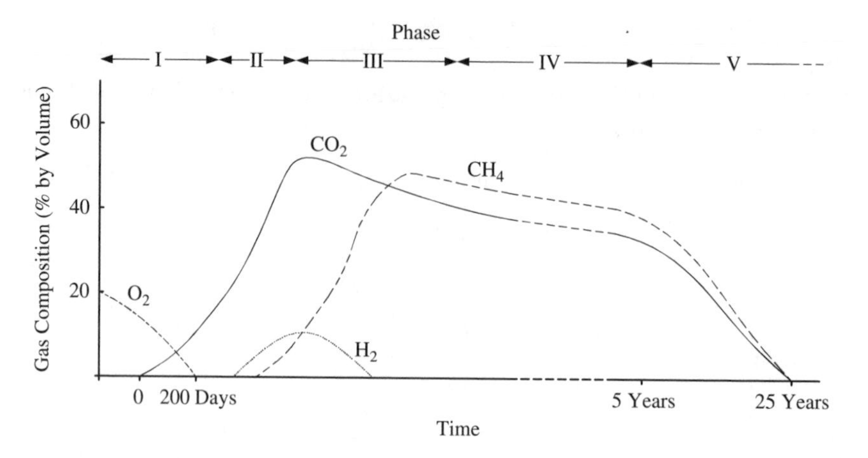

FIGURE 6.3 Sequential production dynamics of gases emitted from a sanitary landfill. (From Beeby, A. 1993. *Applying Ecology*. Chapman & Hall, London. With permission.)

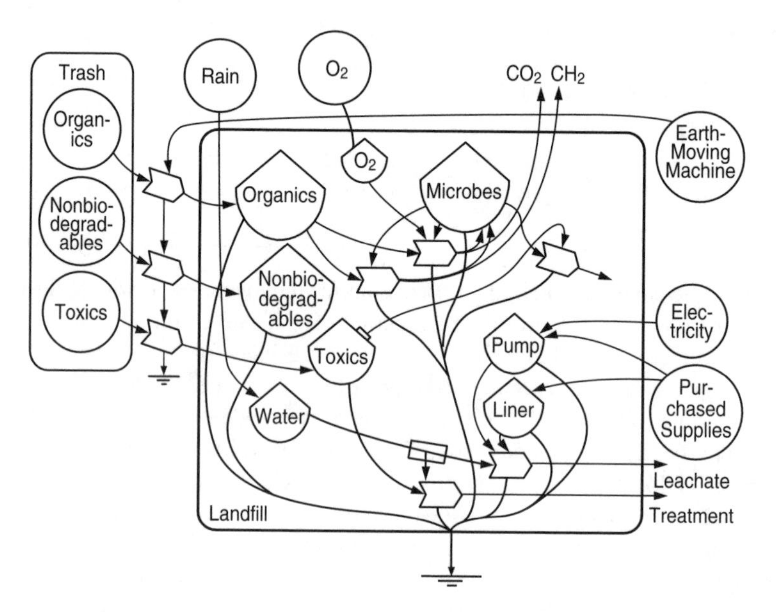

FIGURE 6.4 Energy circuit diagram of a landfill ecosystem.

Optimal sites for landfills are those which are socially acceptable, environmentally safe, and close enough to cities to have reasonable transportation costs. These sites are becoming difficult to locate, especially in highly developed regions such as the eastern U.S. and Europe where solid wastes have accumulated over long time periods. The current trend is to build and operate fewer but larger landfills and to transport wastes greater distances.

The most dangerous components of solid wastes to both humans and the environment are those categorized as hazardous, according to the U.S. Resource Conservation and Recovery Act. These wastes are now regulated and must be separated from other wastes and managed in special facilities. Wastes must exhibit at least one of the following characteristics to be considered hazardous: ignitability, corrosivity, reactivity, and toxicity. Most hazardous wastes come from industries but a small portion is generated in normal household solid wastes. Workers of the Garbage Project at the University of Arizona found a relationship between socioeconomic characteristics of neighborhoods and the composition of hazardous waste: low-income households produce more car-care items (such as motor oil and gas additives); middle-income households produce more items associated with home improvement (such as paints, stains, and varnishes); and upper-income households produce more lawn and garden items (such as pesticides and herbicides) (Rathje and Murphy, 1992). Nuclear waste provides a special case of the highest priority in terms of safety (Weber and Wiltshire, 1985). The proposed approach taken for these materials is geological disposal, where wastes are buried under very thick layers of soil and rock (Carter, 1987). However, risks remain with geological disposal (Shrader-Frechette, 1993), and the technology for nuclear waste management is still evolving. Current debate focuses on the Yucca Mountain, NV, site near Las Vegas which is scheduled to be the world's first geologic repository for high-level nuclear waste (Apted et al., 2002; Ewing and MacFarlene, 2002).

Under current conditions there is little opportunity for ecological engineering to contribute to landfill technology. The systems by necessity are dominated by conventional engineering designs and procedures. The best opportunities may come for leachate treatment, and Mulamoottil et al. (1999) describe constructed wetlands for this purpose. Also, Beeby (1993) suggests that methane (i.e., biogas) production from landfills can be optimized, but "the ecosystem itself has to be managed to favour the methanogenic bacteria." This kind of management theoretically is possible, but it is unlikely to occur for practical reasons.

COMPOSTING ECOSYSTEMS FOR ORGANIC SOLID WASTES

Composting is the process used by humans to break down organic solid wastes into materials that can be reused as soil amendments in agriculture or horticulture (Rechcigl and MacKinnon, 1997). Organic wastes that are composted include food waste, sewage sludge, yard wastes, and animal manures.

Principles of composting are well known and straightforward (Anonymous, 1991; Haug, 1980; Poincelot, 1974) and have been extensively reviewed, especially in the journals entitled *Compost Science* and *Biocycle*. One of the main authorities on technical aspects of composting has been Clarence Golueke (1977, 1991), who has approached the subject as a sanitary engineer.

The primary objective in the design of composting systems is to maximize the decomposition rate of the organic wastes by control of limiting factors. Thus, the goals are to maintain moist, aerobic conditions that are insulated to retain heat and to allow access by decomposer organisms. A wide variety of systems are employed, ranging from large-scale commercial facilities that are highly engineered to small-scale backyard systems used by gardeners. Commercial technology includes two approaches. The more costly mechanical composting involves the use of mechanized enclosed systems that provide control over major environmental factors. The less expensive open or windrow composting involves stacking the raw material in elongated piles (i.e., windrows) in which composting occurs. Because this type results in nonuniform heating of the organic materials, the piles must be mixed or turned periodically so that all of the mass is eventually exposed to the highest temperatures.

The beauty of composting is that it is very easy to see and to understand how a waste product can be converted into something useful. Because of this obvious value and because in most cases it is also safe and easy to do, small-scale composting is popular with gardeners everywhere. A large literature exists for the general public (Martin and Gershuny, 1992) including such titles as *Everyone's Guide to Home Composting* (Bem, 1978) and *Let it Rot!* (Campbell, 1975). The application of compost to soils can increase the organic content and improve the physical structure. Specific benefits that have been reported of compost as a soil amendment include increased aeration, improved moisture and nutrient retention, decreased soil erosion, reduced soil surface crusting, plant disease suppression, and improved tilth. Compost is often used to restore damaged or disturbed soils because of these special qualities.

One of the most important determinants of the rate of decomposition in composting is the chemical quality of the organic materials. The ratio of carbon to nitrogen (C:N) is often used as an index of the chemical quality, and values for various types of organic wastes are listed in Table 6.3. This index is useful because it is composed of two of the most important elements to the microbial decomposers and to living organisms in general: carbon, needed as a source of energy in metabolism, and nitrogen, needed to synthesize protoplasm. The optimal ratio for composting is about 25:1 to 30:1 (Golueke, 1977). The molecular structure of the carbon compounds in the organic wastes is also an important determinant of decomposition rate. Some molecular structures are more resistant to breakdown than others. Highly proteinaceous materials such as food wastes break down quickly and support many kinds of microbes. However, materials with cellulose (such as paper), lignin (such as wood), or aromatics (such as carbon compounds with ring structures) break down slowly because of their resistant chemical configurations and because only a few groups of microorganisms produce the enzymes needed to assimilate these molecular structures. Both indicators of chemical quality listed above (C:N ratio and molecular structure of carbon compounds) are not unique to composting but rather they are generally relevant for decomposition in any kind of ecosystem (Boyd and Goodyear, 1971; Cadisch and Giller, 1997; Enriquez et al., 1993; Jensen, 1929). Russell-Hunter (1970) also provides a review of the C:N ratio in terms of animal nutrition.

Composting is an example of ecological succession because a series of microbial taxa contribute to the breakdown of organic wastes in an organized sequence. Changes in the physical–chemical conditions of the compost occur over time, caused by the metabolic activities of microbes, and different taxa are adapted to only a limited range of these conditions. Composting is an especially interesting example of succession because of the biogenic changes in temperature that are characteristic of the process. Figure 6.5 illustrates temperature and pH changes in a typical composting sequence, with four successional stages (A to D) listed on the time axis.

TABLE 6.3
Carbon (C) to Nitrogen (N) Ratios for Various Kinds of Organic Materials

Material	C:N Ratio
Urine	0.8
Activated sludge	6
Raw sewage sludge	11
Nonlegume vegetable wastes	11–12
Poultry manure	15
Cow manure	18
Mixed grasses	19
Horse manure	25
Potato tops	25
Straw from oats	48
Straw from wheat	128–150
Sawdust	200–500

Source: Adapted from Golueke, C. G. 1977. *Biological Reclamation of Solid Wastes*. Rodale Press, Emmaus, PA.

This kind of succession occurs when organic wastes are gathered up into heaps or piles so that an insulating effect emerges with conservation of heat and a rise in temperature. Heat is a by-product of the metabolic reactions of the microbes as they decompose the organic wastes. The pH also changes over the composting succession, beginning acid and becoming more alkaline over time. The first stage of composting succession is dominated by mesophilic microbes, primarily aerobic and anaerobic bacteria that metabolize simple carbohydrates such as sugars and starches. The optimal temperature for these organisms is about 35°C (95°F). Thermophilic (i.e., "heat loving") microbes dominate next and metabolize proteins and other nitrogenous materials. The optimal temperature for these organisms is about 60°C (140°F). No living organisms can exist above 70°C (175°F), so microbial metabolism stops and a cooling down phase follows the thermophilic stage. During this period actinomycetes and fungi increase in numbers and metabolize cellulose and other more resistant carbon compounds. The final stage shown in Figure 6.5 occurs with the formation of humus, which is the most valuable form of compost.

In a mechanized composting plant the complete succession sequence can take place in about one week, while in an open or windrow operation the sequence can require on the order of a month to complete. Unlike composting systems, in most natural ecosystems organic materials seldom accumulate under aerobic conditions to a sufficient extent for insulation to occur, so temperature remains at ambient

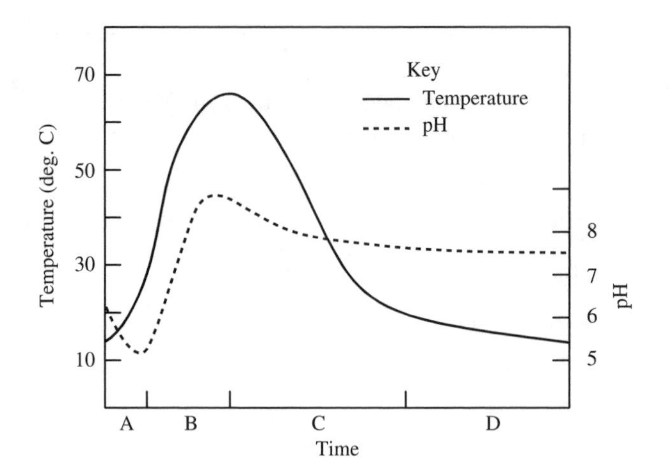

FIGURE 6.5 Patterns of change in temperature and pH over time in a closed composting system. (Adapted from Gray, K. R. and A. J. Biddlestone. 1974. *Biology of Plant Litter Decomposition*. C. H. Dickinson and G. J. F. Pugh (eds.). Academic Press, London.)

levels. Under these conditions decomposition is carried out by mesophilic microbes at moderate temperatures and psychrophilic microbes at lower temperatures. An exception is decomposition in muskrat mounds (see Chapter 2). These mounds are the most obvious construction feature of muskrats in temperate zone marshes and act like compost piles (Figure 6.6) in accelerating the decomposition rate (Berg and Kangas, 1989; Wainscott et al., 1990). The mounds are layered with mud and vegetation similar to a classic Indure compost pile (Martin and Gershuny, 1992). The vegetation used to construct the mound is cut while fresh and alive, and thus it has higher nutrient content than vegetation not used in mound construction, which undergoes physical leaching in a standing dead stage before decompositon. The vegetation, which is used in mound construction, is also "shredded" to some extent by the muskrat and by macroinvertebrates that live inside the mound, which may facilitate colonization by microbial decomposers. Finally, the mound itself is moist but aerobic with some insulation effect as in a compost pile. Other examples of compost pile analogs are the nests built by megapode birds in Australasia (Collias and Collias, 1984). (*Megapode* refers to the big feet, which the birds use to construct large piles of plant materials for their nests.) Alligators in the southeastern U.S. also build similar nests. An open question is who designs the best compost piles: human sanitary engineers or animals such as the muskrat? Perhaps this question could be resolved by careful analysis with heat transfer equations from conventional engineering.

Composting is basically a natural process that is controlled by humans. Most attention has been given to managing or engineering for microbial decomposition of organic wastes. This emphasis is reflected in the standard texts (Golueke, 1977; Insam et al., 2002), which only consider the roles of microbes (see, however, the children's book by Lavies, 1993). Microbes also are the main driving force in decomposition in natural ecosystems but much more biodiversity is involved. Inver-

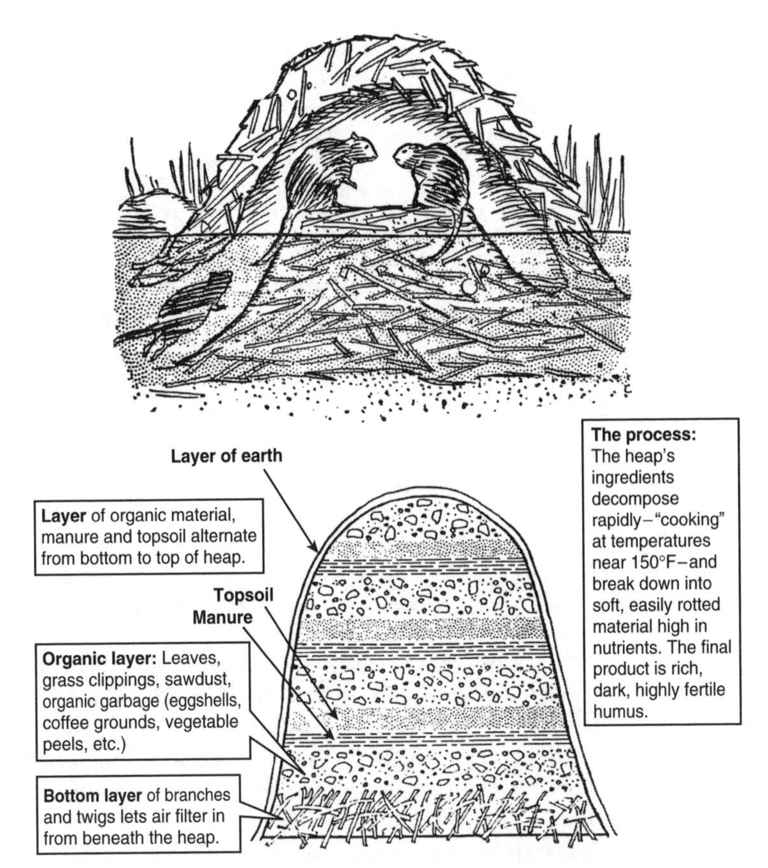

FIGURE 6.6 Comparison of a muskrat mound (above) with a compost pile (below). The top part of the figure is from Hodgson, R.G. 1930. *Successful Muskrat Farming*. Fur Trade Journal of Canada, Toronto. The bottom part of the figure is from United Press International. With permission.

tebrate animals dominate the complex detritus food webs in natural ecosystems, in contrast to the relatively short detritus food chains of microbes found in most human-designed composting systems. Detritus food webs occur primarily in the soil for terrestrial ecosystems and in sediments for wetland and aquatic ecosystems (Anderson and MacFadyen, 1976; Brussaard et al., 1997; Palmer et al., 1997; Snelgrove et al., 1997). Waring and Schlesinger (1985) illustrate the range of invertebrate animal diversity involved in the breakdown of leaf litter detritus with a graph of 50 taxa which spans three orders of magnitude in size — from protozoans at the small end to crayfish and earthworms at the large end of the spectrum. The work of detritus food webs has been called detritus- or leaf-processing which includes physical breakdown, mixing, and consumption of organic matter (Boling et al., 1975; Maltby, 1992; Petersen and Cummins, 1974; Petersen et al., 1989). Successions of different organisms are involved in detritus processing, each with different functional roles (Anderson, 1975; Frankland, 1966; Visser and Parkinson, 1975; Watson et al., 1974). Cousins (1980) also has referred to this kind of processing as a *detritus cascade* with emphasis on the different sizes of organisms that are involved. Cummins (1973;

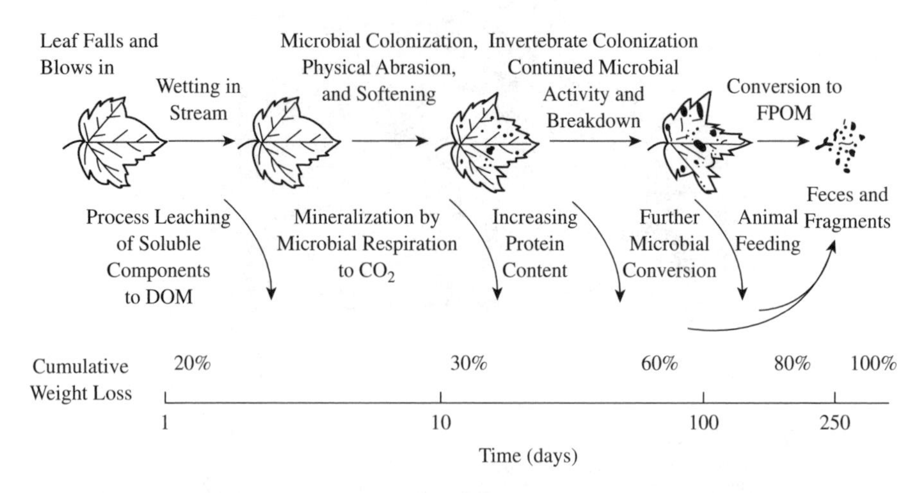

FIGURE 6.7 Leaf litter breakdown in a stream ecosystem. (From Cushing, C. E. and J. D. Allan. 2001. *Streams: Their Ecology and Life*. Academic Press, San Diego, CA. With permission.)

Cummins and Klug, 1979) has developed a classification of functional feeding types of stream invertebrates which illustrates the different roles. His classification includes shredders, collectors, scrapers, and predators, depending on mouthparts and behavior of the animal. The shredder category is particularly important in fragmentation of large pieces of detritus into smaller particles (Anderson and Sedell, 1979; Cummins et al., 1973, 1989; Wallace et al., 1982). Particle size is important in decomposition because it determines the surface area per unit mass exposed to microbial colonization and metabolism. In some commercial-scale composting facilities this kind of fragmentation is carried out with mechanical grinding machines but it is an expensive step that is not always possible. However, in nature it is an inherently important contribution that accelerates the decay process. Figure 6.7 depicts the leaf breakdown process that occurs in freshwater streams. Physical leaching quickly causes an initial weight loss during the first few days. Mineralization by microbes follows after they begin to colonize the leaf surface. Invertebrate animals colonize later and breakdown the main structure of the leaf through their feeding. Complete detritus processing requires approximately 1 year in the temperate zone and follows the exponential decay model described in Chapter 2. A more detailed graphic display of leaf breakdown for a terrestrial forest ecosystem is given by Schaller (1968). Bormann and Likens (1979) identify several fragmentation processes: fenestration, perforation, and deskeletonization and refer to these actions as a kind of "coordinated attack" in the quote given below:

> … it would appear that as soon as soft tissues such as leaves or bud scales fall to the forest floor they are subject to a coordinated attack. … Fungi and bacteria initiate the action but are soon joined by springtails, bark lice, and various larvae which eat or tear holes in the tissue (fenestration), opening it to more rapid microbial attack. Larger larvae and mites bring about further perforation and skeletonization. Large amounts of

> feces, or frass, are produced, which may be consumed again by other fauna. The activities of the soil fauna and microflora are thus closely linked. Chewing, ingestion, and digestion by fauna not only result in decomposition of the organic matter but simultaneously create surface and moisture conditions more favorable to microbial action both within the faunal gut and in the resultant frass. It seems likely that the detritivores obtain their principal energy supplies from the easily decomposable substances within the litter such as sugars, starches, and simple and crude proteins. Exoenzymes of fungi and bacteria not only attack these easily decomposable substances but are largely responsible for the decomposition of the more resistant compounds, such as hemicellulose, cellulose, and lignins, which compose the bulk of the leafy and wood litter.

Thus, invertebrates are important regulators of the decomposition process in natural ecosystems (Anderson and MacFadyen, 1976; Coleman, 1996; Edwards et al., 1970; Lussenhop, 1992; Seastedt, 1984; Visser, 1985). In addition to the physical breakdown or fragmentation which facilitates microbial colonization, invertebrates (1) create zones of active microbial growth through mixing and other actions, (2) select for fast-growing microbial populations through direct grazing, and (3) fertilize microbes through release of nutrients in their excretion. When highly focused, these kinds of control actions have been referred to as "microbial gardening" (Hylleberg, 1975; Reise, 1985; Rhoads et al., 1978) in that the animals directly channel microbial production into their own growth. An example of this interaction can be seen with the leaf-cutter ants (Attine ants: genera *Atta* and *Acromyrmex*) of the tropics that intentionally cultivate fungi for food in their belowground nests with leaves that they cut from the surrounding trees (Lugo et al., 1972; Weber, 1972).

The control of decomposition in soils by earthworms is well known. Charles Darwin (1881) provided some early quantification of the role of earthworms in the last book he wrote before his death in 1882. He felt earthworms were the most important animals on Earth because of their contribution to soil fertility. Earthworms are considered to be keystone species in terrestrial ecosystems because of their (1) physical effects on soils, (2) biogeochemical effects on nutrient cycles, and (3) enhancement of species diversity (Blondel and Aronson, 1995). A huge literature exists on the ecology of earthworms (Edwards, 1998; Satchell, 1983), including popular books with titles such as *Worms Eat My Garbage* (Appelhof, 1997) and *Harnessing the Earthworm* (Barrett, 1947). Earthworm biotechnology (Hartenstein, 1986) includes vermicomposting, where earthworms are managed to accelerate composting, and vermistabilization, where they are managed for sewage sludge processing (Reed et al., 1995).

The paradigm that emerges from this review is that animals manage microbial work in natural ecosystems (Figure 6.8A). In some cases anaerobic microbes occur within the digestive systems of animals (Figure 6.8B). This is an important type of symbiotic relationship to which both taxa contribute. The animals bring food to the microbes and provide anaerobic microenvironments that are necessary for their survival in otherwise aerobic environments. The microbes break down the food with special enzymes that the animals lack. Although many animals have symbiotic gut microbes, the most highly developed example is the herbivore group of ruminants, including cattle, sheep, and deer. These animals have four stomachs and a symbiotic food web

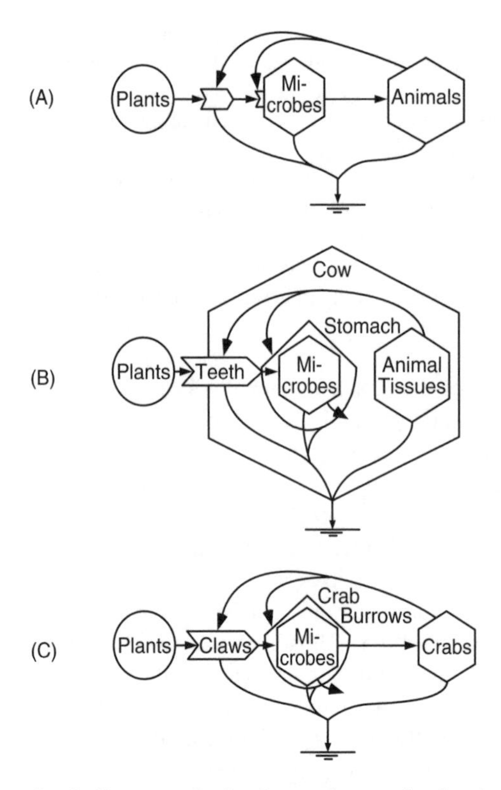

FIGURE 6.8 Energy circuit diagrams of animal control over microbes in various ecosystems. (A) Conceptual view. (B) Digestive microbes in a ruminant system. (C) Soil microbes in crab burrows. (From Odum, H. T. 1983. *Systems Ecology: An Introduction*. John Wiley & Sons, New York.)

of bacteria and protozoans with densities of more than 100 million organisms/ml of stomach solution (Hoshino et al., 1990; Hungate, 1966). Ruminant digestion consists of a sequential set of processes: mastication (chewing), pregastric fermentation (breakdown of cellulose in the first stomach), regurgitation of "cud," and acid hydrolysis in the last stomach. This pattern has been suggested as being comparable to the processes in a sewage treatment plant and even has been an inspiration for the design of bioreactors (Beeby, 1993)! Also relevant here is the work of Penry and Jumars (1986, 1987; Jumars, 2000) on modeling animal guts as chemical reactors. Other animals control microbes in external environments (Figure 6.8C). Examples of this kind of control include shredding and gardening mentioned earlier. Burrowing by animals in soils, sediments, and other materials also provides microenvironments that enhance microbial activities (Meadows and Meadows, 1991). Various invertebrates have been classed as "ecosystem engineers" because of these kinds of roles in aquatic sediments (Levinton, 1995) and soils (Anderson, 1995).

The question for ecological engineering is whether elements of animal control over microbes can be incorporated into composting systems. Does the tremendous untapped biodiversity of animals from natural ecosystems represent an opportunity for improved composting? The challenge is to design and test new composting ecosystems that have higher biodiversity and more effective decomposition efficiency. Many possible systems can be imagined. Can carrion-based food webs (Payne, 1965) be used to accelerate composting of carcasses at animal farm operations (Murphy and Carr, 1991)? Can burrowing clams (Pholadidae) (Komar, 1998) or the high diversity of bioeroders on coral reefs (Glynn, 1997) be used to break down limestone and concrete construction materials? Can marine boring organisms such as gribbles (isopods of the family *Limnoridae*) or shipworms (clams of the family *Teredinidae*) (Ray, 1959) or terrestrial termites (*Termitidae*) (Lee and Wood, 1971) be used to break down wooden construction materials?

Although some commercial-scale municipal composting facilities do operate in the U.S. and elsewhere, the economics is not very favorable. An exception is composting of sewage sludge which is generated by conventional wastewater treatment plants and by septic tank owners. This is a major industry with well-developed technologies (Clapp et al., 1994; Smith, 1996). Stabilization of sewage sludge requires dewatering, after which composting is often utilized. One fairly common practice is to spread the stabilized sludge on agricultural fields for further decomposition and for use as a soil amendment. This mimics the old practice of manuring, whereby animal wastes from farms are applied to fields (Klausner, no date). This kind of composting can have positive benefits as long as proper application rates are used. The application of excess sludge or animal manure can lead to environmental problems such as nonpoint source pollution. This issue brings to focus the sometimes conflicting motivations of composting. On the one hand, composting is (1) a way of disposing of a waste product while, on the other hand, it is (2) a way of producing a useful product. The potential exists for conflicts to arise between these two motivations. For example, sewage sludge or animal manure may be spread on land with the apparent motivation of improving the soil (motivation [2] above) when actually it is done just to dispose of waste materials (motivation [1] above). In this case, excessive applications can easily occur, creating environmental impacts rather than subsidies. This same phenomenon can occur with the use of dredge material for marsh restoration, as was described in Chapter 5. Composting systems for sewage sludge and animal manures may represent the best opportunities for the incorporation of ecological engineering improvements because these operations are large-scale and common. In this regard reed-based wetland systems are widely used for dewatering sewage sludge at the present time.

The compost toilet is a commercially successful composting system that is particularly relevant for ecological engineering (Del Porto and Steinfeld, 1999; Jenkins, 1994; Stoner, 1977; Van der Ryn, 1995). These are toilets which are meant as a substitute for the flush toilet, requiring little or no water. Human feces are collected and stabilized in the composting toilet and, later, used as a soil amendment. Many designs are available but most take the form of an outhouse with storage chambers, aeration with ducts or venting and the addition of a material such as sawdust that acts as an absorbent (Figure 6.9).

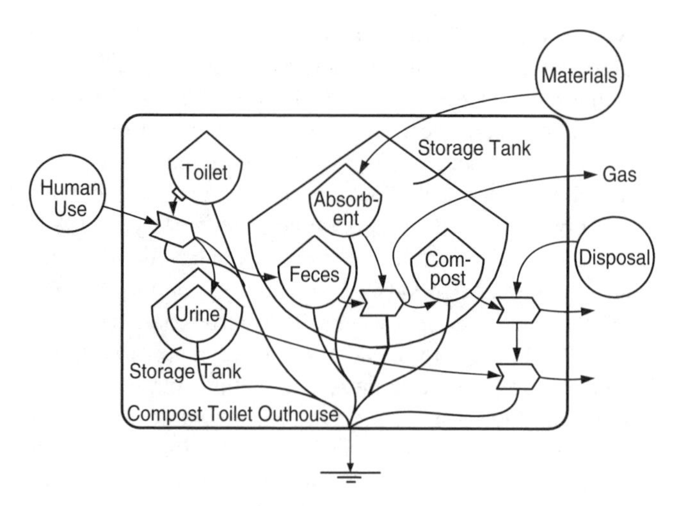

FIGURE 6.9 Energy circuit diagram of a composting toilet.

INDUSTRIAL ECOLOGY

Industrial ecology is an emerging new paradigm that concerns methods for increasing industrial efficiency and improving the relationships between industry and the environment (Graedel and Allenby, 2003; Lifset, 2000; McDonough and Braungart, 2002; Richards, 1997; Richards and Pearson, 1998; Socolow et al., 1994). It is relevant in this chapter on solid waste management because it emphasizes energy and material flows often with a focus on two management methods described earlier: source reduction and recycling (Graedel, 1994). One version of industrial ecology is for a particular industry to become "green" by improving its material cycle (Goldberg and Middleton, 2000; Stevens, 2002). The existing field arose in the 1990s with initiatives from industrial engineers, so it is a very recent development with many ideas and proposals but few working examples. A journal entitled *Industrial Ecology* was begun in 1998, and the International Society for Industrial Ecology was formed in 2001.

One of the most interesting features of industrial ecology in the context of ecological engineering is that it uses ecology as a model for examining industries, as noted in the quote by Richards et al. (1994) below:

> Industrial ecology offers a unique systems approach within which environmental issues can be comprehensively addressed. It is based on an analogy of industrial systems to natural ecological systems …
>
> There are obviously limits to this analogy, but it can help illuminate useful directions in which the system might be changed. Consider, for instance, waste minimization at a scale larger than that of a single unit or facility in light of the biological analogue. A mature natural ecological community operates as a waste minimization system. In general, the waste produced by one organism, or by one part of the community, is not disposed of as waste by the total system as long as it is a source of useful material

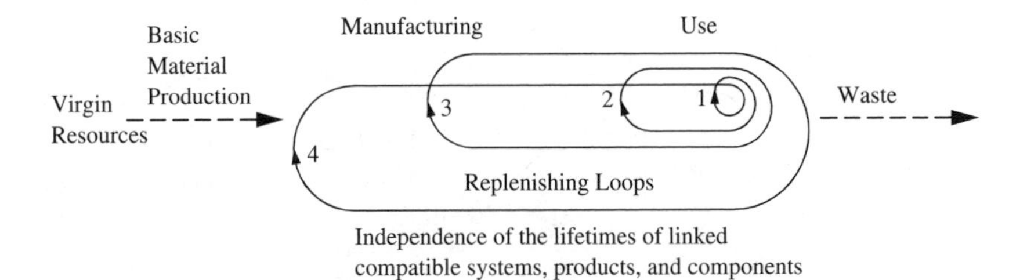

FIGURE 6.10 The spiraling process from industrial ecology, which has similarities with material processing in ecosystems. (From Stahel, W. R. 1994. *The Greening of Industrial Ecosystems*. B. R. Allenby and D. J. Richards (eds.). National Academy Press, Washington, DC. With permission.)

> and energy. Some organism, some part of the ecological system, tends to evolve or adjust to make a living out of any particular waste ...
>
> In an industrial ecology, unit processes and industries are interacting systems rather than isolated components. This view provides the basis for thinking about ways to connect different waste-producing processes, plants, or industries into an operating web that minimizes the total amount of industrial material that goes to disposal sinks or is lost in intermediate processes. The focus changes from merely minimizing waste from a particular process or facility, commonly known as "pollution prevention," to minimizing waste produced by the larger system as a whole.

Industrial ecology uses ecological principles but not actual living organisms to design new systems. David Tilley (personal communication) suggests that in industrial ecology "the ecosystem provides the software for industrial design." Thus, industrial ecology can be thought of as an abstract form of ecological engineering or perhaps as a kind of reverse engineering based on natural ecosystems. For example, Figure 6.10 illustrates some of the strategies of this new field including waste prevention and reduction, product-life extension, and recycling, and is quite reminiscent of the resource spiraling concept from ecology (see Chapter 2). A more concrete example is given by Klimisch (1994) in which the automobile industry is depicted like an ecosystem with trophic levels, material flows, and recycling. Benefits will accrue to both society and the environment if industrial ecology as a field can mature and result in working models. Ecological engineers may be able to help develop the new ideas of industrial ecology since they have backgrounds in both ecology and engineering.

Although industrial ecology is clearly a recent development there are historic precedents for the field. Henry Ford may have been the first true industrial ecologist as evidenced by a number of efforts undertaken from about 1910 until his death in 1948 [though Friedlander (1994) traces origins of the field back to Benjamin Franklin in colonial America]. Ford is best known for developing affordable cars, utilizing mass production techniques, and creating one of the first industrial empires in the U.S. However, he maintained an interest in agriculture throughout his life and continually tried to create compatible system of agriculture and industry (Bryan, 1990; Wik, 1972). For example, his village industry idea has been proposed as a

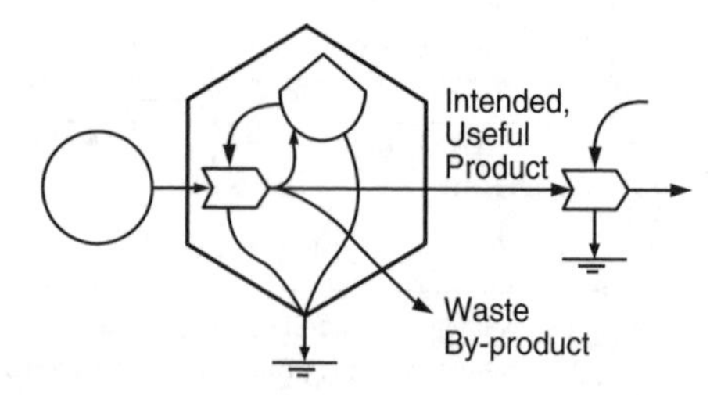

FIGURE 6.11 Energy circuit diagram illustrating the concept of valuation of waste by-products.

model for sustainable development (Kangas, 1997) because it combined farms with local manufacturing plants fueled with renewable energy sources. Ford was also a leader in the "Chemurgy" movement that was popular during the Depression period. *Chemurgy*, which is a term analogous to *metallurgy*, attempted to produce and utilize industrial materials derived from agricultural crops. Soybeans were the focus of a major research effort by the Ford Motor Company. This crop was used to produce a number of items such as oils, paints, varnishes, and plastic-like parts for automobiles. Ford wanted to control all of the raw materials that went into his cars and to utilize as much of the waste products from the automobile as possible. According to Wik (1972), because of Ford's interest in recycling,

> … engineers in his company tried to salvage everything from floor sweepings to platinum. Wood shavings were converted into charcoal briquets, formaldehyde, creosote, and ethyl acetate. Coal derivatives yielded coke, benzol, and ammonium sulfate, while the slag from steel furnaces was used for surfacing roads. In 1925 the company sold coke commercially as well as ammonium sulfate as fertilizer. Eighty-eight gas stations in Detroit sold benzol to auto drivers at the same price as gasoline. Seven tons of Dearborn garbage were distilled daily in the River Rouge plant where it yielded alcohol, refined oil, and gas suitable for heating purposes. Residues were mixed with sand and sold as fertilizer to greenhouses. Tests were made to extract soap from the sewage in Detroit. Sale of the various Ford by-products in 1928 amounted to $20 million. *The New York Times* in 1930 claimed Ford threw nothing away, not even the smoke from his factories.

A thorough review of Henry Ford's work may provide many examples that can inspire modern industrial ecology efforts.

ECONOMIC CONCEPTS AND THE PARADOX OF WASTE

The simple diagram shown in Figure 6.11 illustrates the idea of wastes. A process exists (the work gate on the left) in which energy is transformed to create an intended, useful product, which is subsequently used as an input in another process (the work gate on the right). The product is considered useful because it serves as an input to a productive process. Waste material also is created by the initial process, but because it doesn't serve as an input to a subsequent productive process, it is considered to be useless (or in other words, a waste), and therefore, it is a by-product. While this is logical, a paradox arises when it is realized that the same amount of energy, from the source to the left and from the feedback out of the storage, is required to make both the intended, useful product and the unintended, waste by-product. From the perspective of the energy theory of value (H.T. Odum, 1996; see Chapter 8), both the useful product and the waste by-product can be thought to be equally valuable because they both required the same amount of energy to produce. Kangas (1983a) applied this theory to strip-mined landscapes in central Florida. He proposed that the spoil mounds formed by mining had value in proportion to mining energy inputs and that they should not be leveled for reclamation. Plant communities colonized the mounds and, with sufficient time, succession could produce forests with structure comparable to natural forests. Ultimately, however, the argument was not convincing enough to change mine reclamation policy. Obviously, the energy theory contrasts dramatically with the utility-based value system of market economics, where the balance of supply and demand determines value or price. In human society, which is driven by market economics, wastes occur when no demand exists for their use. In fact, these wastes actually have a negative value because there are costs associated with their disposal.

In natural ecosystems, wastes seldom arise because some population or process evolves to use by-products for a productive purpose. For example, in the basic P–R model of the ecosystem (see Chapter 1), primary productivity (P) produces oxygen as an apparent by-product that is used as an input by respiration (R) and respiration produces carbon dioxide as an apparent by-product that is used as an input by primary productivity. There is no waste in this P–R system because by-products are utilized. The energy theory of value represents this evolutionary perspective of the natural ecosystem by assigning equal values to all outflows from a process. Industrial ecology is striving to achieve this goal in human economics, wherein all waste by-products are used as resources (Allen and Behmanesh, 1994). In a sense, the loss of value that occurs when by-products are wasted, such as in landfilling, is the cost to human society for the evolution of closed material loops characteristic of natural ecosystems.

Of course, some wastes are being recycled but markets are not widely available (Aquino, 1995; Lund, 2001). Many creative uses have been found for certain wastes, often outside of market transactions (Piburn, 1972), such as the use of waste tires for artificial reefs (see Chapter 5), composting examples described earlier in this chapter, and the recycling of oceangoing vessels by "ship breakers" (Langewiesche, 2000). An extreme case occurs when trash is used as art (Greenfield, 1986), which reveals a surprising aesthetic value of waste. Finding productive uses for waste by-products is a goal of ecological engineering, and examples are described throughout the text.

7 Exotic Species and Their Control

INTRODUCTION

The invasion of ecosystems by exotic species is a major environmental problem that has become widely recognized (Culotta, 1991; Mack et al., 2000; Malakoff, 1999). This phenomenon is occurring globally and causing changes to ecosystems, along with associated economic impacts. The most important issue with the invasion of exotics is the replacement of native species, in terms of either reduction of their relative abundance or, in the extreme, their outright extinction. Associated costs to human economies from the invasion of exotics include losses of value derived from the natives they replace, direct damages caused by them, and expenditures for control programs directed at exotics (Pimentel et al., 2000). The invasion of exotic species occurs because of introduction by humans, either intentionally or unintentionally. Of course, intentional introductions are undertaken in an effort to add a useful species to an ecosystem, and there are positive examples of this action such as the introduction of honey bees as a pollinator for crop species. Problems arise, however, when intentionally introduced species take on unintended, expanded, and negative roles in ecosystems or when this occurs with unintentional introductions.

Perhaps because it is an environmental problem caused by excessive growth or "biology gone wrong," the invasion of exotics has become sensationalized by environmentalists and the news media with seemingly good reason. This situation is reflected in titles of news stories about exotics such as "Unstoppable Seaweed Becomes Monster of the Deep" (Simmons, 1997) and other evocative descriptions such as "the Frankenstein effect" (Moyle et al., 1986) and the need to consider exotics as "guilty until proven innocent" (Ruesink et al., 1995; Simberloff and Stiling, 1996). A further example is the announcement of "America's Least Wanted" (Table 7.1), which is a list of the dirty dozen of the country's worst exotics, according to the Nature Conservancy (Flack and Furrlow, 1996). The problem of invasion of exotics has captured the imagination of the public and the scientific community and is receiving greater and greater attention. Figure 7.1 illustrates this growing interest by plotting the number of books published on exotics by decade since World War II (Appendix 1). Although this listing may not be complete, the pattern is clear with relatively little publishing until the 1980s and especially the 1990s when there was an explosion of writing about exotics. This growing literature includes mostly the standard scientific writing but also popular books (e.g., Bright, 1998), books commissioned by the federal government (National Research Council [NRC], 1996a; Office of Technology Assessment, 1993), and even a children's book (Lesinski, 1996). The latter clearly reflects a trickle-down effect and a growing awareness of the issue. This trend is also seen in a growing body of policy and legislation such

TABLE 7.1
List of the Worst Invasive Exotic Species in the U.S.

Zebra mussel	*Dreissena polymorpha*
Flathead catfish	*Pylodictis olivaris*
Purple loosestrife	*Lythrum salicaria*
Hydrilla	*Hydrilla verticillata*
Rosy wolfsnail	*EugCarcinus maenas*
Green crab	*Landina rosea*
Tamarisk	*Tamarix* sp.
Balsam wooly adelgid	*Adelges piceae*
Leafy spurge	*Euphorbia esula*
Brown tree snake	*Boiga irregularis*
Miconia	*Miconia calvescens*
Chinese tallow	*Sapium sebiferum*

Note: This list has been called "America's Least Wanted" and "The Country's Twelve Meanest Environmental Scoundrels."

Source: Adapted from Flack, S. and E. Furlow. 1996. *Nature Conservancy.* 46(6):17–23.

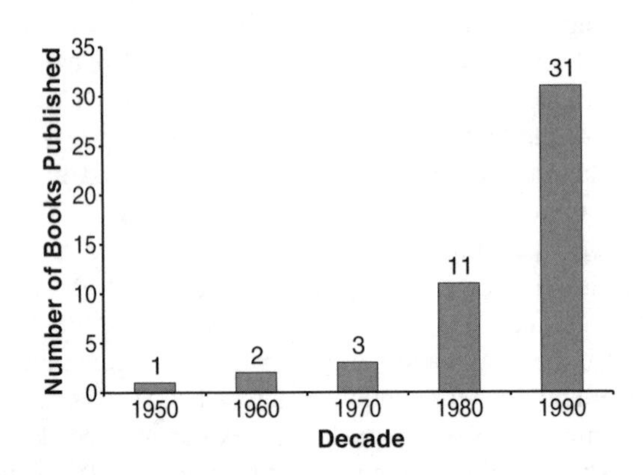

FIGURE 7.1 Exponential increase in the publication of books about exotic species. (See Appendix 1 of this chapter for a list of titles.)

as the National Invasive Species Act of 1996 (Blankenship, 1996) and the proposed Species Protection and Conservation of the Environment Act (Paul, 2002).

Although interest and concern about exotics have recently exploded, the problem is an old one, probably as old as human civilization. For example, Haemig (1978) describes introductions by pre-Colombian people in Mexico several thousand years ago. Modern awareness about exotic species as an environmental impact dates to

Charles Elton's monograph from 1958. Elton defined biological invasions as occurring when species move from an area where they evolved to an area where they did not evolve, and this still may be the best definition of the concept. Although some of the approaches Elton used to explain invasions may be outdated by standards of current ecological theory, his book was clearly far ahead of its time. Recent interest in exotics by ecologists dates to the 1970s when W. E. Odum coined the term *living pollutants* to describe the problem (W. E. Odum, 1974). Also, Courtenay and Robins published what may be the first general paper on exotics in 1975. Finally, Holm et al. (1977) may have presaged the Nature Conservancy's Dirty Dozen list of exotics with their listing of "The World's Worst Weeds."

The greatest fear from exotics for environmentalists, conservation biologists, and natural resource managers is "the homogenization of the world" (Culotta, 1991; Lockwood and McKinney, 2001). In this view a relatively few exotics spread throughout the world's ecosystems reducing native biodiversity. This phenomenon has already occurred with humans, who are exotics in most ecosystems. The fear of homogenization of the world's biodiversity seems real as exotics are clearly occurring as a global environmental problem (Schmitz and Simberloff, 1997; Soule, 1990; Vitousek et al., 1996). This fear cannot be denied but there is still much to understand about the ecology of exotic invasions. For example, MacDonald and Cooper (1995) suggest that alien-dominated ecosystems may be unstable over long time periods and therefore perhaps only a temporary problem. Many new ecosystems, which need to be described and explained, are being formed by the combination of exotics and natives. The prevailing view of exotics as negative additions to ecosystems has been accepted rather uncritically by the scientific majority, and the small amount of published literature on any controversy has been largely ignored (Lugo, 1988, 1990, 1994). Alternative views of exotic species can be imagined (Table 7.2) and some of these are examined in this chapter. The study of exotic species seems to be a wave of the future, and it will be a challenge to ecological theory for some time.

STRATEGY OF THE CHAPTER

A chapter on exotic species is included in this text for several reasons. The systems they come to dominate are not consciously designed by humans, but they are still human-generated systems due to increased dispersal and disturbance. In fact, exotic-dominated ecosystems represent the ultimate in self-organization, one that can become a threat to certain human values. Exotic species often dominate systems because of their high degree of preadaptation to new conditions created by humans. Thus, these species embody several of the important ecological engineering principles introduced in Chapter 1.

Under certain conditions, invasive exotic species provide a significant challenge to environmental managers because of their explosive growth. However, there is potential to take advantage of the successful qualities of these species. It is possible to imagine designs that utilize exotic species under appropriate circumstances, but this use must be carefully employed so as not to increase the problems these species

TABLE 7.2
A Comparison of Different Views Concerning Invasive Exotic Species in Ecosystems

Conventional Thinking	Alternative Hypothesis
Ecosystems infected with exotics are imbalanced systems that must be restored.	Ecosystems infected with exotics are examples of a new class of ecosystems heavily influenced by humans and have value of their own.
Our knowledge of exotics is sufficient to develop management strategies and value judgments on them.	Almost all research on exotics has been at the population scale, with little emphasis on ecosystem relations. More research is needed on ecosystems with high amounts of exoticism (as opposed to endemism).
Exotics are problems that must be exterminated.	Exotic-dominated ecosystems may reveal some aspects of ecology that we have not seen previously; they are a scientific tool for doing ecological theory.
Exotics should not be used in restoration projects; only native species should be used.	Exotics sometime grow faster or have special qualities that may speed up restoration. The key may be to managing exotics. This may be the most effective way of restoring ecosystems.
Ecosystems infected with exotics are less valuable because of their ability to outcompete or harvest to extinction native species.	Exotics may improve certain overall ecosystem parameters such as biomass, production, decomposition, stability, and even diversity.
All exotics should be controlled or kept out of natural systems to reduce their impacts.	The best way to manage exotics may be to add more exotics, so that more control networks (food webs) will arise.
Exotic-free ecosystems are attainable.	There is no way to keep exotics out or to remove them once they have invaded. Exotics may be inevitable. Humans are exotics.

can cause to natural ecosystems (Bates and Hentges, 1976; Ewel et al., 1999). This chapter examines the positive and negative contributions exotics make to biodiversity and outlines the new form of organization they represent. Exotic species provide opportunities to learn about basic ecological structure and function, if viewed objectively, and their success is a challenge to existing ecological knowledge. Finally, ideas of control strategies are reviewed. These strategies vary in their effectiveness and may be better described as management rather than engineering. As a group, exotics are forms of biodiversity that have escaped control by factors that would have regulated their populations. Thus, concepts of control in ecology and engineering are discussed for perspective.

EXOTICS AS A FORM OF BIODIVERSITY

Exotic species affect biodiversity in two opposite ways. On one hand, through their invasion of a community they can reduce biodiversity by reducing populations of native species. On the other hand, through their invasion of a community they increase biodiversity by their own addition to the system. The former process (of exotics' reducing native biodiversity) is often seen as the central problem of the invasions. Reduction in biodiversity is sometimes difficult to attribute solely to exotics because other factors such as pollution, disturbance by humans, and habitat loss also may be involved. However, exotics certainly contribute to declines in native diversity to a greater or lesser extent through competition or predation when they invade natural systems.

The process of exotics' adding biodiversity to communities is much less studied and discussed than their role in causing biodiversity declines. Of course, exotics are biological species as are natives, and they are as intrinsically interesting and valuable as any species taken within an appropriate context. When an exotic invades a community, its addition represents an increase in the community's biodiversity. At least in some cases this process can greatly increase diversity. This phenomenon is especially characteristic of islands which naturally have few species due to dispersal limitations (see the discussion of the theory of island biogeography in Chapters 4 and 5). Fosberg (1987) cites a dramatic example of this situation for an isolated island (Johnson Island) in the central Pacific Ocean. When first visited by a botanist there were only three species of vascular plants on the island. The island became occupied by humans as a military base during World War II, and by 1973 the number of vascular plants had increased to 127. Fosberg (1987) termed this "artificial diversity" because it was attributable to species brought in by humans. He goes on to describe a "pantropical flora" of plants that "... are either commensals with man, cultivated useful or ornamental plants, or what have been called camp-followers, door-yard or garden weeds, or else aggressive pioneer-type plants that produce many long-lived seeds and thrive on disturbed ground, or even in bare mineral soil." This is not a particularly attractive description of biodiversity, but the new communities on Johnson Island and in other locations have higher diversity that deserves to be studied. A continental example for Arizona fishes was described by Cole (1983):

> Thus by constructing artificial waters, we have increased diversity on one hand even as we have decreased it. The overall picture, however, is probably a lessening of diversity. Although the number of fish species in Arizona was originally about 25, exotic introductions have increased the state's fish fauna to more than 100 species (Minckley, 1973). Some of the original native species have disappeared or are endangered because of competition from the new arrivals and alteration of their fragile aquatic habitats.

This quote is instructive because it shows how exotics have increased biodiversity, but the author is quick to qualify the phenomenon by noting possible negative impacts. Ecologists generally have avoided the paradox (though, see Angermeier, 1994), but there is a need to take on the problem of understanding the new systems of exotics and native survivors, which may have more biodiversity than the old systems without exotics. Lugo (1988, 1990, 1994) seems to be the only ecologist who has discussed the problem in any depth. He has tried to take a balanced approach as reflected in the following quote (Lugo, 1988):

> Although conservationists and biologists have an aversion to exotic species such as predatory mammals and pests (with good reason!), this may not be totally justified if the full inventory of exotic fauna and flora and certain ecological arguments are taken into consideration. For example, the growth of exotic plant species is usually an indication of disturbed environments, and under these conditions, exotic species compete successfully (Vermeij, 1986). They accumulate and process carbon and nutrients more efficiently than do the native organisms they replace. In so doing, many exotic species improve soil and site quality and either pave the way for the succession of native species or form stable communities themselves. There is no biological criterion on which to judge *a priori* the smaller or greater value of one species against that of another, and if exotic species are occupying environments that are unavailable to native species, it would probably be too costly or impossible to pursue their local extinction.
>
> The paradox of exotic species invasion of islands with high levels of endemism is discussed by Vitousek (1988) in Chapter 20. He correctly points out that if the invasion of exotic species is at the expense of the extinction of local endemics, the total species richness of the biosphere decreases and the Earth's biota is homogenized since most of the invading exotics are cosmopolitan.

Biodiversity exists at several scales (Whittaker, 1977), and exotics can increase alpha or local (within habitat) diversity. Thus, during the invasion process, a community adds one or more exotics. Biodiversity goes up if there are fewer local extinctions of native species than there are additions of exotics. Beta (between habitats) and gamma (regional) diversity can go down, even while alpha diversity goes up, if local endemic species are driven to extinction. The reductions in beta and gamma diversities with concurrent increase in alpha diversity characterize the homogenization phenomenon mentioned earlier. Although there have been few studies of this phenomenon with sufficient depth to document simultaneous change in diversity at different spatial scales, these kinds of biogeographical surveys are needed. Is homogenization actually happening? How many species have been added through introductions and how many species have gone extinct because of these introductions? If invasions of exotics are proceeding in all geographical directions, perhaps the actual net losses in species diversity are small. For every Asian species that invades North America, is there a North American species that invades Asia? In reality, there seem to be few studies spanning the geographic dimensions of biodiversity (alpha, beta, and gamma) that document changes solely attributable to invasions of exotics. Known losses in biodiversity are perhaps best thought as resulting from cumulative impacts of a number of factors which include exotic invasion, pollution, habitat loss, and others. In this context, it would be interesting to know the contribution of the different factors, especially for decision makers who must allocate scarce resources to mitigate separate impacts, such as invasions of exotic species.

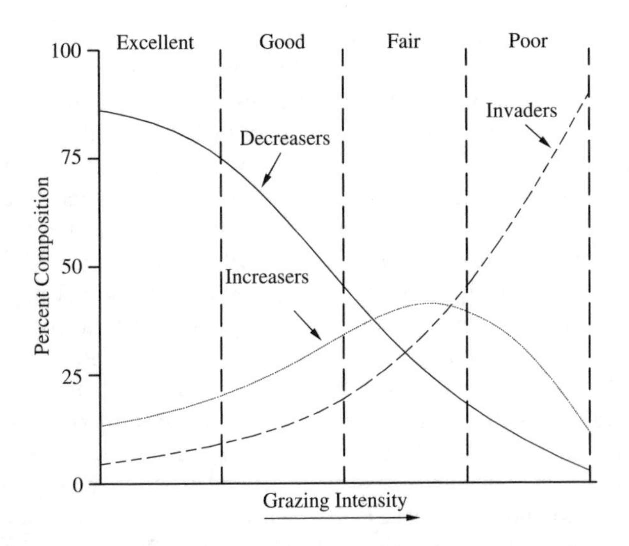

FIGURE 7.2 Classification of rangeland plant species based on adaptation to grazing intensity. Exotic species are like increasers or invaders. (Adapted from Strassmann, B. I., 1986. *Energy and Resource Quality: The Ecology of the Economic Process.* C. A. S. Hall, C. J. Cleveland, and R. Kaufman (eds.). John Wiley & Sons, New York.)

As a form of biodiversity, exotics seem to generally share certain traits, but they are also a diverse group. It is sometimes even difficult to state definitely whether a species is even an exotic (Peek et al., 1987). The problem with defining these kinds of species mirrors the related challenge of defining a "weed." Herbert G. Baker (1965) defined a weed as a plant which grows "entirely or predominantly in situations markedly disturbed by man (without, of course, being deliberately cultivated plants)." The relation between exotics and human disturbance is a key in this definition and it will be explored in more depth in a later section of this chapter. Terminological challenges to defining weeds can be seen in the long lists of alternative definitions given by Harlan (1975) and Randall (1997).

The old range plant terminology (Ellison, 1960) also is instructive for defining exotic biodiversity. Rangeland plants were classified as increasers, decreasers, or invaders depending on their response to grazing. Thus, with increasing grazing intensity, increasers increase in density, decreasers decrease in density, and invaders invade from outside the community (Figure 7.2). This is a common-sense kind of classification that is value-free and that relies on a species response to perturbation.

Exotic species range in size from microbial diseases to wide-ranging wildlife and canopy-level trees. Most are fast growing with wide dispersal capabilities ("*r*-selected," see Chapter 5) but they have other qualities that allow them to be invasive. Some authors have tried to characterize "ideal" invaders (Baker, 1965, 1974, 1986; Ehrlich, 1986, 1989; Mack, 1992; Noble, 1989; Sakai et al., 2001), but many kinds of organisms can take on this role.

One fairly general feature of successful exotic invaders is preadaptation for the conditions of their new community (Allee et al., 1949; Bazzaz, 1986; Weir, 1977).

Preadaptation is a chance feature for unintentional introductions but a conscious choice for those species intentionally introduced by humans. In many cases invasive exotic species are preadapted to the disturbances caused by humans.

A final note on exotics as a form of biodiversity deals with the context of human value judgment. There is an underlying subjective feeling that natural ecosystems should have only native species. In this context, exotic species represent biodiversity in the wrong place. There are anachronistic exceptions such as the feral horses on several U.S. east coast barrier islands (Keiper, 1985), but exotics generally have a negative connotation. In the U.S. this is appropriate for national parks (Houston, 1971; Westman, 1990) where the objective is to preserve natural conditions despite changes in the surrounding landscape. However, in other situations exotics could be viewed with less negative bias. For example, Rooth and Windham (2000) document the positive values of the common reed (*Phragmites australis*) along the eastern U.S. coast, where it is regarded as one of the worst exotic plant species by many workers. These values include marsh animal habitat, water quality improvement, and sediment accumulation, the last of which is especially significant in terms of the impacts caused by the global rising of the sea level. The case for introducing an exotic oyster into Chesapeake Bay for reef restoration provides another case study (Gottlieb and Schweighofer, 1996). Brown (1989) summarizes ideas on value judgments about exotics with the following statement:

> Unless one is a fisherman, hunter, or member of an acclimatization society, there is a tendency to view all exotic vertebrates as "bad" and all native species as "good." For example, most birdwatchers, conservationists, and biologists in North America view house sparrows and starlings with disfavor, if not with outright loathing; they would like to see these alien birds eliminated from the continent if only this were practical. There is a kind of irrational xenophobia about invading animals and plants that resembles the inherent fear and intolerance of foreign races, cultures, and religions. I detect some of this attitude at this conference. Perhaps it is understandable, given the damage caused by some alien species and the often frustrating efforts to eliminate or control them.
>
> This xenophobia needs to be replaced by a rational, scientifically justifiable view of the ecological role of exotic species. In a world increasingly beset with destruction of its natural habitats and extinction of its native species, there is a place for the exotic. Two points are particularly relevant. First, increasing homogenization of the earth's biota is inevitable, given current trends in the human population and land use. …
>
> The second point is that exotic species will sometimes be among the few organisms capable of inhabiting the drastically disturbed landscapes that are increasingly covering the earth's surface. …
>
> It has become imperative that ecologists, evolutionary biologists, and biogeographers recognize the inevitable consequences of human population growth and its environmental impact, and that we use our expertise as scientists not for a futile effort to hold back the clock and preserve some romantic idealized version of a pristine natural world, but for a rational attempt to understand the disturbed ecosystems that we have created and to manage them to support both humans and wildlife. …

The current sentiment among most ecologists and environmentalists is that invasive exotics are "bad" species. However, it must be remembered that this is a subjective assessment. Perspective on the degree of this subjectivity comes from a consideration of a historical case. From the early 1900s until the 1950s, the U.S. government conducted a predator control program on public lands including national parks. Professional hunters and even park rangers were specifically employed in this program to kill wolves, coyotes, and many other mammalian predator species because they were judged to be "bad" species. This situation is described, with an emphasis on national parks, by McIntyre (1996):

> Our country invented the concept of national parks, an idea that represented a new attitude toward nature. In the midst of settling the West, of civilizing the continent, some far-sighted citizens argued for setting aside and preserving the best examples of wild America. Public opinion supported the proposal, and Congress established a system of national parks, including such crown jewels as Yellowstone, Yosemite, Sequoia, Rocky Mountain, Grand Canyon, Glacier, and McKinley. The natural features and wildlife found within these parks would be protected as a trusted legacy, passed on from one generation to another.
>
> But the early managers of these national parks defined preservation and protection in ways that seem incredible today. The contemporary attitude classified wildlife species as either "good" or "bad" animals. Big game species such as elk, deer, moose, bison, and big-horn sheep fell into the favored category. Park administrators felt that national parks existed to preserve and protect those animals. Anything that threatened them, whether poachers, forest fires, or predators, had to be controlled. Based on that premise, predators, especially wolves, became bad animals, and any action that killed them off could be justified.
>
> Besides wolves, many other animals were also blacklisted and shot, trapped, or poisoned during the early decades of the national park system: mountain lions, lynx, bobcats, red foxes, gray foxes, swift foxes, badgers, wolverines, mink, weasels, fishers, otters, martens, and coyotes. Amazingly, rangers even destroyed pelicans in Yellowstone on the premise of protecting trout.
>
> The predator control program in the national parks was just an extension of a national policy to rid the country of undesirable species. …

This control program stopped in the 1950s, and many are questioning its wisdom to the degree that wolves are now being reintroduced to the national parks. Thus, the judgment of these species as being "bad" and needing to be controlled has been reversed as attitudes have changed. Will a similar reversal in attitudes happen with invasive exotics some day? Chase (1986) in his critical review of management policies at Yellowstone National Park labeled the old predator control program as an example of "playing god" with the species. The comparison is striking with current exotic control programs.

EXOTICS AND THE NEW ORDER

Mooney and Drake (1989), in summarizing a text on the ecology of biological invasions, suggested that humans have transformed nature to such a great extent that a "new order" now exists. They list a number of dramatic changes that have occurred due to human population growth and state that the world is now dominated by new systems because of these changes, as is highlighted in the following quote:

> All of these alterations are providing a new landscape with an abundance of disturbed habitats favoring organisms with certain traits. This massive alteration of the biosphere has occurred in conjunction with the disintegration of the great barriers to migration and interchange of biota between continents due to the development by humans of long-distance mass transport systems. The introduction of a propagule of an organism from one region to a distant one has changed from a highly unlikely event to a certainty. The establishment and spread of certain kinds of organisms in these modified habitats, wherever they may occur, is enhanced. The net result of these events is a new biological order. Favored organisms are now found throughout the world and in ever increasing numbers. It is evident that these changes have not yet totally stabilized either in the Old or New World. In the former the success of invading species has changed through time with differing cultural practices and new directions and modes of transport. Old invaders are being replaced by new ones (Heywood, this volume). In the New World additional invading species are still being added.
>
> The kinds of disruptions that non-intentionally introduced invading species can play in natural systems have been outlined above and have been the focus of the SCOPE study. These disruptions may in time stabilize on the basis of a new system equilibrium.

This interpretation might be translated as a kind of algebraic equation for understanding exotic species:

Increased disturbance by humans + Increased dispersal by humans =
New systems with dominance of exotic species

This equation is useful in illustrating the two main causes of exotic invasions but it especially focuses on the idea that the resulting systems are new. To some this is an exciting concept in that these are systems that have never existed previously, and they are new challenges for science to describe and explain. To others this is an environmental disaster that requires remediation or restoration. While the concept is a philosophical statement, there is a definite reality in the new organization of systems with exotic invasion.

Some have focused on the role of disturbance by humans as a key factor in exotic invasions. Elton (1958) was the first to tie exotics to disturbance, as did Baker (1965) in his definition of weeds. More recently others have discussed the connection (Hobbs, 1989; Hobbs and Huenneke, 1992; Horvitz, 1997; Lepart and Debussche, 1991; Orians, 1986). The notion is that invasions are more likely in disturbed ecosystems because resources are available and competition from resident native

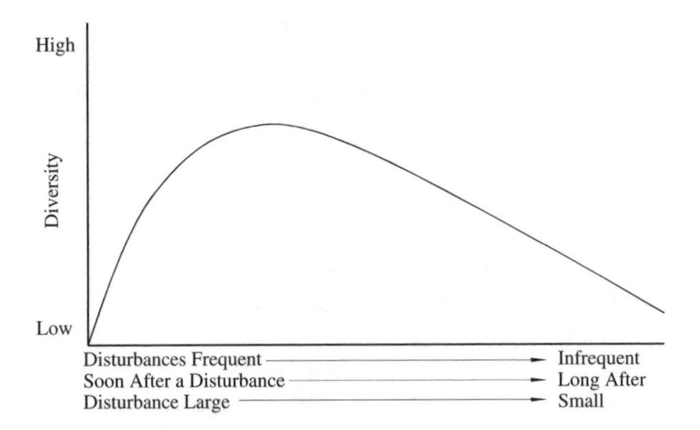

FIGURE 7.3 Graph of the intermediate disturbance hypothesis which suggests that the maximum diversity occurs when the disturbance level is moderate. (Adapted from Connell, J. H., 1978. *Science.* 199:1302–1310.)

species is reduced. This is a promising focus to take for understanding exotic invasions, especially due to the well-developed theory of disturbance in ecology (Clark, 1989; Connell, 1978; Levin and Paine, 1974; Petraitis et al., 1989; Pickett and White, 1985; Pickett et al., 1989; Reice, 2001; Sousa, 1984; Walker, 1999). This theory states that species can adapt to natural disturbances, and in some cases they even use the disturbance as an energy source. As examples, energy from disturbances can be used for accelerating nutrient cycling or dispersing propagules. Connell's (1978) study, which showed that the maximum diversity was found at intermediate levels of disturbance (Figure 7.3), became a benchmark in documenting the important role of disturbance in ecology. The hump-shaped pattern arises for several reasons. At low levels of disturbance, the most adapted species outcompete all of the other species (for example, those that are "*K*-selected," see Chapter 5), which lowers diversity. At high levels of disturbance, only a few species can adapt to the environmental conditions that change so often (for example, those that are "*r*-selected"), which also lowers diversity. The highest diversity occurs at intermediate levels of disturbance because some species adapted to the entire disturbance spectrum are supported. Energy theory provides an alternative explanation: the intermediate levels of disturbance provide the most energy subsidy to the ecosystem, while low levels of disturbances provide less energy subsidy and high levels of disturbance act as stress rather than subsidy (E. P. Odum et al., 1979). Disturbance theory has led ecologists to emphasize nonequilibrium concepts of ecosystems over the earlier ideas of more static "balance-of-nature" concepts. The theory of island biogeography (see Chapters 4 and 5) is an example of an equilibrium model for explaining species diversity. Under equilibrium conditions competitive exclusion can run its course, eliminating inferior competitors and selecting for the species best adapted to a site. However, under nonequilibrium conditions the environment changes frequently enough that competitive exclusion cannot run its course and thus more species are supported on the site. Nonequilibrium theory was first used by Hutchinson (1961) to explain the "paradox of the plankton" or why so many species of phytoplankton are found to coexist in the epilimnion or upper layer of a lake. The epilimnion seemed to offer only one niche for phytoplankton since it was uniformly mixed with constant light intensity. Under these conditions the competitive exclusion principle (see Chapter 1) dictated that only one species of phytoplankton should be found at equilibrium. Yet many species are found there. Hutchinson solved this paradox by suggesting that environmental conditions (such as temperature and nutrient concentrations) actually change with sufficient frequency to preclude the onset of competitive equilibrium, thus allowing many species to coexist. Huston (1979) elaborated and generalized Hutchinson's nonequilibrium concept of species diversity in an important paper published one year after Connell's classic. Since the 1970s nonequilibrium and disturbance theory have become dominant in ecology (Chesson and Case, 1986; DeAngleis and Waterhouse, 1987; Reice, 1994; Wiens, 1984). The shift in emphasis from equilibrium to nonequilibrium perspectives is critically important in ecology, but it does not necessarily imply that the field is without order or predictability. Rather, as noted by Wu and Loucks (1996), "harmony is embedded in the patterns of fluctuation, and ecological persistence is 'order within disorder'."

It is not enough to simply correlate exotic invasion with disturbances caused by humans. Much research is needed for understanding how the various kinds of human disturbances act. Frequency, intensity, and duration have been found to be good descriptors of natural disturbances. Work is needed to quantitatively derive similar descriptors of human disturbances in relation to exotic invasions. For example, no simple relation was found between urbanization as a form of disturbance and degree of exotic invasion by Zinecker (1997) for riparian forest plant species in northern Virginia. The approach of Reeves et al. (1995) in developing "a new human-influenced disturbance regime" might be a good model for the disturbances that facilitate invasion of exotic species.

Relatively less attention has been given to the factor of increased dispersal by humans as the cause of exotic invasions, although it is usually acknowledged as being important. In fact, invasions can occur in systems that are not necessarily disturbed by humans, as long as an invader can reach the system. The invasion of isolated oceanic islands, such as the introduction of goats by explorers in the 1700s, is an example of this situation. However, increased disturbance and dispersal usually occur simultaneously, making it difficult to separate the two factors in most case studies. Increased dispersal of species from one biogeographic province to another is occurring due to increased rates of travel and trade within the global economy. Total amounts of dispersal are seldom known because only successful introductions are recorded (Simberloff, 1981, 1989; Welcomme, 1984). Ship ballast, as a form of increased dispersal for aquatic organisms, is a good example of a well-studied mechanism (Carlton, 1985; Williams, 1988) and has potential for regulation (National Research Council [NRC], 1996a). New syntheses of dispersal by exotic organisms must be based on detailed species-specific studies, such as Carlton's (1993) work on zebra mussels (*Dreissena polymorpha*), which are only now starting to accumulate in the literature. Studies of the dispersal of native species (Bullock et al., 2002; Clobert et al., 2001; Gunn and Dennis, 1976; Howe and Smallwood, 1982; van der Pijl, 1972; Wolfenbarger, 1975) can be models for the syntheses, but there will probably be new elements of preadaptation that help explain increased dispersal rates by exotic organisms.

The final component of the equation given earlier in this section is the concept of new systems with mixes of natives and exotics. Mooney and Drake (1989) emphasize the idea that these are "new," which is a different perspective than one gets from reading most literature on exotics. Rather than thinking of these as natural systems that have been degraded by the introduction of exotic species, they can be seen as new systems that have been reorganized from the old "natural" systems. The value of this perspective is that it allows thinking to be freed from biases to consider new forms of organization (see Chapter 9).

Humans are creating a tremendous number of new habitats that in turn create opportunities for new mixes of species. Cohen and Carlton (1998) describe the San Francisco Bay and Delta ecosystem as having perhaps the highest exotic species diversity of any estuary because the bay is a focal point for transport and, therefore, increased dispersal and because of extensive human disturbance (Nichols et al., 1986; Pestrong, 1974). In another example, Ewel (1986a) describes the new soil conditions of South Florida as being an important factor in the exotic invasion of terrestrial systems in the following quote:

> Substrate modification, such as rock plowing, diking, strip-mining, and bedding, has created soils and topographic features heretofore unknown to Florida. These human-created soils, or anthrosols, are likely to support new ecosystems in which exotic species play dominant roles. The Hole-in-the-Doughnut in Everglades National Park exemplifies this situation. Despite efforts by the National Park Service to restore native vegetation to this rock plowed land, a peppertree/wax myrtle/saltbush ecosystem persists there.

The story of invasion of Gatun Lake in Panama by *cichlid* fish species (Swartzmann and Zaret, 1983; Zaret, 1975; Zaret and Paine, 1973) offers another view of new systems. This is an example that is often used to illustrate the severity of changes that exotic introductions can have on an ecosystem. In this case the *cichlid* is a voracious predator that was introduced into the lake. Changes in the lake's food web over time, which included dramatic reductions in native species and a simplification of the structure of the food web, were documented (Figure 7.4). While this is often used as an example of how much change an exotic can make in a native food web, in fact it may be better explained as an example of a reorganization of a new system because Gatun Lake is a reservoir formed as part of the Panama Canal rather than a natural lake. The original natural system was a river that was subsequently turned into a reservoir when the canal was built. This change in hydrology must have played a significant role in the changes in the food web that Zaret and Paine described. This interpretation is not intended to diminish the importance of Gatun Lake as an example of exotic invasion but rather to highlight the context of the example as a reorganized new system rather than a degraded natural system.

One way to think of the new systems is as examples of alternative stable states. In this concept if a system is perturbed beyond some threshold of resilience, the system may change through succession to a new organization or stable state and not

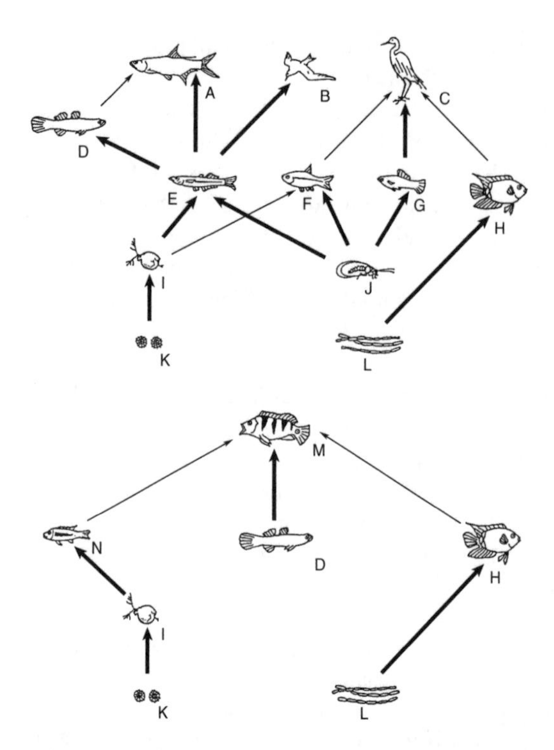

FIGURE 7.4 Comparison of food webs in Gatun Lake, Panama, with and without an exotic fish predator. (A) *Tarpon atlanticus.* (B) *Chlidonias niger.* (C) Several species of herons and kingfishers. (D) *Gobiomorus dormitor.* (E) *Melaniris chagresi.* (F) *Characinidae*, including four common species. (G) *Poeciliidae*, including two common species; on exclusively herbivorous, *Poecilia mexicana*, and one exclusively insectivorous, *Gambusia nicaraguagensis.* (H) *Cichlasoma maculicauda.* (I) Zooplankton. (J) Terrestrial insects. (K) Nannophtoplankton. (L) Filamentous green algae. (M) Adult *Cichla ocellaris.* (N) young *Cichla.* (From Zaret, T.M. and R. T. Paine. 1973. *Science.* 182:449–455. With permission.)

revert back to the old organization (Holling, 1973; May, 1977). Thus, some form of disturbance may push a natural system into a new domain of stability with an entirely new set of species (Figure 7.5). Alternative stable states have been discussed for a number of ecosystems including coral reefs (Done, 1992; Hughes, 1994; Knowlton, 1992), grazing systems (Augustine et al., 1998; Dublin et al., 1990; Laycock, 1991; Rietkerk and Van de Koppel, 1997), mud flats (Van de Koppel et al., 2001), and lakes (Blindow et al., 1993; Scheffer and Jeppesen, 1998). The concept remains controversial but seems to be generally applicable (Carpenter, 2001; Law and Morton, 1993; Sutherland, 1974). Introduction of exotic species can be thought of as an impact that causes the system to change from one stable state to a new one with a reorganized ecosystem structure and function. For example, the invasion of zebra mussels into the Great Lakes has been suggested to cause a shift from a pelagic stable state to a benthic stable state because of the zebra mussels' ability to strip sediments and algae from the water column through suspension feeding (Kay and Regier, 1999; MacIsaac, 1996). With increased dispersal by humans many new mixes

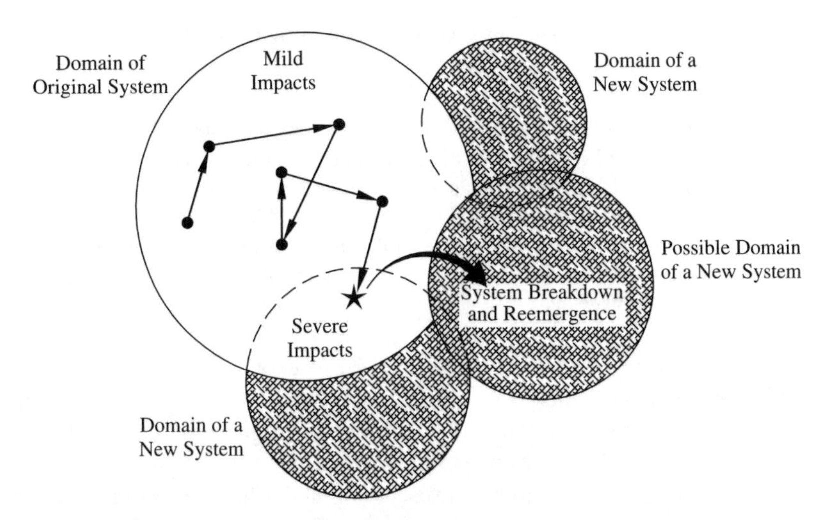

FIGURE 7.5 Theory of alternative stable states in ecology. An ecosystem can be pushed between alternative domains by major disturbances. (Adopted from Bradbury, R.H. et al. 1984. *Australasian Science*. 14(11–12):323–325.)

of species may come together on a site and allow for the creation of new alternative stable states. Perhaps the number of possible alternative stable states is much greater with the accelerated seeding rates of human introductions as compared with what is possible under the old natural conditions. In a sense, genetics is a limiting factor to ecosystem development in natural systems and may be overcome by exotic invasions that add species to the system.

LEARNING FROM EXOTICS

The new order created by exotic invasions is both a challenge and a stimulus for learning about ecology. Marston Bates (1961) made this connection in a relatively early reference:

> The animals and plants that have been accidentally or purposefully introduced into various parts of the world in the past offer many opportunities for study that have hardly been utilized. They can, in a way, be considered as gigantic, though unplanned, experiments in ecology, geography, and evolution, and surely we can learn much from them.

This was also stated by Allee et al. (1949) in their classic text on animal ecology:

> The concept of biotic barriers may be tested by introducing animals and plants from foreign associations and observing the results. In most instances such tests have not been performed consciously. With the advent of modern transportation, many organisms are inadvertently introduced into ancient balanced communities. These unwitting experiments may be studied with profit.

Furthermore, Vitousek (1988) suggested that ecological theory can benefit from studies of exotic invasions:

> Better understanding of biological invasions and their consequences for biological diversity on islands will contribute to the development and testing of basic ecological theory on all levels of biological organization. … An understanding of the effects of invasions on biological diversity in rapidly responding island ecosystems may give us the time and the tools needed to deal with similar problems on continents; it may even contribute to the prediction and evaluation of the effects of environmental releases of genetically altered organisms.

Ecologists are just beginning to explore the use of exotic invasions as unplanned and uncontrolled experiments. Simberloff (1981) used historical records on introductions to examine two relevant ecological theories (equilibrium island biogeography and limiting similarity of competing species). He found little support for the theories in his analysis, and generated discussion about how to use historical data sets on introductions (Herbold and Moyle, 1986; Pimm, 1989). While a few other attempts at using exotics to examine ecological theories have been made (MacDonald and Thom, 2001; Mack, 1985; Ross, 1991), many relevant topics, such as assembly theory, keystone species, and the role of indirect effects, could be examined. Here, two theories are discussed as examples.

Catastrophe theory is a branch of mathematical topology which describes dynamic systems that can exist in alternative stable states and that can dramatically change between states over short periods of time in a discontinuous fashion (Thom, 1975). Although the mathematical basis of the theory was criticized soon after it came out (Kolata, 1977), catastrophe theory has been profitably applied to several kinds of outbreak-type systems including forest insects (Casti, 1982; Jones, 1975; Ludwig et al., 1978), Dutch elm disease (Jeffers, 1978), algal blooms (Beltrami, 1989, 1990), and others (Loehle, 1989; Saunders, 1983). The theory is receiving renewed attention for understanding alternative stable states in ecosystems (Allen, 1998; Scheffer and Jeppeson, 1998) and it may offer a language for understanding invasion and dominance of natural communities by exotic species. For catastrophe theory to apply to exotic takeover, the system must have a certain structure of control variables that results in an equilibrium surface or a map that tracks a periodic outbreak-type of dynamic behavior. Several kinds of maps are described by the theory; most common are the fold and cusp catastrophes, which depend on one and two control variables, respectively. Thus, for catastrophe theory to be useful for understanding exotic invasions, the structure of control variables must be understood. Phelps (1994) suggested that a cusp catastrophe might help explain the invasion of the Potomac River near Washington, DC, by Asiatic clams (*Corbicula fluminea*), and perhaps other exotic invasions can be understood with this approach.

The maximum power principle may also be useful for understanding exotic invasions. This is a systems-level theory that states that systems develop designs that generate the maximum useful power through self-organization (Hall, 1995b; H. T. Odum, 1971, 1983). The concept is based on the premise that "systems that gain more power have more energy to maintain themselves and … to overcome any other shortages or stresses and are able to predominate over competing units" (H. T. Odum, 1983). The general systems design that tends to maximize power is one that develops feedbacks which increase energy inflow during early successional stages or which increase energy efficiency during later successional stages. Feedbacks are performed by species within ecosystems, so the maximum power principle also is a theory about how species composition develops. The theory suggests that those species that are successful and dominate a system must contribute to the system's ability to maximize power. Exotic species that invade a system then should lead to an increase in power flow, if the maximum power principle holds. Thus, exotic invasions may allow a test of the theory by examining power flow or metabolism of systems before invasion and after invasion. For example, the theory predicts that a natural Chesapeake Bay marsh dominated by *Spartina* or *Scripus* would have lower energy flow than the same marsh after invasion by *Phragmites*. This test has not been formally made yet but the work discussed by Vitousek seems to be consistent with the maximum power principle (Vitousek, 1986, 1990; Vitousek et al., 1987) as does the analysis of exotic Spartina marshes in New Zealand (Campbell et al., 1991; H. T. Odum et al., 1983).

Existing ecological theory may not be completely adequate to understand exotic invasions (Abrams, 1996), and entirely new ideas may be needed for their description and explanation. The prospects are good for new theory to be developed from the study of exotics. Much new quantitative modelling has focused on how exotics spread across landscapes (Higgins and Richardson, 1996; Shigesada and Kawasaki, 1997), but the best prospects for new theory may be with invasibility of communities. This subject was first treated by MacArthur and Wilson (1967) in the context of islands using equilibrium approaches to theory. Invasion is the process of species entering an established community. It differs from colonization, which is the process of species entering a community while it is being established. Ewel (1987) noted the importance of this topic when he suggested that invasibility is one of the five most important criteria for assessing newly restored ecosystems. The concept of invasion is receiving increasing attention with empirical studies (Burke and Grime, 1996; Planty-Tabacchi et al., 1996; Robinson and Dickerson, 1984), review articles (Crawley, 1984; Fox and Fox, 1986) and application of existing theory (Hastings, 1986). Elton's (1958) old concept of resistance to invasion is more or less the inverse of invasibility (Orians et al., 1996; Pimm, 1989; Rejmanek, 1989). Resistance of a community to invasion is sometimes found to be proportional to its diversity (Kennedy et al., 2002), but in other cases "invasional meltdowns" can occur where the invasion rate accelerates as more species are added (Ricciardi and MacIsaac, 2000; Simberloff and von Holle, 1999). The invasional meltdown concept has only recently been introduced and may be explained by facilitation interactions between exotic invaders. This is an example of new ecological theory that is being developed to understand exotic invasions.

A final value of exotic invasions as a stimulus to learning would be if knowledge generated from their study can help deal with new problems facing society. The connection between invasions of exotic species and releases of genetically engineered or modified organisms (GMOs) has been made (National Research Council [NRC], 1989b) and similar theories may apply to both problems (Kareiva et al., 1996;

Purrington and Bergelson, 1995). There are many risks associated with the release of GMOs. For example, adding genes for disease resistance to crops is risky because they may pass these genes on to weeds, creating superweeds with enhanced growth potential (Kaiser, 2001b; Snow and Palma, 1997). Moreover, the disease-resistant crops may themselves become weeds (Rissler and Mellon, 1996)! Understanding degrees of weediness in exotic species may help assess the risks associated with GMOs. Products derived from genetically altered food crops have been called "frankenfoods," referring to Mary Shelley's story of Frankenstein. This reference is evocative because in the story the man-made monster escapes and kills his creator. Another issue deals with possible biological cross-contamination caused by extraterrestrial space travel. The concerns are that missions to other planets may infect them with organisms from the Earth and that missions that return from other planets may infect the Earth with alien organisms. Assessment of this risk began with lunar missions in the 1960s and protocols for planetary quarantines were established by NASA (Lorsch et al., 1968). Interest became more intense with planned Mars missions because life on Mars was then thought to be a definite possibility (Pittendrigh et al., 1966). An interesting controversy about the need for quarantines and space craft sterilization developed between some engineers who thought the probabilities of cross-contamination were too remote for concern, and some biologists who understood the ability of living organisms to grow and spread even under harsh environmental conditions. Carl Sagan was a vocal supporter of the need for precautions, and the controversy between engineers and biologists is discussed in depth in one of his biographies (Poundstone, 1999). There is now renewed interest about the issue of cross-contamination because of the chance of false-positive results in planned extraterrestrial life detection experiments caused by Earth organisms (Clarke, 2001) and because of the chance of alien invasion from samples of rocks and soils that are planned to be returned from space (Space Studies Board, 1997, 1998). Perhaps NASA would be well advised to include ecologists specializing in exotic species invasions on committees and advisory boards dealing with planetary cross-contamination.

CONTROL OF EXOTIC SPECIES AND ITS IMPLICATIONS

Control of exotic species is a goal of natural resource managers and conservation biologists. Many methods are available, ranging from quarantining in order to keep them out to eradication so as to remove them once they are established (Dahlsten, 1986; Dahlsten and Garcia, 1989; Groves, 1989; Reichard, 1997; Schardt, 1997; Simberloff, 1997). Eradication in particular is usually difficult and often unpleasant work, but in some cases such as in national parks, it is necessary. As noted by Temple (1990),

> In spite of all that is known about the negative influence of exotics and the obvious conservation benefits of controlling them, their eradication inspires little enthusiasm among most conservationists, the public, or governments. Reasons for this apathy include misconceptions about the nature and magnitude of the problem, fears of the negative public reactions that almost invariably accompany eradication efforts, espe-

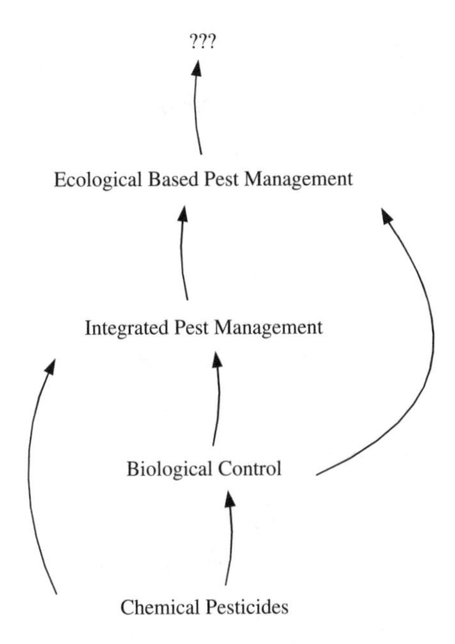

FIGURE 7.6 Succession of pest control paradigms that started after World War II with chemical pesticides.

cially for animals, and intimidation by the inefficient labor-intensive nature of current eradication technologies.

These challenges need to be addressed if exotic control is to be a realistic goal. To meet the challenges Temple (1990) calls for "a better job of educating the public about the threats of exotics," the development of "more palatable methods of eradication that avoid issues of ethics or cruelty," and the recruitment of "scientists whose research will produce new approaches for controlling or eradicating exotic species." This is a call for creative research on control methods that will occupy increasing numbers of applied ecologists in the future.

Foundations of exotic control rest on the long history of pest control, especially in agriculture and forestry in terms of diseases, weeds, and insects. A tremendous amount of knowledge has accumulated on the subject over a long history. However, modern pest management essentially dates from after World War II when agricultural production and pesticide use expanded greatly. A succession of paradigms has emerged (Figure 7.6) but pest problems continue to accelerate. The consensus is that eradication is often impossible, and even control is difficult. At best some form of management is the most reasonable goal (National Research Council [NRC], 1996b). The primary tools for controlling many exotic species are still chemical pesticides, which have positive and negative aspects (Table 7.3).

While the environmental and social costs of pesticides in agriculture and forestry are becoming better understood (Pimentel et al., 1980, 1992), pesticide use continues to increase. Embedded in these pest control systems is an ironic feedback circuit, termed the *pesticide treadmill* (van den Bosch, 1978). In this circuit greater use of

TABLE 7.3
Positive and Negative Aspects of Pesticides

Positive Aspects
Pesticides save lives.
They increase food supplies and lower food costs.
They increase profits for farmers.
They work faster and better than other pest control alternatives.
Safer and more effective pesticides are continually being developed.
Negative Aspects
Development of genetic resistance reduces the effectiveness of pesticides and leads to the pesticide treadmill.
Pesticides kill natural pest enemies and convert minor pest species into major pest species.
Certain persistent pesticides are mobile and can amplify up food chains causing environmental impacts.
There are short-term and long-term threats to human health from pesticide use and manufacture.

Source: Adapted from Miller, G. T., Jr. 1991. *Environmental Science*. Wadsworth, Belmont, CA.

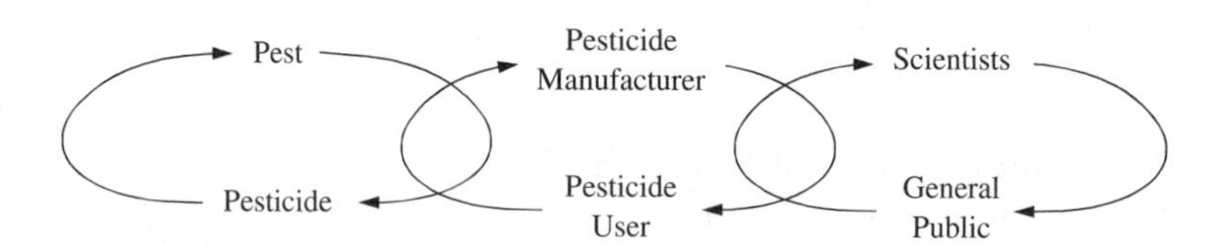

FIGURE 7.7 Linkages in feedback circuits associated with the chemical control of pests. This network creates a cascade of effects when pesticides are used.

pesticides leads to higher levels of pest populations due to the development of increased resistance in pests and due to declines in natural pest predators from pesticide toxicity. The circuit is completed when the resistant pests, which are now released from predation, increase thereby requiring the application of even more pesticides. The numbers of arthropod (insects and mites), plant pathogen, and weed species resistant to chemical pesticides has risen dramatically since World War II (Gould, 1991), and there is no easy solution to the positive feedback circuit. In fact, there are a series of these feedback circuits involved in pest management (Figure 7.7), including pesticide manufacturers who advocate use, farmers and other users, and even extending to scientists and the general public whose perspectives on pesticides are often out of phase (van den Bosch, 1978; Winston, 1997). Narcotics addiction has been used as a metaphor for these feedback circuits by several authors to signify the insidiousness of the problem (DeBach, 1974; Ehrlich, 1978). These circuits are actually interacting coevolutionary games or arms races, such as the "Red Queen relationship" (Van Valen, 1973, 1977) from evolutionary theory. A Red

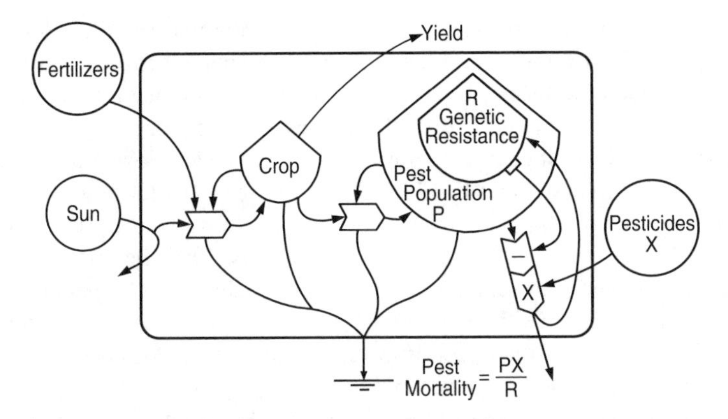

FIGURE 7.8 Energy circuit diagram of the pesticide treadmill concept. Applications of pesticides increase the genetic resistance of the pest population which reduces mortality due to pesticide toxicity.

Queen relationship occurs when any gain in fitness by one species is balanced by losses in fitness by another species. Thus, adaptive success of one species creates selective pressure on the other species to evolve a counter move, which in turn creates selective pressure on the first species, starting the process over again. The Red Queen relationship can occur either between two competing species or between a predator and prey. This kind of coevolution has been named after the Red Queen from Lewis Carroll's *Through the Looking Glass* because she lived in a land where people had to do all the running they could just to stay in the same place; if they actually wanted to go anywhere, they had to run twice as fast as they could. Other examples of the Red Queen type of evolution are given by Clay and Kover (1996), Hauert et al. (2002), and Stenseth (1979). Exotic species and the natural resource managers who try to control them are being drawn into this kind of coevolutionary circuit and they may have to start working as hard as they can to keep up with one another. Figure 7.8 illustrates some aspects of the pesticide treadmill. Genetic resistance reduces mortality of the pest population due to the pesticide applications and resistance to pesticides increases in proportion to pesticide use in this model. These problems force the farmer to use greater doses of pesticides or different types of pesticides to maintain yield. A similar phenomenon is occurring with the development of drug-resistant pathogens, such as the increasing resistance of bacteria to penicillin and other antibiotics. Whole new strategies of dealing with medical wastes are needed to deal with this growing problem (see, for example, Rau et al., 2000).

Frank Egler's work may stand as a model for the kind of creative research that is needed to deal with the problems of exotic species control. Egler was a consummate plant ecologist (Burgess, 1997) who was committed to understanding and using herbicides as part of his research. He published many papers on herbicide effects (Egler, 1947, 1948, 1949, 1950, 1952b), on overviews of the social ecology of pesticides (Egler, 1964, 1979), and on vegetation management with herbicides (Egler, 1958; Egler and Foote, 1975; Pound and Egler, 1953) along with his collaborator, William Niering (Dreyer and Niering, 1986; Niering, 1958; Niering and

Goodwin, 1958). He developed a new kind of ecology that used herbicides as an experimental tool for applied problems. If exotic plants are to be controlled once they have become established, Egler's work on controlling plant community composition may provide lessons on the selective use of herbicides.

Perhaps some kind of ecosystem management (Agee and Johnson, 1988; Boyce and Haney, 1997; Haeuber and Franklin, 1996; Meffe et al., 2002) will be required for exotic species control. The ecosystem scale was examined by traditional pest ecologists (Haynes et al., 1980; Pimentel and Edwards, 1982) before the concept of ecosystem management arose, but most work in agriculture and forestry has focused on the population scale. Although ecosystem management has been criticized for being a philosophy rather than a set of specific techniques, it does present a different context against which exotic species and pests are judged.

A final topic is the economics of exotic control. Economics involves accounting for costs and benefits of exotic control and determines how much control is possible. Unfortunately, economics of exotics control has been overlooked in most assessments of the problem, so it is difficult to know how much control is possible. Studies are needed which evaluate the costs of control (such as purchases of pesticides and labor costs) and relate them to the relative success of control efforts. That such studies have not been published in the many symposium volumes and other texts on exotics is probably a measure of the preliminary stage of the field. Here again, work on pest control in agriculture and forestry can be a guide for the economics of exotic control. As is usually the case in these situations, it may be cheaper to exclude an exotic from a system (i.e., quarantining) rather than trying to eradicate it once established. Detailed studies must confirm this supposition. Exotic control must find a place among other priorities in the budgets of natural resource managers, and new forms of financing may be required. Economics is a reality for managers whose responsibilities it is to control exotics. Will it be possible to control exotic species with the amount of money available? Is there a risk of getting on a coevolutionary treadmill with exotics where more and more money will be required just to maintain levels of invasion? Answers to these kinds of questions will be needed to predict the future of exotics control.

OTHER CONCEPTS OF CONTROL IN ECOLOGY AND ENGINEERING

Considerations of exotic species and their control relates to the broader topic of control in ecology and engineering. Exotic species are often said to cause problems because they have escaped from the natural processes that control or regulate their populations. Human managers of exotic species have attempted to reestablish this control but with uneven success. In this final section of the chapter, discussion of control is expanded because of its importance to ecological engineering in a general sense.

Historically, control in ecology has been discussed in many contexts and often with controversial positions. One of the earliest controversies involved the control or regulation of population sizes. One group led by David Lack (1954) believed in

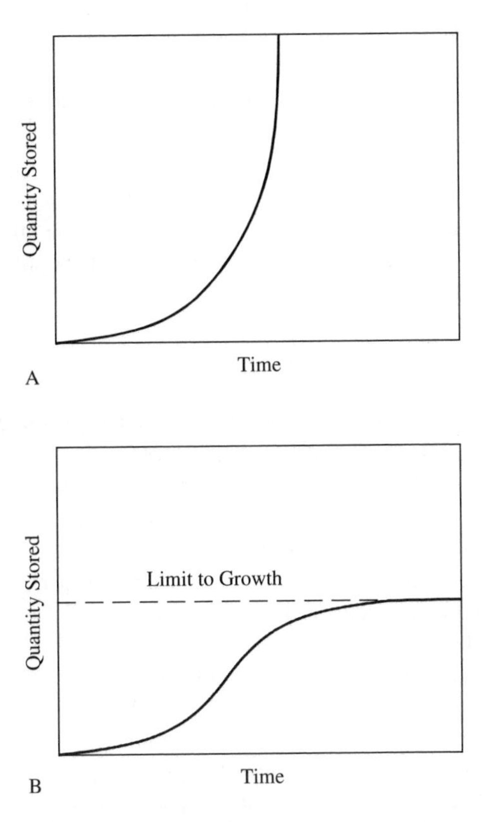

FIGURE 7.9 Patterns of population growth. (A) Exponential growth. (B) Logistic growth.

density dependence in which the severity of mortality factors is correlated with population density (such as for mortality caused by disease, predators or parasites, and food shortage). Another group led by Andrewartha and Birch (1954) believed in density independence in which the severity of mortality is the same at all population densities (such as for mortality caused by extreme weather events). This is a critically important distinction because density dependence allows for a self-regulation mechanism within a population. Cole (1957) reviewed the subject in terms of the search for a "governor" or controlling influence on population size. He showed the governor as a term added to the population growth equation, which converts uncontrolled, exponential growth (Figure 7.9a) into controlled, logistic or limited growth (Figure 7.9b):

$$dN/dt = rN \quad \text{exponential growth equation} \tag{7.1}$$

where

N is the population size
r is the reproductive rate

and

$$dN/dt = g(rN) \quad \text{logistic growth equation} \tag{7.2}$$

where g is the governor

Cole explored various forms for the governor term, the simplest of which has evolved to be the carrying capacity term, which causes the population to be regulated by density dependence:

$$g = (K - N)/N \tag{7.3}$$

where K is the carrying capacity or the maximum size of the population that the environment can support the addition of this term (g) to the exponential growth equation causes population growth to stop and population numbers to level off when $N = K$ (see Equation 3.4). Carrying capacity therefore is an important quality within this elementary theory of population ecology because it causes controlled growth of the logistic equation as opposed to the out-of-control growth of the exponential equation. Much argument occurred between members of the density dependence and density independence schools of thought from the 1950s onward and to some extent the controversy continues (Chitty, 1996). Reviews are given by Krebs (1995) and Tamarin (1978).

At the ecosystem scale, control has been considered to occur either due to resource limitations (i.e., bottom-up control) or due to harvesting by consumers (i.e., top-down control). Bottom-up control of food webs is determined by resources, specifically those resources that are required for primary productivity. This is the process whereby solar energy is transformed into the chemical energy of biomass and is at the base of most food webs (i.e., the bottom). A number of resources are required for primary productivity, such as water, carbon dioxide, and nutrients. Justus Liebig, a German agronomist, proposed his famous Law of the Minimum in the 1800s to describe how resources limit (i.e., control) primary productivity (E. P. Odum, 1971; see also the excerpts of Liebig's publications in Kormondy, 1965; and Pomeroy, 1974). Liebig's law states that the required resource in the least supply will limit production. Thus, resources that limit primary productivity are called *limiting factors*. The primary way to identify a limiting nutrient is with nutrient addition experiments. In this kind of experiment different nutrients are added to a system in controlled locations in order to test for increases in plant growth. Although traditionally it has been thought that only one factor at a time can limit primary production, there is a growing trend of examining how limiting factors are linked or dynamically related. Alfred Redfield was the first to consider this idea in his study of "The Biological Control of Chemical Factors in the Environment" (Redfield, 1958; see also Redfield et al., 1963). In particular, he studied the biogeochemical cycle of the photic zone of the open ocean and found that carbon, nitrogen, and phosphorus cycled in a constant proportion that was roughly equivalent to the ratio of these elements in the biomass of the plankton. This observation indicates that of these three elements, no single one limited production but rather they all simultaneously were limiting. The implication was that the plankton biota had coevolved with the ocean nutrient cycles so that the ratio of elements released by decomposition matched the ratio of elements taken up by primary production. This was judged to be a highly evolved state and the element ratio became known as the Redfield ratio. The coevolution of biota and macronutrient cycles was considered to be possible only in the open ocean where the variable geology of land masses have little influence on chemistry, but even here other micronutrients such as iron may limit primary production (see Chapter 9). H. T. Odum attempted to generalize Redfield's concept with the introduction of the "ecomix" which he defined as "the particular ratio of elemental substances being synthesized into biomass and subsequently released and recirculated" (H. T. Odum, 1960). He suggested that

> Although shortage or excessive accumulation of any one element will stop or retard the system, there is a self-selection for compatibility of the photosynthesis and the regenerative respiration. The characteristic ratio of elements which tends to be stabilized in the average mix of the system is the chemical ecomix .... (H. T. Odum, 1970)

Although Odum's ecomix idea was not picked up by other ecologists, more recently a whole new area of study on ecological stoichiometry has arisen based on Redfield's nutrient ratio approach to understanding bottom-up control in ecosystems (Daufresne and Loreau, 2001; Elser et al., 1996; Hessen, 1997; Lampert, 1999; Lockaby and Conner, 1999; Sterner, 1995).

The top-down control of food webs by consumers has received a great deal of attention in ecology with review articles of field and empirical studies (Chew, 1974; Huntly, 1995; Kitchell et al., 1979; Naiman, 1988; Owen and Wiegert, 1976; Petrusewicz and Grodzinski, 1975; Zlotin and Khodashova, 1980) and with theoretical work (Lee and Inman, 1975; O'Neill, 1976). Consumers make up many categories of organisms including carnivores, herbivores, detritivores, and omnivores along with parasites and even diseases. In each of these categories the consumer consumes different things. When the thing being consumed is living, then the predator–prey theory applies. All predator–prey relationships have the potential for control of prey by predators (see experiments by Gause in Chapter 4), but the strength of the relationship varies significantly. The most dramatic examples are keystone predators which exert strong control over multispecies assemblages (i.e., from the top of the food web). The keystone species concept was introduced by Robert Paine based on his experimental studies of a rocky intertidal food web in Mukkaw Bay, WA. This system is composed of a diverse assemblage of macroscopic attached algae, mussels, barnacles, and a large predatory starfish (*Pisaster ochraceus*). Paine (1966) experimentally removed the starfish from a section of the intertidal zone and compared the dynamics with a control section that contained the starfish. The removal of the predator caused a succession of species to occur with eventual competitive exclusion of other species by the mussel *Mytilus*. This result demonstrated that the predator had diversified the system by regulating the population of an otherwise dominant competitor. Any kind of species can be a keystone species and several are noted throughout this text. The primary way to identify a keystone species is with species removal experiments, as Paine conducted in the rocky intertidal zone. Paine's work is a benchmark in ecology which led to a generation of experimental studies and to the important keystone species concept (Mills et al., 1993; Paine, 1995; Power et al., 1996).

Bottom-up and top-down control are combined in the trophic cascade model (Carpenter et al., 1985; Carpenter and Kitchell, 1993) which uses a food chain approach to describe ecosystem control. These authors studied lake ecosystems and showed how productivity is controlled both by nutrient concentration in the lake water and its effect on phytoplankton, which are at the base of the food chain, and by the effects of the top predator fish species, which are at the top of the food chain. Thus, control can "cascade" either up or down the food chain. One generalization of this model suggests that in certain food chains the direction of control depends on the number of links (Fretwell, 1987) in such a way that top carnivores enhance primary productivity by reducing the intensity of herbivory in odd-numbered chains, while top carnivores reduce primary productivity by enhancing the intensity of herbivory in even-numbered chains. This generalization would be a useful design rule if ecological engineers were able to construct food chains of any significant length. Overall, the trophic cascade is an interesting theory which is much discussed in the literature (Hunter and Price, 1992; Perrson et al., 1996; Strong, 1992).

Another approach to control in ecology has been the application of cybernetics concepts to the ecosystem. Cybernetics as a discipline was first articulated by Norbert Wiener (1948) to cover examples of "control and communication in the animal and the machine." At the heart of cybernetics is an understanding of feedback pathways between components of a system that influence (i.e., control) its behavior. From the start, as envisioned by Wiener, cybernetics involved study of both machine controls developed by human designers and control systems in organisms that have evolved through natural selection, especially in terms of physiology. It was logical for ecologists to apply cybernetics because there seem to be many examples of self-regulation in nature.

Ramon Margalef (1968) was the first ecologist to embrace cybernetics as a foundation for describing control in ecosystems. He set out his ideas in his classic book whose first chapter had the title "The Ecosystem as a Cybernetic System." This chapter is filled with ideas of feedbacks, organization, diversity, stability, and energetics which are presented as general theory in Margalef's unique writing style. In his view ecosystems are composed of many feedback circuits mediated by species which collectively result in macroscopic behavior. E. P. Odum (1971) also added cybernetics to the introductory chapter on ecosystems in the third edition of his text. He uses the example of the heating of a room with a thermostat-controlled furnace to illustrate feedback and control concepts, but he also discusses several more complex ecological applications which contrast with the thermostat-heating system. For example, he states:

> … control mechanisms operating at the ecosystem level include those which regulate the storage and release of nutrients and the production and decomposition of organic substances. The interplay of material cycles and energy flows in large ecosystems generates a self-correcting homeostasis with no control or set-point required.

The concept of homeostasis mentioned in E. P. Odum's quote has come to be important in cybernetics and has been applied to ecology. In a homeostatic system adjustments take place in components so that some property of the system remains constant despite changes in the surrounding environment. For example, a thermostat maintains room temperature by adjusting the amount of fuel used in heating. Adjustments occur because of feedback about the system property that is being homeostatically controlled. The mechanisms of adjustment in ecology are diverse. Wynne-Edwards (1970) spoke of the "homeostatic machinery" involved in population regulation as including "changes in mutual tolerance or aggression, in territory size, the amount of emigration, the age of sexual maturation, changes in fertility and reproductive success, in cannibalism and other socially-promoted forms of mortality both of young stages and adults." Hardin (1993) spoke of a "demostat," using the thermostat as an analogy, for similar mechanisms. Levins (1998) listed systems-level factors of homeostasis including "the redundancy of the set of variables (if they are species, niche overlap expresses this property), self-damping of the variable, positive and negative feedbacks among variables, long and short pathways in the system, the connectivity of the network, time delays and sinks, the heterogeneity of flows and interactions and the 'shapes' of functional relationships." However, one problem with applying the concept of homeostasis to ecosystems is identifying the system property that is maintained at a constant level. This relates to the ecological notions of stability mentioned earlier (see Chapter 4). Stability has many meanings and there is no consensus among ecologists about which aspects are most important. Species composition and community structure of an ecosystem are not good candidates because species can move across landscapes and change relative abundances through time. More likely, species are the components that adjust within the homeostatic process. A microcosm experiment conducted by Copeland (1965) indicated homeostasis of ecosystem metabolism. He moved a turtle grass (*Thalassia testudinum*) microcosm that was at steady-state with a light regime of 1,500-ft candles to a new light regime of 230-ft candles. Under the lower light environment the turtle grass died back and was replaced by blue-green algae as the dominant primary producer. Ecosystem metabolism declined initially after the drop in light intensity, but it returned to the previous level within 3 months. This is a remarkable experiment that ought to be repeated. Schultz (1964, 1969) provides another example concerning the Arctic tundra as a homeostatic system. A feedback model of homeostasis is given which includes levels of the lemming population, plant biomass and nutrient content, and soil temperature. The overall system oscillates but the composition remains stable.

H. T. Odum (1971) developed ideas of cybernetics and homeostasis by differentiating between power and control circuits in ecosystems, as noted in the following quote:

> In very highly organized natural systems the flows of power are much divided among the species circuits, but they can be roughly separated into power circuits and control circuits. Thus, if oak trees process 50 percent of the power budget of a forest system, they constitute a power circuit. The squirrels of that forest may be processing much less than 1 percent of the forest budget. Their procedures for gathering and planting acorns may, however, serve as a control on the patterns of the oaks. Thus, we must

distinguish between the power flow in a circuit and the power being controlled by a circuit. ... Power circuits must be large and sluggish, whereas control circuits with small energies are easily insulated and can perform delicate operations and provide a directive influence on the power circuits. These principles are the same in electrical power distribution, in the forest, or in the complex industrial systems of man.

These generalizations were applied to a specific example of the grazing control system in the Puerto Rican rainforest study (H. T. Odum and Ruiz-Reyes, 1970). He further illustrated his ideas on control in terms of the maximum power theory. In this context work of consumer organisms in the forest increases total system energy flow through their dispersed control actions. In the following quote (H. T. Odum, 1978b), he suggests a demonstration of this aggregate control action by comparing a forest plantation, which lacks most of the complex consumer diversity, with the rain forest that grows side by side in the Luquillo Mountains of eastern Puerto Rico:

A tropical forest plantation of Cadam trees in Puerto Rico has a productive net yield of photosynthesis 20g/m^2/day (80 Calories/m^2/day wood equivalents) as a monoculture without many consumers. In contrast, a fully developed ecosystem nearby (with fully developed consumers feeding back in an organized manner) showed an increase in this basic primary production. An increase of 7 g/m^2/day (28 Calories/m^2/day), most of which was used by the consumers without any net energy, was measured. The system with consumers contained more energy flow (power) than the same system without consumers. Most of the web of producer-consumer interaction was required to maximize power.

In addition to the more or less classic approaches to cybernetics in ecology listed above, many other studies are noteworthy. Knight (1983) developed H. T. Odum's consumer-control hypothesis and made experimental tests with microcosms in Silver Springs, FL. He concluded that the stable population levels of aquatic organisms seen by tourists through the famous glass bottom boats of the springs were maintained by "a harmonious system of feedback controls." Mattson and Addy (1975) provided a review of insect herbivory with many examples of similarities between insects and cybernetic regulators. Further examples of cybernetics in ecology are Montague's (1980) model of feedback actions of fiddler crabs in temperate saltmarshes, Gutierrez and Fey's (1980) discussion of feedback and ecosystem succession, and the review of DeAngelis et al. (1986).

Other ecologists have criticized the application of cybernetics. Perhaps the most vehement has been Lawrence Slobodkin (1993) who provided the quote below in a review of a book on the Gaia hypothesis:

The idea of feedback and cybernetics was born in engineering and imported into environmental sciences and biology so long ago that there is a tendency to forget that organisms have not been constructed by, or even for, engineers. Biological systems may indeed be represented by diagrams that look like those of a cybernetic engineer, but they do not have the properties of engineered cybernetic systems.

For example, compare a temperature control system for a living room and the processes that regulate the number of animals in a population. The thermostat–furnace–air con-

> ditioner system is designed by an engineer with the express purpose of keeping a comfortable temperature for humans despite ambient changes outside the living room. It has an engineered purpose. Population size, on the other hand, is an epiphenomenal consequence of the environment and the properties of the organisms in the environment. The properties of the organisms are an outcome of an evolutionary process in which absolute population size has no particularly important meaning except for the trivial stipulation that the organisms' ancestors did not die before they reproduced. Therefore, although population regulation may appear to be comparable to an engineered cybernetic system, the appearance is deceiving. But the deceptive appearance of effective feedback in nature extends well beyond the scale of a single population.
>
> It was already clear 30 years ago that some of the processes carried out by organisms on Earth tended to negate deviations from existing properties that are of importance for organisms. The carbon cycle, for example, is portrayed in elementary biology texts as a set of boxes representing plants, animals, and microbes and little else. The boxes are connected by arrows representing flow. They are wired together as if they were a diagram of a negative feedback system, maintaining constancy of atmospheric oxygen and carbon dioxide. … Despite the cybernetic appearance of the block-and-arrow diagrams, there is no guarantee that the carbon cycle actually is an effective feedback system, particularly in the context of anthropogenic returns of buried carbon to the atmosphere … .

Earlier in his career Slobodkin (1964, 1968) was a bit more generous towards cybernetics and he suggested that the optimal strategy for species was to maximize homeostatic ability. The context for this suggestion was a model of evolution as an "existential game" that he developed. Slobodkin contended that the only measure of success of a species playing the existential game of evolution was persistence, which was proportional to homeostatic ability. Conrad's (1995) model of the ecosystem as an "existential computer" is a systems-level expression of this same kind of behavior.

Other critiques of cybernetics in ecology are given by O'Neill et al. (1986) and DeAngelis (1995). An interesting dialogue involved a critique by Engelberg and Boyarsky (1979) that elicited a number of rebuttals (Jordan, 1981; Knight and Swaney, 1981; McNaughton and Coughenour, 1981; Patten and Odum, 1981). These arguments involved much semantics with the positive result of recording a number of opinions on the cybernetic nature of ecosystems.

Unlike ecology, control theory in engineering is noncontroversial and straightforward. Control theory is a technical field common to all engineering disciplines. Traditionally, controls were small machines (sometimes called servo-mechanisms) which used feedback information to regulate larger processes. These devices date back to nearly 5000 BC in Egypt with many applications (Mayr, 1970). One old version was called a "governor," because it governed the rate at which a larger machine (such as a steam engine) operated. As noted earlier in this section, Cole (1957) used the "governor" as a metaphor for the regulatory mechanism in population dynamics. Perhaps the most widely known example of a mechanical control is the thermostat, which is part of the heating system of all modern homes and in many types of industries. Cornelius Drebbel, a Dutch engineer, is credited with inventing the thermostat in the 17th century for controlling temperature in his alchemy exper-

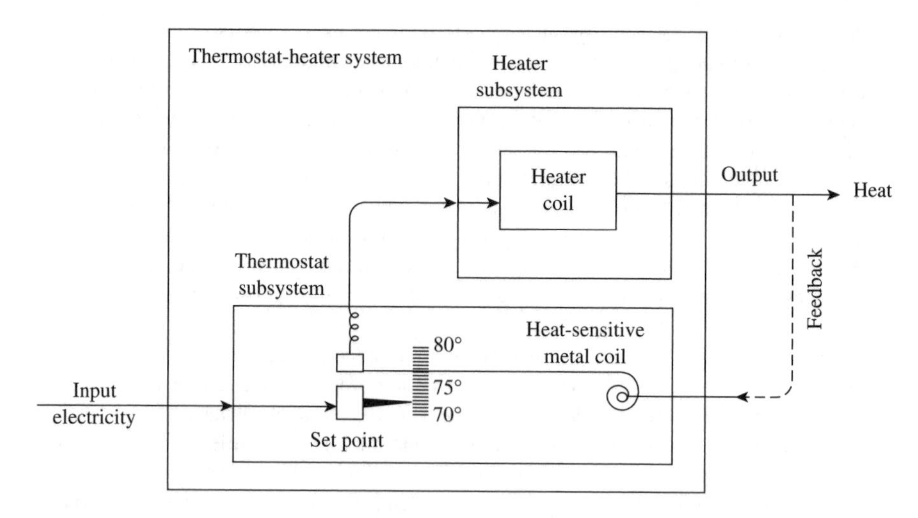

FIGURE 7.10 Details of a typical thermostat. (From Sutton, D. B. and N. P. Harmon, 1973. *Ecology: Selected Concepts*. John Wiley & Sons, New York. With permission.)

iments (Angrist, 1973). The modern thermostat is a device which senses temperature with a bimetallic strip and opens or closes an electrical circuit to a fuel source with reference to a set-point temperature (Figure 7.10). It is therefore a small engineered device that processes information (i.e., about temperature) and controls the operation of a larger system. Figure 7.11 shows an engineering block diagram of a temperature control system along with a translation in the energy circuit language. In the block diagram the thermostat is shown as the comparator unit while in the energy circuit diagram it is shown with the switch symbol. Basically, the thermostat compares the temperature in the building against the set-point value and turns on the furnace if the building temperature is below the set-point or turns off the furnace if the building temperature is above the set-point. This operation stabilizes the entire system by maintaining a constant indoor temperature even if the outdoor temperature changes dramatically.

Mathematical techniques describing control in engineering have developed especially within the last 100 years to provide a quantitative basis for design. These techniques are standardized and described in a number of textbooks, variously titled *Control Engineering* (Murphy, 1959), *Feedback Control System Analysis and Synthesis* (D'Azzo and Houpis, 1960), etc. Analysis is performed on sets of equations that describe the system, and designs are modified to ensure stable performance. Frequency response is one example where system performance is evaluated against variations in input conditions, followed by design modification. These kinds of techniques along with others represent the powerful tools that have allowed engineers to design, build, and operate the amazing array of technologies characteristic of modern society.

Engineering control theory has been successfully applied in physiology (Grodins, 1963; Milhorn, 1966; Milsum, 1966; Toates, 1975). This is not surprising since there are many examples of self-regulation of physiological processes where steady-

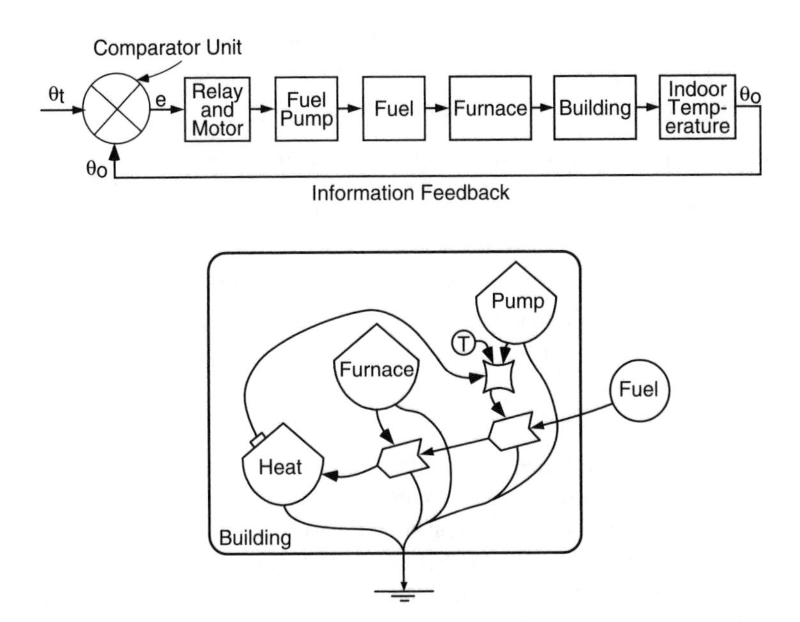

FIGURE 7.11 Comparison of an engineering block diagram of a thermostat controlled system with a translation in the energy circuit language. The thermostat is the comparator unit in the block diagram above and the switch symbol in the energy diagram below. Note the role of feedback in each model.

state conditions are maintained, as in body temperature, breathing rates, etc. (Langley, 1965, 1973). In fact, Walter Cannon, a physiologist and medical doctor, coined the important cybernetic term *homeostasis* to describe these systems in his classic work in 1932. However, the application of engineering control theory to ecology has not been nearly as successful. Some attempts were made in the 1960s and 1970s (Lowes and Blackwell, 1974; Mulholland and Sims, 1976), but the applications did not lead to advancements in ecological understanding. Clearly, ecological circuits do not behave very much like human-designed circuits such as the thermostat-furnace system. For example, on the one hand, if the thermostat is removed from the system in Figure 7.11, as Paine did in his experiment of the keystone predator of the rocky intertidal zone, then the heating system becomes unstable and basically stops functioning. On the other hand, removal of the keystone species in Paine's ecosystem changed the system dramatically but it continued to function. One hypothesis to explain the difference between ecological and engineering control systems may be that ecosystems are more complex. Hill and Wiegert (1980) indicate the difference in control mechanisms between ecosystems and human-designed systems in the following quote:

> Applying feedback control theory to engineered system is often much more successful than applying it to ecosystems. One reason is that engineered systems and control theory are eminently compatible because they have coevolved.

TABLE 7.4
Comparison of Machine Analogies in Ecology

Machine Analogy	Ecological System Modelled	Reference
Block-and-springs	Freshwater plankton food chain	Leavitt, 1992
Conveyor belt	Population dynamics	Oster, 1974
Pin ball machine	Population dynamics	Pearson, 1960
Connected gears	Marine plankton food chain	Clarke, 1946; Margalef, seen in Odum, 1983
Chemical reactor	Oceanic biogeochemistry	Siever, 1968
Cannon-ball catcher	Euphotic zone of ecosystems	H.T Odum et al., 1958

Thus, engineering control theory was specifically developed for systems that have been designed by humans. Because engineering control theory does not apply to ecosystems, fundamental differences must exist between these two kinds of systems. Berryman et al. (1992) also suggest that engineering control theory "should evolve to meet the needs and terminology of the ecologist."

Even though engineering ideas of control have not provided much new insight in ecology, they are useful in order to contrast the degree of difference in understanding about control between the two fields. There is no general theory of control in ecology, unlike in engineering. Perhaps new generations of ecological engineers with balanced training and experience in both disciplines will be able to make contributions to the academic field of ecology about the nature of control in an ecosystem context. Possible directions are outlined by Conrad (1976) and Hannon (1986; Hannon and Bentsman, 1991). E. P. Odum (1997) suggested that ecologists should shift emphasis from the tight control homeostasis of to homeorhesis, which is a looser form of control, for ecological studies above the organismal level of organization.

An interesting aside to the discussion of control ideas in ecology and engineering is the role of the machine analogy. Machines or mechanical devices, which have obvious relations to engineering, have long been used as analogies to help understand complex living systems. Calow (1976), Channell (1991), and Grmek (1972) reviewed the history of this subject which is filled with fascinating examples such as the importance of the development of the mechanical clock to theories of biology during the Renaissance. Unlike these historical references, mechanical analogies have seldom been useful in explaining ecology. Examples are given in Table 7.4, but none are well known. Clarke's (1946) gear diagram is perhaps most noteworthy (Figure 7.12). Here gears in the mechanism represent different trophic levels, which are scaled to turn at different speeds depending on production rates. H. T. Odum (1950) elaborated on the gear analogy in a section of his dissertation entitled "A Biosphere of Cogwheels," citing Clarke's paper (see also Figure 8.7 in H. T. Odum and E. C. Odum, 1976). Even though he also once used a gear model of a food web in his earlier work (see H. T. Odum, 1983, Figures 15–24), Margalef (1985) offers an explanation for this general lack of success of the mechanical analogy in ecology in the following quote:

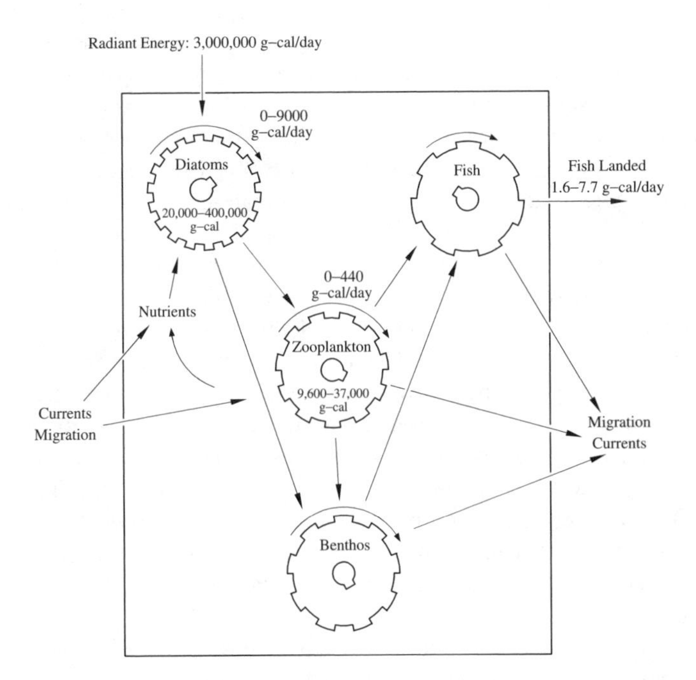

FIGURE 7.12 Example of a machine analogy from ecology. (From Clarke, G. L. 1946. *Ecological Monographs.* 16:321–335. With permission.)

> The essence of the ecosystem is the pattern that links its components. The a priori freedom of behavior of each of the parts is more or less diminished when the parts join the system. But the elements retain some flexibility or elasticity, because of time delays at the junctions. Ecosystems behave in just this way, being made up of individuals of a certain size, of behavior more less unpredictable, and out of equilibrium most of the time.
>
> The ecosystem is not a rigid machine made of gears and levers, but involves more costly transmission of information across a fluid (and turbulent) environment. In one sense it is comparable to a Turing machine, an automaton able to read information from a tape, and to use it for any purpose, including writing it on to a new tape, or for building the machines. However, it makes little sense to keep separate the instructions to build the machines, to operate them, or to process and pass information to other machines. The blurring of proper distinctions between operative parts and memories makes organisms and ecosystems quite different from computers, and this has to be kept in mind in formulating ecosystem models. They cannot be rigid clockwork.

Ulanowicz (1993, 1997) provides an even stronger critique of the machine analogy in ecology. He even suggests that mechanistic (machinelike) explanations are inadequate, and he calls for a post-Newtonian ecology with alternative concepts of causality. While these criticisms have merit and need to be explored, ecological engineering brings a renewed interest to the machine analogy. John Todd's living machines (see Chapter 2) and Robert Kadlec's (1997) "biomachine" treatment wetlands are hybrid systems. Todd's work in particular goes beyond the use of machine as analogy in development of design knowledge for hybrid systems. His work on living machines is somewhat reminiscent of the work of Franz Reuleaux, the German engineer who is credited with developing the fundamental theory of machine design in mechanical engineering in the late 1800s (O'Brien, 1964). At an elemental level, design of complex machines consists of combinations of the five basic "simple machines": the lever, the wheel and axle, the pulley, the wedge, and the screw. Reuleaux developed principles and algorithms for machine design based on *kinematics*, which is the science of motion. Thus, the simple machines are combined in such a way that their coordinated motion results in the transformation of energy and the output of useful work. Todd's design of living machines represents a kind of kinematics for ecological engineering that is both effective and elegant. Concepts of both machine and ecosystem are needed for the design of the new living machines. Ideas of the ecosystem as a computer and succession as a form of computation (see Chapter 5) are possible examples of new directions for the machine analogy in ecology. Technoecosystems are discussed as a future direction in Chapter 9. An interesting nexus may emerge as the field of ecological engineering develops among the machine analogs in ecology, the living machine concept, and the theory of self-reproducing machines (see Chapter 3).

At least two management fields exist which involve actual ecological control through manipulation of species. The field of biological control is well established in agriculture (Batra, 1982; Debach, 1974; Murdoch and Briggs, 1996; National Research Council, 1996), involving the introduction of predators, parasites, or diseases for pest population control. This is a population scale approach that is implemented to reduce the use of chemical pesticides. Although well established, the field is coming under greater scrutiny because of the risks of biological control agents becoming invasive (Simberloff and Stiling, 1996). Biomanipulation is an ecosystem scale approach to control of eutrophication in lakes (Kitchell, 1992; Reynolds, 1994). This field was first outlined by Shapiro et al. (1975), and it involves manipulating piscivore (fish predators) abundance to reduce phytoplankton abundance and to improve water clarity. Biomanipulation is based on the trophic cascade model discussed earlier and, like this model, remains controversial (Carpenter and Kitchell, 1992; DeMelo et al., 1992). However, the advocates of biomanipulation are ambitious as noted in the quote by Carpenter et al. (1985) given below:

> The concept of cascading trophic interactions links the principles of limnology with those of fisheries biology and suggests a biological alternative to the engineering techniques that presently dominate lake management. Variation in primary productivity is mechanistically linked to variation in piscivore populations. Piscivore reproduction and mortality control the cascade of trophic interactions that regulate algal dynamics. Through programs of stocking and harvesting, fish populations can be managed to regulate algal biomass and productivity.

Drenner and Hambright (1999) reviewed 41 biomanipulation trials and found that 61% were successful at improving water quality.

The fields of biological control and biomanipulation represent examples of the concept of "ecological engineering through control species" that was introduced by H. T. Odum (1971). Berryman et al. (1992) also advocate ecological engineering through species manipulation in the following quote:

> Natural ecosystems contain a plethora of feedbacks between their biotic components; for example, negative feedbacks between predators and prey and positive feedbacks between competing species. The ecological control engineer can, theoretically, manipulate these feedbacks in an attempt to regulate the system at desired steady states or to amplify certain components.

Unlike other examples of ecological engineering, manipulation of control species involves no familiar hardware such as pumps and pipes or electrical circuits. It is more analogous to the new discipline of software engineering, with manipulations of information rather than energy and materials. This may be the most sophisticated form of ecological engineering and will require much more experience in order to achieve success, as opposed to other applications discussed in earlier chapters where successful progress is established and growing.

In conclusion, invasive exotic species are an example of biology that is "out of control." On the one hand this may be thought of negatively as in the story of Frankenstein, where exotics represent the monster turned loose on the innocent villagers. It is interesting to note that this same fear occurs with certain forms of modern technology that exceed their intended functions (Winner, 1977). Tenner (1997) describes many of these examples in his book entitled *Why Things Bite Back: Technology and the Revenge of Unintended Consequences*." Bill Joy presents an even stronger case for the potential dangers of genetic engineering nanotechnology and robotics, which he refers to as the GNR technologies because these technologies are capable of self-reproduction, Joy (2000) warns that "they can spawn whole new classes of accidents and abuses" (see Crichton, 2002 for a science fiction interpretation). On the other hand, the idea of being "out of control" may be an example of a higher level phenomenon where new forms of order emerge out of old systems (Kelly, 1994). Rodney Brooks has also written on the positive aspects of being out of control. Control mechanisms require extra energy input for maintenance. If a system can be designed that does not require control (i.e., one that is out of control), then energy can be saved and used for other productive purposes. Brooks (2002; Brooks and Flynn, 1989) has explored this concept by desigining and building simple robots that achieve complex tasks through collective, emergent behavior (See Chapter 3). In a sense, the design of all ecologically engineered systems is "out of control" to some extent because of the contributions of self-organization. For example, the restoration ecologist may try to achieve a certain species composition in a restored marsh through intentional plantings but the final plant community is different because of natural selection and the addition of volunteer species that disperse in from the surrounding landscape. Thus, ecological engineers must be able to give up some control over their designs in order to create them. This represents a new kind of design paradigm for engineering, which is actively evolving as noted by the many examples given in this text. How much control are ecological engineers willing and able to give up in order to "design" new systems? The self-organization that is taking place with invasive exotic species may be a guide. In fact, the best way to conserve biodiversity may be to maximize dispersal and invasion globally. Instead of allowing a subset of highly preadapted species to homogenize the biosphere, more species may be supported by accelerating the dispersal processes. For example, species endangered in the U.S. might flourish in China and vice versa. In this way humans give up control in order to create a more diverse planet. Perhaps humanity needs both preserves (i.e., national parks), where the old ecosystems without exotics are maintained with care and at high cost, and new systems, where exotic species and native species are actively and intentionally mixed together. These new systems might be called mixing zones where self-organization is encouraged in order to save species and to create useful ecosystem designs. This may happen anyway, whether or not humans wish it to happen. Perhaps ecological engineers are best prepared by their balanced training to study these ideas and to be leaders in encouraging a new order of biodiversity.

APPENDIX 1: LIST OF BOOKS PUBLISHED ON EXOTIC SPECIES USED TO PRODUCE FIGURE 7.1

1950s

Elton, C. S. 1958. *The Ecology of Invasions*. Methuen, London.

1960s

Laycock, G. 1966. *The Alien Animals*. Ballantine Books, New York.

Mead, A. R. 1961. *The Giant African Snail: A Problem in Economic Malacology*. University of Chicago Press, Chicago, IL.

1970s

Mann, R. (ed.). 1979. *Exotic Species in Mariculture*. MIT Press, Cambridge, MA.

Roots, C. 1976. *Animal Invaders*. Universe Books, New York.

Silverstein, A. and V. Silverstein. 1974. *Animal Invaders*. Atheneum, New York.

1980s

Courtenay, W. R., Jr. and J. R. Stauffer, Jr. (eds.). 1984. *Distribution, Biology, and Management of Exotic Fishes*. Johns Hopkins University Press, Baltimore, MD.

Dahlsten, D. L. and R. Garcia (eds.). 1989. *Eradication of Exotic Pests*. Yale University Press, New Haven, CT.

Drake, J. A., H. A. Mooney, F. di Castri, R. H. Groves, F. J. Kruger, M. Rejmanek, and M. Williamson (eds.). 1989. *Biological Invasions: A Global Perspective*. John Wiley & Sons, Chichester, U.K.

Druett, J. 1985. *Exotic Intruders: The Introduction of Plants and Animals into New Zealand*. Heinemann, Auckland, New Zealand.

Groves, R. H. and J. J. Burdon (eds.). 1986. *Ecology of Biological Invasions*. Cambridge University Press, Cambridge, U.K.

Hengeveld, R. 1989. *Dynamics of Biological Invasions*. Chapman & Hall, London.

Long, J. L. 1981. *Introduced Birds of the World*. Universe Books, New York.

Macdonald, I. A. W., F. J. Kruger, and A. A. Ferrar (eds.). 1986. *The Ecology and Management of Biological Invasions in Southern Africa*. Oxford University Press, Cape Town, South Africa.

Mooney, H. A. and J. A. Drake (eds.). 1986. *Ecology of Biological Invasions of North America and Hawaii*. Springer-Verlag, New York.

Thomas, L. K., Jr. 1980. *The Impact of Three Exotic Plant Species on a Potomac Island*. National Park Service Scientific Monograph Series No. 13, U.S. Dept. of the Interior, Washington, DC.

Wilson, C. L. and C. L. Graham (eds.). 1983. *Exotic Plant Pests and North American Agriculture*. Academic Press, New York.

1990s

Bright, C. 1998. *Life Out of Bounds: Bioinvasion in a Borderlesss World*. W. W. Norton, New York.

Cox, G. W. 1999. *Alien Species in North America and Hawaii*. Island Press, Washington, DC.

Cronk, Q. C. B. and J. L. Fuller. 1995. *Plant Invaders*. Chapman & Hall, London.

Devine, R. 1998. *Alien Invasion*. National Geographical Society, Washington, DC.

de Waal, L. C., L. E. Child, P. M. Wade, and J. H. Brock (eds.). 1994. *Ecology and Management of Invasive Riverside Plants*. John Wiley & Sons, Chichester, U.K.

Di Castri, F., A. J. Hansen, and M. Debussche (eds.). 1990. *Biological Invasions in Europe and the Mediterranean Basin*. Kluwer Academic, Dordrecht, The Netherlands.

D'Itri, F. M. (ed.). 1997. *Zebra Mussels and Aquatic Nuisance Species*. Ann Arbor Press, Chelsea, MI.

Goldschmidt, T. 1996. *Darwin's Dreampond.* S. Marx-Macdonald (Trans.). MIT Press, Cambridge, MA.

Groves, R. H. and F. di Castri (eds.). 1991. *Biogeography of Mediterranean Invasions.* Cambridge University Press, Cambridge, U.K.

Jaffe, M. 1997. *And No Birds Sing: The Story of an Ecological Disaster in a Tropical Paradise.* Barricade Books, New York.

Johnston, R. F. and M. Janiga. 1995. *Feral Pigeons.* Oxford University Press, New York.

Lesinski, J. M. 1996. *Exotic Invaders.* Walker and Co., New York.

Lever, C. 1994. *Naturalized Animals: The Ecology of Successfully Introduced Species.* T & A D Poyser, London.

Luken, J. O. and J. W. Thieret (eds.). 1997. *Assessment and Management of Plant Invasions.* Springer, New York.

Mayer, J. J. and I. Lehr Brisbin, Jr. 1991. *Wild Pigs of the United States.* University of Georgia Press, Athens, GA.

McKnight, B. N. (ed.). 1993. *Biological Pollution.* Indiana Academy of Science, Indianapolis, IN.

Meinesz, A. 1999. *Killer Algae.* D. Simberloff (Trans). University of Chicago Press, Chicago, IL.

Nalepa, T. F. and D. W. Schloesser (eds.). 1993. *Zebra Mussels.* Lewis Publishers, Boca Raton, FL.

National Research Council. 1996. *Stemming the Tide.* National Academy Press, Washington, DC.

Office of Technology Assessment (OTA). 1993. *Harmful Non-Indigenous Species in the United States.* OTA-F-565. U.S. Government Printing Office, Washington, DC.

Pitcher, T. J. and P. J. B. Hart (eds.). 1995. *The Impact of Species Changes in African Lakes.* Chapman & Hall, London.

Pysek, P., K. Prach, M. Rejmanek, and M. Wade (eds.). 1995. *Plant Invasions: General Aspects and Special Problems.* SPB Academic Publ., Amsterdam, The Netherlands.

Ramakrishnan, P. S. 1991. *Ecology of Biological Invasions in the Tropics.* International Scientific Publications, New Delhi, India.

Randall, J. M. and J. Marinelli (eds.). 1996. *Invasive Plants.* Brooklyn Botanic Garden, New York.

Rodda, G. H., Y. Sawai, D. Chiszar, and H. Tanaka (eds.). 1999. *Problem Snake Management, the Habu and the Brown Treesnake.* Comstock Publ. Associates, Ithaca, NY.

Rosenfield, A. and R. Mann (eds.). 1992. *Dispersal of Living Organisms into Aquatic Ecosystems.* Maryland Sea Grant Publ., College Park, MD.

Shigesada, N. and K. Kawasaki. 1997. *Biological Invasions: Theory and Practice.* Oxford University Press, Oxford, U.K.

Simberloff, D., D. C. Schmitz, and T. C. Brown (eds.). 1997. *Strangers in Paradise.* Island Press, Washington, DC.

Stone, C. P., C. W. Smith, and J. T. Tunison (eds.). 1992. *Alien Plant Invasions in Native Ecosystems of Hawaii.* University of Hawaii Cooperative National Park Resources Studies Unit, Honolulu, HI.

Williams, D. F. (ed.). 1994. *Exotic Ants.* Westview Press, Boulder, CO.

Williamson, M. 1996. *Biological Invasions.* Chapman & Hall, London.

8 Economics and Ecological Engineering

INTRODUCTION

Economics plays an important role in any engineering field, primarily as an aid in making design decisions. There is always a need to find the least expensive way to solve a problem and, at the most basic level, economics provides a system for this accounting. The typical approach is to generate alternative solutions or designs and to evaluate these alternatives with economic criteria. The application of economics to engineering is a traditional subdiscipline called engineering economics (Grant and Ireson, 1964; Sepulveda et al., 1984).

In a sense economics reveals which alternatives are realistic in terms of implementation. Some may be too costly and are thus not realistic. However, reality in this context depends on the accounting system that is used for evaluation. As will be discussed in this chapter, conventional economics has some limitations, especially in terms of being capable of evaluating aspects of the environment. To deal with these limitations, new forms of economics are being developed and applied to ecological engineering in order to improve decision making (Maxwell and Costanza, 1989; H. T. Odum, 1994b; Van Ierland and deMan, 1996). An accounting system is needed for a variety of special issues in ecological engineering. For example, it is often stated that pollution is cheaper to prevent than to clean up. This is an economic generalization that requires the capability of full accounting of costs of pollution treatment technologies, costs of pollution impacts on the environment, and costs of pollution cleanup which might include site remediation or even ecological restoration. Hazard evaluation with microcosms is another example. How much funding is appropriate for adequate testing of potential toxins that are to be released into the environment? This decision requires costs of testing with microcosms and mesocosms vs. costs of potential environmental impacts of the toxins.

In practice engineers usually become involved in a project after a certain stage of decision making. Often, they are not asked whether the project should be done, but rather they are asked how best to implement the project. For example, the engineer is asked where to build a dam or what kind of dam to build, not whether the dam should be built. Thus, engineers do not usually go beyond the typical uses of economic accounting. However, ecological engineering implies a wider scale of thinking. Ecological engineering designs are specifically intended to combine nature with human technology, which requires a complete accounting system. In this chapter alternative accounting systems are presented with recommendations for those best suited to the field of ecological engineering.

A related issue concerning economics is the large-scale question of the future of society. Some believe that growth will continue without limit, while others believe society is already or will soon be limited. These limits come from declining amounts of fossil fuels that are the driving force supporting society and from the carrying capacity constraints of the biosphere. Classical economic theory suggests that forms of human capital can substitute for natural resources such as fossil fuels through technology. In this view the future depends on the power of technology to overcome limits (Ausubel, 1996). A dichotomy has evolved between people who believe technology will continue to develop fast enough to compensate for lost or spent natural resources and people who believe that the planet's capacity to absorb society's wastes and provide raw materials and energy is finite and limited. Costanza (1989) referred to these groups as technological optimists and technological pessimists, respectively, and he suggests that policy makers need to carefully weigh their perspectives in making decisions. No definite resolution is possible to the dichotomy at this time because both sides can present evidence to support their beliefs, but fossil fuels are definitely becoming more expensive and limits to humanity are becoming apparent.

A relevant question is where the field of ecological engineering falls along the gradient of opinions. On one hand ecological engineering designs are among the most advanced forms of technology by combining conventional engineering with living ecosystems in a symbiotic coupling, making them consistent with the beliefs of the technological optimists. On the other hand, by relying on renewable energies, by reducing costs, and by emphasizing natural ecosystems, ecological engineering designs are best adapted to a future with limited resources, making them consistent with the belief of the technological pessimists. Thus, ecological engineering has a dual conception that makes it correct and appropriate for either the technological optimist or the technological pessimist position.

STRATEGY OF THE CHAPTER

Issues of economic evaluation and assessment are covered in this chapter. All ecological engineering projects are concerned with economics, usually in several contexts, making this a subject of general relevance. The conventional economic approaches are covered first with a survey of cost–benefit analysis and assessments based on market valuation. Ecological engineering designs can save money especially because they use more free, renewable energies and less purchased energies than traditional alternatives. They also can produce by-products that add value to their assessment. Limitations of conventional economic analysis are discussed with focus on environment. While a number of approaches have evolved to include environmental issues and values in conventional economics, some aspects are still not adequately considered. Alternative policy and accounting systems, such as the new field of ecological economics, have been developed in response to these limitations. Important topics stimulated by the development of this field are discussed, including ecosystem services, carrying capacity, natural capital, and sustainability. A new approach, emergy analysis, is presented in some depth as an example of an accounting system that is appropriate for ecological engineering. The chapter concludes with brief considerations of relevant economic policies and issues: financing, patenting, regulatory permitting, and ethics.

CLASSICAL ECONOMICS PERSPECTIVES ON ECOLOGICAL ENGINEERING

One of the primary roles of economics is in assessments of costs and benefits of a system or project. The most practical applications are financial assessments that deal with classical microeconomics of market values. A market is a self-organizing economic system that balances supply of and demand for goods and services. Theoretically, the market controls production of goods and services to match the demand by consumers so that there is no excess in terms of extra supply or unmet demand. In this context, the price of a good or service is the measure of its value. Financial analyses deal only with values of costs and benefits that are determined by markets. This is the day-to-day reality of the business world in which most decisions are made.

A financial goal of ecological engineering designs is to reduce costs of a project by substituting free renewable energies, through use of natural or constructed ecosystems, for some of the purchased energies that dominate conventional alternative designs. Thus, a goal of ecological engineering is to save money. This saving may occur at any or all of the stages of the design–build–operate sequence of a project. However, it is at the operating stage of a project that savings are most likely to occur because it is here that free, renewable energies are used to drive the long-term dynamics of the ecosystem part of the design.

Perhaps the best demonstration of financial savings comes from the field of treatment wetlands which is the most advanced application of ecological engineering. Table 8.1 compares financial aspects of a conventional treatment system with a treatment wetland. Although this example leaves off the design costs, the treatment wetland is cheaper for both the construction and operating costs. Other examples of financial cost savings from treatment wetlands are given by Breaux et al. (1995), Campbell and Ogden (1999), Cueto (1993), Ko et al. (2000), and Petersen (1991), though savings do not always occur (Latchum and Kangas, 1996).

Another quality of at least some ecological engineering designs is by-products which have market value. In other words, ecologically engineered systems often generate beneficial goods that have value as by-products of the normal operation of the system. Two examples of living machines with by-product values are shown in Figure 8.1. The direct purpose of these systems is to treat wastewater and produce clean water that can be discharged back into the environment. However, they also have the ability to generate by-products that can be sold to add value to the system. The Frederick, MD, living machine was a demonstration project that treated a small portion of the domestic sewage from the local urban area (Josephson et al., 1996a,b). In this system ornamental plants and aquarium fish were produced in the tanks near the end of the living machine and sold to local businesses. At one point more than $1,000/month was generated from these sales, which was an indirect benefit of treating the wastewater. The Henderson, NV, living machine treats wastewater from

TABLE 8.1
Comparison of Budgets for Wastewater Treatment Alternatives

Cost Category	Conventional Sewage Treatment Plant	Treatment Wetland System
	Construction Costs ($)	
Mobilization and administration	95,000	91,000
Earthwork	381,000	1,336,000
Wetland Planting	0	309,000
Other Sitework (electrical, controls, and piping)	728,000	1,720,000
Conventional primary	639,900	0
Conventional activated sludge	698,000	0
Sludge handling	687,000	0
Biological nitrification	476,000	0
Chlorination and outfall	208,000	208,000
Total	4,112,000	3,664,000
	Operating Costs ($/year)	
Personnel	63,000	24,000
Utilities	23,000	5,000
Chemicals	23,000	11,000
Equipment and supplies	47,000	5,000
Total	156,000	45,000

Source: Adapted from Kadlec, R. H. and R. L. Knight. 1996. *Treatment Wetlands*. CRC Press, Boca Raton, FL.

a chocolate factory (Shaw, 1999). In this system some of the clean water produced by the treatment process is used as irrigation water for the company's landscape plants. This generates value because the company is located near Las Vegas in the arid southwest U.S. where the value of irrigation water is considerable. A savings is realized by the company because they produce their own irrigation water and do not have to buy an equivalent amount of water for the landscaping. This system also produces vegetables (Figure 8.2), though these are not sold but rather distributed to the operators of the living machine. Federal laws prohibit sale of food products grown in treated sewage waters, but because the Henderson living machine is not treating sewage, it is possible to raise food products.

There are many opportunities to develop valuable by-products from ecologically engineered systems because of the production capacity of ecosystems. Although several system-specific examples have been discussed (Devik, 1976; National Research Council, 1981), the topic of by-product generation is underdeveloped. To some extent knowledge of business and marketing is essential along with ecological engineering to develop these opportunities. The future will likely include more

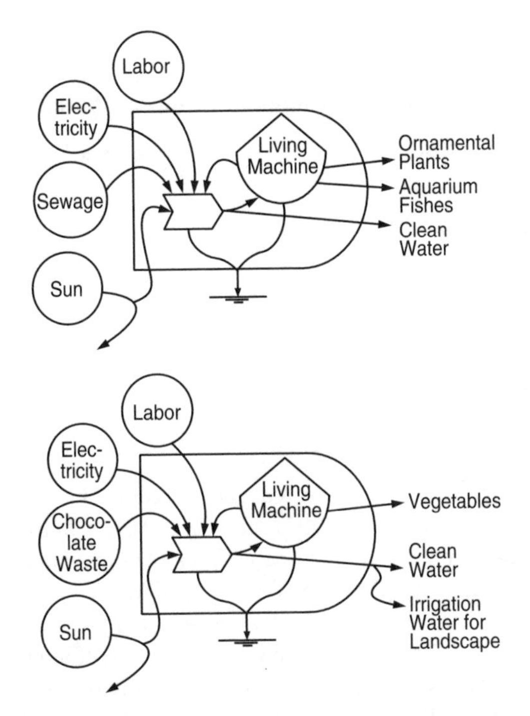

FIGURE 8.1 Comparison of two living machines with different by-product values: Frederick, MD (top) and Henderson, NV (bottom).

FIGURE 8.2 Tomatoes are being grown in the living machine in Henderson, NV, which treats food wastewater rather than sewage.

multipurpose uses of ecological engineered systems, and policies that regulate by-products should be critically examined.

The by-products produced from a system can be added up to assess total indirect benefits. This approach has also been applied in a related context for assessing the value of natural ecosystems. It has been termed the *component value method* and

TABLE 8.2
Summary of Use Values for Michigan's Coastal Wetlands for 1977

Use Category	Economic Value ($/acre/year)
Sport fishing	286.00
Nonconsumptive recreation	138.24
Waterfowl hunting	31.23
Trapping of furbearers	30.44
Commercial fishing	3.78
Total	489.69

Source: Adapted from Raphael, C. N. and E. Jaworski. 1979. *Coastal Zone Management Journal.* 5:181–194.

one of the earliest applications was in wetland valuation. Table 8.2 shows an early example by Raphael and Jaworski (1979) for Great Lakes wetlands. Using market values they estimated benefits to local economies from wetlands for five different uses. They then extrapolated their unit value ($489.69/acre/year) across the total acreage of Great Lakes wetlands (105,855 acres or 42,870 ha) to make a total assessment of $51,836,135/year. This exercise is useful to communicate the concept of value of natural ecosystems, especially in units (dollars) that are widely understood. More recently, the component value method has been applied to tropical rainforest conservation by showing the value of nontimber products that can be harvested from intact forests. Peters et al. (1989) produced the first comprehensive assessment of these nontimber forest products, and their estimations indicated that more value could be generated to local economies from intact forests than from conversion of the forest to other land uses such as ranching or tree plantations. Although controversial, assessments of value of nontimber forest products are an important conservation strategy for the tropics (Nepstad and Schwartzman, 1992; Plotkin and Famolare, 1992). Balmford et al. (2002) extend this type of analysis to a global scale.

The most comprehensive form of financial assessment is cost–benefit analysis, in which all costs and benefits of a project are considered. This has been the standard technique used for choosing between alternative designs in the field of engineering economics. In this analysis the costs and benefits of each alternative design for a project are evaluated in the same units, usually dollars, and then summed. Annual values of costs and benefits over a given life cycle of the project generally are divided by a discount rate to calculate net present value. After this calculation, the alternative with the highest ratio of benefits to costs is considered to be the best choice for implementing the project. The strength of this approach is in the logic of summing costs and benefits to determine the best alternative. The particular alternative with the best cost–benefit ratio represents the best investment opportunity for either a private or public (i.e., government) funded initiative. McAllister (1980) reviews this and other evaluation approaches. Reviews of methods for incorporating environmental values into economic cost–benefit analyses are given by Loomis and Walsh (1986) and Schulze (1991).

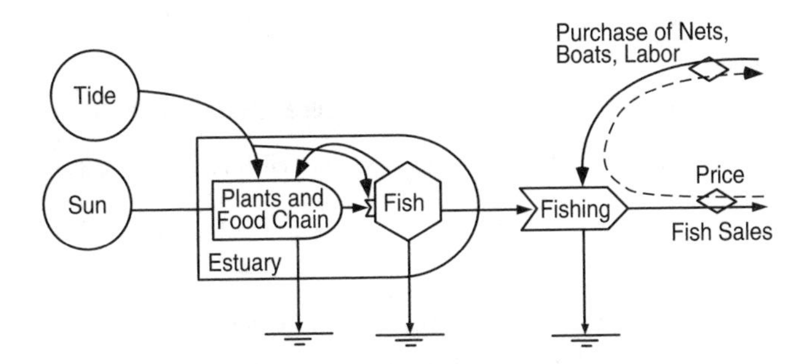

FIGURE 8.3 Energy circuit model of a fisheries system. The solid lines represent energy flows and the dashed lines represent flows of money. This model illustrates how money flows account only for the work done by the fisherman and are not connected to the actual production of fish by the estuary. (From Odum, H. T. and E. C. Odum. 1976. *Energy Basis for Man and Nature*. McGraw-Hill, New York. With permission.)

PROBLEMS WITH CONVENTIONAL ECONOMICS

Problems arise with classical economics when it extends beyond the market system. Classical economics is intentionally anthropocentric; it was developed to deal with issues between humans and especially about goods and services that humans can provide, make, sell, and own. However, this is only a subset of human concerns because the environment enters into human affairs in many ways. The environment, which consists of natural energy sources and ecosystems, provides to humans many goods and services that are not accounted for by classical economics. Figure 8.3 illustrates this fact for an estuary where fish are harvested and sold by fishermen. The estuarine ecosystem produces fish through interactions with an energy signature of tide and sun. Fishing is a process that removes fish from the estuary through interaction with purchased inputs from the fisherman. Money, shown with the dashed line, flows into the system in proportion to sales of fish, and it flows out in proportion to the inputs used by the fisherman. The problem here is that the process of fishing is based on inputs both from the estuary and from the fisherman, but money only goes to compensate the fisherman and not the estuary. Thus, the inputs from the estuary are considered to be free and are not accounted for in the economic transaction. The market, which determines the price of fish, only considers part of the actual system that produces fish. This problem has serious consequences because the accounting system used for decision making (conventional economics) does not properly account for all of the value. Overfishing inevitably occurs in the case shown in Figure 8.3 because of an inadequate accounting of value, and it ultimately leads to the collapse of the fishery to the detriment of both the estuary and the fisherman! This is an example of the "tragedy of the commons" in which the estuary is a common property resource (Hardin, 1968).

Another example of the failure of classical economics to account for value is shown in Table 8.3. This table lists values to humans from wetland ecosystems in a hierarchical ranking. The market system adequately accounts for only the population level of values. This is the realm considered in the analysis by Raphael and

TABLE 8.3
Listing of Wetland Values According to Hierarchical Level

Hierarchical Level	Value Category	Economic Values as Percent Total	
		Market System	Nonmarket Accounting
Population	Fish and wildlife and other component values	100%	5%
Ecosystem and global	Hydrological values Productivity values Waste assimilation values Atmospheric values Life-support values	0%	95%

Source: Adapted from Odum, E. P. 1979a. *Wetland Function and Values: The State of Our Understanding*. P. E. Greeson, J. R. Clark, and J. E. Clark (eds.). American Water Resources Association, Minneapolis, MN.

Jaworski for the Great Lakes wetlands described earlier (Table 8.2). Higher level values, at the ecosystem and global levels, are not accounted for by the market system and thus are largely considered to be free by humans. These values can be considered to be free as long as humans do not drain the system. However, when the drain on the systems becomes too great, the values must be accounted for and managed, or there is a threat of collapse due to overuse.

This problem of valuation is well known as are other criticisms of classical economics. Particularly interesting discussions are given by ecologists who provide a fresh perspective on these problems (Farnworth et al., 1981; Hall, 1990; Hall et al., 2001; Maxwell and Costanza, 1989; E. P. Odum, 1979a). Economists have attempted to deal with the problems in various ways. To some extent these are termed "market imperfections," which occur because people have incomplete knowledge of the basis of true value when making decisions. Various concepts have been developed to deal with these "imperfections," such as externalities for common property resources and shadow prices for adjusting values derived from the market. Whole subdisciplines have evolved with environmental economics dealing with the use of the environment as a sink for waste products from the human economy and natural resource economics dealing with the use of natural resources as inputs into the human economy. The next section deals with new approaches for integrating ecology and economics more fully.

However, even though conventional economics has limitations, in terms of assessing the environment, it is useful. It does account for things that have markets, which allows for financial analyses, and it is still the language of decision makers, which gives it practical utility. Thus, conventional economics should not be abandoned but new ways of thinking are needed to improve its utility during times of growing environmental impacts and resource shortages. Public policy decisions are

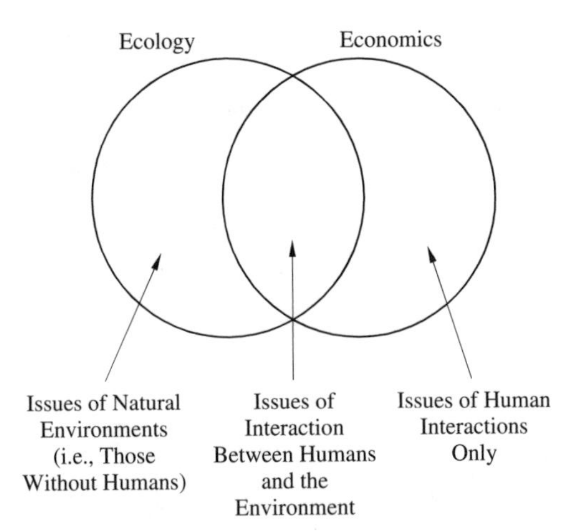

FIGURE 8.4 Venn diagram showing the differences and similarities between the fields of ecology and economics.

determined by conventional economics for the most part and ecologists should at least develop "diplomatic relations" with economists (Roughgarden, 2001).

ECOLOGICAL ECONOMICS

A new field that has developed to deal with the problems of accounting for the environment is ecological economics. This is a somewhat radical departure from classical economics that attempts to address "the relationships between ecosystems and economic systems in the broadest sense" (Costanza, 1989). The intention of workers in this new field is to reinvent economics with connections to ecology, rather than simply trying to correct "market imperfections" as with natural resource economics or environmental economics.

At a very basic level of comparison, ecology and economics have some commonalities or similarities that support the development of the new field. Both fields are named with the same prefix, *eco* from the Greek *oikos*, which refers to household. Thus, they appear to deal with similar systems. Ecology deals with natural environments and with environments that include human use and impact. Economics deals with interactions between humans and between humans and the environment. Thus, there is some overlap between the fields (Figure 8.4). Ecological economists hope to expand this overlap so that methods are developed that can be used to make better decisions about the environment.

Further evidence of the similarities between ecology and economics is the sharing of concepts and techniques by analogy. This sharing has been especially influential in ecology, which went through a phase during which economic tools and approaches were applied to a number situations (see reviews given by Bernstein, 1981, and Rapport and Turner, 1977). This same phenomenon has occurred in economics but apparently to a much lesser extent. Table 8.4 lists some examples of these cross-disciplinary analogies.

TABLE 8.4
Cross-Disciplinary Analogies between Ecology and Economics

Topic	Reference
Ecology Applied to Economics	
Natural selection	Winter, 1964
Niche	Hardesty, 1975; Lloyd et al., 1975; Mark et al., 1985
Diversity patterns	Golley, 1966
Resource partitioning	Kangas and Risser, 1979
Economics Applied to Ecology	
Consumer behavior	Tullock, 1971; Covich, 1972, 1974
Cost–benefit analysis	Roughgarden, 1975; Orians and Solbrig, 1977; Solbrig and Orians, 1977; Givnish et al., 1984; Riessen, 1992; Matsuda and Shimada, 1993
The theory of the firm	Bloom et al., 1985
Input–output analysis	Richey et al., 1978
Equilibrium concepts	Tschirhart, 2000

Ecological economics goes beyond analogies in order to remake classical economics with emphasis on philosophy as much as on actual accounting techniques. The field emerged in the 1980s from several starting points. On the one hand there were a number of established economists who began discussing fundamental problems of their field around the time of the first Earth Day in 1970. These included E. F. Schumacher (1968, 1973) with his reference to Buddhist economics, N. Georgescu-Roegen (1971, 1977) with his reference to "the entropy law and the economic process," and K. E. Boulding (1966, 1972, 1973, 1978) who made many connections between ecology and economics. The most influential of these workers trained as classical economists has been Herman Daly (1968, 1973, 1977, 1996; Daly and Towsend, 1993), a student of Georgescu-Roegen, whose early contribution was the idea of an economy based on steady-state rather than growth. Daly is one of the founders of ecological economics, and he continues to add original ideas to the field.

On the other hand founders of ecological economics came from ecology-based training. H. T. Odum was a forerunner in this effort and his emergy analysis will be described in the next section. Perhaps the most influential person in ecological economics has been Robert Costanza, who was a student of H. T. Odum. Starting with the publication of his dissertation research (Costanza, 1980), Costanza has been at the center of developments in the field. For example, he has edited many conference proceedings (Costanza, 1991; Costanza and Daly, 1987; Costanza and O'Neill, 1996) and the new journal called *Ecological Economics*. He was the first president of the International Society of Ecological Economics and the lead author on the first text on the subject (Costanza et al., 1997a). The agenda Costanza outlined with coworkers (Costanza et al., 1991) continues to identify areas of work in ecological economics: (1) valuation of natural resources and natural capital, (2) ecological economic system accounting, (3) sustainability, (4) developing innovative instruments for environmental management, and (5) ecological modelling at all scales. Several of these areas are discussed in relation to ecological engineering below.

Life-Support Valuation of Ecosystem Services

One of the earliest quantitative measures in ecological economics was the calculation of the value of an ecosystem based on its contribution to the overall life-support system that the biosphere provides to humans. The concept states that ecosystems produce clean water and air through the biogeochemistry of their normal metabolism. These actions collectively constitute a life-support system for humans that is not valued by the economic system. However, it is expensive to reproduce by technological means, as is evidenced for example in the case of a physio–chemical life-support system for manned space flight (see the discussion of closed systems in Chapter 4). H. T. Odum (1971) first discussed and quantified the concept of life-support valuation by using the gross primary productivity (GPP) of the ecosystem and an energy-to-dollar ratio, which he estimated by dividing the total energy flow of the U.S. by the gross national product (GNP). This was fundamentally an ecological economics calculation because it combined ecology (through the use of GPP) and economics (through the use of an energy-to-dollar ratio) to quantify the value of an ecosystem. GPP is the appropriate measure to use for energy flow in the ecosystem because it integrates the metabolism (photosynthesis and respiration) of all biological populations within most systems. This calculation was a significant breakthrough that allowed for natural ecosystems to be evaluated with dollar values. H. T. Odum (1971) extended the concept by suggesting that humans have a constitutional right to a life-support system, as noted below:

> Basic to many of the legal battles underway and developing in the defense of the environment is a long ignored constitutional freedom — the human right to a safe life-support system. There can be no more fundamental right to an individual than his opportunity to breathe, drink water, eat, and move about with safety. Long taken for granted, these rights are not free but are paid for daily by the metabolic works of the life-support system processing the wastes and by-products. The water and mineral cycles, the complex of complicated organisms that process varied chemicals, and the panorama of ecological subsystems that organize and manage the earth's surface are not the property of individuals, but are part of the essential basic right, the life-support system. A fundamental flaw in the legal systems allowed owners of land to assume special rights to the public life-support means.

TABLE 8.5
An Early Comparison of Different Wetland Values

Value Category		Annual Return ($/acre/year)	Capitalized Value (at 5% interest rate) ($/acre)
Commercial and sports fisheries		100	2000
Aquaculture potential	Moderate oyster culture	350	7000
	Intensive oyster culture	900	18,000
Waste treatment	Secondary only	280	5600
	Tertiary	2500	50,000
Maximum noncompetitive summation of values from above:			
Commercial and sports fisheries + Tertiary waste treatment		2600	52,000
Intensive oyster culture + Tertiary waste treatment		3400	68,000
Total life-support value		4150	83,000

Source: Adapted from Gosselink, J. G., E. P. Odum, and R. M. Pope. 1973. The value of the tidal marsh. Working Paper No. 3. Urban and Regional Development Center, University of Florida, Gainesville, FL.

H. T. Odum even suggested how the calculation could be used to estimate losses in the life-support system due to pollution. Using a human population figure from the late 1960s and an average global ecosystem metabolism, he calculated that

> Each human's portion of the earth's life-support system is 1.7 × 10E5 g oxygen or 4 times this amount of energy (6.8 ¥ 10E5 kcal) processing by the system daily. Every time someone discharges about 380 pounds of organic waste per day he has diverted the life-support fraction of one person. If the substances are toxic, the amplifier destructive action is much greater. Large storages of oxygen and carbon in air and sea protect us from immediate difficulty with them, but we use their flows as an index to our disturbance of nature.

The life-support calculation approach was used in a classic paper on valuation of coastal wetlands (Gosselink et al., 1973). This paper was important because it provided the first comparison of different methods for calculation of ecosystem values (Table 8.5). The life-support approach provided the highest estimate of value. Of particular interest, the life-support value was higher than the component summations of individual values. This result attracted criticism from economists (i.e., King et al., 1979) and, in particular, an interesting exchange of opinions is recorded among Shabman and Batie (1978, 1980) and H. T. Odum (1979) and E. P. Odum (1979b). H. T. Odum and Hornbeck (1997) provide a review of the issue of saltmarsh valuation and update the early calculations by incorporating emergy analysis. The basic difference of opinion seems to be that some workers feel that the life-support

TABLE 8.6
Listing of Ecosystem Services

Purification of air and water

Mitigation of floods and droughts

Detoxification and decomposition of wastes

Generation and renewal of soil and soil fertility

Pollination of crops and natural vegetation

Control of the vast majority of potential agricultural pests

Dispersal of seeds and translocation of nutrients

Maintenance of biodiversity, from which humanity has derived key elements of its agricultural, medicinal, and industrial enterprise

Protection from the sun's harmful ultraviolet rays

Partial stabilization of climate

Moderation of temperature extremes and the force of winds and waves

Support of diverse human cultures

Providing of aesthetic beauty and intellectual stimulation that lift the human spirit

Source: Adapted from Daily, G. C. (ed.). 1997. *Nature's Services*. Island Press, Washington, DC.

approach is an overestimate because it implicitly includes values beyond what society is normally prepared to acknowledge. Other applications of the life-support calculation, all of which happen to involve coastal wetlands, are given by Lugo and Brinson (1979), Faber and Costanza (1987), and Costanza et al. (1989).

Probably for various reasons attention has turned in a different direction, and there has been no recent application of the life-support approach to ecosystem valuation. The discussion by ecological economists has evolved to focus on what are termed *ecosystem services* (Daily, 1997; Daily et al., 1997, 2000; Dakers, 2002; Ehrlich and Mooney, 1983; van Wilgen et al., 1996; Westman, 1977), which essentially constitute life-support functions (Table 8.6). The approach is to identify individual services that ecosystems provide to society and to estimate their value based on methods that are more consistent with classical economics, such as the "willingness-to-pay" approach. A fundamental difference between life-support and ecosystem services revolves around using one number as a measure of value (i.e., ecosystem metabolism) or breaking down a number of individual measures of value (various ecosystem services). The most extensive calculation of ecosystem services is given by Costanza et al. (1997b) who estimated a biosphere value of $16 to 54 trillion ($10^{12}$) per year with an average of $33 trillion per year for global ecosystem services, which is greater than the global gross national product of strictly economic flows at $18 trillion per year.

These approaches are very relevant to ecological engineering, which involves the construction of new ecosystems to solve problems. These new ecosystems will contribute life-support values and ecosystem services to society beyond their intended purposes. Thus, each new constructed ecosystem, such as a treatment wetland or even a microcosm, adds to the life-support capacity of the environment. Some feasibility studies in ecological engineering are being undertaken to account for this kind of value and the interest can only be expected to grow in the future.

A related valuation approach is to calculate the value of an ecosystem as the cost required to replace it. This is also very relevant to ecological engineering in regard to restoration ecology and associated fields, which seek the least expensive method of ecosystem creation.

Natural Capital, Sustainability, and Carrying Capacity

Ecological economics includes many other ideas that relate to ecological engineering. One example is the concept of natural capital, which is analogous to human capital traditionally considered by economists (Prugh et al., 1995). In a sense, natural capital is the structure of the biosphere's economy from which ecosystem services flow. The concept is described below by Costanza et al. (1997a):

> Thinking of the natural environment as "natural capital" is in some ways unsatisfactory, but useful within limits. We may define capital broadly as a stock of something that yields a flow of useful goods or services. Traditionally capital was defined as produced means of production, which we call here human-made capital, as distinct from natural capital which, though not made by man, is nevertheless functionally a stock that yields a flow of useful goods and services. We can distinguish renewable from nonrenewable natural capital, and marketed from nonmarketed natural capital, giving four cross-categories. Pricing natural capital, especially nonmarketable natural capital, is so far an intractable problem, ... All that need be recognized for the argument at hand is that natural capital consists of physical stocks that are complementary to human-made capital.

Although this concept has been developed by ecological economists, Wes Jackson, an environmentalist and agroecologist, introduced the term *ecological capital* in his proposed revision of modern agriculture (Jackson, 1980). Jackson elaborated his conception in terms of topsoil, with concern for erosion. Figure 8.5 illustrates the data he provides as a simple mass balance analogous to a bank account. The storage of topsoil represents natural or ecological capital, which is produced slowly by biogeochemical processes and soil management procedures but drained relatively quickly by agricultural erosion. Jackson's proposed switch to perennial plant species for crop production, along with other techniques such as no-till cultivation, reduces erosion and allows for greater accumulations of natural capital (in terms of topsoil storage) by agricultural systems. Thus, Jackson presaged the ecological economics concept of natural capital with his metaphor about the value of topsoil.

The natural capital concept has led to a macroeconomics perspective for ecological economics. Developments include the incorporation of natural resources into national-scale accounting (Repetto et al., 1989, 1999) and alternative indices such as the index of sustainable economic welfare (Costanza et al., 1997a; Daly and Cobb, 1989) that provide different perspectives from traditional measures, like gross national product.

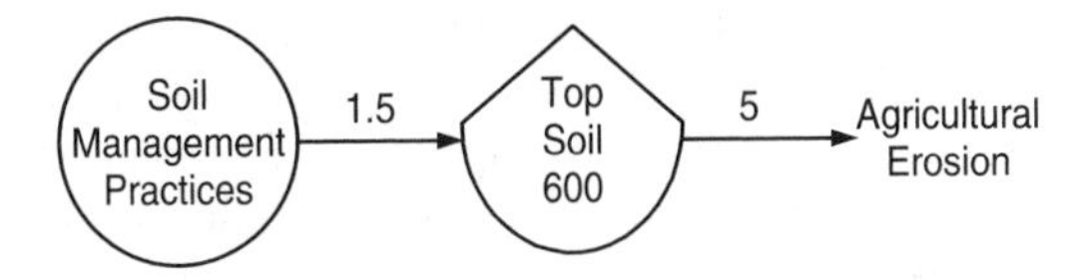

FIGURE 8.5 Energy circuit diagram of the system used by Jackson (1980) to discuss the concept of "ecological capital," shown by the storage of topsoil on a farm. Flows are in units of tons/acre/year and the storage is in units of tons/acre, assuming a 4-in. topsoil layer. These values are typical of U.S. agriculture.

Another important topic in ecological economics is the creation of new kinds of economies that are sustainable over long time periods. Ecological engineering can help society move towards sustainability by reducing costs and by utilizing natural, renewable energy sources. The concept of sustainable development covers many adaptations of society for long-term survival, of which ecological engineering is one of several recent advancements. In this larger context, ecological engineering can play an important role for society as a whole. Two definitions of sustainable development are given below:

1. To live on renewable income and to not deplete natural capital
2. To provide for the needs of the present generation without sacrificing the ability of future generations to meet their needs

A significant contribution from ecological economics has been to differentiate between aspects of growth and development in thinking about sustainable development of an economy (Costanza and Daly, 1992). As noted by Costanza et al. (1997a),

> Improvement in human welfare can come about by pushing more matter-energy through the economy, or by squeezing more human want satisfaction out of each unit of matter-energy that passes through. These two processes are so different in their effect on the environment that we must stop conflating them. Better to refer to throughput increase as *growth*, and efficiency increase as *development*. Growth is destructive of natural capital and beyond some point will cost us more than it is worth — that is, sacrificed natural capital will be worth more than the extra man-made capital whose production necessitated the sacrifice. At this point growth has become anti-economic, impoverishing rather that enriching. Development, or qualitative improvement, is not at the expense of natural capital. There are clear economic limits to growth, but not to development.

The great challenge of sustainability is a kind of social engineering. Ecological economists not only must help design new systems of resource use but also must find ways to change people's attitudes so that they can change from consumptive lifestyles to sustainable lifestyles. This is a major challenge and the long-term fate of global civilization may depend on its outcome.

One approach to sustainability is to establish the carrying capacity of a system for humans. Carrying capacity is the maximum number of individuals of a population that can be stably maintained in a given environment. It is an important ecological concept that has developed from both mathematical population biology and wildlife management. In mathematical population biology, carrying capacity is a constant (K) developed for the logistic growth equation (see Chapters 3 and 7) that represents the equilibrium population size. It is an asymptote that a population grows up toward when starting from low initial conditions. The mathematical concept was first used by Pierre Verhulst in the early 1800s, but it was "rediscovered" in the early 1900s by Raymond Pearl who incorporated it into modern population biology (Kingsland, 1985). In wildlife management, carrying capacity was defined as the "maximum density of wild game which a particular range is capable of carrying" (Leopold, 1933). It is usually related to the amount of food, water, and cover available to the animals. While this basic definition is quite simple and straightforward, the concept has been used in different ways (Edwards and Fowle, 1955). If a carrying capacity for humans could be established, then the limits to sustainability could be known. This is a critical and controversial subject (Cohen, 1995; Daly, 1995; Hardin, 1986; H. T. Odum, 1976; Sagoff, 1995). One of the latest developments along this line of thought is ecological footprint analysis which attempts to calculate the land and water areas required to support human communities (Wackernagel and Rees, 1996; Wackernagel et al., 1999).

EMERGY ANALYSIS

One example of a new form of economics, related to ecological economics, is termed *emergy analysis* (H. T. Odum, 1996). Emergy (short for "energy memory") is a measure of embodied energy in a product or process which in turn is a measure of its value. Emergy analysis is an accounting system in which everything is accounted for with energy units rather than money. In this way contributions from nature and environmental impacts can be assessed with the same units as traditional economic values. Emergy analysis is an analytical technique that calculates values that can be used for making decisions. This makes it one of only a few existing types of accounting systems. The concept and method were developed by H. T. Odum, based on his earlier work on ecological energetics. Recently, Mark Brown, who was a student of H. T. Odum, has become a leader in applying emergy analysis to a number of problems (Brown and Herendeen, 1996; Brown and Ulgiati, 1997, 1999; Brown et al., 1995).

The approach of emergy analysis is to convert everything to one unit which is then used in decision-making algorithms, such as cost–benefit analysis, and others, such as the investment ratio, that have been developed especially for this approach (H. T. Odum, 1996). Thus, two major steps are involved. First, all flows and storages relevant to a problem are quantified and converted to emergy, using published conversion factors called *transformities*. Then, the emergy values are used in algorithms to make assessments and to provide perspective for decision making. Emergy analysis was applied to the energy signature of a mesocosm in Chapter 4. Several other examples are discussed below to illustrate the approach.

Figure 8.6 shows an assessment of the estuarine fishery discussed earlier. The fish harvest is shown as input from the environment to the economic process of the fishery. It is valued as the emergy flow from the estuarine ecosystem that produces the fish. This flow (3×10^6 coal equivalent calories) is then divided by an energy-

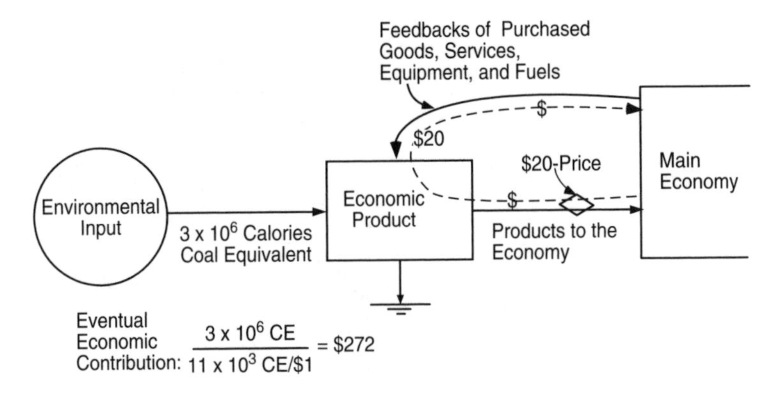

FIGURE 8.6 Energy circuit model of a fishery evaluated with emergy analysis. The economic contribution of the estuary is equivalent to $272/acre/year whereas the economy only recognizes a value of $20/acre/year, based on the work of the fisherman. This type of analysis documents the undervaluing of nature by conventional economics. (From Odum, H. T. and E. C. Odum. 1976. *Energy Basis for Man and Nature*. McGraw-Hill, New York. With permission.)

TABLE 8.7
Comparison of Costs for a Treatment Wetland vs. a Conventional Wastewater Treatment Plant Using Emergy Analysis

Energy Input (fossil fuel Kcal/gal)	Treatment Wetland	Conventional Treatment Plant
Fossil fuel energy input	3.28	25.3
Natural energy input	3.30	0

Note: The inputs are expressed in equivalent units so that direct comparisons can be made between the different energy types.

Source: Adapted from Mitsch, W. J. 1977. *Proceedings of the International Conference on Energy Use Management*. R. Fazzolari and C. B. Smith (eds.). Pergamon Press, Oxford, UK.

to-dollar ratio (11,000 coal equivalent calories to the dollar) to calculate the value of the fish in dollars. Thus, fishes are valued at $272 through emergy analysis, while the traditional economic analysis established the value at $20. This is an example of the common result that traditional economics undervalues nature.

Table 8.7 is an example of emergy analysis applied to the wetland option of wastewater treatment. The alternative of treating wastewater in a cypress wetland is compared with conventional technology in terms of two kinds of emergy. The conventional treatment alternative requires more fossil fuel energy and uses essentially no natural energy compared with the wetland alternative. Thus, the rational choice would be to choose the wetland alternative over the conventional technology alternative for wastewater treatment.

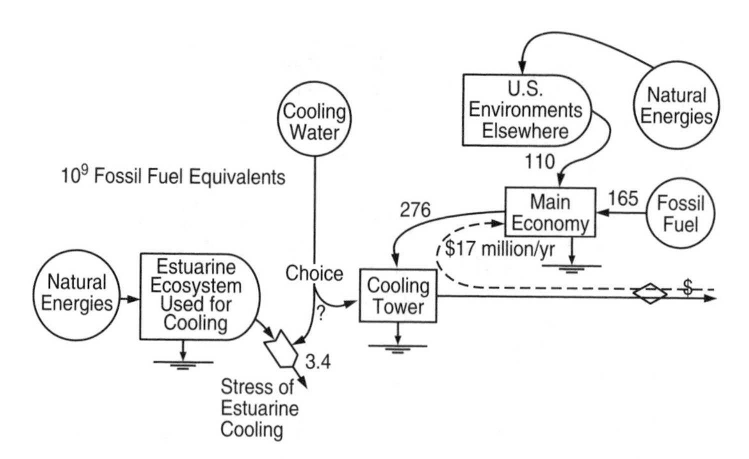

FIGURE 8.7 Energy circuit diagram showing alternative choices for treating thermal pollution from cooling water. Using the estuary for cooling drains much less energy from the entire system than using a cooling tower, and therefore it is the optimal choice for treatment from the perspective of emergy analysis. (From Odum, H. T., W. Kemp, M. Sell, W. Boynton, and M. Lehman. 1977. *Environmental Management.* 1:297. Springer-Verlag GmbH & Co. KG, Heidelberg, Germany. With permission.)

While the results of emergy analysis from Table 8.7 are consistent with the results from traditional economic analysis (Table 8.1), surprises do occur when emergy analysis is applied. Figure 8.7 illustrates a case for alternative methods of treating thermal pollution from a power plant (H. T. Odum et al., 1977b). Water can be cooled by either discharging it into an estuary or by passing it through constructed cooling towers. Both alternatives are evaluated on the diagram in terms of the emergy: for estuarine discharge the environmental impact is quantified, and for the cooling towers the total cost of construction and operation is shown. The results indicate a relatively small environmental impact (3.4×10^9 fossil fuel equivalents) due to thermal pollution vs. a larger load put on the economy by the cooling tower (276×10^9 fossil fuel equivalents), which causes more environmental impact elsewhere. This is consistent with the nature of thermal pollution, which is not highly disruptive to natural systems, especially in comparison with other kinds of pollution. The best choice then may be to discharge the heated water into the natural estuary even though intuition based on environmentalism might suggest that the cooling towers be built in order to avoid any impact to the environment.

The examples mentioned above are just a small sample of the set of problems and issues that can be addressed with emergy analysis (H. T. Odum, 1996). However, difficulties arise from several directions when attempting to implement emergy analysis for public policy. It is a radical new form of economics because it is based on a completely different currency than humans are familiar with. Emergy analysis represents a kind of physical theory of value rather than a social theory of value, which seems to make it objectionable to some people. Many problems arise because this currency is a physical quantity (i.e., a form of energy) rather than an information marker (i.e., money). Part of the problem also is that existing fields such as physics have other ideas of energy which easily get confused with emergy. Exergy is one example (Jorgenson, 1982, 2000), which is based on a form of mechanical energy. Unfortunately, people unfamiliar with these fields and concepts quickly get confused and turn away from the approach.

Major disagreements exist between emergy analysis and conventional economics (Lavine and Butler, 1981). In large part the disagreement concerns the issue of accounting for both human-centered and ecosystem-centered values. Conventional economics does not do this completely, even with various kinds of adjustments mentioned earlier; emergy analysis does, at least theoretically. This goal is achieved in emergy analysis by utilizing a quantity, emergy, which can be calculated for both human-centered and ecosystem-centered quantities, making it a "common denominator." Conventional economics is computationally sophisticated enough to account for both these quantities, but it philosophically denies that certain kinds of values exist (such as fishes in Figure 8.3). The philosophy and analytical techniques of conventional economics evolved when human populations were at low densities and the ecological life-support system seemed limitless. Under these conditions it was possible, and it even made sense, to exclude certain things from the value system. Thus, the ecological life-support system was taken for granted, and it was assumed that it didn't need to be accounted for. As populations grew, however, the environment became more important to humans and new approaches have evolved to recognize this importance, sometimes through accounting techniques with conventional economics and at other times through social instruments such as regulations and public policies. However, even though the human–environment relationship has changed over time, the philosophical basis of conventional economics has not changed, only the details of some of the accounting techniques. Thus, there may always be disagreement between those who believe in conventional economics and those who believe in emergy analysis because of the philosophical differences between the approaches.

Should society as a whole question the accounting systems used for public policy? Which of the different approaches (conventional economics, ecological economics, emergy analysis) is correct and should be used by society as a guide for making decisions? Is it just a matter of personal preference or belief? The answers to these questions are very important, but they are beyond the scope of this text. However, because ecological engineering involves systems of both man and nature, the more holistic approaches seem to be necessary and appropriate for evaluation, assessment, and design activities.

RELATED ISSUES

A number of topics related to ecological engineering economics are presented below. Many steps are required for a project to be implemented, and the project can be stopped for any of a number of reasons. While this text focuses more on the technical and specifically the ecological dimensions of ecological engineering, various economic, business, and policy concerns deserve attention.

FINANCING

Financing involves developing the capital necessary for paying for a project. This action obviously involves economic and business-oriented information but politics can also be a critical factor. Ecological engineering projects can be financed with either public or private funding sources. Some examples, such as domestic wastewater treatment, are financed as public projects with traditional methods (Green, 1932). Monies raised through taxes or similar means are available for these types of projects, and engineering firms submit bids to undertake them. Usually, the firm with the lowest bid wins the contract and conducts the project. Ecological restoration is being funded in this fashion, though these projects must compete for public funds with other projects. In this situation the government decision makers decide the allocation of the monies. Thus, the financing of restoration projects is often a political decision.

Other financing methods are used to fund ecological engineering projects from private sources. Interesting examples come from situations where a private company causes pollution or environmental impact, requiring an ecological engineering system for cleanup or restoration. In this case the company must pay for damages, which are assessed through legal means. Large-scale examples are cleanup of superfund sites and of the Exxon Valdez oil spill, which have been dramatic and contentious, but many other examples are small-scale.

An exceptional example of private funding of an ecological engineering project was Biosphere 2 in southern Arizona. In this case a wealthy individual became convinced of the merits of the project and provided funding. A company was formed, named Space Biosphere Ventures, Inc. (SBV), to develop technologies from the construction and operation of Biosphere 2, in particular for space travel and for the eventual colonization of Mars. This was an example of a venture capital business. SBV ultimately failed, as do many of these kinds of businesses that involve high financial risk.

A final example of private finance is the situation with strip-mine restoration or wetland mitigation. In this case mining companies or developers pay money into a fund in proportion to their activities of clearing land. These monies are later used to reclaim or restore the land that was disturbed in the case of mining, or to create new ecosystems elsewhere in the case of wetland mitigation. This is similar to the situation where a polluter must pay for environmental damages. Costanza and Perrings (1990) have developed this idea further as an assurance bonding model.

REGULATION

Ecological engineering projects are regulated by government agencies, in various cases from city or county, to state, to federal scales. Regulation is necessary (1) to document and maintain system performance, (2) to protect the environment, and (3) to ensure human health and safety. Wastewater treatment systems have perhaps the most developed regulatory system because environment and human welfare depend on their effective operation; they are used as an example below.

Regulation usually begins with evaluation of plans or designs for a proposed treatment system. These documents must be signed by registered professional engineers as a first step in validating the proposed project. Plans and designs are examined for adequacy by the government agency and if they are found to be satisfactory, a permit for construction and operation is issued. Permits usually stipulate performance ratings that must be achieved for continued operation of the system and a monitoring program to provide information to the regulatory agency that verifies that the performance ratings are being met.

Regulations establish standards or criteria of performance and hold the operators of the treatment systems accountable based on these ratings. The absolute values of the standards or criteria are very important because if they are not met, then the system must be upgraded or closed. Thus, regulatory standards and criteria can become contentious with large amounts of money at stake from the perspective of the treatment plant operator who suffers the costs of plant closure or upgrade, or with environmental and human health at stake from the perspective of the regulator who represents the public interest. Unfortunately, there is often insufficient information available to establish standards or criteria for ecological systems and many problems occur as a consequence. For wastewater treatment systems these standards are concentrations of chemicals or other materials in the discharge waters released from the system. Regulatory criteria for restoration or mitigation projects may be levels of vegetation coverage or the presence of particular plant species that indicate overall ecosystem character.

Because ecological engineering alternatives are relatively new and therefore relatively unknown, there is a resistance to them by potential clients and regulators. This is a general phenomenon with any new technology (Bauer, 1995). Clients and regulators are comfortable with conventional technologies that are known and reliable, and they naturally resist new alternatives because of their risk of failure. The resistance is natural and prudent up to a point. However, at some point in the development of a new technology, resistance becomes nonadaptive and can cause resources to be wasted. This occurs when the new technology has been tested and proven effective in a number of trials. The state of the art of at least some examples of ecological engineering seems to be at the threshold of overcoming resistance. More data and studies are needed on these systems to help convince clients and regulators on the attributes of ecologically engineered alternatives. Figure 8.8 presents a general model showing the role of resistance, which comes from several sources, in reducing the flow of technology for the solution of a problem. See Chapter 2 for a discussion of resistance to the development of treatment wetland technology.

PATENTS

The U.S. Patent System consists of a federal unit within the Department of Commerce and a set of laws that govern invention rights. Patents are granted for inventions, which provide a property right to the inventor. The patent system is important not just because it protects the rights of inventors but also because it promotes the progress of technology at the larger scale. At the scale of the inventor the patent gives the patentee the right to exclude others from practicing the invention and thus

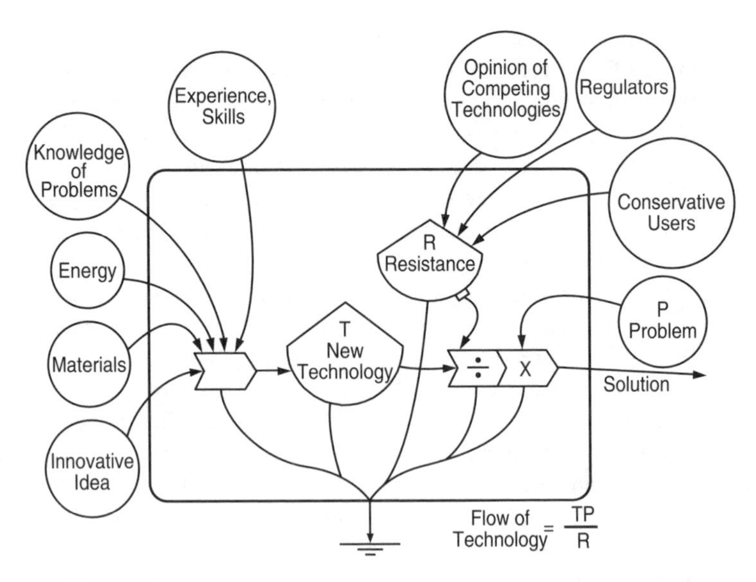

FIGURE 8.8 Energy circuit model of the flow of technology in problem solving. The flow of new technology to the solution, shown on the right-hand side of the diagram, is high if the technology is effective (high T) or if the problem is urgent (high P), but it is reduced in proportion to the resistance (R).

gives the patentee rights to profits from the sale of the invention for a limited period of time (20 years). After the allotted time period the exclusion rights expire and the invention enters the public domain. The patent system promotes technology at a larger scale by the disclosure to the public of the invention in the patent document which can stimulate the creation of other inventions. According to the legal statute, any person who "invents or discovers any new or useful process, machine, manufacture, or composition of matter, or any new and useful improvement thereof, may obtain a patent." An invention must exhibit three qualities to be patentable: novelty, utility, and nonobviousness. Descriptions of the patent system are given by Anonymous (1997), Gordon and Cookfair (1995), and Tuska (1947).

The patent application includes a complete written description and drawings that illustrate the invention. Once the patent is granted these materials are published through the Patent Office which constitutes disclosure. Thus, a patent is both a technical publication and a legal document. In a sense, patents represent one of the best available sources of information on current technological developments. Moreover, patents take the place of the traditional, academic publications in scientific and engineering journals for some inventors. For example, Thomas Edison published little on any of his inventions, but he was granted 1093 patents over his lifetime. Buckminster Fuller, perhaps best known for inventing the geodesic dome, similarly emphasized patents over academic publication for his technical work (Fuller, 1983; Robertson, 1974).

Some ecological engineering designs have been patented and examples are given in Table 8.8. These are interesting because they represent constructed ecosystems,

TABLE 8.8
Examples of Ecological Engineering Patents

Patent Title	Inventor	Patent Number
Method for treating wastewater using microorganisms and vascular aquatic plants	W. C. Wolverton	4,415,450
Water purification system and apparatus	W. H. Adey	4,966,096
Algal turf scrubber	W. H. Adey	4,333,263
Ecological fluidized bed method for the treatment of polluted water	J. H. Todd J. M. Shaw	5,486,291
Method for treating water	J. H. Todd B. Silverstein	5,389,257
Solar aquatic apparatus for treating waste	J. H. Todd B. Silverstein	5,087,353

which are part living in terms of biological populations and part nonliving in terms of containment structure, pumps, pipes, etc. "Products of nature" are not supposed to be patentable, but in the case of ecological engineering systems, constructed ecosystems are patentable. There are similarities here with patenting of genetic engineering or biotechnology designs (Adler, 1984), where patents are assigned to organisms whose genetic code has been altered by humans for useful purposes. However, it may be even more interesting or contentious in ecological engineering because of the role of self-organization in creating designs. Should the patent rights for a constructed ecosystem be given to a human inventor if nature is responsible for a significant portion of the design? Perhaps a legal statute is needed for at least a sharing of profits from an ecological engineering design with some monies going to the human inventor and some going back to nature in the form of feedbacks supporting biodiversity.

The general method of ecological engineering is not patented and it can be used by anyone to construct a useful ecosystem. Basically, the method is to construct a containment system which includes the problem to be solved (eroded shoreline or waste stream) and to over-seed it with biodiversity. Self-organization will create an appropriate ecosystem for the given boundary conditions that will solve the problem over time. Self-organization can be accelerated by the human designer by seeding with species preadapted to the specific problem. The role of self-organization in this process is in selecting useful species for the identified problem. This is similar to the role ascribed to indigenous peoples in selecting species useful to pharmaceutical companies in medical drug production (Cunningham, 1991; Greaves, 1994).

Ethics

It may seem strange to conclude a chapter on economics with the subject of ethics; however, like economics, ethics is a guide to decision making. Ethics is a system of beliefs that provide self-imposed limitations on the freedom to act. Formal codes of ethics or conduct exist for engineers who become members of professional societies or who become licensed by engineering boards. These codes provide guidance to engineers, especially in terms of understanding the consequences of their actions on the health and safety of humans. While this role for ethics in engineering is well established, some believe that the ethical boundaries need to be expanded to include the environment (Gunn and Vesilind, 1986). The need for sustainability requires development of ethics that may be very difficult to achieve in the present-day society, which is often oriented towards growth and short-term objectives.

The new field of ecological engineering must develop its own code of ethics, from its own unique perspective. This will probably include traditional concerns for human health and safety, environmental ethics, sustainability, and perhaps a new respect for biodiversity, which provides an important component of ecological engineering designs. In terms of a concern for biodiversity, engineers might look to ideas on Biophilia or the philosophical connections between humans and all other forms of life (Wilson, 1984). The list of "78 reasonable questions to ask about any technology" given by Mills (1997) might be a good starting point for ethical developments in any engineering discipline. An ecological engineering code of ethics based on ethics of computer hackers is suggested in Chapter 9.

9 Conclusions

These ecosystems, as we may call them, are of the most various kinds and sizes.

— **A. G. Tansley, 1935**

THE EMERGENCE OF NEW ECOSYSTEMS

A central theme of this book has been the development of the concept that new ecosystems can be designed, constructed, and operated for the benefit of humanity through ecological engineering. The concept of new ecosystems was introduced in Chapter 1 and was elaborated in subsequent chapters that focused on particular case studies. New ecosystems originate through human management, along with the self-organizational properties of living systems. The mix of engineered design with nature's self-design makes these ecosystems unique. The study of new ecosystems is often marked with surprises because they are not yet fully understood (Loucks, 1985; O'Neill and Waide, 1981). Like genetically engineered organisms, these ecosystems have never existed previously. Those who design, construct, and operate the new ecosystems are therefore exploring new possibilities of ecological structure and function. In this sense, ecological engineering is really a form of theoretical ecology. This book is an introduction to the new ecosystems that are emerging all around us through self-organization in different contexts.

Humans have been creating new ecosystems for thousands of years, but it is only in the last 30 years or so that these ecosystems have been recognized as objects for study by ecologists. Some of these ecosystems have been intentionally created while others have developed for various unintended reasons. Agriculture is probably the best example of a system that has been intentionally created. The origin of agriculture, on the order of 10,000 years ago, consisted of domesticating certain wild plants and animals and creating production systems from these species in modified natural ecosystems. Thus, plants were raised on cropland and grazing animals were raised on pastures or rangeland. Early agriculture differed little from natural ecosystems, but the modifications increased over time with greater uses of energy subsidies. Although the agricultural system is dominated by domesticated species, a variety of pest species has self-organized as part of the system. Management of agricultural land involves inputs of energy to channel production to humans and away from pests, and to reduce losses due to community respiration. In their modern forms, agricultural systems differ greatly from natural ecosystems, often with very low diversity (i.e., monocultures), large inputs of fossil fuel-based energies (i.e., mechanized tillage, fertilizers, etc.), and regular, orderly spatial patterns of component units (i.e., row crops arrangements).

The idea that agricultural systems actually were ecosystems evolved in the early 1970s. This occurred concurrently with the wide use of the ecosystem concept in the International Biological Program. Previously, ecologists almost exclusively studied natural ecosystems or their components. During this time agricultural systems themselves were studied by applied scientists with narrow focus in agronomy, entomology, or animal science. The ecosystem concept allowed ecologists to "discover" agriculture as systems of interest and for the applied scientists to expand their view to a more holistic perspective. Antecedent ecological studies of agricultural crops had been undertaken, with emphasis on primary production and energy flow (Bray, 1963; Bray et al., 1959; Gordon, 1969; Transeau, 1926), but this work had relatively little influence on the science of ecology. After the early 1970s, however, whole system studies of agriculture by ecologists became common (Cox and Atkins, 1975; Harper, 1974; Janzen, 1973; Loucks, 1977) and similar studies by the traditional agricultural scientists followed soon after. In fact, a journal named *Agroecosystems* was initiated in 1974 as a special outlet for ecological studies of agricultural systems. This line of research is very active with many useful contributions on nutrient cycling (Hendrix et al., 1986; Peterson and Paul, 1998; Stinner et al., 1984), conservation biology (Vandermeer and Perfecto, 1997), and the design of sustainable agroecosystems (Altieri et al., 1983; Ewel, 1986b).

Around this same time period the ecosystem concept was applied to other new systems. For example, Falk (1976, 1980) studied suburban lawn ecosystems near Washington, DC. Lawns are heavily managed ecosystems that provide aesthetic value to humans. Falk identified food chains, measured energy flows, and documented management techniques using approaches developed for natural grassland systems. This work was an in-depth study of a new ecosystem type that later was expanded on by Bormann et al. (1993). Much more significant has been research on urban ecosystems. This work began in the 1970s (Davis and Glick, 1978; Stearns and Montag, 1974) and steadily increased, especially in Europe (Bernkamm et al., 1982; Gilbert, 1989; Tangley, 1986). Urban areas include many fragments of natural habitats along with entirely new habitats (Kelcey, 1975) and have unique features as noted by Rebele (1994):

> … there are some special features of urban ecosystems like mosaic phenomena, specific disturbance regimes, the processes of species invasions and extinctions, which influence the structure and dynamics of plant and animal populations, the organization and characteristics of biotic communities and the landscape pattern as well in a different manner compared with natural ecosystems. On behalf of the ongoing urbanization process, urban ecosystems should attract increasing attention by ecologists, not only to solve practical problems, but also to use the opportunity for the study of fundamental questions in ecology.

Much research is currently being carried out on urban ecosystems (Adams, 1994; Collins et al., 2000; Pickett et al., 2001; Platt et al., 1994; Rebele, 1994), including significant projects funded by the National Science Foundation at two long-term ecological research sites in Baltimore, MD, and Phoenix, AZ (Parlange, 1998). In addition, a journal named *Urban Ecosystems* was begun in 1996 for publishing the growing research on this special type of new system.

In a sense, then, there has been a paradigm shift in ecology since the 1970s with ecologists embracing the idea that humans have created new ecosystems. Most ecologists probably still prefer to study only natural systems, but research is established and growing on agroecosystems and urban ecosystems. This work is not necessarily considered to be applied research, though it is certainly an easy and logical connection to make. Rather, there are a number of ecologists who are studying agriculture and urban areas as straightforward examples of ecosystems. These are new systems with basic features (energy flow, nutrient cycling, patterns of species distributions, etc.) common to all ecosystems but with unique quantitative and qualitative characteristics that require study to elucidate. Ludwig (1989) called these *anthropic ecosystems* because of their strong human influence and proposed an ambitious program for their study.

There are many examples of new ecosystems beyond those mentioned above and throughout this book. Hedgerows, fragmented forests, brownfields, rights-of-way, and even cemeteries (Thomas and Dixon, 1973) are examples of new terrestrial systems, and there are many aquatic examples as well. H. T. Odum originally began referring to polluted marine systems as new ecosystems and developed a classification system that can be generalized to cover all ecosystem types. His ideas developed from research along the Texas coast in the late 1950s and early 1960s. This work involved ecosystem metabolism studies of natural coastal systems and those altered by human influences. The latter included brine lagoons from oil well pumping, ship channels, harbors receiving seafood industry waste discharges, and bays with multiple sources of pollution. H. T. Odum first referred to these systems as "abnormal marine ecosystems" (H. T. Odum et al., 1963), then as "new systems associated with waste flows" (H. T. Odum, 1967), and finally as "emergent new systems coupled to man's influence" (H. T. Odum and Copeland, 1972). The concept of emergent new systems is best articulated in the classification system developed for U.S. coastal systems (Copeland, 1970; H. T. Odum and Copeland, 1969, 1972). This system classified ecosystems by their energy signatures with names associated with the most prominent feature or, in other words, the one that had the greatest impact on the energy budget of the ecosystem. A whole category in this classification was given to new ecosystems (Table 9.1) with examples of all major types of human-dominated estuarine systems. This is a philosophically important conceptualization. Although H. T. Odum acknowledged that these ecosystems were "unnaturally" stressed by humans, he chose to refer to them as new systems rather than stressed systems. This distinction may at first seem subtle, but it is not. It carries with it a special notion of ecosystem organization.

The concept of new ecosystems implies that the human influence is literally a part of the system and therefore an additional feature to which organisms must adapt (Figure 9.1A). Thus, human pollution is viewed the same as natural stressors such as salt concentration or frost, and ecosystems exposed to pollution reorganize to accommodate it. The tendency to consider humans and their stressors as being outside of the ecosystem is common in modern thought. This conception generally holds that human influence, such as pollution, leads to a degraded ecosystem (Figure

TABLE 9.1
Classification of New Estuarine Ecosystems

Name of Type	Characteristic Energy Source or Stress
Sewage waste	Organic and inorganic enrichment
Seafood wastes	Organic and inorganic enrichment
Pesticides	An organic poison
Dredging spoil	Heavy sedimentation by man
Impoundment	Blocking of current
Thermal pollution	High and variable temperature discharges
Pulp mill waste	Wastes of wood processing
Sugarcane waste	Organics, fibers, soils of sugar industry wastes
Phosphate wastes	Wastes of phosphate mining
Acid waters	Release or generation of low pH
Oil shores	Petroleum spills
Piling	Treated wood substrates
Salina	Brine complex of salt manufacture
Brine pollution	Stress of high salt wastes and odd element ratios
Petrochemicals	Refinery and petrochemical manufacturing wastes
Radioactive stress	Radioactivity
Multiple stress	Alternating stress of many kinds of wastes in drifting patches
Artificial reef	Strong currents

Source: Adapted from Odum, H. T. and B. J. Copeland. 1972. *Environmental Framework of Coastal Plain Estuaries*. The Geological Society of America, Boulder, CO.

9.1B). However, is it appropriate only to think of an ecosystem as degraded when a source of pollution is added to the energy signature? What actually happens is that the ecosystem reorganizes itself in response to the new pollution source. Thus, degradation (Figure 9.1B) is really reorganization of a new ecosystem (Figure 9.1A). This seems like a contradiction because degradation carries a negative connotation while reorganization has a more positive sense. Both views in Figure 9 are valid. What is advocated here is the straightforward notion that ecosystem identity (i.e., elements of structure and function) is determined by the energy signature, and if the energy signature is changed, then a new ecosystem is created.

In another sense the concept of emergent new systems attempts to reduce value judgment in ecosystem classification. Rather than considering ecosystems with human pollution as degraded natural systems, the classification labels them as new systems. The value-free approach frees thinking so that the organization of new

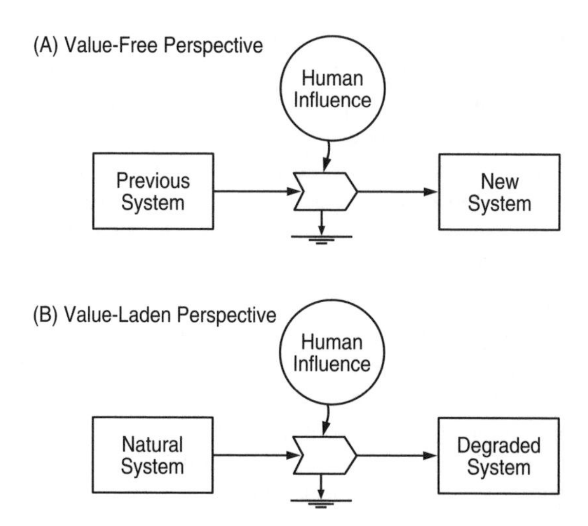

FIGURE 9.1 Comparison of philosophical positions or interpretations of the effects of human influence on ecosystems. (A) View focusing on change to a new system. (B) View focusing on degradation.

systems can be more clearly understood. Of course, the trick is to not throw out the value-laden thinking. It is important to understand and account for human influences which society judges to be negative. Some new systems are "good" (cropland agriculture dominated by domesticated exotic species) and some are "bad" (forest invaded by exotic species), but this distinction is determined by human social convention, not by ecological structure or function.

Consider another application of this way of thinking. A distinction is made between native species and exotic species in ecosystems as discussed in Chapter 7. Native species are those that are found in a particular location naturally or, in other words, without recent human disturbance, while exotic species are those that evolved in a distant biogeographical region but have invaded the particular location under discussion. The reference point in the distinction between natives and exotics is location. However, in the energy theory of ecosystems the reference point is the energy signature that exists at the location, not the location itself. A causal relationship is implied which matches a set of energy sources to ecosystem components. Thus, if the energy signature of a location changes, then the species native to the location may no longer be as well adapted to it as compared with exotic species that invade. Under these circumstances nature favors the exotic species which are preadapted to the new energy signature, while human policy favors the old native species due to an inappropriate respect for location. Exotics are said to be the problem, when really the problem is that the energy signature has changed. Clear examples of this circumstance are the tree species that invade where hydrology has changed dramatically as in the southwest U.S. with salt cedar (*Tamarix* sp.) and in South Florida with melaleuca (*Melaleuca quinquenervia*). Tree-of-heaven (*Ailanthus altissima*) is another example of an exotic tree species which occupies urban areas and roadside edges (Parrish, 2000). These habitats have different energy signatures as compared with the surrounding forests in the eastern U.S. and tree-of-heaven can dominate under these new conditions. Humans are everywhere changing old energy signatures and creating new ones that never existed previously, and the results are changing ecosystems. The issue is how to choose reference points to interpret changes. This requires a philosophical position and the position advocated here is that new ecosystems are being created which have few or no reference points for comparison in the past. Thus, the future will require new ways of thinking about the new ecosystems that are being created as humans change the biosphere. The concept of new ecosystems may be especially useful for the ecological engineer who designs ecosystems. What criteria will be used to judge the new systems? Will new designs be limited to native species that are no longer fully adapted or can exotic species be used? Can humans allow nature to perform some of the design, even if it results in unanticipated or undesirable species compositions? What are the limits to ecological structure and function that can be achieved through design?

THE ECOLOGICAL THEATER AND THE SELF-ORGANIZATIONAL PLAY

Study of the new systems that are emerging unintentionally is especially instructive. These systems demonstrate the process of self-organization, and their study can be a guide to the intentional engineering of new systems. The two main classes of unintentional new systems are (1) those ecosystems exposed to human stresses, in one form or another, for which they have no adaptational history and (2) those ecosystems with mixes of species that didn't evolve together (i.e., native and exotic species). These kinds of unintentional new systems are coming to dominate landscapes, and therefore, they deserve study even independent of ecological engineering. A very interesting common feature of these systems is that the traditional Darwinian evolution concept no longer provides the fullest context for understanding them. This common feature comes from the fact that the new systems lack direct or explicit adaptations for some features of their current situation because humans have changed conditions faster than evolution can occur. New systems differ from what are normally considered to be natural systems in which a more or less stable set of associated species has evolved together, in the Darwinian sense, over a long period of time with a given external environment. G. E. Hutchinson described the natural situation as the "ecological theater and the evolutionary play" (Hutchinson, 1965), in which ecology and evolution act together to produce organization in ecosystems. This is a wonderful metaphor that captures the way that nature consists of multiple, simultaneous time scales. Populations interact over the short-term in the "ecological theater" while simultaneously being subjected to natural selection over the long-term in the "evolutionary play." However, in the view presented here for the new unintentional systems, the conventional concept of evolution is becoming less important, and perhaps a new evolutionary biology will be required.

This is a strong statement that requires elaboration. First, consider those ecosystems stressed by human influences that never existed in the natural world. There

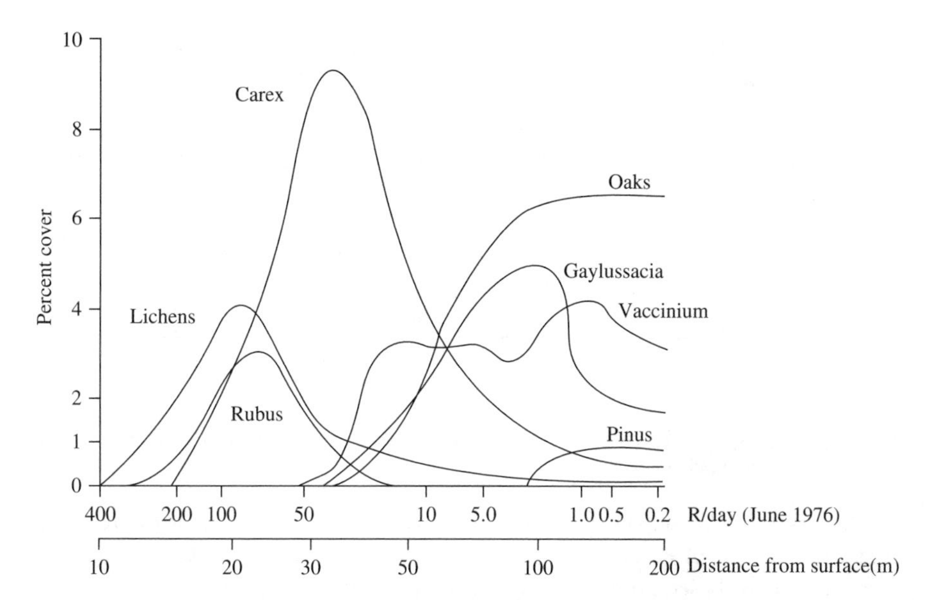

FIGURE 9.2 Patterns of vegetation extending out from a radiation source in the temperate forest at Brookhaven, New York. (From Woodwell, G. M. and R. A. Houghton. 1990. *The Earth in Transition: Patterns and Processes of Biotic Impoverishment.* G. M. Woodwell (ed.). Cambridge University Press, Cambridge, U.K. With permission.)

are, of course, many kinds of pollution that have been created by humans; many new kinds of habitats have also been created, especially in agricultural and urban landscapes. A whole new field of stress ecology has arisen to understand these systems with many interesting generalizations (Barrett and Rosenberg, 1981; Barrett et al., 1976; Lugo, 1978; E. P. Odum, 1985; Rapport and Whitford, 1999; Rapport et al., 1985). These references indicate that many changes in natural ecosystems caused by human impacts are similar and predictable, such as simplification (reductions in diversity) and shifts in metabolism (increased production or respiration). A good example is the set of experiments done in the 1960s which exposed ecosystems to chronic irradiation from a 137 Cs source, such as at Brookhaven National Laboratory in New York. These experiments were conducted to help understand the possible consequences of various uses of atomic energy by society. In these studies point sources of radiation were placed in forests for various lengths of time and ecosystem responses were studied. At Brookhaven, "the effect was a systematic dissection of the forest, strata being removed layer by layer" (Woodwell, 1970). Thus, a pattern of concentric zones of impact emerged outward from the radiation source, perhaps best characterized by these vegetation zones (Figure 9.2):

1. Central zone with no higher plants (though with some mosses and lichens)
2. Sedge zone of *Carex pennsylvanica*
3. Shrub zone with species of *Vaccinium* and *Gaylussacia*
4. Zone of tolerant trees (*Quercus* species)
5. Undisturbed forest

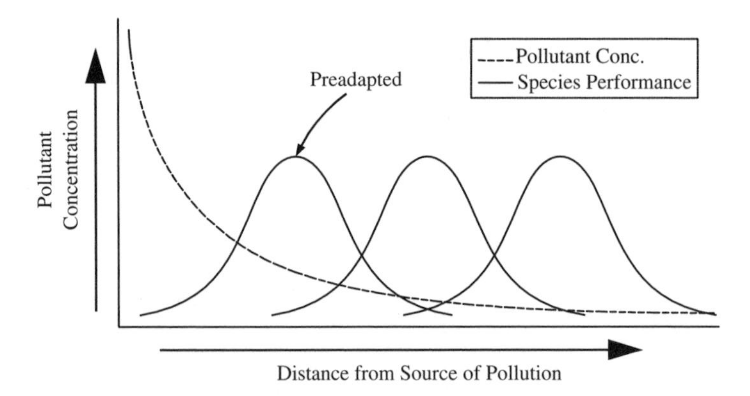

FIGURE 9.3 Model of longitudinal succession caused by a pollutant source, illustrating the position of preadapted species.

In this case the ecosystem had no adaptational history to the stress but self-organization took place in the different zones of exposure, resulting in viable but simpler systems based on genetic input from the surrounding undisturbed forest. It is interesting to note that Woodwell (1970) found similarities between the new stress of radiation and the "natural stress" of fire. Some species in this forest were adapted to fire, and there was a direct correspondence in species adaptation between fire frequency and radiation exposure. Thus, with high fire frequency *Carex pennsylvanica* dominates vegetation just as it does with relatively high radiation exposure. This is an example of preadaptation, which has been noted as being important in stress ecology by Rapport et al. (1985). A general model for the special case described above is shown in Figure 9.3. Concentration of the pollutant declines away from a point source along a linear transect in the model. Associated with the decline in pollutant concentration is a longitudinal succession of species, shown by the series of bell-shaped species performance curves. Each curve represents the ability of a species to exploit resources within the context of the pollution gradient (see Figure 1.8). This pattern of species is characteristic of a variety of ecological gradients and Robert Whittaker developed an analytical procedure for studying the pattern called gradient analysis (Whittaker, 1967). When there is no adaptational history for the pollutant, then the species closest to the point source can be said to be preadapted to the pollutant. In the classic river pollution model (Figure 2.3) the species closest to the sewage outfall are classified as tolerant. Using an alternative line of reasoning, these species are preadapted to the high sewage concentrations, and the proximity of the peak in their performance curves to the point source is an index of the degree of preadaptation. The decline in pollutant concentration in the model is due to various biogeochemical processes. When species have a role to play in the decline, then ecological engineering is possible to enhance treatment capacity of the pollution. To some extent the sequential design of John Todd's living machines (see Chapter 2) corresponds with the species patterns shown in Figure 9.3. Perhaps an adaptation of Whittaker's gradient analysis can be used as a tool for living machine design (see the upcoming section on a universal pollution treatment ecosystem).

The other class of unintentional system is the system dominated by exotic species. The situation here is that species with no common evolutionary history are being mixed together by enhanced human dispersal at rates faster than evolution. The results, as described in Chapter 7, are new viable communities with some exotic and some native species.

In both cases of unintentional systems then, evolution does not provide full understanding or predictive value of the new systems. There are a few examples of evolution taking place in the new systems, such as resistance to pesticides in insect pests or to antibiotics by bacteria and tolerance to heavy metals by certain plants (Antonovics et al., 1971; Bradshaw et al., 1965), but these are exceptions. Certain species with fast turnover can adapt to rapid changes caused by humans (Hoffmann and Parsons, 1997), but this is not possible for all species. Soule's (1980) discussion of "the end of vertebrate evolution in the tropics" is a dramatic commentary on the inability of some species with low reproductive rates to adapt, in this case, to loss of habitat due to tropical deforestation. The idea that Soule refers to is loss of genetic variability in vertebrate populations due to declining population sizes. Natural selection operates on genetic variability to produce evolution, so with less genetic variability there is less evolution.

Thus, the new systems are being organized at least in part by new processes. Janzen (1985) discussed this situation and proposed the term *ecological fitting* for these processes. Self-organization is proposed as the general process organizing new systems in this book. To address this new situation, Hutchinson's classic phrase may need to be reworded as "the ecological theater and the self-organizational play."

A key feature of the organization of new systems is preadaptation. The new systems are often dominated by preadapted species, whether they be native species that are tolerant of the new conditions or exotic species that evolved in a distant biogeographical region under conditions similar to the new system. There appear to be two avenues of preadaptation: those species that are preadapted through physiology and those that are preadapted through intelligence or the capacity to learn.

The best example of physiological preadaptation is for species that have been used as indicator organisms. These species indicate or identify particular environmental conditions by their presence or absence, or by their relative abundance. Indicator organisms can be either tolerant, (i.e., those present and/or abundant under stressful conditions) or intolerant, (i.e., those absent or with reduced abundance under stressful conditions). Only tolerant organisms are preadapted and they indicate the existence of new systems. Tolerant indicator organisms have been widely used in water quality assessments, dating back to the German *Saprobien* system in the early 1900s. A large literature exists in this field (Bartsch, 1948; Cairns, 1974; Ford, 1989; Gaufin, 1973; Patrick, 1949; Rosenberg and Resh, 1993; Wilhm and Dorris, 1968), and it can be an important starting point to developing an understanding of preadaptation as a phenomenon. Hart and Fuller (1974) provide a tremendous amount of information about the adaptations and preadaptations of freshwater invertebrates in relation to pollution. Another example of indicator organisms is plant species found on soils with unusual mineral conditions. Methods of biogeochemical prospecting have been developed by identifying particular indicator species of plants

(Brooks, 1972; Cannon, 1960; Kovalevsky, 1987; Malyuga, 1964); this approach could be important in selecting species for phytoremediation of waste zones in the future (Brown, 1995). The study of tolerant organisms for the purpose of understanding preadaption is similar to the approach of genetic engineers who study "super bugs" or microbes adapted to extreme environmental conditions (Horikoshi and Grant, 1991). These microbes have special physiological adaptations that the genetic engineers hope to exploit when designing microbes for new applications. Species can be found with adaptations for high (thermophilic) and low (psychrophilic) temperature, high salt concentrations (halophilic), low (acidophilic) and high (alkaliphilic) pH, and other extreme environments.

The other avenue of preadaptation involves intelligence or the capacity to learn. This is primarily found in vertebrate species with sophisticated nervous systems. Intelligence or the capacity to learn allows organisms to react to new systems. A. S. Leopold (1966) provided a discussion of this kind of preadaptation in the context of habitat change. Animals that can learn are able to adjust to new systems by avoiding stressful or dangerous conditions and by taking advantage of additional resources or habitats. Many examples exist including urban rats and suburban deer, along with a variety of bird species, which take advantage of new habitats: falcons in cities (Frank, 1994), gulls at landfills (Belant et al., 1995), terns on roof tops (Shea, 1997), and crows in a variety of situations (Savage, 1995).

Although some empirical generalities exist such as those from the field of stress ecology or from the long history of use of indicator organisms, little or no theory exists to provide an understanding of the organization of new emerging ecosystems. As mentioned earlier (see Chapter 1), preadaptation is little discussed in the conventional evolutionary biology literature, yet it is a major source of species that become established and dominate in the new systems through self-organization. More research on preadaptation is clearly needed. Can there be a predictive theory of preadaptation? Or is it simply based on chance matching of existing adaptations with new environmental conditions? Is a new evolutionary biology possible based on preadaptation?

One interesting topic from ecology that offers possibilities for an explanation of new systems is the theory of alternative stable states (see Chapter 7). This theory suggests that alternative equilibria or states, in terms of species composition, exist for ecosystems and that a system may move between these alternatives through bifurcations caused by environmental changes. Several authors have suggested possible views of alternative stable states in terms of human impact (Bendoricchio, 2000; Cairns, 1986b; Margalef, 1969; Rapport and Regier, 1995; Regier et al., 1995) and Gunderson et al. (2002) propose a theory called "panarchy" to explain how systems can shift between alternative states. This theory describes system dynamics across scales of hierarchy (hence the name panarchy) with a four-phase cycle of adaptive renewal. One view of the alternative stable state concept is shown in Figure 7.5 with a Venn diagram in which different sets represent alternative states. A system moves within a set due to normal environmental variations, but can jump to another set, representing a new system in the terminology of this chapter, due to some major environmental change (Parsons, 1990). The states differ qualitatively in their basic species compositions, but within a state a similar species composition exists, though in quantitatively different combinations. The alternative stable-state concept involves folded equilibria from dynamical systems theory, which may provide a foundation for understanding the new emerging systems of human impact and exotic species. Can we predict new alternative states that have never been recorded previously? Can we create alternative states through ecological engineering?

Ecological engineers will be interested in the new emerging systems for several reasons. First, these systems will be sources of organisms to seed into their new designs. Species from the new emerging systems will be variously preadapted to human-dominated conditions so that they may also be successful in interface ecosystems. For example, biodiversity prospecting is taking place at Chernobyl (where the nuclear reactor disaster took place in 1989) for microbes that might have special value due to mutations. Ecological engineers also can learn from the new systems as in reverse engineering. What kinds of patterns of ecological structure and function exist in communities of preadapted species? Useful design principles may arise from the study of the new emerging ecosystems, and the engineering method may be a helpful vantage point for study, as discussed in the next section.

EPISTEMOLOGY AND ECOLOGICAL ENGINEERING

The inherent qualities of ecological engineering — the combination of science and engineering and the goal of designing and studying ecosystems that have never existed before — lead to a consideration of methods and ways of knowing, which is the subject of a branch of philosophy termed *epistemology*. Here the orientation used is that given by Gregory Bateson (1979) who defines epistemology as "the study of the necessary limits and other characteristics of the processes of knowing, thinking, and deciding." While science, as the application of the scientific method, is philosophically well understood as a way of knowing, methods of engineering are not well articulated as noted in Chapter 1. For this reason the methods of ecological engineering are considered in the context of ecology, which is a scientific discipline, rather than in the context of engineering. Moreover, from this perspective, ecological engineering can be seen to offer a new way of knowing about ecology, which can be a significant contribution to the science.

Ecologists have not formally examined epistemology very deeply and only a few references have even mentioned the branch of philosophy (Kitchell et al., 1988; Scheiner et al., 1993; Zaret, 1984). Most ecologists seem to consider only the scientific method of hypothesis testing as the way of knowing about nature (Loehle, 1987, 1988). Although standard hypothesis testing is an excellent method, it is not the only approach available for studying ecosystems. For example, Norgaard (1987) discusses how certain indigenous peoples use different thinking processes compared with the traditional Western worldview in dealing with agroecosystems. Also, the complexity found in ecosystems creates challenges to the conventional philosophy of science as discussed by Morowitz (1996) and Weaver (1947). It is proposed here that the new discipline of ecological engineering should utilize a distinct, alternative

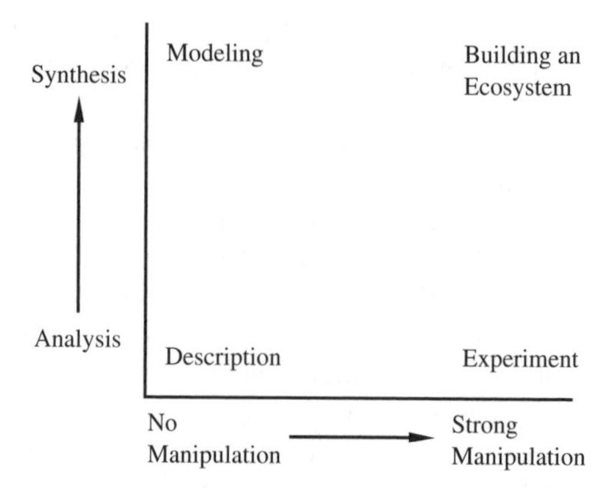

FIGURE 9.4 Spectrum of methods for ecology. Note the important new approach of building an ecosystem which is the main activity in ecological engineering.

method of epistemology that arises from the fundamental basis of engineering as a way of knowing.

Figure 9.4 provides a view of the methods used to develop knowledge in ecology along two axes. The horizontal axis represents the degree to which a method involves manipulation of the environment. The vertical axis represents the degree to which a method relies on dissecting a system into parts and mechanisms (i.e., analysis) vs. synthesizing parts into a whole system (i.e., synthesis). The space enclosed by these axes allows for different methods to be contrasted by their relative positions. By moving outward from the ordinate along either axis, a historical track of scientific development in ecology is outlined. Thus, ecology began with simple descriptions of populations and processes and advanced by focusing on experiments (movement along the horizontal axis) or by focusing on modelling (movement along the vertical axis). Each of the four methods shown in Figure 9.4 is a fundamental approach to developing knowledge, and each has a special contribution to make.

Description is the most basic approach in any discipline. It involves observations of systems, which usually lead to classifications of component parts and their behaviors. This approach is highly empirical and is the foundation of any of the other approaches shown in Figure 9.4. It also is the least respected method because the kinds of knowledge that can be generated from pure description are limited. As a science, ecology was in a descriptive phase from its origins around the turn of the century until after World War II when more advanced methods came to dominate the field.

Modelling refers to the mathematical description and prediction of interacting component parts of a system. At minimum, some knowledge of the component parts and how they interact is needed to create a model, and this knowledge comes from description, though other methods can also contribute. Modelling is primarily an act of synthesis as opposed to analysis because the emphasis is on connecting components in such a way as to capture their collective behavior. Although there is continuing interest in the parts, the focus of the modelling method is on the interaction of the parts and the building up of networks of interaction. The construction of the model requires a very systematic and precise description with mathematical relationships. This effort often identifies missing data, which leads to more description or to additional experiments. Once the model is built, it can be analyzed by various techniques. In this sense the model itself becomes an object of description, and the work can be thought to move back down the axis from synthesis to analysis. The models also can be simulated to study their dynamic behavior. This work can lead to a better understanding of the system being modelled and/or to predictions of how the system will behave under some new conditions. A somewhat extreme position on the heuristic value of models was given by H. T. Odum who taught that "you don't really understand a system until you can model it." Model-building itself involves no manipulation of the environment but, once constructed, a model is often "validated" in relation to the systems being modelled through a comparison of predictions with data gathered from the environment.

Experimentation, as shown in Figure 9.4, refers to the traditional scientific method of hypothesis testing. In this sense an experiment is a test of hypotheses. This is of importance in the philosophy of science since, as noted by Frankel and Soule (1981), "human science evolves by the natural selection of hypotheses." Hypotheses are statements about how component parts or whole systems behave, and an experiment is an event in which the validity of a hypothesis is checked. Experiments are carefully designed so that only one variable changes with a treatment, as described by the hypothesis in question. In this way a causal link is established between the treatment and the change in the variable. The method is thus analytical because only one variable at a time is studied while all others are held constant. The critical goal of this method is to disprove a hypothesis rather than to prove it. This is necessary because it is never possible to prove something is always true, but it is possible to demonstrate that something is definitely false. Experiments involve manipulating the environment through various treatments so that the consequences of hypotheses can be examined. Experimentation is the dominant method used in the present state of ecology (Resetarits and Bernardo, 1998; Roush, 1995).

The final method shown in Figure 9.4 is most important to the present discussion because it relates to ecological engineering. Building ecosystems is the defining activity of ecological engineering, whether it be a treatment wetland for absorbing stormwater runoff, a microcosm for testing toxicity of a pollutant, or a forest planted to restore strip-mined lands. Each constructed ecosystem is a special kind of experiment from which the ecological engineer "learns by building." This action is at once a form of strong manipulation of the environment and a form of synthesis so that the method occupies the extreme upper right-hand portion of Figure 9.4. Moreover, the method of building an ecosystem occupies a critical position in the plot because the science of ecology has no approach for developing knowledge in this region of space in the diagram. Building ecosystems is inherently an engineering method but it represents a whole new epistemology for ecology. In a sense it represents one of the "existential pleasures of engineering" described by Florman (1976). Through the process of designing, building, and operating objects, engineers have always utilized this approach to learning as noted in Chapter 1. It is essentially a kind of trial-and-error method in which each trial (a design) is tested for performance. The test provides a feedback of information to the designer, which represents learning. Engineers search for successful designs or, in other words, things that work. Errors provide a large feedback but, in a sense, they are not really looked upon as problems as much as opportunities to learn, as described by MacCready (1997) in the following quote:

> In a new area, where you can't do everything by prediction, it's just so important to get out there and make mistakes: have things break, not work, and learn about it early. Then you're able to improve them. If your first test in some new area is a success, it is rarely the quickest way to get a lasting success, because something will be wrong. It's much better to get quickly to that point where you're doing testing.
>
> You must tailor the technique to the job. Breaking and having something seem like it's going wrong in a development program is not bad. It's just one of the best ways to get information and speed the program along. If you've had nothing but success in a development program, it means that you shot too low, and were too cautious, and that you could've done it in half the time. Pursuing excellence is not often a worthy goal. You should pursue good enough, which in many cases, requires excellence, but in other cases is quick and dirty. The pursuit of excellence has infected our society. Excellence is not a goal; good enough is a goal. Nature just worries about what is good enough. What succeeds enough to pass the genes down and have progeny.

Several other authors have discussed the philosophical view of errors as being an inherent part of the learning process (Baldwin, 1986; Dennett, 1995; Petroski, 1982, 1997b). This kind of trial-and-error is not a blind, random process, but rather it is always informed by past experience. In this way it is self-correcting. Thomas Edison used a variation of this approach, which he called the "hunt-and-try method," as the basis for his inventions. Edison's approach blended theory and systematic investigation of a range of likely solutions. As noted by Millard (1990), "in Edison's lab it was inventing by doing, altering the experimental model over and over again to try out new ideas."

The emphasis of the engineering method is on testing a design to demonstrate that it works. In this way, it differs fundamentally from the scientific method of hypothesis testing described earlier. In hypothesis testing the goal is to disprove a hypothesis, while in the engineering method the goal is to prove that a design works. Philosophically this difference arises because in science there is only one correct answer to a question, and its method works by systematically removing incorrect answers from consideration. However, in engineering many designs are possible solutions to a problem, and its method works by systematically improving designs with continual testing (see Figure 1.4).

Several ecologists have begun to declare the value of building an ecosystem as an epistemological method. In terms of restoration Bradshaw (1987b) called it "an acid test for ecology" and Ewel (1987) added the following quote:

> Ecologists have learned much about ecosystem structure and function by dissecting communities and examining their parts and processes. The true test of our understanding of how ecosystems work, however, is our ability to recreate them.

Ecological engineering, then, may increasingly become important as a method for understanding nature, as well as an active, applied field that adds to the conservation value of society as a whole. All of the methods listed in Figure 9.4 should be utilized. A special emphasis on description of the new systems that are emerging both intentionally and unintentionally may be necessary because they may have patterns and behaviors that have not been seen previously. Finally, Aldo Leopold's (1953) famous quote (which, interestingly, implies a machine analogy of nature — see Chapter 7) is particularly relevant to a consideration of the ecological engineering method:

> If the biota, in the course of aeons, has built something we like but do not understand, then who but a fool would discard seemingly useless parts? To keep every cog and wheel is the first precaution of intelligent tinkering.

Ecological engineers are doing "intelligent tinkering" when they design, build, and operate new constructed ecosystems.

FUTURE DIRECTIONS FOR DESIGN

Ecological engineering is a growing field with many possible future directions. Most existing technologies, such as described in Chapters 2 through 6, are of relatively recent origin, and they can be expected to be improved upon. Whole new paths of developments also can be expected, especially as more young people are educated in the field. However, although the future appears to be promising, there is much to be done to bring ecological engineering into the mainstream of societal, academic, and professional arenas. The field does not yet even appear in the vocabulary of the U.S. National Environmental Technology Strategy (National Science and Technology Council, 1995), though several related applications such as bioremediation and restoration ecology are becoming widely recognized. Mitsch (1998b) has summarized the recent accomplishments of the field and has posed a number of questions about the future (Table 9.2). He concludes with several recommendations and a call for ecologists and engineers to work together for continued development of the field.

One critical fact about the future is that environmental problems will continue to grow and to multiply. These problems include global climate change and sea level rise, along with declining levels of freshwater availability, agricultural land and fossil fuels, and increasing levels of pollution. These pressures may lead society to focus on ecological engineering designs that "do more with less," that utilize natural energies and biodiversity, and that convert by-product wastes into resources. Several examples of possible directions are outlined below. These are selected to illustrate various dimensions such as size extremes from molecular to planetary and applications of biodiversity, technology, and social action. Some directions rely on futures

TABLE 9.2
Questions for the Future of Ecological Engineering

What is the rationale for ecological engineering and what are its goals?
What are the major concepts of ecological engineering?
What are the boundaries of ecological engineering?
What are the measures of success of ecological engineering projects?
What are the linkages of ecological engineering to the science of ecology?
How do we balance theory vs. empiricism?
At what scale do we approach ecological engineering?
What tools are available for analyzing ecological engineering?
What are the ramifications of ecological engineering in developing countries with differing values and cultures?
How do we institutionalize ecological engineering education?
How will we integrate the ecological and the engineering paradigms?
Under what conditions will ecological engineering flourish or disappear?

Source: Adapted from Mitsch, W. J. 1998. *Ecological Engineering.* 10:119–130.

with expanding energy resources (technoptimism) while others require less energy (technopessimism).

Ecological Nanotechnology

The smallest size ecological engineering application may be in nanotechnology, which has been called the last frontier of miniaturization. Nanotechnology is molecular engineering or "the art and science of building complex, practical devices with atomic precision" (Crandall, 1999). It involves working at the scale of billionths of a meter with microscopic probes. This field was first articulated by physicist Richard Feynman in 1959 and has been championed by futurist Eric Drexler (1986, 1990). While nanotechnology is very early in its development (Stix, 1996), small-scale engineering applications are arising (for examples, see Caruso et al., 1998; Singhvi et al., 1994). There are probably many possible uses of nanotechnology in ecological engineering, such as the construction of molecular machines that cleanse polluted sediments or regulate biofilms, but this kind of design must wait for future developments in the field. Several speculative environmental applications are listed by Chesley (1999) and Lampton (1993). To be truly ecological, these applications need to affect interactions between species or biogeochemical pathways. A molecular machine, for example, that improves phosphorus sequestering in a treatment wetland might significantly increase overall performance.

Beyond speculation, however, there is already an interesting connection between ecological engineering and nanotechnology. Both fields rely on self-organization as the basis for design. In ecological engineering, species populations and abiotic

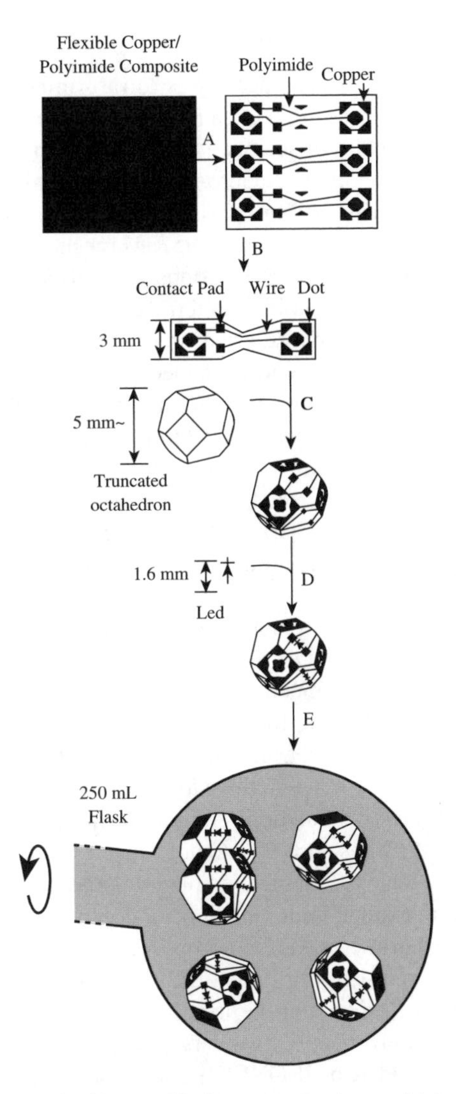

FIGURE 9.5 An example of self-assembly in nanotechnology, which occurs in step E of the diagram. Led: Light emitting diode. (From Gracias, D. H., J. Tien, T. L. Breen, C. Hsu, and G. M. Whitesides. 2000. *Science* 289:1170. With permission.)

components self-organize into ecosystems that provide some service to humans. In nanotechnology, molecular self-assembly is used to create desired products and functions (Rietman, 2001; Service, 1995; Whitesides, 1995; Whitesides et al., 1991). Chemical molecules and their environments are manipulated to facilitate the self-organization of devices in this form of engineering (Figure 9.5). Perhaps in the future, engineers from these widely different scales may be able to share ideas about self-organization as an engineering design approach.

Terraforming and Global Engineering

The largest scale of ecological engineering is terraforming, which is the modification of a planetary surface so that it can support life (Fogg, 1995). While this application is still in the realm of science fiction, it is receiving credible attention. Some interesting theory about biosphere-scale ecological engineering is being discussed, especially in terms of Mars (Allaby and Lovelock, 1984; Haynes and McKay, 1991; McKay, 1999; Thomas, 1995). Mars has a thin atmosphere and probably has water frozen in various locations. The principal factor limiting life seems to be low temperature. One idea to terraform Mars is to melt the polar ice cap in order to initiate a greenhouse effect that would raise temperature (Figure 9.6). Then, living populations would be added, perhaps starting with microbial mats from cold, dry regions of the earth that might be preadapted to the Martian surface. The mats are dark-colored and would facilitate planetary warming by lowering the albedo and absorbing solar radiation. These actions are envisioned to set up climate control, as described by the Gaia hypothesis on earth (Margulis and Lovelock, 1989). Arthur C. Clarke (1994), the famous science fiction author, has extended the theory with many imaginative views of the stages of succession involved in terraforming Mars.

While actual terraforming may not be expected to be possible for hundreds of years in the future, some practical applications are being debated for engineering at this scale on the earth. There is much interest in understanding feedbacks between the biota and climate systems (see, for example, Woodwell and MacKenzie, 1995). Some applied planetary engineering has been suggested to deal with the present climate change in the form of tree plantings to absorb and sequester carbon dioxide (Booth, 1988), though these calculations are not promising as a long-term solution to the greenhouse effect (Vitousek, 1991). A more uncertain plan is ocean fertilization with iron as a planetary scale CO_2 mitigation plan. John Martin (1992) first suggested the "iron hypothesis" to explain limitation of open ocean primary productivity based on small-scale bottle experiments. He later proposed that large-scale iron fertilization could generate a significant sink for global CO_2 and boldly stated, "give me a half a tanker of iron and I will give you the next ice age" (Dopyera, 1996)! Since his proposal (and his untimely death), two large-scale experiments (Transient Iron Addition Experiment I and II or IRONEX I and II) in the southern Pacific Ocean have basically confirmed Martin's hypothesis. Proposals about commercial iron fertilization for CO_2 mitigation are currently being debated (Chisholm et al., 2001; Johnson and Karl, 2002; Lawrence, 2002).

From Biosensors to Ecosensors

Biosensors are a growing form of technology becoming widely used in medical applications (Schultz, 1991). As noted by Higgins (1988)

> a biosensor is an analytical device in which a biological material, capable of specific chemical recognition, is in intimate contact with a physico-chemical transducer to give an electrical signal.

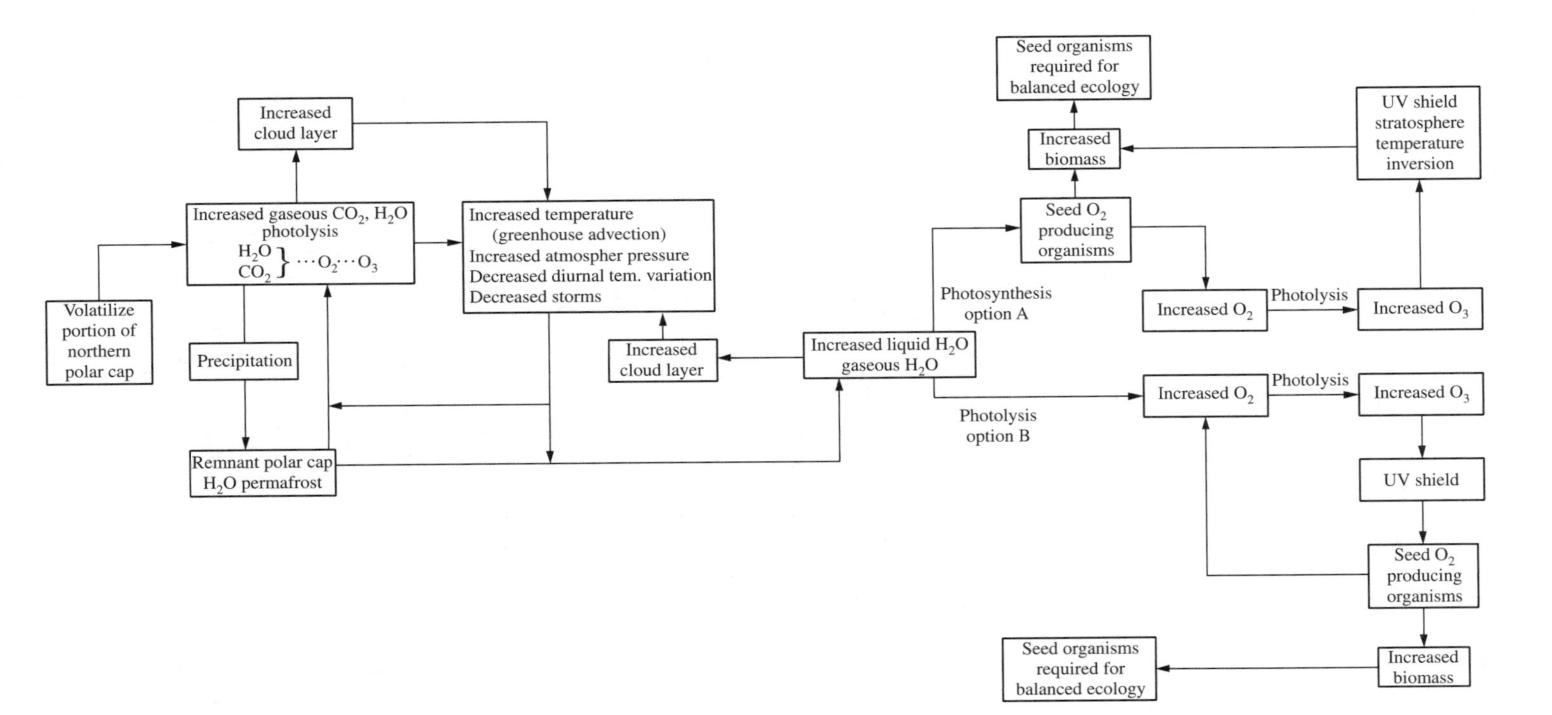

FIGURE 9.6 Hypothetical sequence of events caused during terraforming on Mars, initiated by volatilization of the northern polar ice cap. (Adapted from Wharton, R. A., Jr., D. T. Smeroff, and M. M. Averner. 1988. *Algae and Human Affairs*. C. A. Lembi and J. R. Waaland (eds.). Cambridge University Press, Cambridge, U.K.)

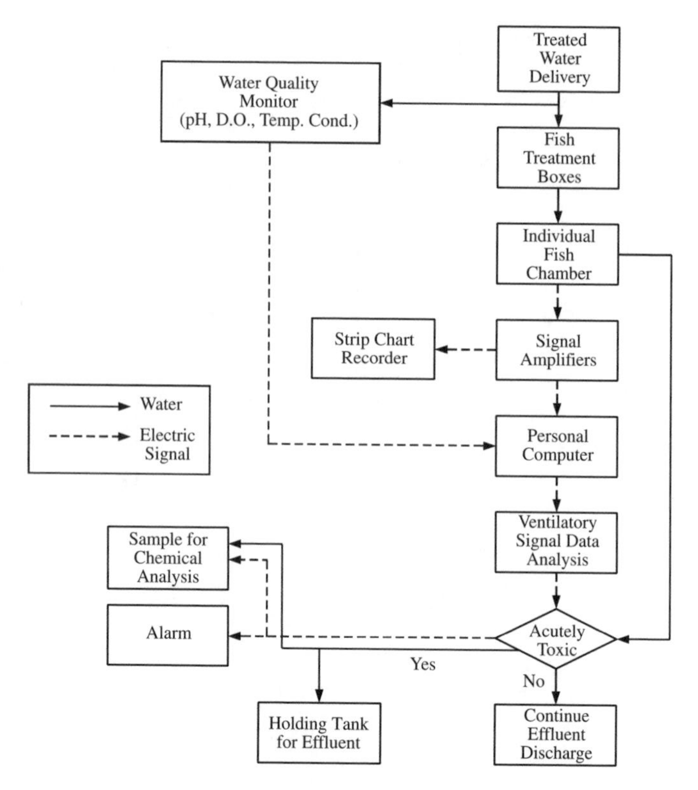

FIGURE 9.7 An example of a system for toxicity assessment with continuous monitoring sensors. (Adapted from American Society for Testing and Materials. 1996. *Annual Book of ASTM Standards*. American Society for Testing and Materials, West Conshohocken, PA.)

Biological materials offer unique capabilities in specificity, affinity, catalytic conversion, and selective transport, which make them attractive alternatives to chemical methods of sensing. This is an interesting area that involves the interfacing of biology with electronics. The three basic components of a biosensor are (1) a biological receptor, (2) a transducer, such as an optical fiber or electrode, and (3) associated signal processing electronics. Environmental applications of biosensors have focused on continuous monitoring for water quality evaluation (Grubber and Diamond, 1988; Harris et al., 1998; Rawson, 1993; Riedel, 1998). An example employing respiratory behavioral toxicity testing with fish (American Society for Testing and Materials, 1996) is shown in Figure 9.7. In this case gill movements are sensed with electrodes placed in the fish tank and related to pollutant concentrations in the water. The system can predict toxicity of a water stream with associated interfacing. In the future, biosensors may be able to be scaled up to ecosensors by ecological engineers. As noted by Cairns and Orvos (1989), most environmental uses of biosensors rely on single-species indicators of pollution stress that may not be adequate for all purposes. Ecosensors could be devised that utilize information on multispecies community composition or on ecosystem metabolism, as mentioned in the next section on technoecosystems.

TECHNOECOSYSTEMS

H. T. Odum (1983) defined *technoecosystems* as "homeostatically coupled" hybrids of living ecosystems and hardware from technological systems. This is a vision of a living machine but with added control. The simplest version would be the turbidostat (Myers and Clark, 1944; Novick, 1955) which is a continuous culture device for studying suspended populations of algae or bacteria. In this device, turbidity of the suspension is proportional to density of the microbial population. A photocell senses turbidity and is connected to a circuit that controls a valve to a culture media reservoir. If the turbidity is higher than a given threshold, then the circuit remains off, leaving the valve to the reservoir closed. However, if the turbidity is lower than the threshold, then the circuit opens the valve which adds culture media to the suspension. The added media causes growth of the population, which in turn causes an increase in turbidity. The increased turbidity thus causes the circuit to turn off, halting the addition of media. In this fashion the turbidostat provides for density dependent growth of the microbial population. The key to the turbidostat and other technoecosystems is feedback through a sensor circuit which allows for self-control. This action is similar to the concept of biofeedback from psychobiology (Basmajian, 1979; Schwartz, 1975). Biofeedback allows humans or other animals to control processes such as heart rate, blood pressure, or electrical activity of the brain when provided with information from a sensor about their physiological function.

A variety of simple technoagroecosystems have been developed including irrigation systems that sense soil water status (Anonymous, 2001), aquacultural systems that sense growth conditions for fishes (Ebeling, 1994), and computerized greenhouses (Goto et al., 1997; Hashimoto et al., 1993; Jones, 1989). Ecological engineers may design more complex technoecosystems. For example, studies by R. Beyers and J. Petersen were described in Chapter 4 for microcosms which sensed ecosystem metabolism and regulated light inputs. Wolf (1996) constructed a similar system which regulated nutrient fertilizer inputs for experimental bioregeneration. Robert Kok of McGill University envisioned even more complicated hardware interfaces in his "Ecocyborg Project." Along with his students and colleagues Kok published many designs and analyzes for ecosystems with artificial intelligence control networks (Clark et al., 1996, 1998, 1999; Kok and Lacroix, 1993; Parrott et al., 1996). Blersch (in preparation) has built this kind of design around a wetland soil microcosm (Figure 9.8). The microcosm is part of a hardware system that attempts to maximize denitrification in the microcosm by controlling limiting factors. Based on a sensing of the change in the microcosm's redox potential, either nitrogen or carbon is added to accelerate microbial metabolism. Denitrification is monitored as the rate of consumption of nitrogen addition, and microbial metabolism is monitored as the rate of decline in redox potential. Artificial intelligence is being investigated with a logic system that evaluates inputs from the redox probe in the actual microcosm and inputs from a simulation model of the system that is run simultaneously with the microcosm. The goal is to achieve the maximum denitrification rate through the use of the decision algorithm to optimize the input of elements that stimulate microbial metabolism.

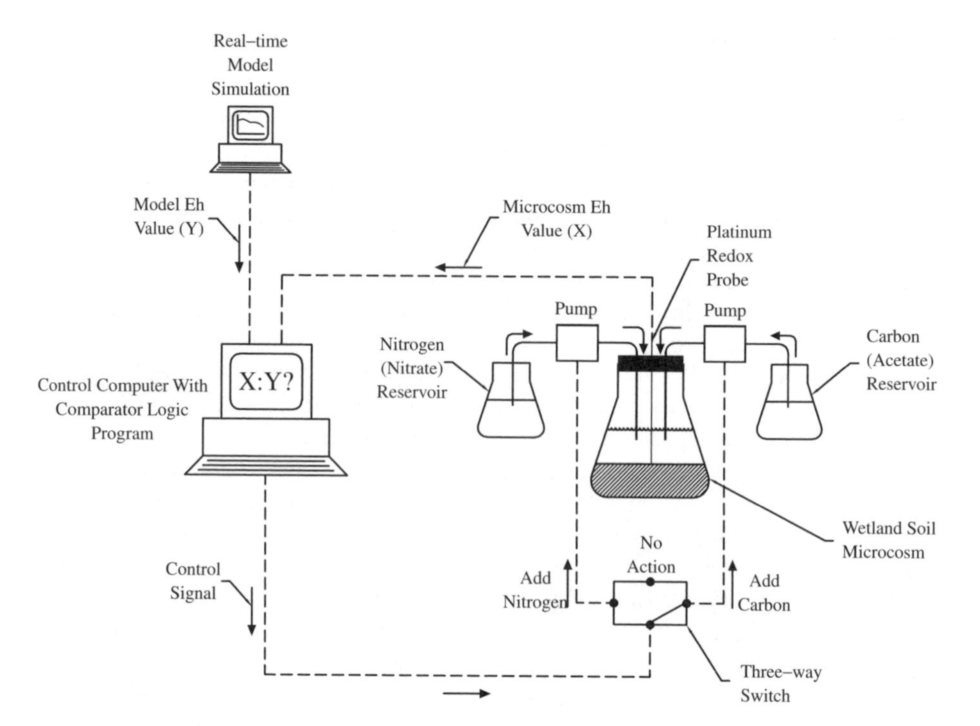

FIGURE 9.8 Diagram of a redox microcosm with artificial control from a simulation model. (From Blersch, in preparation. With permission.)

A Universal Pollution Treatment Ecosystem

The main component elements of ecological engineering designs are species populations, and the designs themselves are ecosystems. If ecological engineering was similar to other fields such as chemical, electrical, or civil engineering, it would be possible to build up designs from component elements that are well known in terms such as capacity, conductance, and reliability. However, species populations are not so well known. A million species have been discovered in nature and even for the common, widely occurring species, knowledge isn't complete. Agricultural species are best known, and the discipline of agriculture involves design of production systems with these species. Ecological engineering seeks to use the much greater biodiversity of wild species for its designs. Attempts have been made to summarize information on wild species, but these efforts have always been incomplete. The closest examples to a handbook as exists in other engineering disciplines are those produced by the Committee on Biological Handbooks in the 1960s (see, for example, Altman and Dittmer, 1966), which are composed of hundreds of tables of data. While these are interesting compilations, the ecological engineer needs different information to design networks of species. Needed are lists of who eats what and whom, chemical compositions of excretion, behaviors, tolerances, performance ranges, adaptations to successional sequences, and much other information (i.e., the species

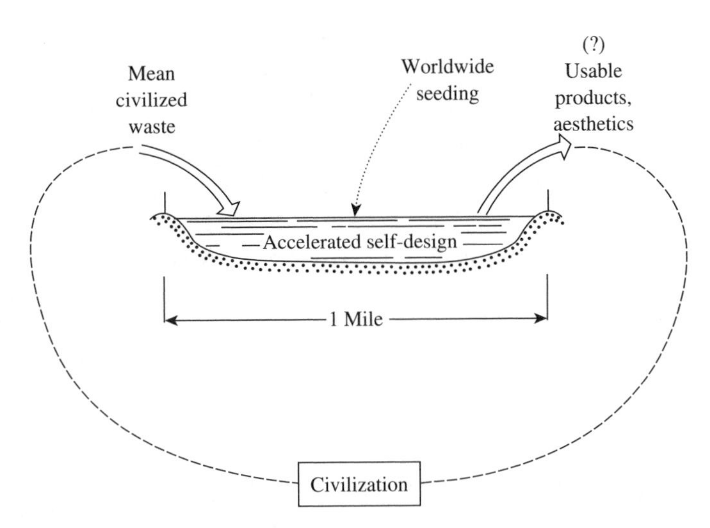

FIGURE 9.9 H. T. Odum's concept for a universal pollution treatment ecosystem from 1967. (From Odum, H. T. 1967. *Pollution and Marine Ecology.* T. A. Olson and F. J. Burgess (eds.). John Wiley & Sons, New York. With permission.)

niche). The closest existing examples may be the work on national biotic inventories, such as exists for Costa Rica (Gamez et al., 1993; Janzen, 1983) or the records on species used for biological control of agricultural pests (Clausen, 1978). However, the conventional engineer would be disappointed even in these extensive sources.

In place of handbooks on component elements, the ecological engineer utilizes the self-organizing properties of nature. H. T. Odum envisioned an example of what might be a universal pollution treatment ecosystem based on this principle that has yet to be intentionally tried (Figure 9.9 and Figure 9.10). Basically, the design would mix a variety of pollutants together in a large circulating impoundment that would be seeded with as much aquatic biodiversity as possible. H. T. Odum projected that the result would be a treatment ecosystem that could absorb and cleanse any pollution source. The growth of treatment wetland technology as described in Chapter 2 demonstrates that such ecosystems are possible. Furthermore, species from around the world can self-organize into new networks, as discussed in Chapter 7. Thus, H. T. Odum's vision for a universal treatment ecosystem may be possible. The critical aspect seems to be size of the impoundment necessary for self-organization to transcend adaptation to any particular pollutant and result in more universal treatment capacity. Size relates to spatial heterogeneity, which improves ecosystem qualities such as diversity and stability. Interestingly, H. T. Odum expanded the size of his design from a maximum diameter of 1 mi (1.6 km) in 1967 (Figure 9.9) to 5 mi (8 km) in 1971 (Figure 9.10). Perhaps his experience at Morehead City, NC, with self-organization of marine ponds and domestic sewage in the late 1960s suggested to H. T. Odum that his conception needed enlargement.

H. T. Odum's design may be equivalent to a living machine consisting of a very large number of tanks connected in series (see Chapter 2). The hypothesis is that any pollution source can be treated, given a long enough set of tanks filled with

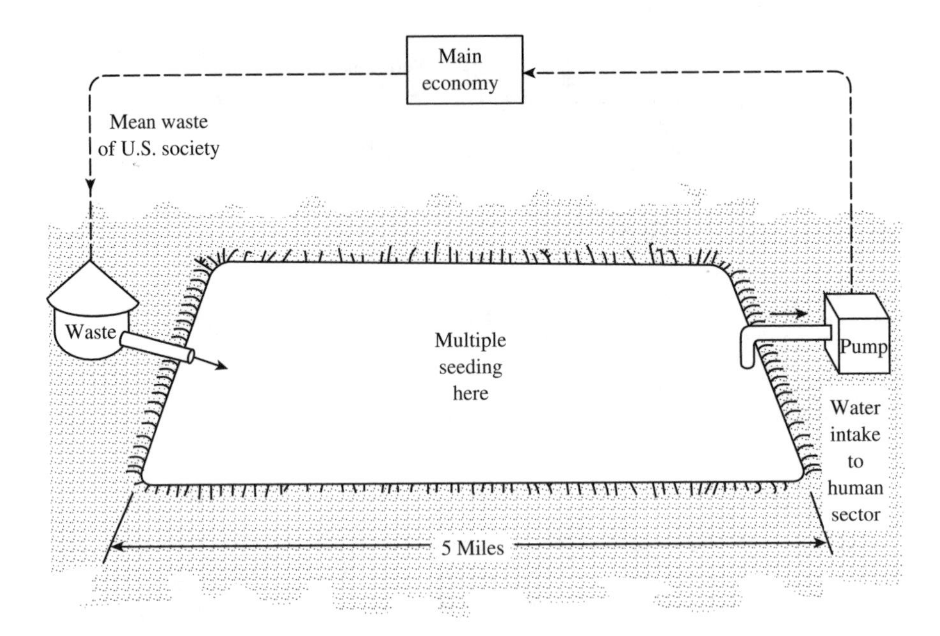

FIGURE 9.10 H. T. Odum's concept for a universal pollution treatment ecosystem from 1971. (From Odum, H. T. 1971. *Environment, Power, and Society.* John Wiley & Sons, New York. With permission.)

different biota. A quantitative expression for the treatment capacity of a living machine is given below:

$$T = \sum_{i=1}^{n} \left(P - C_i + B_i\right) \tag{9.1}$$

where

T = total treatment capacity of the living machine
$P - C_i$ = physical–chemical treatment capacity of tank i
B_i = biological treatment capacity of tank i
n = number of tanks in the living machine

Treatment capacity is increased by increasing the number of tanks (n in the equation). In an analogous sense, the digestive system of a ruminant is an example of this principle. Three extra stomachs are found in ruminants which aid in digestion of plant material with low nutritive value (see Chapter 6). Each stomach has a different function in the digestion process, and recycle is even included in the regurgitation of cud.

FIGURE 9.11 The example of Buckminister Fuller's Dymaxion house on display at the Henry Ford Museum in Dearborn, MI.

While the experiments described above may be as much science fiction as terraforming, they also may be happening inadvertently in polluted bays and harbors around the world today. For example, see the discussion of San Francisco Bay in Chapter 7 for a possible candidate. Intentional ecological engineering of the design would increase progress, which may require "a national project of self-design" as proposed by H. T. Odum more than 30 years ago.

Ecological Architecture

Strong ties already exist between architecture and ecological engineering. Architecture deals with design of human environments, and many architects have evolved approaches that are responsive to, or even inspired by, nature (Zeiher, 1996). Well-known examples are philosophies of organic or living architecture (Wright, 1958) and the idea of "design with nature" as a guide to landscape architecture (McHarg, 1969). McHarg's famous phrase actually may have been derived from Olgyay's (1963) treatise on bioclimatic architecture that was titled "*Design with Climate*."

The design process is somewhat different in architecture as compared with traditional engineering, and ecological engineers can learn much from the contrast. Often times, architects seem to open up new lines of thinking by creating bold designs that are unconstrained by practical limitations. Buckminister Fuller's "Dymaxion" house is an example of this creative approach to design from the 1920s. The Dymaxion house included many features that were completely unconventional but farsighted. For example, (1) it was made out of aluminum and could be mass-produced, and (2) it had a circular floor plan and was suspended on a central mast which made maximum use of space and facilitated climate control. Also, the amount of material used per unit floor space was minimized, which reflects Fuller's motto of "doing more with less." Although the Dymaxion house was never commercially produced, it generated new thinking that was influential (Baldwin, 1996). An example of an actual Dymaxion house is on exhibit at the Henry Ford Museum in Dearborn, MI (Figure 9.11). Paolo Soleri's "Arcosanti" is another example of an extremely visionary form of architecture. Soleri (1973) developed a unique philosophy of the future building environment and ecology of humans. His approach is to design and build huge skyscrapers of highly integrated living and working spaces, in this way concentrating the built environment onto a small footprint and leaving as much as possible of the surrounding open space for agriculture and nature preservation. An actual model of Soleri's Arcosanti exists and is growing as an experiment in the desert grassland north of Phoenix, AZ.

Of course, at another extreme architecture can be eminently practical, as demonstrated by Butler (1981) in his book on how to build an "ecological house" based on principles of energy efficiency. John and Nancy Todd's work on bioshelters is another expression of ecological architecture that also emphasizes food production and wastewater treatment designs (Todd and Todd, 1984). Reviews of these approaches to architecture are given by Steele (1997) and Stitt (1999).

Several initiatives of ecological architecture represent new directions for collaboration between architects and ecological engineers. Plant-based systems are being integrated with architecture in innovative ways, such as for air quality improvement in interior environments (see below) and roof gardens (Kohler and Schmidt, 1990) for stormwater management in external environments. Although these applications involved straightforward horticulture, ecological engineering may be important especially if treatment function is to be optimized. For example, Golueke and Oswald (1973) describe a plan for a home that uses an algal regenerative system for multiple functions. Another direction for collaboration may be in terms of the recycling of buildings and their materials (Brand, 1994). Here, ideas of industrial ecology such as life cycle analysis and network accounting of material flows may be appropriate. The term *construction ecology* recently has been used to describe some or all the applications listed above (Kibert et al., 2002).

Biofiltration and Indoor Environmental Quality

The quality of the indoor environment is directly related to ecological architecture but at a smaller scale. Many aspects are involved (Godish, 2001), though air quality with respect to human occupation and activity has received the greatest attention (Meckler, 1991). Human health problems can arise inside buildings due to the accumulation of toxic chemicals or organisms. These accumulations are facilitated by the static atmosphere that occurs inside buildings, especially in those buildings that are tightly sealed for energy efficiency. Furthermore, because humans spend a high percentage of their time indoors, exposure levels can become critical. When the cause of the health problem is diagnosable, the condition is known as "building related illness." When the cause of the health problem cannot be diagnosed, the condition is known as "sick building syndrome." Together these kinds of problems are serious enough to require costly treatment or even the abandonment of whole buildings. Causes include such factors as volatile organic chemicals, dust fibers, asbestos, carbon monoxide, and molds.

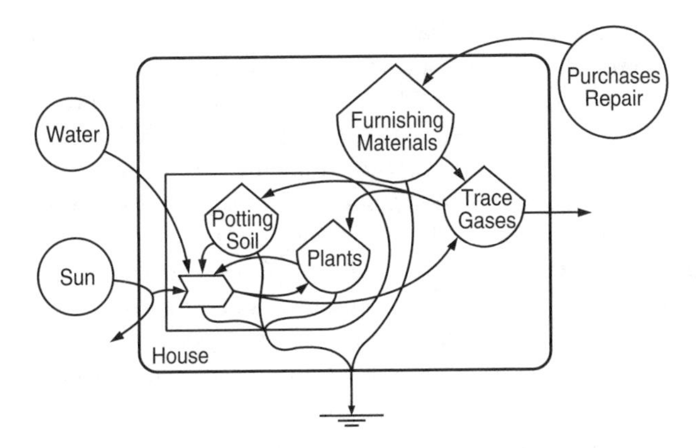

FIGURE 9.12 Energy circuit diagram of the role of house plants in air quality improvement in a house environment.

Biofiltration is an ecological engineering approach that has been taken to solve these problems. The approach is to pass contaminated air through chambers containing media and organisms (i.e., biofilters) in order to remove the contaminants. Soil beds have a long history of use for this purpose, for example, in treating odors from sewage treatment plants (Bohn, 1972; Carlson and Leiser, 1966). However, recent trends are to explore more sophisticated designs (Darlington et al., 2000, 2001; Leson and Winer, 1991; Wood et al., 2002). Most biofilters rely on microbes for biological treatment processes such as oxidation of volatile organic chemicals. However, higher plants also are used. These applications can involve common houseplants that contribute to the air processing in buildings (Figure 9.12). B. C. Wolverton has been a leader in this approach, building on his early experience in treatment wetland design (see Chapter 2). Much of his extensive published research on use of houseplants in biofiltration is summarized in a text entitled *How to Grow Fresh Air* (Wolverton, 1996).

One direction for future design is to begin thinking of the indoor environment as an ecosystem. A significant amount of biodiversity can be found in houses, even though these species are largely considered pests. For example, Ordish (1981) lists more than 150 taxa as being found in a review of the history of a bicentennial house, and he even gives graphs of relative abundances and estimates of animal metabolism. These organisms are links in food webs and in biogeochemical cycles within the houses. Conscious design of house animal food chains might be able to control mold populations, when practically no other solutions are available. The ecosystem approach used by ecological engineers could aid in future solutions to indoor environmental quality problems.

Ecology and Aquacultural Design

Aquaculture is the controlled production of aquatic organisms for human use. This is an important field that has the potential to provide a significant source of food to the world's growing population (Bardach et al., 1972; Brown, 1980; Limburg, 1980; Naylor et al., 2001). Designs in aquaculture range from commercial scale, energy intensive, indoor tank systems to backyard-scale, low energy, outdoor pond systems. Engineering aspects of aquaculture, especially on a commercial scale, are well developed (Wheaton, 1977) and mostly concern problems such as temperature and oxygen controls, water filtration, and waste disposal. Thus, aquacultural engineering largely involves controlling and maximizing conditions for growth of particular species of fish and other organisms within constructed environments. The economic basis of commercial scale aquaculture is tenuous because both capital and operating costs are high and markets are often uncertain. Some systems such as catfish production in the southern U.S. are established and economically viable, while many others require stronger markets and/or further technological development.

Ecological engineering probably can make little contribution to commercial scale aquaculture where emphasis is on producing large amounts of food product from monocultures of single species. Work on intermediate scale or backyard-type aquaculture systems is more appropriate for ecological engineering. Systems that rely on a polyculture of multiple species and/or an integration of multiple uses of water may be able to be improved with ecological engineering knowledge. Much is already known about lower-energy aquaculture (Chakroff, 1976; Logsdon, 1978), but some new initiatives may be possible. Swingle's (1950; Swingle and Smith, 1941) old work on fishing ponds actually represents an early version of ecological engineering. He developed much design knowledge from experimentation with artificial impoundments at the Alabama Agricultural Experiment Station. For example, he found that the ratio of forage fish to game fish should be about 4 to 1 for optimal productivity of a fish community. One approach would be to try to incorporate aspects of Chinese fish culture, which focuses on a polyculture production system (Jingson and Honglu, 1989; Lin, 1982), into the Western world. The Chinese systems are remarkable for their diversity and they incorporate features based on ecological principles. Creative designs based on these models are possible and could be constructed widely in rural and suburban environments (Todd, 1998). Other models also are possible. For example, Pinchot (1966, 1974) designed an aquacultural system based on natural oceanic upwelling ecosystems. These ecosystems are usually found on the west coasts of continents where wind patterns cause a localized circulation that brings deep ocean water to the surface (i.e., upwelling). The water from the deep ocean is nutrient rich, because of the dominance of decomposition processes there. Thus, upwellings are naturally eutrophic sites with high production and short food chains due to the fertilization effect of nutrient rich water being brought to the surface where sunlight levels are highest (Boje and Tomczak, 1978; Cushing, 1971). Pinchot's design for an artificial upwelling involved pumping deep ocean water into a coral reef lagoon with power from windmills (Figure 9.13). Whales would then be raised in the lagoon where they could be easily harvested for food because of their confinement in the enclosure of surrounding reef. This is a rather imaginative design, which represents an ecological engineering approach, but it probably would not be feasible at present due to animal rights concerns about the welfare of whales. Other examples of integrated systems of aquaculture and sewage treatment are described in Chapter 2.

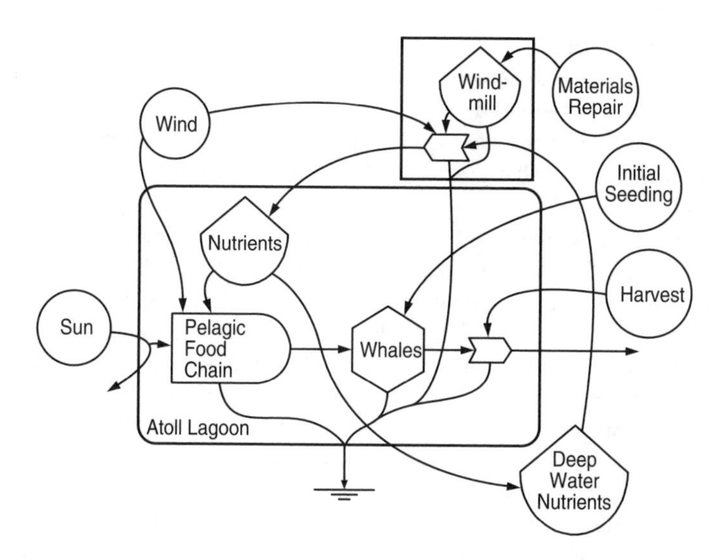

FIGURE 9.13 Energy circuit diagram of a hypothetical whale aquaculture system proposed by Pinchot.

The last area in which ecological engineering may possibly contribute to aquaculture involves pest biodiversity. Outdoor aquacultural systems often attract natural predators that feed on the species being cultured. For example, fish-eating birds, such as pelicans and cormorants, are responsible annually through their feeding actions for millions of dollars of damage to outdoor aquaculture. An ecologically engineered design for reducing this energy flow in a humane fashion would be a significant achievement.

Biotechnology and Ecological Engineering

Biotechnology or genetic engineering, as it is sometimes referred to, involves the creation of organisms with new properties through various forms of genetic manipulation. The field has a long history of producing useful organisms that contribute to the well-being of humans, such as the centuries-old examples of microorganisms that function in the leavening of bread and the fermentation of grapes to make wine. Recombinant DNA technology, which developed in the mid-1970s, is only the most recent form of biotechnology. In all forms of biotechnology the genetic makeup of organisms is modified to produce an organism (or a product) that is different from the starting organism. However, significant differences exist between classical techniques such as controlled breeding and the new molecular techniques. The recombinant DNA technology is faster, can deal with many more kinds of genes, and is more precise than the classic methods. Because of these developments, a biotechnology revolution has been envisioned with great advances expected in medicine, agriculture, and other fields (see, for example, Koshland, 1989). Many kinds of improvements can be imagined and are being studied such as engineering crop species that use less water or that have enhanced resistance to disease, or microbes which metabolize hazardous wastes. The potential benefits of biotechnology seem endless, but there are risks associated with releasing genetically engineered organisms into the environment (Flanagan, 1986; Gillett, 1986; Pimentel et al., 1989; Tiedje et al., 1989). The risks are similar to those described in Chapter 7 for exotic species, but with a somewhat greater degree of uncertainty about impacts (Drake et al., 1988).

While there are fundamental differences between genetic and ecological engineering (Mitsch, 1991), the two fields might find areas of collaboration (Forcella, 1984). Ecological engineering could contribute the perspective of multispecies networks to genetic engineering which otherwise focuses on individual species. Perhaps biotechnology could involve pairs or sets of interacting species in a kind of coevolutionary genetic engineering. Thus, predator–prey pairs might be coengineered rather than just a single species. This might build more security into releases because a predator or parasite would be simultaneously designed for a specific purpose. In this way a predator would already be available to control the new engineered organism if its populations caused unintended impacts. Biotechnologists would need to collaborate with their ecological counterparts to ensure success of the pair of species. Other kinds of interspecific interactions could also be explored such as symbiotic systems for decomposition and nutrient cycling. Contributions from ecology in the relationship could be seen as amplifying biotechnology in the long run, though in the short run more genetic–ecological engineering design would be required than presently occurs. This would be the opposite of the typical negative role that ecology has had with biotechnology in terms of simply regulating releases of genetically engineered organisms.

Biocultural Survey for Alternative Designs

Most designs discussed in this text have been the product of Western thinking. Possible examples from the Orient were mentioned in Chapter 3 but alternative designs might be employed by many different cultures. Perhaps biocultural surveys could be undertaken to search for these alternatives. An analogous approach exists for useful plants, called *ethnobotany*, in which anthropologists conduct studies of the ways different cultures utilize plants of economic importance. This is a well-known approach to learning about indigenous knowledge that can then be adapted to Western society. Interfacing with other cultures is a complicated enterprise that involves human rights issues (Rubin and Fish, 1994; Schmink et al., 1992), and biocultural surveys must be conducted with care and respect.

Alternate ways of thinking about ecological engineering can be expected to occur because different human societies are known to develop unique material cultures, such as house form (Rapoport, 1969) or agriculture (Gliessman, 1988). Todd (1996b) has warned against the disappearance of regional and local technologies and the spread of a "technological monoculture" found in Western society. Thus, there may be a need to search out and "salvage" cultural approaches to ecological engineering before they are lost (Cox, 2000). Balinese irrigation engineering may be one such example as shown in Lansing's studies (1987, 1991). In

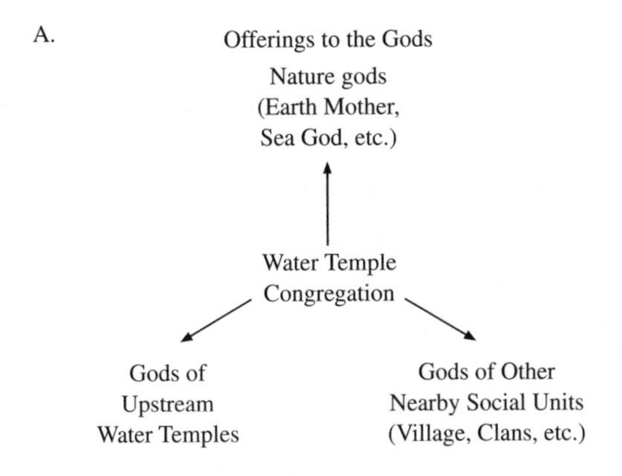

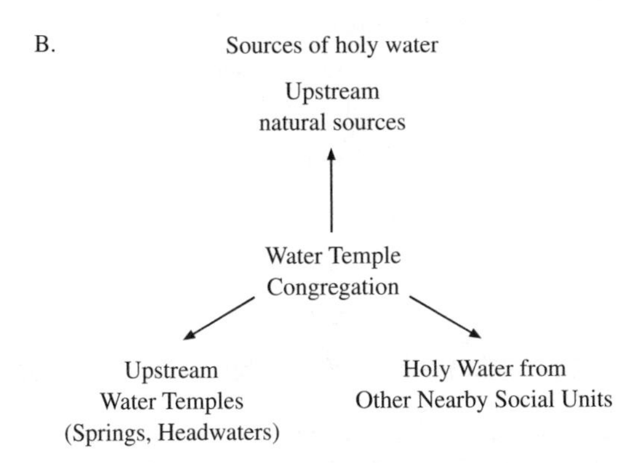

FIGURE 9.14 Two views of Balinese water temples in terms of (A) religion and (B) hydrology. (Adapted from Lansing, J. S. 1991. *Priests and Programmers: Technologies of Power in the Engineered Landscape of Bali*. Princeton University Press, Princeton, NJ.)

this case, social organizations and irrigation structures as well as management are highly integrated into a form of "ritual technology" (Figure 9.14). Kremer and Lansing (1995) constructed simulation models of this system of water management and used them as a way of facilitating communication between the indigenous culture and officials from the Westernized government of Bali. Further studies such as these are needed to know whether a kind of ethnoecological engineering would be an instructive activity. Perhaps similar philosophies to McHarg's "design with nature" or Buckminster Fuller's "do more with less" can be found embedded in some Amazon Indian cosmology (Reichel-Dolmatoff, 1976); or perhaps some new technology such as a novel living fence (Steavenson et al., 1943) tucked away in a Mayan field in the highlands of Guatemala can be discovered and incorporated into Western ecological engineering. Can the sacred groves of India (Gadgil and Vartak, 1976; Marglin and Mishra, 1993; Reddy, 1998) provide designs for socially integrated urban rain gardens and riparian buffers?

TABLE 9.3
Some of the Educational Challenges of Ecological Engineering

Ecological Engineering Curricula
Should curricula be offered at the graduate or undergraduate level?
What balance should there be between ecology courses and engineering courses in the curricula?
What particular ecology courses should be required?
What particular engineering courses should be required?
Should there be ABET review of curricula?
Ecological Engineering Course
At what level (graduate, senior, junior, etc.) should an ecological engineering course be taught?
What prerequisites should there be for the course?
Should the course be lecture only or should it be taught with a lab?
How many credits should be given for the course?
What coverage of topical material should be given in the course?

Note: Questions are arranged according to two scales: the entire curriculum for a major and an individual course on the subject.

ECOLOGICAL ENGINEERING EDUCATION

Education in ecological engineering will need to develop for significant advancements in the field to occur. In the present state of the art there are very few academic programs in ecological engineering, and most of these are of recent origin. Existing practitioners are largely self-taught with backgrounds that combine training in ecology and some established engineering discipline, along with on-the-job experience. Ecological engineering is probably more interdisciplinary than any traditional engineering field, and this presents challenges for the evolution of academic programs. Many important questions exist about how to provide training in this new hybrid field of ecology and engineering (Table 9.3). Perhaps the self-organization of curricula and courses taking place at universities in the U.S. and elsewhere will lead to one educational model through natural selection, and it is premature to generalize yet on the best approach for education. In this section some issues are discussed and educational initiatives are explored.

CURRICULA

Obviously, an academic program in ecological engineering must consist of some kind of balance of ecology and environmental science with engineering principles and technology. However, many paths are possible and no agreed-upon solutions have yet arisen. Whole new educational approaches are needed and, as discussed in

Chapter 1, no existing disciplines in ecology or engineering can serve as exact models. One possible approach would be to require ecological engineering projects to be carried out by teams of specialists with training in traditional disciplines, but most workers agree that it is possible to educate individuals in this new interdisciplinary field. H. T. Odum, who is the founder of ecological engineering, has written about his experiences in offering an "informal" academic program for more than 30 years (H. T. Odum, 1989, 1994). He suggests that ecological engineering requires at least education through a master's level degree (typically with 4 years of an undergraduate degree and 2 years of a master's degree) to cover all of the necessary coursework. The optimal program may be the combination of an undergraduate engineering degree and a graduate degree that emphasizes ecology and environmental science, though he also notes the success of individuals who entered his graduate program with education in a biologically based undergraduate degree program.

A workshop was held in 1999 at Ohio State Univeristy in Columbus to discuss educational options in ecological engineering, and several different curricula were discussed (Mitsch and Kangas, 1999). These included stand-alone undergraduate degrees modelled after existing engineering disciplines, and graduate degrees that would accept undergraduates with training in either ecology or engineering. The special challenge of a stand-alone undergraduate degree is fitting enough coursework into an approximately 4-year time period characteristic of a bachelor's degree. This coursework must include enough depth in both engineering and ecology to produce a professional who might enter the job market. This will require sacrifices that might result in training that is too weak in either ecology or engineering for an individual to be able to carry out ecological engineering design, construction, and operation. The special challenge of a graduate degree is overcoming the biases of specialized undergraduate training in a relatively short time period, usually 2 years for a master's degree. Also, different tracks of graduate coursework would be required for those entering with an undergraduate engineering degree vs. those entering with an undergraduate science degree, such as biology.

Some of these issues can be addressed by having any educational program formally accredited by an external board of experts. Undergraduate engineering degrees are accredited by the American Board for Engineering and Technology (ABET), which reviews programs periodically and confers a standardized certification of adequate educational depth and quality. This kind of accreditation allows entrants into an academic program to know they will receive a proper education and allows employers who hire graduates from a program to know that they are employing people with appropriate backgrounds. Accreditation by ABET is accepted as the best approach by those advocating a stand-alone undergraduate degree in ecological engineering, but much work will have to be done to achieve this goal. ABET accreditation requires a certain number of existing professionals to be available to establish review criteria and to carry out the routine accreditation activities. Also, current rules dictate that a field must have a minimum of 50 graduates per year across all universities before accreditation can begin. Because ecological engineering is such a new field, ABET accreditation may have to wait until a critical mass of practitioners is available to meet the formal requirements of the process.

Another process from traditional engineering that may address educational issues is professional certification. This kind of certification is bestowed on individuals who meet certain criteria and who pass a standardized exam. Currently, professional engineering exams in the U.S. are administered by state boards that have different requirements. Most require that an applicant have graduated from an ABET-accredited academic program and have worked in the field for several years to develop practical experience. Those who pass the exam are thus formally certified as engineers. Certification documents that the recipient has a given level of knowledge and ability. This option also requires a critical mass of practitioners before exams can be designed and administered and, at least as presently conducted, it requires that ABET-accredited academic programs exist. Other relevant certification models have been developed, such as with the Ecological Society of America and the Society of Wetland Scientists. These are less formal and rigorous than certification as a professional engineer but they could be modified and implemented by a professional society, such as the existing International Society for Ecological Engineering or the new American Society for Ecological Engineering.

While discussions about accreditation and certification need to continue, they may be premature because both require a critical mass of practitioners. This group either does not yet exist or, at least, has not yet emerged from the many disciplines related to ecological engineering (see Table 1.6). Indications from the workshop on education (Mitsch and Kangas, 1999) and from the growth of the journal *Ecological Engineering* suggest that many universities are developing academic programs in ecological engineering and that these programs will generate a critical mass of practitioners soon. The existing programs combine training in ecology and engineering with traditional university core requirements of humanities, social sciences, history, and other disciplines. Most curricula include several ecology and/or environmental science courses, such as general ecology, population and ecosystem ecology, and applied ecology, along with electives from fields such as aquaculture, bioremediation, and restoration ecology. Required engineering background includes thermodynamics, fluids, and principles of design, along with the associated mathematics, physics, and chemistry characteristic of traditional programs. Specialized requirements include a basic course in ecological engineering, ecological modelling, some form of economics for design evaluation, and a practicum course that involves a group experience in an actual design. Examples of design seminars are given by Biermann et al. (1999) and Yaron et al. (2000). This kind of curriculum requires participation by a number of faculty with different skills. As noted in Chapter 1 the best environment for this kind of curriculum may be in agricultural or biological engineering departments in which existing faculty have training in engineering with biology. Environmental engineering departments are another logical location for an ecological engineering academic program. Other situations are also possible such as the more liberal arts approach advocated by Orr (1992a, 1992b, 2002), but there is a growing consensus for academic programs in ecological engineering to take on some modification of a traditional engineering curriculum.

The Ecological Engineering Laboratory of the Future

What is the best learning environment to train ecological engineers? Because of the special interdisciplinary nature of the field, a new kind of lab may be required that generates both ecology and engineering experiences. The ecology side must provide knowledge and skills dealing with biodiversity as the building blocks of design. It must also provide a whole system perspective necessary to create domestic and interface ecosystems that perform useful functions. The engineering side must provide access to existing technologies in terms of machines and electronics, with an emphasis on the design process itself. Considered below are two historical and two existing models for perspective on the ecological engineering laboratory of the future.

Thomas Edison's "Invention Factory"

Thomas Edison, sometimes referred to as the "Wizard of Menlo Park," was one of the great American inventors. Although best known for his work on the incandescent light, he produced many useful devices in his lifetime, as witnessed by his more than 1,000 patents. These devices were the products of his "invention factory" which evolved as a kind of institution through three physical complexes over his adult lifetime of 55 years. It was a place where he developed a method of invention that involved the organized application of scientific research to commercial ends. This work was especially significant because it took place before engineering formally emerged and broke into academic and professional disciplines, and it became the forerunner of modern industrial research and development. At its peak, Edison's invention factory realized his prediction of "a minor invention every ten days and a big thing every six months or so." Because Edison was privately supported, either through venture capitalists or companies he created himself, his inventions had to make a profit. Thus, his institution was half research lab and half factory.

Edison's work began at Menlo Park, NJ, where he operated from 1876 to 1881. Work here focused on early telephone designs, the phonograph and, of course, the electric lighting system. He temporarily left the invention field to develop electric lighting as a commercial industry but returned in 1887 when he built a larger lab in West Orange, NJ. Emphasis in this lab was on five product lines: musical phonographs, dictating machines, primary batteries, storage batteries, and cement (Millard, 1990). Edison worked in the West Orange lab until his death in 1931, but he also established a small lab at his winter home in Fort Myers, FL. This was the "green laboratory" where he worked on alternate sources of natural rubber, primarily in the years just before his death (Thulesius, 1997). Remarkably, each of these labs exists as a public museum: the Menlo Park lab was moved to southeastern Michigan and reconstructed by Henry Ford at Greenfield Village in Dearborn (Figure 9.15), the West Orange lab became a national park, and the green laboratory became a museum administered by the City of Fort Myers.

The Menlo Park lab was where Edison achieved the fame that continued to develop throughout his life. It also was where he established a method of invention

FIGURE 9.15 Views of Menlo Park at the Henry Ford Museum in Dearborn, MI.

that was the product of his creative genius (Pretzer, 1989). The Menlo Park lab began in 1876 with a simple two-story frame structure that contained an office, machine shop, and lab. It was expanded to six buildings in 1878 to work on the electric lighting system. These included the main lab building, a separate and enlarged machine shop, the office/library, and three small buildings that housed essential skills and materials that were constantly needed (a glasshouse, carpenters' shop, and the carbon shed). Most of the principal work took place in the main lab which was filled with chemicals and mechanisms used to conduct experimental projects in the spacious, second-floor work room. The machine shop was also a critical part of the overall lab in producing experimental devices for the continual process of testing and redesign. As noted by Israel (1989),

> By adapting the machine shop solely to inventive work Edison and his assistants could rapidly construct, test, and alter experimental devices, thus increasing the rate at which inventions were developed. In this way the laboratory became a true invention factory.

Moreover, Edison adopted the machine shop culture into his invention process. This was a unique work culture "that stemmed from craft traditions of the pre-industrial era, traditions that stressed the skill of the worker and preserved the dignity and independence of his work" (Millard, 1990). The American machine shop was an innovative institution that evolved in the early 19th century with the industrial revolution, as noted in the following quote.

> Most of the early shops worked almost entirely on special order rather than for a broad, competitive market requiring mass production and standardization. This was true partly because the designs of the products made, steam engines, machine tools, and locomotives, had not yet fully evolved. The shop frequently was an experimental laboratory which developed and perfected industrial and mechanical processes and equipment (Calvert, 1967).

Machine shops supplied machines, advice, designs, and repair services. They also provided a unique social environment in which information was shared between shops and within shops as a kind of educational network. Included in the machine shops were lathes, drills, milling machines, and planers used to cut and shape iron and steel with great precision. These were machines that made machines, and they were the backbone of America's industrial revolution.

Connections between Edison's invention factory and a modern ecological engineering lab are through analogies. The important connection is Edison's spirit of invention and his method, which can be directly applied. New ecosystems must be invented to provide specific functions in ecological engineering. It can be contended that the Edison spirit of invention exists in several leading ecological engineers such as John Todd and Walter Adey, and an educational goal is to instill this quality in the next generation of students.

The New Alchemy Institute

The New Alchemy Institute was initiated in 1969 through the shared discussions of John Todd, his wife Nancy Jack Todd, and William McLarney. It became a small, nonprofit organization dedicated to research and education on renewable resource technologies. Although the official goal was grandiose, "To Restore the Lands, Protect the Seas, and Inform the Earth's Stewards," the institute produced a number of very practical technologies and provided education to many through workshops, tours, and publications. As a formal organization, New Alchemy lasted for more than 20 years; it has evolved into new organizations with similar goals.

In the early 1970s the Institute became centered at a 12-acre (4.8 ha) farm on Cape Cod, MA, where work focused on development of technologies that support low-cost, year-round food production and energy-efficient shelter design (Figure 9.16). Examples of projects included intensive agriculture, aquaculture, tree crops, and renewable energy alternatives using solar and wind power. These were full-scale experimental projects in alternative technologies. The living machine, which was described in Chapter 2, is a good example of these designs. Another important invention was the bioshelter which is a solar-heated building that links a variety of

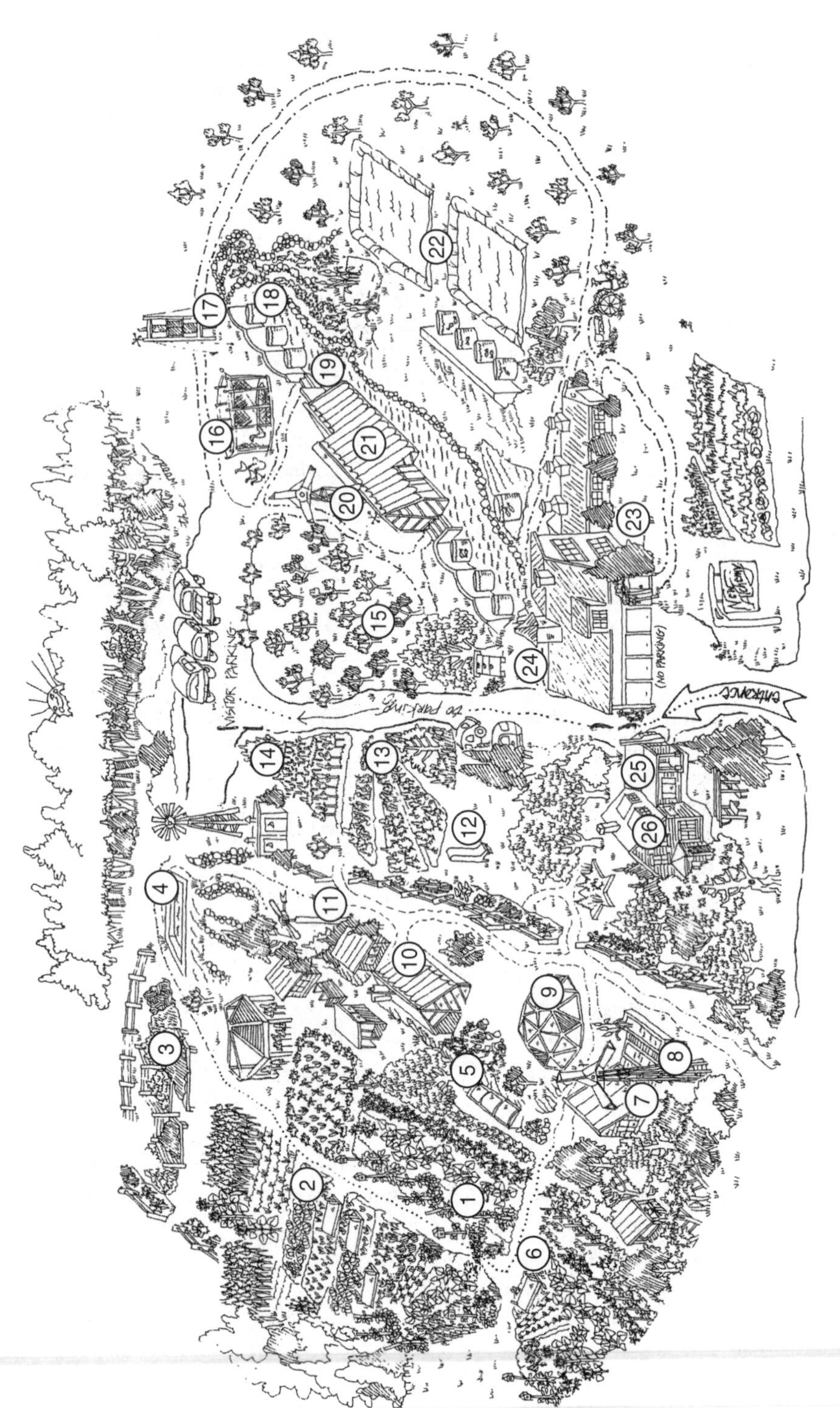

FIGURE 9.16 View of the New Alchemy Institute on Cape Cod, MA. (From John Todd, Ocean Arks Inc. Burlington, VT. With permission.)

biological elements together for food and energy production and biological waste treatment. Large bioshelters were called *arks* and several were built in different locations.

In many ways the New Alchemy Institute was an "invention factory" in producing ecological engineering designs. The living machines of New Alchemy are analogous to the machine tools of Edison's labs, which were described earlier as "machines that made machines." Also, New Alchemy fostered a social organization somewhat analogous to the machine shop culture, with a sharing of information and value systems among participants. A final similarity is dominance by a big thinker. Although New Alchemy had an egalitarian structure involving many individuals, John Todd does emerge as a dominant figure that was somewhat analogous to Edison in providing inspiration and organizational skills. Todd left the Institute in the early 1980s and started a new organization called Ocean Arks International in which he continues some of the work started in the New Alchemy Institute. However, unlike Edison's operations which were driven by profit and capitalism, the New Alchemy Institute was a nonprofit organization driven by the goal of fostering sustainable development.

The Waterways Experiment Station

The U.S. Army Corps of Engineers Waterways Experiment Station (WES) in Vicksburg, MS, is a national lab with many activities related to ecological engineering (Anonymous, no date). Although the scale of WES is much larger than could be achieved at an individual university, it provides an existing model for perspective on the ecological engineering lab of the future. WES consists of five engineering labs along with various administrative and technical support units (Figure 9.17). It was established in 1929 with emphasis on hydraulics, after the disastrous 1927 flood on the Mississippi River. Various missions were added to the lab over the years, including a significant military role after World War II.

The environmental lab at WES already is involved with a variety of research lines that relate to ecological engineering such as aquatic plant control, simulation modelling, and wetland creation. The association of these activities with the other engineering labs, especially the geotechnical and hydraulics units, makes WES an ideal location for the development of ecological engineering technologies. Somewhat of a paradigm shift may be required, however, for the Corps of Engineers to become heavily involved in this field. Past work of the Corps has not always demonstrated the holistic thinking inherent in ecology, though as noted in Chapter 5 they have initiated several programs that are moving in this direction. It remains to be seen if the Corps of Engineers will become a leader in the field of ecological engineering, which it is clearly capable of, or continue to generate environmental problems that require ecological engineering solutions.

The Olentangy River Wetland Research Park

Perhaps the closest model for the ecological engineering lab of the future at a university is the Olentangy River Wetland Research Park on the Ohio State Univer-

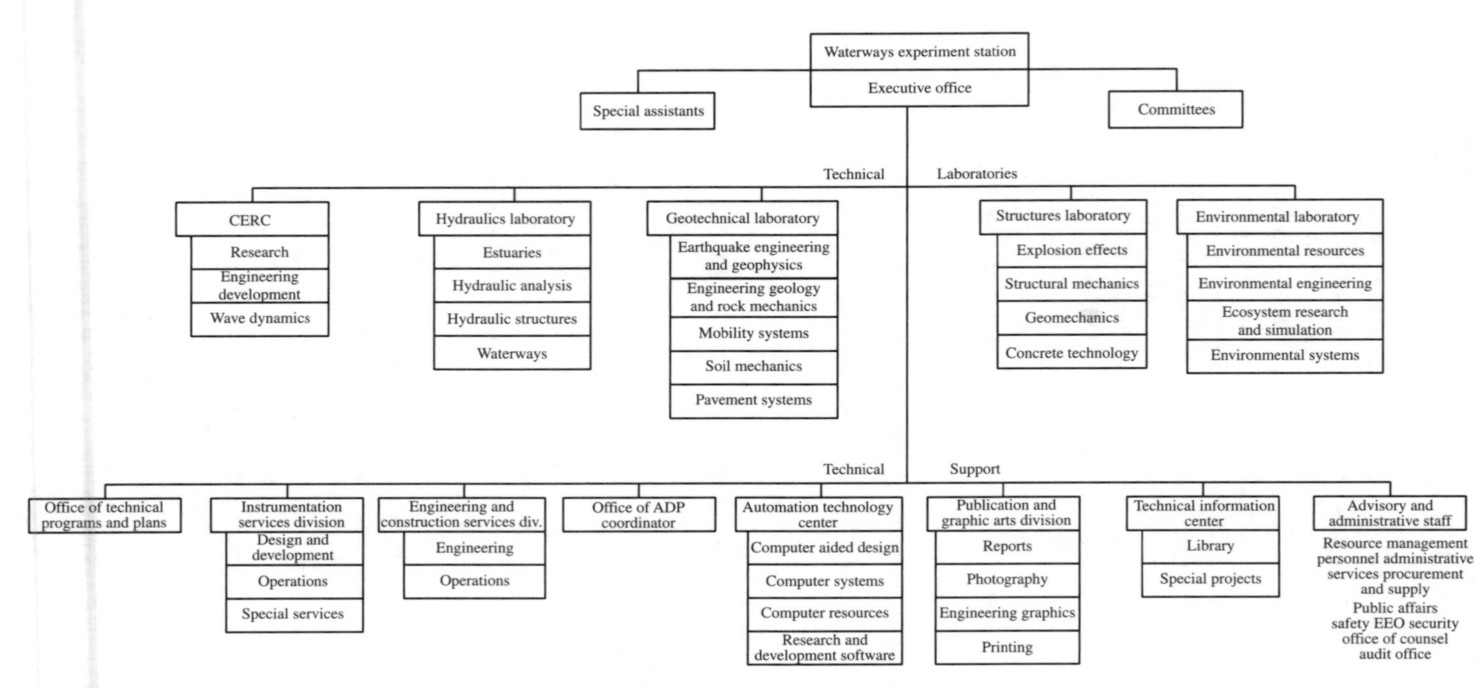

FIGURE 9.17 Organizational chart of the Corps of Engineers at the Waterways Experiment Station in Vicksburg, Mississippi. (Adapted from Public Affairs Office. No date. *Summary of Capabilities*. Waterways Experiment Station, U.S. Army Corps of Engineers, Vicksburg, MS.)

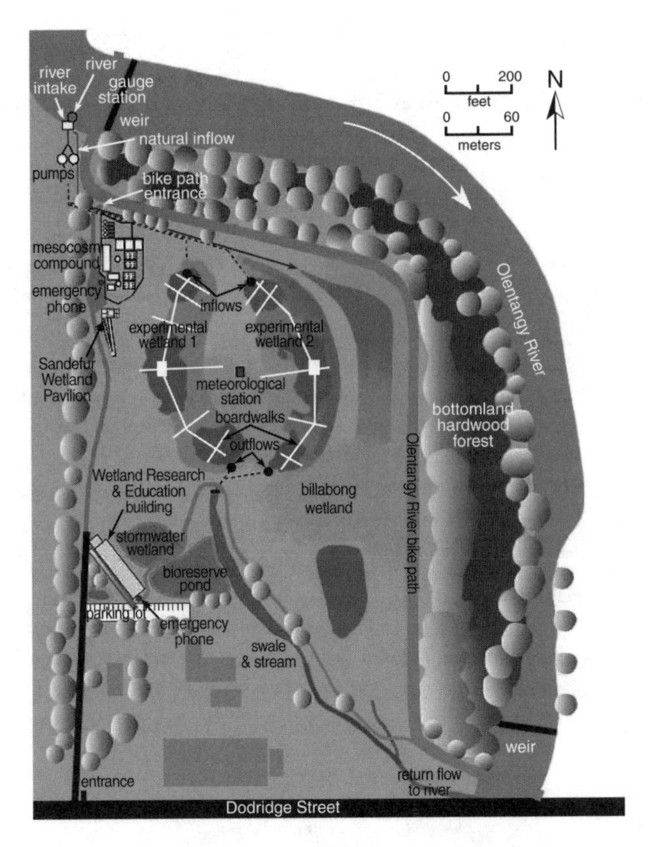

FIGURE 9.18 View of the Wetlands Field Lab at the Ohio State University. (From William Mitsch. Ohio State University, Columbus, OH. With permission.)

sity campus in Columbus. This is a 307-acre (123-ha) outdoor lab composed of a set of natural and constructed wetlands that was initiated in 1994 by William Mitsch. The purpose of the lab is to foster research and education on wetland ecology and engineering. Development of the lab is in progress, and it will ultimately include a number of features (Figure 9.18). Presently, the site is dominated by two 2.5 acre (1 ha) kidney-shaped constructed marshes that are supported by a flow-through pumping of the adjacent Olentangy River. Water is pumped continuously into the two wetlands and then flows by gravity back into the river. An experiment in planting strategies is being undertaken with one wetland receiving an intentional planting effort of known species while the other wetland received no intentional plantings but is self-designing a plant community from volunteer species. Preliminary results demonstrate many similarities between the wetlands (Mitsch et al., 1998), and the plan is to continue this study indefinitely with contributions by OSU students and faculty. The site also includes a number of replicated wetland mesocosms for small-scale research and several other experimental wetland cells. A variety of educational uses are made of the site, which is facilitated by its location on the OSU campus (Mitsch, 1998c). The strengths of the Olentangy River Wetland Research Park are

TABLE 9.4
A Code of Ethics for Ecological Engineering Based on the Hacker Code of Ethics

Hacker's Ethic	Analogous Ecological Engineering Ethic
Access to computers – and anything which might teach you something about the way the world works – should be unlimited and total. Always yield to the Hands-On Imperative!	Access to ecosystems – and anything which might teach you something about the way the world works – should be unlimited and total. Always yield to the Hands-On Imperative!
All information should be free.	All species should be free and be able to be used in ecological engineering design (including invasive exotics).
Mistrust Authority – Promote Decentralization.	Mistrust Environmental Regulators who give permits and Funding Agencies who give research grants — Promote Self-Organization.
Hackers should be judged by their hacking, not bogus criteria such as degrees, age, race, or position.	Ecological engineers should be judged by their ability to create ecosystems, not bogus criteria such as a degree in engineering, passing the P.E. exam, or ABET certification.
You can create art and beauty on a computer.	You can create art and beauty with a constructed ecosystem.
Computers can change your life for the better.	Ecosystems can change your life for the better.

Note: The ethics listed on the left are direct quotes from Chapter 2 of Levy's book while the ethics listed on the right are the analogous equivalents for ecological engineering.

Source: Adapted from Levy, S. 1984. *Hackers: Heroes of the Computer Revolution*. Dell, New York.

the variety of experimental settings available and the hands-on research experiences made available to students. A wetland research and education building on the site has recently opened which will enhance the utility of the park.

The ideal ecological engineering lab of the future would combine the strengths of the facilities described above. Emphasis would be on the design and construction of experimental living machines and other practical systems. An association with commercial companies would be desirable to keep research practical. Symbiotic relationships should be tried in which research is conducted for companies by faculty and students, and the companies provide equipment, guest lectures, and grants to the lab. The best learning situation is in which students build experimental systems themselves; perhaps a new generation of "hackers" might be envisioned in ecological engineering design that would be analogous to the young people who learned to program computers in the 1960s. Levy (1984) describes the early hackers as "heroes of the computer revolution." Through their passion for programming the old mainframe computers, these young people are credited with developing the software and hardware of personal computers and even with initiating the Internet. Perhaps a new generation of ecological engineering hackers is needed — young people beginning to program living machines to develop new appropriate technologies. Levy recorded the original hacker ethics which guided their productive work, and Table 9.4 offers a possible corresponding ethic for ecological engineering hackers. Providing an educational environment that facilitates this kind of spirited learning will accelerate the development of the discipline. To some extent this is already happening in academic departments of a few universities where experiments in ecological engineering education are taking place.

References

Abbe, T. B. and D. R. Montgomery. 1996. Large woody debris jams, channel hydraulics and habitat formation in large rivers. *Regulated Rivers: Research and Management.* 12:201–221.

Abbey, E. 1975. *The Monkey Wrench Gang*. Avon Books, New York.

Abbott, W. 1966. Microcosm studies on estuarine waters. I. The replicability of microcosms. *Journal of the Water Pollution Control Federation.* 38:258–270.

Abrams, P. A. 1996. Evolution and the consequences of species introduction and deletions. *Ecology.* 77:1321–1328.

Abrams, P. and C. J. Walters. 1996. Invulnerable prey and the paradox of enrichment. *Ecology.* 77:1125–1133.

Abrams, P., B. A. Menge, G. G. Mittelbach, D. Spiller, and P. Yodzis. 1996. The role of indirect effects in food webs. pp. 371–395. In: *Food Webs: Integration of Patterns and Dynamics*. G. A. Polis and K. O. Winemiller (eds.). Chapman & Hall, New York.

Adams, J. L. 1991. *Flying Buttresses, Entropy, and o-Rings: The World of an Engineer.* Harvard University Press, Cambridge, MA.

Adams, L. W. 1994. *Urban Wildlife Habitats*. University of Minnesota Press, Minneapolis, MN.

Adamus, P. R. 1988. Cmriteria for created or restored wetlands. pp. 369–372. In: *The Ecology and Management of Wetlands*, Vol. 2. D. D. Hook et al. (eds.). Timber Press, Portland, OR.

Adey, W. H. 1973. Temperature control of reproduction and productivity in a subarctic coralline algae *Phycologia.* 12:111–118.

Adey, W. H. 1978. Algal ridges of the Caribbean Sea and West Indies. *Phycologia.* 17:361–367.

Adey, W. 1983. The microcosm: a new tool for reef research. *Coral Reefs.* 1:193–201.

Adey, W. H. 1987. Marine microcosms. pp. 133–149. In: *Restoration Ecology.* W. R. Jordan, III, M. E. Gilpin, and J. D. Aber (eds.). Cambridge University Press, Cambridge, U.K.

Adey, W. H. 1995. Controlled ecologies. pp. 413–423. In: *Encyclopedia of Environmental Biology,* Vol. 1. W. A. Nierenberg (ed.). Academic Press, San Diego, CA.

Adey, W. H. and R. Burke. 1976. Holocene bioherms (algal ridges and bank barrier reefs) of the eastern Caribbean. *Bulletin of the American Geological Society.* 87:95–109.

Adey, W. H. and T. Goertemiller. 1987. Coral reef algal turfs: master producers in nutrient poor seas. *Phycologia.* 26:374–386.

Adey, W. H. and J. M. Hackney. 1989. The composition and production of tropical marine algal turf in laboratory and field experiments. pp. 1–106. In: *The Biology, Ecology, and Mariculture of* Mithrax spinosissimus *Utilizing Cultural Algal Turfs*. W. H. Adey (ed.). The Mariculture Institute, Washington, DC.

Adey, W. H. and K. Loveland. 1991. *Dynamic Aquaria*, 1st ed. Academic Press, San Diego, CA.

Adey, W. H. and K. Loveland. 1998. *Dynamic Aquaria*, 2nd ed. Academic Press, San Diego, CA.

Adey, W. H. and R. S. Steneck. 1985. Synergistic effects of light, wave action and geology on coral reef primary production. pp. 163–187. In: *The Ecology of Coral Reefs*. M. L. Reaka (ed.). *NOAA Symp. Ser. Undersea Res.* Vol. 3, No. 1. Washington, DC.

Adey, W. H., M. Finn, P. Kangas, L. Lange, C. Luckett, and D. M. Spoon. 1996. A Florida Everglades mesocosm — model veracity after four years of self-organization. *Ecological Engineering.* 6:171–224.

Adey, W. H., C. Luckett, and K. Jensen. 1993. Phosphorus removal from natural waters using controlled algal production. *Restoration Ecology.* 1:29–39.

Adey, W. H., C. Luckett, and M. Smith. 1996. Purification of industrially contaminated groundwaters using controlled ecosystems. *Ecological Engineering.* 7:191–212.

Adler, R. G. 1984. Biotechnology as an intellectual property. *Science.* 224:357–363.

Agee, J. K. and D. R. Johnson (eds.). 1988. *Ecosystem Management for Parks and Wilderness.* University of Washington Press, Seattle, WA.

Ahn, C. and W. J. Mitsch. 2002. Scaling considerations of mesocosm wetlands in simulating large created freshwater marshes. *Ecological Engineering.* 18:327–342.

Alexander, M. 1965. Biodegradation: problems of molecular recalcitrance and microbial fallibility. pp. 35–80. In: *Advances in Applied Microbiology.* Academic Press, New York.

Alexander, M. 1973. Microorganisms and chemical pollution. *BioScience.* 23:509–515.

Alexander, M. 1981. Biodegradation of chemicals of environmental concern. *Science.* 211:132–138.

Alexander, M. 1994. *Biodegradation and Bioremediation.* Academic Press, San Diego, CA.

Alford, R. A. and H. M. Wilbur. 1985. Priority effects in experimental pond communities: competition between Bufo and Rana. *Ecology.* 66:1097–1105.

Allaby, M. and J. Lovelock. 1984. *The Greening of Mars.* Warner Books, New York.

Allee, W. C., A. E. Emerson, O. Park, and T. Park. 1949. *Principles of Animal Ecology.* W. B. Saunders, Philadelphia, PA.

Allen, A. W. and R. D. Hoffman. 1984. Habitat suitability index models: muskrat. FWS/OBS-82/10.46, U.S. Department of the Interior, Washington, DC.

Allen, D. T. and N. Behmanesh. 1994. Wastes as raw materials. pp. 69–89. In: *The Greening of Industrial Ecosystems.* B. R. Allenby and D. J. Richards (eds.). National Academy Press, Washington, DC.

Allen, G. H. and R. L. Carpenter. 1977. The cultivation of fish with emphasis on salmonids in municipal wastewater lagoons as an available protein source for human beings. pp. 479–527. In: *Wastewater Renovation and Reuse.* F. M. D'Itri (ed.). Marcel Dekker, New York.

Allen, H. H. and J. W. Webb. 1993. Bioengineering methods to establish saltmarsh on dredged material. pp. 118–132. In: *Coastlines of the Gulf of Mexico.* S. Laska and A. Puffer (eds.). American Society of Civil Engineers, New York.

Allen, J. 1973. Sewage farming. *Environment.* 15(3):36–41.

Allen, J. 1991. *The Human Experiment.* Penguin Books, New York.

Allen, L. A. 1944. The bacteriology of activated sludge. *Journal of Hygiene.* 43:424–431.

Allen, T. F. H. 1977. Scale in microscopic algal ecology: a neglected dimension. *Phycologia.* 16:253–258.

Allen, T. F. H. 1998. Community ecology. pp. 315–383. In: *Ecology.* Oxford University Press. New York.

Allen, T. F. H. and T. W. Hoekstra. 1992. *Toward a Unified Ecology.* Columbia University Press, New York.

Allenby, B. R. and D. J. Richards (eds.). 1994. *The Greening of Industrial Ecosystems.* National Academy Press, Washington, DC.

Aller, J. Y., S. A. Woodin, and R. C. Aller. (eds.). 2001. *Organism-Sediment Interactions.* University of South Carolina Press, Columbia, SC.

Alling, A. and M. Nelson. 1993. *Life Under Glass: The Inside Story of Biosphere 2.* The Biosphere Press, Oracle, AZ.

Alling, A., M. Nelson, L. Leigh, T. MacCallum, N. Alvarez-Romo, and J. Allen. 1993. Experiments on the closed ecological system in the Biosphere 2 test module. pp. 463–479. Appendix C. In: *Ecological Microcosms.* R. J. Beyers and H. T. Odum (eds.). Springer-Verlag, New York.

Allred, M. 1986. *Beaver Behavior.* Naturegraph Publishers, Happy Camp, CA.

Allsopp, D., R. R. Colwell, and D. L. Hawksworth (eds.). 1993. *Microbial Diversity and Ecosystem Function.* CAB International, Tucson, AZ.

Alper, J. 1998. Ecosystem engineers shape habitats for other species. *Science.* 280:1195–1196.

Altieri, M. A., D. K. Letourneau, and J. R. Davis. 1983. Developing sustainable agroecosystems. *BioScience.* 33:45–49.

Altieri, M. A. 1994. *Biodiversity and Pest Management in Agroecosystems.* Food Products Press, New York.

Altman, P. L. and D. S. Dittmer (eds.). 1966. *Environmental Biology.* Federation of American Societies for Experimental Biology, Bethesda, MD.

American Society for Testing and Materials (ASTM). 1998. Standard Guide for Assessment of Wetland Functions. ASTM designation: E 1983–98. ASTM, West Conshohocken, PA.

Amundson, R. and H. Jenny. 1997. On a state factor model of ecosystems. *BioScience.* 47:536–543.

Andersen, A. N. and G. P. Sparling. 1997. Ants as indicators of restoration success: relationship with soil microbial biomass in the Australian seasonal tropics. *Restoration Ecology.* 5:109–114.

Anderson, D. L., R. L. Siegrist, and R. J. Otis. 1985. Technology Assessment of Intermittent Sand Filters. Office of Municipal Pollution Control (WH-546), U.S. Environmental Protection Agency, Washington, DC.

Anderson, J. M. 1975. Succession, diversity and trophic relationships of some soil animals on decomposing leaf litter. *Journal of Animal* Ecology. 44:475–495.

Anderson, J. M. 1995. Soil organisms as engineers: microsite modulation of macroscale processes. pp. 94–106. In: *Linking Species and Ecosystems.* C. G. Jones and J. H. Lawton (eds.). Chapman & Hall, New York.

Anderson, J. M. and A. Macfadyen (eds.). 1976. *The Role of Terrestrial and Aquatic Organisms in Decomposition Processes.* Blackwell Science, Oxford, U.K.

Anderson, N. H. and J. R. Sedell. 1979. Detritus processing by macroinvertebrates in stream ecosystems. *Annual Review of Entomology.* 24:351–377.

Anderson, R. C. 1982. An evolutionary model summarizing the roles of fire, climate, and grazing animals in the origin and maintenance of grasslands: an end paper. pp. 297–308. In: *Grasses and Grasslands.* J. R. Estes, R. J. Tyrl, and J. N. Brunken (eds.). University of Oklahoma Press, Norman, OK.

Andrewartha, H. G. and L. C. Birch. 1954. *The Distribution and Abundance of Animals.* University of Chicago Press, Chicago, IL.

Andrus, C. W., B. A. Long, and H. A. Froehlich. 1988. Woody debris and its contribution to pool formation in a coastal stream 50 years after logging. *Canadian Journal of the Fisheries and Aquatic* Sciences. 45:2080–2086.

Angel, K. and D. T. Wicklow. 1974. Decomposition of rabbit faeces: an indication of the significance of the coprophilous microflora in energy flow schemes. *Journal of Ecology.* 62:429–437.

Angermeier, P. L. 1994. Does biodiversity include artificial diversity? *Conservation Biology.* 8:600–602.

Angrist, S. W. 1973. *Closing the Loop: The Story of Feedback.* Thomas Y. Crowell, New York.

Anonymous. no date. Summary of Capabilities. U. S. Army Corps of Engineers Waterways Experiment Station, Vicksburg, MS.

Anonymous. 1975. The Illinois Natural History Survey Reports 152.

Anonymous. 1991. *The Biocycle Guide to the Art & Science of Composting.* The JG Press, Inc., Emmaus, PA.

Anonymous. 1995a. Mosquitos in constructed wetlands — a management bugaboo? *Watershed Protection Techniques.* 1:203–207.

Anonymous. 1995b. Coconut rolls as a technique for natural streambank stabilization. *Watershed Protection Techniques.* 1:176–178.

Anonymous. 1997. General Information Concerning Patents. U.S. Patent and Trademark Office, U.S. Government Printing Office, Washington, DC.

Anonymous. 2001. A wireless system saves water. *Resource.* 8(5):4.

Antonovics, J., A. D. Bradshaw, and R. G. Turner. 1971. Heavy metal tolerance in plants. *Advances in Ecological Research.* 7:2–86.

Appelhof, M. 1997. *Worms Eat My Garbage.* Flower Press, Kalamazoo, MI.

Appenzeller, T. 1994. Biosphere 2 makes a new bid for scientific credibility. *Science.* 263:1368–1369.

Apted, M. J., D. Langmuir, D. W. Moeller, and J. Ahn. 2002. Yucca Mountain: should we delay? *Science.* 296:2333–2335.

Aquino, J. T. (ed.). 1995. *Waste Age/Recycling Times' Recycling Handbook.* Lewis Publishers, Boca Raton, FL.

Armenante, P. M. 1993. Bioreactors. pp. 65–112. In: *Biotreatment of Industrial and Hazardous Waste.* M. A. Levin and M. A. Gealt (eds.). McGraw-Hill, New York.

Armstrong, N. E. and H. T. Odum. 1964. Photoelectric ecosystem. *Science.* 143:256–258.

Aronson, J., S. Dhillion, and E. Le Floc'h. 1994. On the need to select an ecosystem of reference, however imperfect: a reply to Pickett and Parker. *Restoration Ecology.* 3:1–3.

Arrhenius, O. 1921. Species and area. *Journal of Ecology.* 9:95–99.

Ashton, G. D. 1979. River ice. *American Scientist.* 67:38–45.

Asimov, I. 1950. *I, Robot.* Doubleday, Garden City, NY.

ASTM. 1996. Standard guide for ventilatory behavioral toxicology testing of freshwater fish. E 1768-95. *Annual Book of ASTM Standards.* American Society for Testing and Materials, West Conshohocken, PA.

Astro, R. 1973. *John Steinbeck and Edward F. Ricketts: The Shaping of a Novelist.* University of Minnesota Press, New Berlin, MN.

Atlas, R. M. 1993. Bioaugmentation to enhance microbial bioremediation. pp. 19–37. In: *Biotreatment of Industrial and Hazardous Waste.* M. A. Levin and M. A. Gealt (eds.). McGraw-Hill, New York.

Atwood, G. 1975. The strip-mining of Western coal. *Scientific American.* 233(6):23–29.

Augustine, D. J., L. E. Frelich, and P. A. Jordan. 1998. Evidence for two alternative stable states in an ungulate grazing system. *Ecological Applications.* 8:1260–1269.

Ausubel, J. H. 1996. Can technology spare the earth? *American Scientist.* 84:166–178.

Avise, J. C. 1994. The real message from Biosphere 2. *Conservation Biology.* 8:327–329.

Axelson, G., M. Morris, and A. Gettman. 2000. Righting an ecological wrong: saltmarsh restoration on the North Atlantic coast. *Land and Water.* 44(4):10–14.

Azam, F., T. Fenchel, J. G. Field, J. G. Gray, L. A. Meyer-Reil, and F. Thingstad. 1983. The ecological role of water-column microbes in the sea. *Marine Ecology Progress Series* 10:257–263.

Ayres, Q. C. 1936. *Soil Erosion and Its Control.* McGraw-Hill, New York.

Bada, J. L. and A. Lazcano. 2003. Prebiotic soup – revisiting the Miller experiment. Science. 300:745–746.

Bagchi, A. 1990. *Design, Construction, and Monitoring of Sanitary Landfill.* John Wiley & Sons, New York.

Bak, P. 1996. *How Nature Works: The Science of Self-Organized Criticality.* Springer-Verlag, New York.

Baker, H. G. 1965. Characteristics and modes of origin of weeds. pp. 147–172. In: *The Genetics of Colonizing Species.* H. G. Baker and G. L. Stebbins (eds.). Academic Press, New York.

Baker, H. G. 1974. The evolution of weeds. *Annual Reviews of Ecology and Systematics.* 5:1–24.

Baker, H. G. 1986. Patterns of plant invasion in North America. pp. 44–57. In: *Ecology of Biological Invasions of North America and Hawaii.* H. A. Mooney and J. A. Drake (eds.). Springer-Verlag, New York.

Baldwin, A. D., Jr., J. de Luce, and C. Pletsch (eds.). 1994. *Beyond Preservation: Restoring and Inventing Landscapes*. University of Minnesota Press, Minneapolis, MN.

Baldwin, J. 1986. Born to fail. pp. 263–266. In: *Ten Years of CoEvolution Quarterly: News That Stayed News*. A. Kleiner and S. Brand (eds.). North Point Press, San Francisco, CA.

Baldwin, J. 1996. *Bucky Works: Buckminster Fuller's Ideas for Today.* John Wiley & Sons, New York.

Baldwin, J. 1997. On tools. pp. 122–125. In: *Design Outlaws on the Ecological Frontier.* C. Zelov and P. Cousineau (eds.). Knossus Publishing, Philadelphia, PA.

Balmford, A. et al. 2002. Economic reasons for conserving wild Nature. *Science.* 297:950–953.

Bang, M. No date. *Chattanooga Sludge*. Harcourt Brace, San Diego, CA.

Barash, D. P. 1973. The ecologist as zen master. *American Midland Naturalist.* 89:214–217.

Bardach, J. E., J. H. Ryther, and W. O. McLarney. 1972. *Aquaculture*. John Wiley & Sons, New York.

Barker, A. N. 1946. The ecology and function of protozoa in sewage purification. *Annals of Applied* Biology. 33:314–325.

Barrett, T. J. 1947. *Harnessing the Earthworm.* Bruce Humphries Publishing, Boston, MA.

Bartsch, A. F. 1948. Biological aspects of stream pollution. *Sewage Works Journal.* 20:292–302.

Bartsch, A. F. 1961. Algae as a source of oxygen in waste treatment. *Journal of the Water Pollution Control Federation.* 33:239–249.

Bartsch, A. F. 1970. Water pollution — an ecological perspective. *Journal of the Water Pollution Control Federation.* 42:819–823.

Bartsch, A. F. 1971. Accelerated eutrophication of lakes in the U.S. *Environmental Pollution.* 1:133–140.

Bartsch, A. F. and M. O. Allum. 1957. Biological factors in treatment of raw sewage in artificial ponds. *Limnology and Oceanography.* 2:77–84.

Bascom, W. 1964. *Waves and Beaches*. Anchor Books, Garden City, NY.

Basmajian, J. V. (ed.). 1979. *Biofeedback: Principles and Practice for Clinicians*. Williams & Wilkins, Baltimore, MD.

Bastian, R. K. (ed.). 1993. *Constructed Wetlands for Wastewater Treatment and Wildlife Habitat*. EPA832-R-93-005. U.S. Environmental Protection Agency, Washington, DC.

Bastian, R. K. and D. A. Hammer. 1993. The use of constructed wetlands for wastewater treatment and recycling. pp. 59–68. In: *Constructed Wetlands for Water Quality Improvement*. G. A. Moshiri (ed.). Lewis Publishers, Boca Raton, FL.

Bada, J. L. and A. Lazcano. 2003. Prebiotic soup – revisiting the Miller experiment. *Science.* 300:745–746.

Bates, M. 1961. *Man in Nature*. Prentice Hall, Englewood Cliffs, NJ.

Bates, R. P. and J. F. Hentges, Jr. 1976. Aquatic weeds — eradicate or cultivate? *Economic Botany.* 30:39–50.

Bateson, G. 1979. *Mind and Nature: A Necessary Unity.* Bantam Books, Toronto, Canada.

Batra, S. W. T. 1982. Biological control in agroecosystems. *Science.* 215:134–139.

Bauer, M. 1995. "Technophobia": a misleading conception of resistance to new technology. pp. 97–122. In: *Resistance to New Technology.* M. Bauer (ed.). Cambridge University Press, Cambridge, U.K.

Bazzaz, F. A. 1986. Life history of colonizing plants: some demographic, genetic, and physiological features. pp. 96–110. In: *Ecology of Biological Invasions of North America and Hawaii*. H. A. Mooney and J. A. Drake (eds.). Springer-Verlag, New York.

Beakley, G. C. 1984. *Careers in Engineering and Technology*. Macmillan, New York.

Beck, R. E. 1994. The movement in the United States to restoration and creation of wetlands. *Natural Resources Journal*. 34:781–822.

Beeby, A. 1993. *Applying Ecology*. Chapman & Hall, London.

Beeby, A. and A.-M. Brennan. 1997. *First Ecology*. Chapman & Hall, London.

Beecher, W. 1942. *Nesting birds and the vegetation substrate*. Chicago Ornithology Society. Chicago, IL.

Beeton, A. M. and W. T. Edmondson. 1972. The eutrophication problem. *Journal of Fisheries Research Broad of Canada*. 29:673–682.

Belant J. L., T. W. Seamans, S. W. Gabrey, and R. A. Dolbeer. 1995. Abundance of gulls and other birds at landfills in northern Ohio. *American Midland Naturalist*. 134:30–40.

Beltaos, S. (ed.). 1995. *River Ice Jams*. Water Resources Publications, Highlands Ranch, CO.

Beltrami, E. J. 1989. A mathematical model of the brown tide. *Estuaries*. 12:13–17.

Beltrami, E. J. 1990. Brown tide dynamics as a catastrophe model. pp. 307–315. In: *Novel Phytoplankton Blooms*. E. M. Cosper, V. M. Bricelj, and E. J. Carpenter (eds.). Springer-Verlag, Berlin.

Bem, R. 1978. *Everyone's Guide to Home Composting*. Van Nostrand Reinhold, New York.

Ben-Ari, E. T. 1999. Not just slime. *BioScience*. 49:689–695.

Ben-Ari, E. T. 2001. What's new at the zoo? *BioScience*. 51:172–177.

Bender, M. 1995. The prairie as a model for the Sunshine Farm. pp. 199–205. In: *Fourteenth North American Prairie Conference*. D. C. Hartnett (ed.). Kansas State University, Manhattan, KS.

Bendoricchio, G. 2000. Continuity and discontinuity in ecological systems. pp. 395–409. In: *Handbook of Ecosystem Theories and Management*. S. E. Jorgensen and F. Muller (eds.). Lewis Publishers, Boca Raton, FL.

Bennett, H. H. and W. C. Lowdermilk. 1938. General aspects of the soil-erosion problem. pp. 581–608. In: *Soils and Men: Yearbook of Agriculture*. U.S. Department of Agriculture, Washington, DC.

Benyus, J. M. 1997. *Biomimicry*. William Morrow, New York.

Berg, K. M. and P. C. Kangas. 1989. Effects of muskrat mounds on decomposition in a wetland ecosystem. pp. 145–151. In: *Freshwater Wetlands and Wildlife*. R. R. Sharitz and J. W. Gibbons (eds.). DOE Symposium Series N. 61, USDOE Office of Scientific and Technical Information, Oak Ridge, TN.

Berger, J. J. 1985. *Restoring the Earth*. Knopf, New York.

Berner, T. 1990. Coral-reef algae. pp. 253–264. In: *Coral Reefs, Ecosystems of the World*, Vol. 25. Z. Dubinsky (ed.). Elsevier, Amsterdam, the Netherlands.

Bernstein, B. B. 1981. Ecology and economics: complex systems in changing environments. *Annual Reviews of Ecology and Systematics*. 12:309–330.

Berry, J. F. and M. S. Dennison. 1993. Wetland mitigation. pp. 278–303. In: *Wetlands, Guide to Science, Law, and Technology*. M. S. Dennison and J. F. Berry (eds.). Noyes Publications, Park Ridge, NJ.

Berryman, A. A. 1992. The origins and evolution of predator-prey theory. *Ecology*. 73:1530–1535.

Berryman, A. A., M. A. Valenti, M. J. Harris, and D. C. Fulton. 1992. Ecological engineering — an idea whose time has come? *Trends in Ecology and Evolution*, 7:268–270.

Beven, K. and P. Germann. 1982. Macropores and water flow in soils. *Water Resource Research*. 18:1311–1325.

Beyers, R. J. 1963. Balanced aquatic microcosms — their implications for space travel. *American Biology Teacher.* 25:422–428.

Beyers, R. J. 1963. The metabolism of twelve aquatic laboratory microecosystems. *Ecological Monographs.* 33:255–306.

Beyers, R. J. 1963. A characteristic diurnal metabolic pattern in balanced microcosms. *Publ. Inst. Mar. Sci.* Texas. 9:19–27.

Beyers, R. J. 1964. The microcosm approach to ecosystem biology. *American Biology Teacher.* 26:491–498.

Beyers, R. J. 1974. Report of Savannah River Laboratory, University of Georgia, Aiken, SC. (Seen in Odum, 1983).

Beyers, R. J. and H. T. Odum. 1993. *Ecological Microcosms.* Springer-Verlag, New York.

Bhatla, M. N. and A. F. Gaudy, Jr. 1965. Role of protozoa in the diphasic exertion of BOD. *Journal of Sanitary Engineering Division*, ASCE. 91(3):63–87.

Bick, H. and H. P. Muller. 1973. Population dynamics of bacteria and protozoa associated with the decay of organic matter. *Bulletin of Ecology Research* Committee (*Stockholm*). 17:379–386.

Biermann, E., C. Streb, B. Jessup, P. May, J. Schaafsma, and P. Kangas. 1999. The development of an ecological engineering design seminar. *Annals of Earth.* 17(1):17–19.

Bilby, R. E. and G. E. Likens. 1980. Importance of organic debris dams in the structure and function of stream ecosystems. *Ecology* 61:1107–1113.

Bird, E. C. F. 1996. *Beach Management.* John Wiley & Sons, Chichester, U.K.

Bitter, S. D. and J. K. Bowers. 1994. Bioretention as a water quality best management practice. *Watershed Protection Techniques.* 1:114–116.

Blank, J. L., R. K. Olson, and P. M. Vitousek. 1980. Nutrient uptake by a diverse spring ephemeral community. *Oecologia.* 47:96–98.

Blankenship, K. 1996. Invasion by sea. *Bay Journal* (Alliance for the Chesapeake Bay). 6(2):1, 8–9.

Blankenship, K. 1997. Algae crops up in nutrient removal system. *Bay Journal* (Alliance for the Chesapeake Bay). 7(8):1, 6.

Blindow, I., G. Andersson, A. Hargeby, and S. Johansson. 1993. Long-term pattern of alternative stable states in two shallow eutrophic lakes. *Freshwater Biology.* 30:159–167.

Bloesch, J., P. Bossard, H. Buhrer, H. R. Burgi, and U. Uehlinger. 1988. Can results from limnocorral experiments be transferred to *in situ* conditions? *Hydrobiologia.* 159:297–308.

Blondel, J. and J. Aronson. 1995. Biodiversity and ecosystem function in the Mediterranean Basin: human and non-human determinants. pp. 43–120. In: *Mediterranean-Type Ecosystems: The Function of Biodiversity.* Springer, New York.

Bloom, A. L. 1969. *The Surface of the Earth.* Prentice Hall, Englewood Cliffs, NJ.

Bloom, A. J., F. S. Chapin, III, and H. A. Mooney. 1985. Resource limitation in plants — an economic analogy. *Annual Reviews of Ecology and Systematics.* 16:363–392.

Bohn, H. L. 1972. Soil adsorption of air pollutants. *Journal of Environmental Quality.* 1:372–377.

Bohnsack, J. A. 1991. Habitat structure and the design of artificial reefs. pp. 412–426. In: *Habitat Structure.* S. S. Bell, E. D. McCoy, and H. R. Mushinsky (eds.). Chapman & Hall, London.

Bohnsack, J. A., D. L. Johnson, and R. F. Ambrose. 1991. Ecology of artificial reef habitats and fishes. pp. 61–107. In: *Artificial Habitats for Marine and Freshwater Fisheries.* W. Seaman, Jr. and L. M. Sprague (eds.). Academic Press, San Diego, CA.

Boje, R. and M. Tomczak (eds.). 1978. *Upwelling Ecosystems.* Springer-Verlag, Berlin.

Boling, R. H., Jr., E. D. Goodman, J. A. Van Sickle, J. O. Zimmer, K. W. Cummins, R. C. Petersen, and S. R. Reice. 1975. Toward a model of detritus processing in a woodland stream. *Ecology.* 56:141–151.

Bonabeau, E., M. Dorigo, and G. Theraulaz. 1999. *Swarm Intelligence: From Natural to Artificial Systems.* Oxford University Press, New York.

Boon, P. J. 2000. Bacterial biodiversity in wetlands. pp. 281–310. In: *Biodiversity in Wetlands: Assessment, Function and Conservation*, Vol. 1. Backhuys Publishers, Leiden, the Netherlands.

Booth, T. 1977. Muskox dung: its turnover rate and possible role on Truelove Lowland. pp. 531–545. In: *Truelove Lowland, Devon Island, Canada: A High Arctic Ecosystem.* L. C. Bliss (ed.). University of Alberta Press, Edmonton, Canada.

Booth, W. 1988. Johnny Appleseed and the greenhouse. *Science.* 242:19–20.

Bormann, F. H. and G. E. Likens. 1967. Nutrient cycling. *Science.* 155:424–429.

Bormann, F. H. and G. E. Likens. 1979. *Pattern and Process in a Forested Ecosystem.* Springer-Verlag, New York.

Bormann, F. H., D. Balmori, and G. T. Geballe. 1993. *Redesigning the American Lawn: A Search for Environmental Harmony.* Yale University Press, New Haven, CT.

Botkin, D. B., S. Golubic, B. Maguire, B. Moore, H. J. Morowitz, and L. B. Slobodkin. 1979. pp. 3–12. In: *Life Sciences and Space Research,* Vol. XVII. R. Holmquist (ed.). Pergamon Press, Oxford, U.K.

Boulding, K. E. 1966. The economics of the coming spaceship earth. pp. 3–14. In: *Environmental Quality in a Growing Economy.* H. Jarrett (ed.). Johns Hopkins Press, Baltimore, MD.

Boulding, K. E. 1972. Economics as a not very biological science. pp. 357–375. In: *Challenging Biological Problems.* J. A. Behnke (ed.). Oxford University Press, New York.

Boulding, K. E. 1973. The economics of ecology. pp. 27–30. In: *Managing the Environment.* A. Neuschatz (ed.). EPA-600/5-73-010, U.S.E.P.A., Washington, DC.

Boulding, K. E. 1978. *Ecodynamics: A New Theory of Societal Evolution.* Sage Publications, Beverly Hills, CA.

Boulton, A. J. and P. I. Boon. 1991. A review of methodology used to measure leaf litter decomposition in lotic environments: time to turn over an old leaf? *Australian Journal of Marine Freshwater Research.* 42:1–43.

Bovbjerg, R. V. and P. W. Glynn. 1960. A class exercise on a marine microcosm. *Ecology.* 41:229–232.

Bowers, K. 1993. Bioengineering and the art of landscaping. *Land and Water.* 37(1):14–17.

Boyce, M. S. 1991. Natural regulation or the control of nature? pp. 183–208. In: *The Greater Yellowstone Ecosystem: Redefining America's Wilderness Heritage.* R. B. Keiter and M. S. Boyce (eds.). Yale University Press, New Haven, CT.

Boyce, M. S. and A. Haney (eds.). 1997. *Ecosystem Management.* Yale University Press, New Haven, CT.

Boyd, C. E. 1970. Vascular aquatic plants for mineral nutrient removal from polluted waters. *Economic Botany.* 24:95–103.

Boyd, C. E. and C. P. Goodyear. 1971. Nutritive quality of food in ecological systems. *Archives of Hydrobiology.* 69:256–270.

Boyt, F. L., S. E. Bayley, and J. Zoltek, Jr. 1977. Removal of nutrients from treated municipal wastewater by wetland vegetation. *Journal of the Water Pollution Control Federation.* 48:789–799.

Bradshaw, A. D. 1983. The reconstruction of ecosystems. *Journal of Applied Ecology.* 20:1–17.

Bradshaw, A. D. 1987a. The reclamation of derelict land and the ecology of ecosystems. pp. 53–74. In: *Restoration Ecology: A Synthetic Approach to Ecological Research.* W. R. Jordan III, M. E. Gilpin, and J. D. Aber (eds.). Cambridge University Press, Cambridge, U.K.

Bradshaw, A. D. 1987b. Restoration: an acid test for ecology. pp. 23–29. In: *Restoration Ecology. A Synthetic Approach to Ecological Research.* W. R. Jordan III, M. E. Gilpin, and J. D. Aber (eds.). Cambridge University Press, Cambridge, U.K.

Bradshaw, A. D. 1995. Alternative endpoints for reclamation. pp. 165–185. In: *Rehabilitating Damaged Ecosystems.* J. Cairns, Jr. (ed.). Lewis Publishers, Boca Raton, FL.

Bradshaw, A. D. 1997a. What do we mean by restoration? pp. 8–16. In: *Restoration Ecology and Sustainable Development.* K. M. Urbanska, N. R. Webb, and P. J. Edwards (eds.). Cambridge University Press, London.

Bradshaw, A. D. 1997b. The importance of soil ecology in restoration science. pp. 33–64. In: *Restoration Ecology and Sustainable Development.* K. M. Urbanska, N. R. Webb, and P. J. Edwards (eds.). Cambridge University Press, London.

Bradshaw, A. D., T. S. McNeilly, and R. P. G. Gregory. 1965. Industrialization, evolution and the development of heavy metal tolerance in plants. pp. 327–343. In: *Ecology and the Industry Society.* G. T. Goodman, R. W. Edwards, and J. M. Lambert (eds.). John Wiley & Sons, New York.

Brand, S. 1994. How Buildings Learn: *What Happens after They're Built.* Viking, New York.

Brand, S. 1997. Sitting at the counterculture. pp. 68–70. *Design Outlaws on the Ecological Frontier.* C. Zelov and P. Cousineau (eds.). Knossus Publishing, Philadelphia, PA.

Brandon, R. N. 1990. *Adaptation and Environment.* Princeton University Press, Princeton, NJ.

Branson, F. A. 1975. Natural and modified plant communities as related to runoff and sediment yields. pp. 157–172. In: *Coupling of Land and Water Systems.* A. D. Hasler (ed.). Springer-Verlag, New York.

Braun, E. L. 1916. The Physiographic Ecology of the Cincinnati Region. Bulletin #7, Ohio Biological Survey, Colombus, OH.

Bray, J. R. 1963. Root production and estimation of net productivity. *Canadian Journal of Botany.* 41:65–72.

Bray, J. R., D. B. Lawrence, and L. C. Pearson. 1959. Primary production in some Minnesota terrestrial communities of 1957. *Oikos.* 10:46–49.

Breaux, A. M. and J. W. Day. 1994. Policy considerations for wetland wastewater treatment in the coastal zone — a case study for Louisiana. *Coastal Management.* 22:285–307.

Breaux, A., S. Farber, and J. Day. 1995. Using natural coastal wetlands systems for wastewater treatment: an economic benefit analysis. *Journal of Environmental Management.* 44:285–291.

Brechignac, F. and R. MacElroy (eds.). 1997. *Life Sciences: Artificial Ecosystems.* The Committee on Space Research, Pergamon Press, Oxford, U.K.

Bright, C. 1998. *Life Out of Bounds: Bioinvasion in a Borderless World.* W. W. Norton, New York.

Brinson, M. M. and L. C. Lee. 1989. In-kind mitigation for wetland loss: statement of ecological issues and evaluation of examples. pp. 1069–1085. In: *Freshwater Wetlands and Wildlife.* R. R. Sharitz and J. W. Gibbons (eds.). CONF-8603101, DOE Symposium Series No. 61, USDOE Office of Scientific and Technical Information, Oak Ridge, TN.

Brinson, M. M. and R. Rheinhardt. 1996. The role of reference wetlands in functional assessment and mitigation. *Ecological Applications.* 6:69–76.

Brinson, M. M., H. D. Bradshaw, and E. S. Kane. 1984. Nutrient assimilative capacity of an alluvial floodplain swamp. *Journal of Applied Ecology.* 21:1041–1057.

Brock, M. A. and D. L. Britton. 1995. The role of seed banks in the revegetation of Australian temporary wetlands. pp. 183–188. In: *Restoration of Temperate Wetlands*. B. D. Wheeler, S. C. Shaw, W. J. Fojt, and R. A. Robertson (eds.). John Wiley & Sons, Chichester, U.K.

Brooks, R. R. 1972. *Geobotany and Biogeochemistry in Mineral Exploration*. Harper & Row, Publishers, New York.

Brooks, R. A. 1991. New approaches to robotics. *Science*. 253:1227–1232.

Brooks, R. A. 2002. *Flesh and Machines: How Robots Will Change Us*. Pantheon Books, New York.

Brooks, R. A. and A. M. Flynn. 1989. Fast, cheap and out of control: a robot invasion of the solar system. *Journal of the British Interplanetary Society*. 42:478–485.

Brooks, R.R. 1972. Geobotany and Biogeochemistry in Mineral Exploration. Harper & Row Publishers, New York.

Brooks, R. R. (ed.). 1998. *Plants That Hyperaccumulate Heavy Metals: Their Role in Phytoremediation, Microbiology, Archaeology, Mineral Exploration and Phytomining*. CAB International Publishing, New York.

Brooks, W. K. 1891. *The Oyster*. The Johns Hopkins University Press, Baltimore, MD.

Broome, S. W. 1990. Creation and restoration of tidal wetlands of the southeastern United States. pp. 37–72. In: *Wetland Creation and Restoration: The Status of the Science*. J. A. Kusler and M. E. Kentula (eds.). Island Press, Washington, DC.

Broome, S. W., E. D. Seneca, and W. W. Woodhouse, Jr. 1986. Long-term growth and development of transplants of the salt-marsh grass *Spartina alterniflora*. *Estuaries*. 9:63–74.

Broome, S. W., E. D. Seneca, and W. W. Woodhouse, Jr. 1988. Tidal saltmarsh restoration. *Aquatic Botany*. 32:1–22.

Brown, A. H. 1966. Regenerative systems. pp. 82–119. In: *Human Ecology in Space Flight*. D. H. Calloway (ed.). The New York Academy of Sciences, New York.

Brown, D. S. and S. C. Reed. 1994. Inventory of constructed wetlands in the United States. *Water Science Technology* 29:309–318.

Brown, E. E. 1980. *Fish Farming Handbook*. AVI Publishing Co., Inc., Westport, CT.

Brown, J. 1989. Patterns, modes and extents of invasions by vertebrates. pp. 85–109. In: *Biological Invasions: A Global Perspective*. J. A. Drake, H. A. Mooney, F. di Castri, R. H. Groves, F. J. Kruger, M. Rejmanek, and M. Williamson (eds.). John Wiley & Sons, Chichester, U.K.

Brown, J. H. 1971. Mammals on mountaintops: non-equilibrium insular biogeography. *American Naturalist*. 105:467–478.

Brown, J. S. 1994. Restoration ecology: living with the Prime Directive. pp. 355–380. In: *Restoration of Endangered Species*. M. L. Bowles and C. J. Whelan (eds.). Cambridge University Press, Cambridge, U.K.

Brown, K. S. 1995. The green clean. *BioScience*. 45:579–582.

Brown, L. R. 1984. The global loss of topsoil. *Journal of Soil Water Conservation*. 39:162–165.

Brown, M. T. and R. A. Herendeen. 1996. Embodied energy analysis and emergy analysis: a comparative view. *Ecological Economics*. 19:219–235.

Brown, M. T. and S. Ulgiati. 1997. Emergy-based indices and ratios to evaluate sustainability: monitoring economies and technology toward environmentally sound innovation. *Ecological Engineering*. 9:51–69.

Brown, M. T. and S. Ulgiati. 1999. Emergy evaluation of the biosphere and natural capital. *Ambio*. 28:486–493.

Brown, M. T., H. T. Odum, R. C. Murphy, R. A. Christianson, S. J. Doherty, T. R. McClanahan, and S. E. Tennenbaum. 1995. Rediscovery of the world: developing an interface of ecology and economics. pp. 216–250. In: *Maximum Power: The Ideas and Applications of H. T. Odum*. C. A. S. Hall (ed.). University Press of Colorado, Niwot, CO.

Brown, S., M. M. Brinson, and A. E. Lugo. 1979. Structure and function of riparian wetlands. pp. 17–31. In: Strategies for Protection and Management of Floodplain Wetlands and Other Riparian Ecosystems. General Technical Report WO-12, USDA, Forest Service, Washington, DC.

Brussaard, L. et al. 1997. Biodiversity and ecosystem functioning in soil. *Ambio*. 26:563–570.

Bryan, F. R. 1990. *Beyond the Model T: The Other Ventures of Henry Ford*. Wayne State University Press, Detroit, MI.

Bucciarelli, L. L. 1994. *Designing Engineers*. MIT Press, Cambridge, MA.

Buckley, G. P. 1989. *Biological Habitat Reconstruction*. Belhaven Press, London.

Buddhavarapu, L. R. and S. J. Hancock. 1991. Advanced treatment for lagoons using duckweed. *Water, Environment, and Technology* 3(3):41–44.

Bullock, J. M., R. E. Kenward, and R. S. Hails. (eds.). 2002. *Dispersal Ecology*. Blackwell Publishing, Malden, MA.

Burgess, R. L. 1997. Resolution of respect, Frank Edwin Egler 1911–1996. *Bulletin of the Ecological Society of America*. 78:193–194.

Burke, G. W., Jr. 1970. Engineering responsibilities in protecting the marine environment. pp. 99–107. In: *Bioresources of Shallow Water Environments*. W. G. Weist, Jr. and P. E. Greeson (eds.). American Water Resources Association, Urbana, IL.

Burke, M. J. W. and J. P. Grime. 1996. An experimental study of plant community invasibility. *Ecology*. 77:776–790.

Burnor, D. 1980. Ed Ricketts: "From the Tidepool to the Stars." *CoEvolution Quarterly*, 28:14–20.

Butler, D. R. 1995. *Zoogeomorphology*. Cambridge University Press, Cambridge, U.K.

Butler, R. B. 1981. *The Ecological House*. Morgan & Morgan, Dobbs Ferry, NY.

Cadisch, G. and K. E. Giller (eds.). 1997. *Driven by Nature, Plant Litter Quality and Decomposition*. CAB International, Wallingford, U.K.

Cairns, J., Jr. 1974. Indicator species vs. the concept of community structure as an index of pollution. *Water Resources Bulletin*. 10:338–347.

Cairns, J., Jr. 1976. Heated waste-water effects on aquatic ecosystems. pp. 32–38. In: *Thermal Ecology II*. G. W. Esch and R. W. McFarlane (eds.). CONF-750425, Technical Information Center, Energy Research and Development Administration, Washington, DC.

Cairns, J., Jr. (ed.). 1982. Artificial Substrates. Ann Arbor Science Publishers, Inc., Ann Arbor, MI.

Cairns, J., Jr. (ed.). 1980. *The Recovery Process in Damaged Ecosystems*. Ann Arbor Science, Ann Arbor, MI.

Cairns, J., Jr. 1983. Are single species toxicity tests alone adequate for estimating environmental hazard? *Hydrobiolgia*. 100:47–57.

Cairns, J., Jr. (ed.). 1985. *Multispecies Toxicity Testing*. Pergamon Press, Oxford, U.K.

Cairns, J., Jr. (ed.). 1986a. *Community Toxicity Testing*. ASTM, Philadelphia, PA.

Cairns, J., Jr. 1986b. Restoration, reclamation, and regeneration of degraded or destroyed ecosystems. pp. 465–484. In: *Conservation Biology*. M. E. Soule (ed.). Sinauer Associates, Sunderland, MA.

Cairns, J. Jr. 1988. Putting the eco in ecotoxicology. *Regulatory Toxicology Pharmaceuticals*. 8:226–238.

Cairns, J., Jr. 1995a. Future trends in ecotoxicity. pp. 217–222. In: *Ecological Toxicity Testing*. J. Cairns, Jr. and B. R. Niederlehner (eds.). Lewis Publishers, Boca Raton, FL.

Cairns, J., Jr. (ed.). 1995b. *Rehabilitating Damaged Ecosystems*. Lewis Publishers, Boca Raton, FL.

Cairns, J., Jr. 1998. The zen of sustainable use of the planet: steps on the path to enlightenment. *Population and Environment*. 20:109–123.

Cairns, J., Jr. 2000. The genesis and future of the field of ecotoxicology. pp. 1–13. In: *Integrated Assessment of Ecosystem Health*. K. M. Scow, G. E. Fogg, D. E. Hinton, and M. L. Johnson (eds.). Lewis Publishers, Boca Raton, FL.

Cairns, J., Jr. and D. Orvos. 1989. Ecological consequence assessment: predicting effects of hazardous substances upon aquatic ecosystems using ecological engineering. pp. 409–442. In: *Ecological Engineering: An Introduction to Ecotechnology*. W. J. Mitsch and S. E. Jørgensen (eds.). John Wiley & Sons, New York.

Cairns, J., Jr. and J. A. Ruthven. 1970. Artificial microhabitat size and the number of colonizing protozoan species. *Transaction of the American Microscopy Society*. 89:100–109.

Cairns, J., Jr., K. L. Dickson, and E. E. Herricks (eds.). 1977. *Recovery and Restoration of Damaged Ecosystems*. University Press of Virginia, Charlottesville, VA.

Calabrese, E. J. and L. A. Baldwin. 1999. Reevaluation of the fundamental dose–response relationship. *BioScience*. 49:725–732.

Calaway, W. T. 1957. Intermittent sand filters and their biology. *Sewage Works Journal*. 29:1–5.

Callahan. J. T. 1984. Long-term ecological research. *BioScience*. 34:363–367.

Callicott, J. B. and R. T. Ames. 1989. *Nature in Asian Traditions of Thought: Essays in Environmental Philosophy*. State University of New York Press, Albany, NY.

Calow, P. 1976. *Biological Machines: A Cybernetic Approach to Life*. Edward Arnold, London.

Calvert, M. A. 1967. *The Mechanical Engineer in America, 1830–1910*. Johns Hopkins Press, Baltimore, MD.

Camazine, S., J.-L. Deneubourg, N. R. Franks, J. Sneyd, G. Theraulaz, and E. Bonabeau. 2001. *Self-Organization in Biological Systems*. Princeton University Press, Princeton, NJ.

Campbell, C. S. and M. H. Ogden. 1999. *Constructed Wetlands in the Sustainable Landscape*. John Wiley & Sons, New York.

Campbell, D. E., H. T. Odum, and G. A. Knox. 1991. Organization of a new ecosystem: exotic Spartina saltmarsh in New Zealand. pp. 24–25. In: *Spartina Workshop Record*. T. F. Mumford, Jr., P. Peyton, J. R. Sayce, and S. Harbell (eds.). Washington Sea Grant Program, University of Washington, Seattle, WA.

Campbell, S. 1990. *Let It Rot! The Gardener's Guide to Composting*. Storey Communications, Pownal, VT.

Campos, J. A. and C. Gamboa. 1989. An artificial tire-reef in a tropical marine system: a management tool. *Bulletin of Marine Sciences*. 44:757–766.

Candle, R. D. 1985. Scrap tires as artificial reefs. pp. 293–302. In: *Artificial Reefs: Marine and Freshwater Applications*. F. M. D'Itri (ed.). Lewis Publishers, Chelsea, MI.

Cannon, H. L. 1960. Botanical prospecting for ore deposits. *Science*. 132:591–598.

Cannon, W. B. 1932. *The Wisdom of the Body*. W. W. Norton, New York.

Capra, F. 1991. *The Tao of Physics*. Shambhala, Boston, MA.

Capra, F. 1995. Deep ecology, A new paradigm. pp. 19–25. In: *Deep Ecology for the 21st Century*. G. Sessions (ed.). Shambhala, Boston, MA.

Carlson, D. A. and C. P. Leiser. 1966. Soil beds for the control of sewage odors. *Journal of the Water Pollution Control Federation*. 38:829–840.

Carlson, D. B., P. D. O'Bryan, and J. R. Rey. 1994. The management of Florida's (USA) saltmarsh impoundments for mosquito control and natural resource enhancement. pp. 805–814. In: *Global Wetlands: Old World and New*. W. J. Mitsch (ed.). Elsevier Publishing, Amsterdam.

Carlton, J. M. 1974. Land-building and stabilization by mangroves. *Environmental Conservation*. 1:285–294.

Carlton, J. T. 1993. Dispersal mechanisms of the zebra mussel (*Dreissena polymorpha*). pp. 677–697. In: *Zebra Mussels*. T. F. Nalepa and D. W. Schloesser (eds.). Lewis Publishers, Boca Raton, FL.

Carlton, J. T. 1985. Transoceanic and interoceanic dispersal of coastal marine organisms: the biology of ballast water. *Annual Review of Oceanography and Marine Biology*. 23:313–371.

Carlson, R. E. 1977. A trophic state index for lakes. *Limnology and Oceanography*. 22:361–369.

Caron, D. A. and J. C. Goldman. 1990. Protozoan nutrient regeneration. pp. 283–306. In: *Ecology of Marine Protozoa*. G. M. Capriulo (ed.). Oxford University Press, New York.

Carpenter, S. R. 1996. Microcosm experiments have limited relevance for community and ecosystem ecology. *Ecology*. 77:677–680.

Carpenter, S. R. 2001. Alternate states of ecosystems: evidence and some implications. pp. 357–383. In: *Ecology: Achievement and Challenge*. M. C. Press, N. J. Huntly, and S. Levin (eds.). Blackwell Science. Oxford, U.K.

Carpenter, S. R. and J. F. Kitchell. 1992. Trophic cascade and biomanipulation: interface of research and management — a reply to the comment by DeMelo et al. *Limnology and Oceanography*. 37:208–213.

Carpenter, S. R. and J. F. Kitchell (eds.). 1993. *The Trophic Cascade in Lakes*. Cambridge University Press, Cambridge, U.K.

Carpenter, S.R., N. F. Caraco, D. L. Correll, R. W. Howarth, A. N. Sharpley, and V. H. Smith. 1998. Nonpoint pollution of surface waters with phosphorus and nitrogen. *Issues in Ecology 3*. Ecological Society of American, Washington, DC.

Carpenter, S. R., J. F. Kitchell, and J. R. Hodgson. 1985. Cascading trophic interactions and lake productivity. *BioScience*. 35:634–639.

Carson, R. 1962. *Silent Spring*. Houghton Mifflin, Cambridge, MA.

Carter, L. J. 1987. *Nuclear Imperatives and Public Trust, Dealing with Radioactive Waste*. Resources for the Future, Washington, DC.

Carter, M. R., L. A. Burns, T. R. Cavinder, K. R. Dugger, P. L. Fore, D. B. Hicks, H. L. Revells, and T. W. Schmidt. 1973. Ecosystems analysis of the Big Cypress swamp and estuaries. U.S. Environmental Protection Agency, Atlanta, GA.

Caruso, F., R. A. Caruso, and H. Mohwald. 1998. Nanoengineering of inorganic and hybrid hollow spheres by colloidal templating. *Science*. 282:1111–1114.

Casti, J. 1982. Catastrophes, control and the inevitability of spruce budworm outbreaks. *Ecological Modelling*. 14:293–300.

Center, T. D., J. H. Frank, and F. A. Dray, Jr. 1997. Biological control. pp. 245–263. In: *Strangers in Paradise: Impact and Management of Nonindigenous Species in Florida*. D. Simberloff, D. C. Schmitz, and T. C. Brown (eds.). Island Press, Washington, DC.

Chadwick, M. J. and G. T. Goodman (eds.). 1975. *The Ecology of Resource Degradation and Renewal*. John Wiley & Sons, New York.

Chakroff, M. 1976. *Freshwater Fish Pond Culture and Management*. VITA, Arlington, VA.

Chambers, J. C. and J. A. MacMahon. 1994. A day in the life of a seed: movements and fates of seeds and their implications for natural and managed systems. *Annual Review of Ecology and Systematics.* 25:263–292.

Channell, D. F. 1991. *The Vital Machine.* Oxford University Press, New York.

Chapman, P. F. and S. J. Maund. 1996. Considerations for the experimental design of aquatic mesocosm and microcosm experiments. pp. 657–673. In: *Techniques in Aquatic Toxicology.* G. K. Ostrander (ed.). Lewis Publishers, Boca Raton, FL.

Chapman, R. N. 1931. *Animal Ecology.* McGraw-Hill, New York.

Characklis, W. G. and K. C. Marshall. (eds.). 1990. *Biofilms.* John Wiley & Sons, New York.

Chase, A. 1986. *Playing God in Yellowstone: The Destruction of America's First National Park.* The Atlantic Monthly Press, Boston, MA.

Chase, E. S. 1964. Nine decades of sanitary engineering. *Water Works and Wastes Engineering.* 1(4):57, 98.

Chen, C.-C., J. E. Petersen, and W. M. Kemp. 1997. Spatial and temporal scaling of periphyton growth on walls of estuarine mesocosms. *Marine Ecology Progress Series.* 155:1–15.

Cheremisinoff, P. N. 1994. *Biomanagement of Wastewater and Wastes.* Prentice Hall, Englewood Cliffs, NJ.

Cherrett, J. M. 1988. Key concepts: the results of a survey of our members' opinions. pp. 1–16. In: *Ecological Concepts.* J. M. Cherrett (ed.). Blackwell Scientific, Oxford, U.K.

Chesley, H. 1999. Early applications. pp. 89–105. In: *Nanotechnology: Molecular Speculations on Global Abundance.* B. C. Crandall (ed.). MIT Press, Cambridge, MA.

Chesson, P. L. and T. J. Case. 1986. Overview: nonequilirium community theories: chance, variability, history, and coexistence. pp. 229–239. In: *Community Ecology.* J. Diamond and T. J. Case (eds.). Harper & Row, New York.

Chew, R. M. 1974. Consumers as regulators of ecosystems: an alternative to energetics. *Ohio Journal of Science.* 74:359–370.

Chisholm, S. W., P. G. Falkowski, and J. J. Cullen. 2001. Discrediting ocean fertilization. *Science.* 294:309–310.

Chitty, D. 1996. *Do Lemmings Commit Suicide*? Oxford University Press, New York.

Chow, V. T. (ed.). 1964. *Handbook of Applied Hydrology.* McGraw-Hill, New York.

Christian, R. R., L. E. Stasavich, C. R. Thomas, and M. M. Brinson. 2000. Reference is a moving target in sea-level controlled wetlands. pp. 805–825. In: *Concepts and Controversies in Tidal Marsh Ecology.* M. P. Weinstein and D. A. Kreeger (eds.). Kluwer Academic, Dordrecht, the Netherlands.

Chung, C.-H. 1989. Ecological engineering of coastlines with salt-marsh plantations. pp. 255–289. In: *Ecological Engineering.* W. J. Mitsch and S. E. Jørgensen (eds.). John Wiley & Sons, New York.

Church, M. 1992. Channel morphology and typology. pp. 126–143. In: *The Rivers Handbook, Hydrological and Ecological Principles*, Vol. I. P. Calow and G. E. Petts (eds.). Blackwell Science Oxford, U.K.

Clapham, W. B., Jr. 1981. *Human Ecosystems.* Macmillan, New York.

Clapp, C. E., W. E. Larson, and R. H. Dowdy (eds.). 1994. *Sewage Sludge: Land Utilization and the Environment.* SSSA Miscellaneous Publication, American Society of Agronomy, Inc. Madison, WI.

Clark, E. H. II, J. A. Haverkamp, and W. Chapman. 1985. *Eroding Soils.* The Conservation Foundation, Washington, DC.

Clark, J. 1977. *Coastal Ecosystem Management.* John Wiley & Sons, New York.

Clark, J. S. 1989. Ecological disturbance as a renewal process: theory and application to fire history. *Oikos.* 56:17–30.

Clark, M. J. 1997. Ecological restoration — the magnitude of the challenge: an outsider's view. pp. 353–377. In: *Restoration Ecology and Sustainable Development*. K. M. Urbanska, N. R. Webb, and P. J. Edwards (eds.). Cambridge University Press, Cambridge, U.K.

Clark, O. G. and R. Kok. 1998. Engineering of highly autonomous biosystems: review of the relevant literature. *International Journal of Intelligent Systems*. 13:749–783.

Clark, O. G., R. Kok, and R. Lacroix. 1996. Engineering intelligent ecosystems. NABEC Paper No. 9662. Presented at the Northeast Agricultural/Biological Engineering Conference, Canaan Valley, WV, August 4–7, 1996.

Clark, O. G., R. Kok, and R. Lacroix. 1999. Mind and autonomy in engineered biosystems. *Engineering Applications of Artificial Intelligence*. 12:389–399.

Clarke, A. C. 1994. *The Snows of Olympus: A Garden on Mars*. W. W. Norton, New York.

Clarke, G. L. 1946. Dynamics of production in a marine area. *Ecological Monographs*. 16:321–335.

Clarke, T. 2001. The stowaways. *Nature*. 413:247–248.

Clausen, C. P. (ed.). 1978. *Introduced Parasites and Predators of Arthropod Pests and Weeds: A World Review*. Agriculture Handbook No. 480, Agricultural Research Service, USDA, Washington, DC.

Clay, K. and P. X. Kover. 1996. The red queen hypothesis and plant/pathogen interactions. *Annual Review of Phytopathology*. 34:29–50.

Clewell, A. and J. P. Rieger. 1997. What practitioners need from restoration ecologists. *Restoration Ecology*. 5:350–354.

Clobert, J., E. Danchin, A. A. Dhondt, and J. D. Nichols (eds.). 2001. *Dispersal*. Oxford University Press, Oxford, U.K.

Clymer, F. 1960. *Treasury of Early American Automobiles, 1877–1925*. Bonanza Books, New York.

Coffman, L. S. and D. A. Winogradoff. 2001. *Design Manual for Use of Bioretention in Stormwater Management*. Watershed Protection Branch, Department of Environmental Resources, Prince George's County, MD.

Coffman, W. P., K. W. Cummins, and J. C. Wuycheck. 1971. Energy flow in a woodland stream ecosystem. I. Tissue support trophic structure of the autumnal community. *Archives of Hydrobiology*. 68:232–276.

Cohen, A. N. and J. T. Carlton. 1998. Accelerating invasion rate in a highly invaded estuary. *Science*. 279:555–558.

Cohen, J. E. 1971. Mathematics as metaphor. *Science*. 172:674–675.

Cohen, J. E. 1978. *Food Webs and Niche Space*. Princeton University Press, Princeton, NJ.

Cohen, J. E. 1995. Population growth and Earth's human carrying capacity. *Science*. 269:341–348.

Cohen, J. E. and D. Tilman. 1996. Biosphere 2 and biodiversity: the lessons so far. *Science*. 274:1150–1151.

Cohen, J. E., F. Briand, and C. M. Newman. 1990. *Community Food Webs: Data and Theory*. Springer-Verlag, Berlin.

Cohn, J. P. 2002. Biosphere 2: turning an experiment into a research station. *BioScience*. 52:218–223.

Cole, G. A. 1983. *Textbook of Limnology*. C. V. Mosby, St. Louis, MO.

Cole, L. C. 1957. Sketches of general and comparative demography. pp. 1–15. In: *Population Studies: Animal Ecology and Demography*, Cold Spring Harbor Symposia on Quantitative Biology, Vol. XXVII. The Biological Laboratory, Cold Spring Harbor, Long Island, NY.

Cole, S. 1998. The emergence of treatment wetlands. *Environmental Science and Technology.* May, 1998:218A–223A.

Coleman, D. C. 1985. Through a ped darkly: an ecological assessment of root-soil-microbial-faunal interactions. pp. 1–22. In: *Ecological Interactions in Soil: Plants, Microbes and Animals.* A. H. Fitter (ed.). Blackwell Scientific, Oxford, U.K.

Coleman, D. C. 1996. Energetics of detritivory and microbivory in soil in theory and practice. pp. 39–50. In: *Food Webs, Integration of Patterns and Dynamics.* G. A. Polis and K. O. Winemiller (eds.). Chapman & Hall, New York.

Colinvaux, P. 1993. *Ecology* 2. John Wiley & Sons, New York.

Collias, N. E. and E. C. Collias. 1984. *Nest Building and Bird Behavior.* Princeton University Press, Princeton, NJ.

Collins, J. P., A. Kinzig, N. B. Grimm, W. F. Fagan, D. Hope, J. Wu, and E. T. Borer. 2000. A new urban ecology. *American Scientist.* 88:416–425.

Confer, S. R. and W. A. Niering. 1992. Comparison of created and natural freshwater emergent wetlands in Connecticut (USA). *Wetland Ecology Management.* 2:143–156.

Connell, J. H. 1978. Diversity in tropical rain forests and coral reefs. *Science.* 199:1302–1310.

Connor, J. L. and W. H. Adey. 1977. The benthic algal composition, standing crop, and productivity of a Caribbean algal ridge. *Atoll Research Bulletin.* 211.

Conquest, L. L. and F. B. Taub. 1989. Repeatability and reproducibility of the standardized aquatic microcosm: statistical properties. pp. 159–177. In: *Aquatic Toxicology and Hazard Assessment*, Vol. 12. ASTM, Philadelphia, PA.

Conrad, M. 1976. Patterns of biological control in ecosystems. pp. 431–456. In: *Systems Analysis and Simulation in Ecology*, Vol. IV. B. C. Patten (ed.). Academic Press, New York.

Conrad, M. 1995. The ecosystem as an existential computer. pp. 609–622. In: *Complex Ecology: The Part-Whole Relation in Ecosystems.* B. C. Patten and S. E. Jørgensen (eds.). Prentice Hall, Englewood Cliffs, NJ.

Conrad, M. and H. H. Pattee. 1970. Evolution experiments with an artificial ecosystem. *Journal of Theoretical Biology.* 28:393–409.

Cooke, G. D. 1967. The pattern of autotrophic succession in laboratory microcosms. *BioScience.* 17:717–721.

Cooke, G. D. 1968. Erratum. *BioScience.* 18:305.

Cooke, G. D. 1971. Ecology of space travel. pp. 498–509. In: *Fundamentals of Ecology.* E. P. Odum. W. B. Saunders, Philadelphia, PA.

Cooke, G. D., R. J. Beyers, and E. P. Odum. 1968. The case for the multispecies ecological system, with special reference to succession and stability. pp. 129–139. In: *Bioregenerative Systems.* NASA, Washington, DC.

Cooke, G. D., E. B. Welch, S. A. Peterson, and P. R. Newroth. 1993. *Restoration and Management of Lakes and Reservoirs*, 2nd ed. Lewis Publishers, Boca Raton, FL.

Cooke, W. G. 1959. Trickling filter ecology. *Ecology.* 40:273–291.

Cookson, J. T., Jr. 1995. *Bioremediation Engineering: Design and Application.* McGraw-Hill, New York.

Cooper, S. D. and L. A. Barmuta. 1993. Field experiments in biomonitoring. pp. 399–441. In: *Freshwater Biomonitoring and Benthic Macroinvertebrates.* D. M. Rosenberg and V. H. Resh (eds.). Chapman & Hall, New York.

Copeland, B. J. 1965. Evidence for regulation of community metabolism in a marine ecosystem. *Ecology.* 46:563–564.

Copeland, B. J. 1970. Estuarine classification and responses to disturbances. *Transactions of American Fisheries Society.* 99:826–835.

Corbitt, R. A. 1990. *Standard Handbook of Environmental Engineering*. McGraw-Hill, New York.

Cornwell, D. A., J. Zoltek, C. D. Patrinely, T. de S. Furman, and J. I. Kim. 1977. Nutrient removal by water hyacinths. *Journal of the Water Pollution Control Federation*. 49:57–65.

Costa-Pierce, B. A. 1998. Preliminary investigation of an integrated aquaculture-wetland ecosystem using tertiary-treated municipal wastewater in Los Angeles County, California. *Ecological Engineering*. 10:341–354.

Costanza, R. 1980. Embodied energy and economic valuation. *Science*. 210:1219–1224.

Costanza, R. 1989. What is ecological economics? *Ecological Economics*. 1:335–362.

Costanza, R. (ed.). 1991. *Ecological Economics*. Columbia University Press, New York.

Costanza, R. and H. E. Daly. 1987. Toward an ecological economics. *Ecological Modelling*. 38:1–8.

Costanza, R. and H. E. Daly. 1992. Natural capital and sustainable development. *Conservation Biology*. 6:37–46.

Costanza, R. and R. V. O'Neill. 1996. Introduction: ecological economics and sustainability. *Ecological Applications*. 6:975–977.

Costanza, R. and C. H. Perrings. 1990. A flexible assurance bonding system for improved environmental management. *Ecological Economics*. 2:57–76.

Costanza, R., J. Cumberland, H. Daly, R. Goodland, and R. Norgaard. 1997a. *An Introduction to Ecological Economics*. St. Lucie Press, Boca Raton, FL.

Costanza, R., H. E. Daly, and J. A. Bartholomew. 1991. Goals, agenda, and policy recommendations for ecological economics. pp. 1–20. In: *Ecological Economics*. R. Costanza (ed.). Columbia University Press, New York.

Costanza, R., R. d'Arge, R. de Groot, S. Farber, M. Grasso, B. Hannon, K. Limburg, S. Naeem, R. V. O'Neill, J. Paruelo, R. G. Raskin, P. Sutton, and M. van den Belt. 1997b. The value of the world's ecosystem services and natural capital. *Nature*. 387:253–260.

Costanza, R., S. C. Farber, and J. Maxwell. 1989. The valuation and management of wetland ecosystems. *Ecological Economics*. 1:335–361.

Cottam, G. 1987. Community dynamics on an artificial prairie. pp. 257–270. In: *Restoration Ecology: A Synthetic Approach to Ecological Research*. W. R. Jordan, III, M. E. Gilpin, and J. D. Aber (eds.). Cambridge University Press, Cambridge, U.K.

Cottingham, K. L. 2002. Tackling biocomplexity: the role of people, tools, and scale. *BioScience*. 52:793–799.

Couch, L. K. 1942. Trapping and Transplanting Live Beavers. Conservation Bulletin 30, U.S. Department of the Interior, Washington, DC.

Courtenay, W. R., Jr. and C. R. Robins. 1975. Exotic organisms: an unsolved, complex problem. *BioScience*. 25:306–313.

Cousins, S. H. 1980. A trophic continuum derived from plant structure, animal size and a detritus cascade. *Journal of Theoretical Biology*. 82:607–618.

Covich, A. 1972. Ecological economics of seed consumption by Peromyscus. *Transactions of Connecticut Academy of Arts and Sciences*. 44:71–93.

Covich, A. 1974. Ecological economics of foraging among coevolving animals and plants. *Annuals Missouri Botanical Garden*. 61:794–805.

Cowan, J. 1998 A Constructed Wetland for Wastewater Treatment in Maryland: Water Quality, Hydrology, and Amphibian Dynamics. M.S. thesis, University of Maryland, College Park, MD.

Cowardin, L. M., V. Carter, F. C. Golet, and E. T. LaRoe. 1979. *Classification of Wetlands and Deepwater Habitats of the United States*. U.S. Fish and Wildlife Service Pub. FWS/OBS-79/31, Washington, DC.

Cowell, P. J. and B. G. Thom. 1994. Morphodynamics of coastal evolution. pp. 33–86. In: *Coastal Evolution*. R. W. G. Carter and C. D. Woodroffe (eds.). Cambridge University Press, Cambridge, U.K.

Cowles, H. C. 1900. The physiographic ecology of northern Michigan. *Science*. 12:708–709.

Cowles, H. C. 1901. The physiographic ecology of Chicago and vicinity: A study of the origin, development, and classification of plant societies. *Botanical Gazette*. 31:73–108, 145–182.

Cox, G. W. and M. D. Atkins. 1975. Agricultural Ecology. *Bulletin of the Ecological Society of America*. 56:2–6.

Cox, P. A. 2000. Will tribal knowledge survive the millennium? *Science*. 287:44–45.

Craggs, R. J., W. H. Adey, B. K. Jessup, and W. J. Oswald. 1996. A controlled stream mesocosm for tertiary treatment of sewage. *Ecological Engineering*. 6:149–169.

Crandall, B. C. 1999. Molecular engineering. In: *Nanotechnology: Molecular Speculations on Global Abundance*. B. C. Crandall (ed.). MIT Press, Cambridge, MA.

Crawley, M. J. 1984. What makes a community invasible? pp. 429–453. In: *Colonization, Succession and Stability*. A. J. Gray, M. J. Crawley, and P. J. Edwards (eds.). Blackwell Science Oxford, U.K.

Crichton, M. 2002. *Prey*. HarperCollins Publishers, Inc., New York.

Crisp, D. J. 1965. The ecology of marine fouling. pp. 99–118. In: *Ecology and the Industrial Society*. G. T. Goodman, R. W. Edwards, and J. M. Lambert (eds.). John Wiley & Sons, New York.

Crites, R. and G. Tchobanoglous. 1998. *Small and Decentralized Wastewater Management Systems*. WCB McGraw-Hill, Boston, MA.

Cronk, J. K. and M. S. Fennessy. 2001. *Wetland Plants, Biology and Ecology*. Lewis Publishers, Boca Raton, FL.

Cueto, A. J. 1993. Development of criteria for the design and construction of engineered aquatic treatment units in Texas. pp. 99–105. In: *Constructed Wetlands for Water Quality Improvement*. G. A. Moshiri (ed.). Lewis Publishers, Boca Raton, FL.

Cullen, A. H. 1962. *Rivers in Harness: The Story of Dams*. Chilton Books, Philadelphia, PA.

Culley, D. D. and E. A. Epps. 1973. Use of duckweeds for waste treatment and animal feed. *Journal of the Water Pollution Control Federation*. 45:337–347.

Culotta, E. 1991. Biological immigrants under fire. *Science*. 254:1444–1447.

Culver, D. C. 1970. Analysis of simple cave communities. I. Caves as islands. *Evolution*. 24:463–474.

Cummins, K. W. 1973. Trophic relations of aquatic insects. *Annual Reviews of Entomology*. 18:183–206.

Cummins, K. W. 1974. Structure and function of stream ecosystems. *BioScience*. 24:631–641.

Cummins, K. W. and M. J. Klug. 1979. Feeding ecology of stream invertebrates. *Annual Reviews of Ecology and Systematics*. 10:147–172.

Cummins, K. W., R. C. Petersen, F. O. Howard, J. C. Wuycheck, and V. I. Holt. 1973. The utilization of leaf litter by stream detritovores. *Ecology*. 54:336–345.

Cummins, K. W., M. A. Wilzbach, D. M. Gates, J. B. Perry, and W. B. Taliaferro. 1989. Shredders and riparian vegetation. *BioScience*. 39:24–30.

Cunningham, A. B. 1991. Indigenous knowledge and biodiversity. *Cultural Survival Quarterly*. 15(3):4–8.

Cunningham, S. 2002. *The Restoration Economy: The Greatest New Growth Frontier*. Berrett-Koehler Publishers, San Francisco, CA.

Curds, C. R. 1975. Protozoa. pp. 203–268. In: *Ecological Aspects of Used-Water Treatment. Vol. 1 — The Organisms and Their Ecology.* C. R. Curds and H. A. Hawkes (eds.). Academic Press. London.

Curds, C. R., A. Cockburn, and J. M. Vandyke. 1968. An experimental study of the role of the ciliated protozoa in the activated-sludge process. *Water Pollution Control.* 67:312–327.

Curtis, J. T. and M. L. Partch. 1948. Effects of fire on the competition between blue grass and certain prairie plants. *American Midland Naturalist.* 39:437–443.

Cushing, C. E. and J. D. Allan. 2001. *Streams: Their Ecology and Life*. Academic Press, San Diego, CA.

Cushing, D. H. 1971. Upwelling and the production of fish. *Advances in Marine Biology.* 9:255–334.

Cushing, J. M., R. F. Costantino, B. Dennis, R. A. Desharnais, and S. M. Henson. 2003. *Chaos in Ecology: Experimental Nonlinear Dynamics*. Academic Press, Amsterdam.

Czapowskyj, M. M. 1976. Annotated Bibliography on the Ecology and Reclamation of Drastically Disturbed Areas. USDA Forest Service Gen. Tech. Rept. NE-21, Northeastern Forest Experiment Station, Upper Darby, PA.

D'Azzo, J. J. and C. H. Houpis. 1960. *Feedback Control System Analysis and Synthesis.* McGraw-Hill, New York.

Dacey, J. W. H. 1981. Pressurized ventilation in the yellow waterlily. *Ecology.* 62:1137–1147.

Dahlsten, D. L. 1986. Control of invaders. pp. 275–302. In: *Ecology of Biological Invasions of North America and Hawaii*. H. A. Mooney and J. A. Drake (eds.). Springer-Verlag, New York.

Dahlsten, D. L. and R. Garcia (eds.). 1989. *Eradication of Exotic Pests: Analysis with Case Histories*. Yale University Press, New Haven, CT.

Daily, G. C. (ed.). 1997. *Nature's Services*. Island Press, Washington, DC.

Daily, G. C. et al. 2000. The value of nature and the nature of value. *Science.* 289:395–396.

Daily, G. C., S. Alexander, P. R. Ehrlich, L. Goulder, J. Lubchenco, P. A. Matson, H. A. Mooney, S. Postel, S. H. Schneider, D. Tilman, and G. M. Woodwell. 1997. Ecosystem services: benefits supplied to human societies by natural ecosystems. *Issues in Ecology No. 2*. Ecological Society of America, Washington, DC.

Dakers, A. 2002. Ecosystem services, their use and the role of ecological engineering: state of the art. pp. 101–126. In: *Understanding and Solving Environmental Problems in the 21st Century.* R. Costanza and S. E. Jørgensen (eds.). Elsevier Science. Oxford, U.K.

Dale, P. E. R. and K. Hulsman. 1991. A critical review of saltmarsh management methods for mosquito control. *Critical Reviews in Aquatic Sciences.* 3:281–311.

Daly, H. E. 1968. On economics as a life science. *Journal of Political Economy.* 76:392–406.

Daly, H. E. (ed.). 1973. *Toward a Steady-State Economy.* W. H. Freeman, San Francisco, CA.

Daly, H. E. 1977. *Steady State Economics*. W. H. Freeman, San Francisco, CA.

Daly, H. E. 1995. Reply to Mark Sagoff's "Carrying capacity and ecological economics." *BioScience.* 45:621–624.

Daly, H. E. 1996. *Beyond Growth*. Beacon Press, Boston, MA.

Daly, H. E. and J. Cobb. 1989. *For the Common Good: Redirecting the Economy Towards Community, the Environment, and a Sustainable Future*. Beacon Press, Boston, MA.

Daly, H. E. and K. N. Townsend. (eds.). 1993. *Valuing the Earth.* MIT Press, Cambridge, MA.

Dame, R. F. 1996. *Ecology of Marine Bivalves: An Ecosystem Approach.* CRC Press, Boca Raton, FL.

Dame, R. F. 2001. Benthic suspension feeders as determinants of ecosystem structure and function in shallow coastal waters. pp. 11–37. In: *Ecological Comparisons of Sedimentary Shores*. K. Reise (ed.). Springer-Verlag, Berlin.

Danell, K. 1982. Muskrat. pp. 202–203. In: *CRC Handbook of Census Methods for Terrestrial Vertebrates*. D. E. Davis (ed.). CRC Press, Boca Raton, FL.

Dangerfield, J. M., T. S. McCarthy, and W. N. Ellery. 1998. The mound-building termite *Macrotermes michaelseni* as an ecosystem engineer. *Journal of Tropical Ecology.* 14:507–520.

Darlington, A. B., J. F. Dat, and M. A. Dixon. 2001. The biofiltration of indoor air: air flux and temperature influences the removal of toluene, ethylbenzene, and xylene. *Environment Science Technology.* 35:240–246.

Darlington, A., M. Chan, D. Malloch, C. Pilger, and M. A. Dixon. 2000. The biofiltration of indoor air: implications for air quality. *International Journal of Indoor Air Quality and Climate.* 10:39–46.

Darnell, R. M. 1961. Trophic spectrum of an estuarine community, based on studies of Lake Pontchartrain, Louisiana. *Ecology.* 42:553–568.

Darnell, R. M. 1964. Organic detritus in relation to secondary production in aquatic communities. *Verhandlungen Internationale Vereinigung für Theoretische und Angewandte Limnologie.* 15:462–470.

Darnell, R. M. 1967. Organic detritus in relation to the estuarine ecosystem. pp. 376–382. In: *Estuaries*. G. H. Lauff (ed.). Publ. No. 83, American Association for the Advancement of Science, Washington, DC.

Darnell, R. M. 1971. *Organism and Environment: A Manual of Quantitative Ecology.* W. H. Freeman, San Francisco, CA.

Darwin, C. 1881. *The Formation of Vegetable Mould Through the Action of Worms with Observations on their Habits*. Murray Publishing, London.

Dasgupta, D. (ed.). 1999. *Artificial Immune Systems and Their Applications*. Springer-Verlag, Berlin.

Daufresne, T. and M. Loreau. 2001. Ecological stoichiometry, primary producer-decomposer interactions, and ecosystem persistence. *Ecology.* 82:3069–3082.

Davidson, A. K. 1983. *The Art of Zen Gardens: A Guide to Their Creation and Enjoyment.* G. P. Putnam's Sons, New York.

Davies, J. M. and J. C. Gamble. 1979. Experiments with large enclosed ecosystems. *Philosophical Transactions of the Royal Society of London B.* 286:523–544.

Davis, D. G. 1989. No net loss of the nation's wetlands: a goal and a challenge. *Water Environment & Technology.* 1:512–514.

Davis, J. H. 1940. The ecology and geologic role of mangroves in Florida. *Carnegie Institute Washington Publication.* 517:303–412.

Davis, L. no date. *A Handbook of Constructed Wetlands*, Vol. 2, Domestic Wastewater. USDA — Natural Resources Conservation Service, Washington, DC.

DeAngelis, D. L. 1992. *Dynamics of Nutrient Cycling and Food Webs.* Chapman & Hall, London.

DeAngelis, D. L. 1995. The nature and Significance of feedback in ecosystems. pp. 450–467. In: *Complex Ecology: The Part-Whole Relation in Ecosystems.* B. C. Patten and S. E. Jørgensen (eds.). Prentice Hall, Englewood Cliffs, NJ.

DeAngelis, D. L. and J. C. Waterhouse. 1987. Equilibrium and nonequilibrium concepts in ecological models. *Ecological Monographs.* 57:1–21.

DeAngelis, D. L., W. M. Post, and C. C. Travis. 1986. *Positive Feedback in Natural Systems.* Springer-Verlag, Berlin.

DeBach, P. 1974. *Biological Control by Natural Enemies*. Cambridge University Press, Cambridge, U.K.

Deevey, E. S. 1970. Mineral cycles. *Scientific American.* 223(3):148–158.

Deevey, E. S., Jr. 1984. Stress, strain, and stability of lacustrine ecosystems. pp. 203–229. In: *Lake Sediments and Environmental History.* E. Y. Haworth and J. W. G. Lund (eds.). University of Minnesota Press, Minneapolis, MN.

Del Porto, D. and C. Steinfeld. 1999. *The Composting Toilet System Book*. The Center for Ecological Pollution Prevention, Concord, MA.

Deland, M. R. 1992. No net loss of wetlands: a comprehensive approach. *Natural Resources & Environment.* 7:3–55.

Delgado, M., M. Biggeriego, and E. Guardiola. 1993. Uptake of Zn, Cr, and Cd by water hyacinths. *Water Research.* 27:269–272.

DeMelo, R., R. France, and D. J. McQueen. 1992. Biomanipulation: hit or myth. *Limnology and Oceanography.* 37:192–207.

Dennett, D. C. 1995. How to make mistakes. pp. 137–144. In: *How Things Are: A Science Tool-Kit for the Mind.* J. Brockman and K. Matson (eds.). William Morrow, New York.

Devik, O. (ed.). 1976. *Harvesting Polluted Waters*. Plenum Press, New York.

Diamond, J. M. 1974. Colonization of exploded volcanic islands by birds, the supertramp strategy. *Science.* 184:803–806.

Diamond, J. M. 1975. The island dilemma: Lessons of modern biogeographic studies for the design of natural reserves. *Biological Conservation.* 7:129–46.

Diamond, J. 1986. Overview: laboratory experiments, field experiments, and natural experiments. pp. 3–22. In: *Community Ecology.* J. Diamond and T. J. Case (eds.). Harper & Row, New York.

Diamond, J. M. and R. M. May. 1976. Island biogeography and the design of natural reserves. pp. 163–186. In: *Theoretical Ecology.* R. M. May (ed.). W. B. Saunders, Philadelphia, PA.

Diaz, M. C. and K. Rutzler. 2001. Sponges: an essential component of Caribbean coral reefs. *Bulletin of Marine Science.* 69:535–546.

Dickerson, J. 1995. Introduction to Soil Bioengineering. Unpublished Course Materials. USDA-NRCE, Syracuse, NY.

Dickerson, J. E., Jr. and J. V. Robinson. 1985. Microcosms as islands: a test of the MacArthur-Wilson equilibrium theory. *Ecology.* 66:966–980.

Dickson, K. L., T. Duke, and G. Loewengart. 1985. A synopsis: workshop on multispecies toxicity tests. pp. 248–253. In: *Multispecies Toxicity Testing*. J. Cairns, Jr. (ed.). Pergamon Press, New York.

Dillon, P. J. and F. H. Rigler. 1975. A simple method for predicting the capacity of a lake for development based on lake trophic status. *Journal of Fisheries Research Board of Canada.* 32:1519–1531.

Dinges, R. 1982. *Natural Systems for Water Pollution Control*. Van Nostrand Reinhold Co., New York.

Dobson, A. P., A. D. Bradshaw, and A. J. M. Baker. 1997. Hopes for the future: restoration ecology and conservation biology. *Science.* 277:515–522.

Dolan, T. J., S. E. Bayley, J. Zoltek, Jr., and A. J. Hermann. 1981. Phosphorus dynamics of a Florida freshwater marsh receiving treated wastewater. *Journal of Applied Ecology.* 18:205–219.

Done, T. J. 1992. Phase shifts in coral reef communities and their ecological significance. *Hydrobiologia.* 247:121–132.

Donkin, R. A. 1979. *Agricultural Terracing in the Aboriginal New World*. University of Arizona Press, Tucson, AZ.

Dopyera, C. 1996. The iron hypothesis. *Earth.* 5(5):26–33.

Dorigo, M. and L. M. Gambardella. 1997. Ant colonies for the traveling salesman problem. *BioSystems.* 43:73–81.

Drake, J. A. 1990a. The mechanics of community assemble and succession. *Journal of Theoretical Biology.* 147:213–233.

Drake, J. A. 1990b. Communities as assembled structures: Do rules govern pattern? *Trends in Ecology and Evolution.* 5:159–164.

Drake, J. A. 1991. Community-assembly mechanics and the structure of an experimental species ensemble. *American Naturalist.* 137:1–26.

Drake, J. A., T. E. Flum, G. J. Witteman, T. Voskuil, A. M. Hoylman, C. Creson, D. A. Kenny, G. R. Huxel, C. S. Larue, and J. R. Duncan. 1993. The construction and assembly of an ecological landscape. *Journal of Animal Ecology.* 62:117–130.

Drake, J. A., G. R. Huxel, and C. L. Hewitt. 1996. Microcosms as models for generating and testing community theory. *Ecology.* 77:670–677.

Drake, J. A., D. A. Kenny, and T. Voskuil. 1988. Environmental biotechnology. *BioScience.* 38:420.

Drengson, A. and Y. Inoue (eds.). 1995. *The Deep Ecology Movement: An Introductory Anthology.* North Atlantic Books, Berkeley, CA.

Drenner, R. W. and K. D. Hambright. 1999. Biomanipulation of fish assemblages as a lake restoration technique. *Archives of Hydrobiology.* 146:129–165.

Drenner, R. W., D. J. Day, S. J. Basham, J. D. Smith, and S. I. Jensen. 1997. Ecological water treatment system for removal of phosphorus and nitrogen from polluted water. *Ecological Applications.* 7:381–390.

Drew, R. D. and N. S. Schomer. 1984. An ecological characterization of the Caloosahatchee River/Big Cypress watershed. FWS/OBS-82/58.2. U.S. Dept. of the Interior. Washington, DC.

Drexler, K. E. 1986. *Engines of Creation: The Coming Era of Nanotechnology.* Doubleday, New York.

Drexler, K. E. 1992. *Nanosystems: Molecular Machinery, Manufacturing, and Computation.* John Wiley & Sons, New York.

Dreyer, G. D. and W. A. Niering. 1986. Evaluation of two herbicide techniques on electric transmission rights-of-way: development of relatively stable shrublands. *Environmental Management.* 10:113–118.

Drury, W. H. and I. C. T. Nisbet. 1971. Inter-relations between developmental models in geomorphology, plant ecology, and animal ecology. *General Systems.* 14:57–68.

Dublin, H. T., A. R. E. Sinclair, and J. McGlade. 1990. Elephants and fire as causes of multiple stable states in the Serengeti-Mara woodlands. *Journal of Animal Ecology.* 59:1147–1164.

Dudzik, M., J. Harte, A. Jassby, E. Lapan, D. Levy, and J. Rees. 1979. Some considerations in the design of aquatic microcosms for plankton studies. *International Journal of Environmental Studies.* 13:125–130.

Dunne, T. and L. B. Leopold. 1978. *Water in Environmental Planning.* W. H. Freeman, San Francisco, CA.

Dunstan, W. M. and K. R. Tenore. 1972. Intensive outdoor culture of marine phytoplankton enriched with treated sewage effluent. *Aquaculture.* 1:181–192.

Durall, D. M., W. F. J. Parsons, and D. Parkinson. 1985. Decomposition of timothy (Phleum pretense) litter on a reclaimed surface coal mine in Alberta, Canada. *Canadian Journal of Botany.* 63:1586–1594.

During, H. J., A. J. Schenkeveld, H. J. Verkaar, and J. H. Willems. 1985. Demography of short-lived forbs in chalk grassland in relation to vegetation structure. pp. 341–370. In: *The Population Structure of Vegetation*. J. White (ed.). Junk Publishing, Dordrecht, the Netherlands.

Ebeling, J. M. 1994. Monitoring and control. pp. 307–324. In: *Aquaculture Water Reuse Systems: Engineering Design and Management*. M. B. Timmons and T. M. Losordo (eds.). Elsevier, Amsterdam, the Netherlands.

Eckart, P. 1994. *Life Support and Biospherics*. Herbert Utz Publishers, Munchen, Germany.

Eddy, S. 1928. Succession of protozoa in cultures under controlled conditions. *Transactions of the American Microscopy Society.* 47:283–319.

Edie, L. C. 1974. Flow theories. pp. 1–108. In: D. C. Gazis (ed.). *Traffic Science*. John Wiley & Sons, New York.

Edmondson, A. C. 1992. *A Fuller Explanation: The Synergetic Geometry of R. Buckminster Fuller.* Van Nostrand Reinhold, New York.

Edmondson, W. T. 1991. *The Uses of Ecology: Lake Washington and Beyond.* University of Washington Press, Seattle, WA.

Edwards, C. A. (ed.). 1998. *Earthworm Ecology.* CRC Press, Boca Raton, FL.

Edwards, C. A., D. E. Reichle, and D. A. Crossley, Jr. 1970. The role of soil invertebrates in turnover of organic matter and nutrients. pp. 147–172. In: *Analysis of Temperate Forest Ecosystems*. D. E. Reichle (ed.). Springer-Verlag, Berlin.

Edwards, R. W. and J. W. Densem. 1980. Fish from sewage. pp. 221–270. In: *Applied Biology*, Vol. V. T. H. Coaker (ed.). Academic Press, London.

Edwards, R. Y. and C. D. Fowle. 1955. The concept of carrying capacity. *Transactions of the North American Wildlife Conference.* 20:589–602.

Egan, D. and E. A. Howell (eds.). 2001. *The Historical Ecology Handbook: A Restorationist's Guide to Reference Ecosystems*. Island Press, Washington, DC.

Egler, F. E. 1947. Effects of 2,4-D on woody plants in Connecticut 1946. *Journal of Forestry.* 45:449–452.

Egler, F. E. 1948. 2,4-D effects in Connecticut vegetation, 1947. *Ecology.* 29:382–386.

Egler, F. E. 1949. Herbicide effects in Connecticut vegetation, 1948. *Ecology.* 30:248–256.

Egler, F. E. 1950. Herbicide effects in Connecticut vegetation, 1949. *Botanical Gazette.* 112:76–85.

Egler, F. E. 1952a. Southeast saline everglades vegetation, Florida and its management. *Vegetation.* 3:213–265.

Egler, F. E. 1952b. Herbicide effects in Connecticut vegetation, 1950. *Journal of* Forestry. 50:198–204.

Egler, F. E. 1958. Science, industry, and the abuse of rights of way. *Science.* 127:573–580.

Egler, F. E. 1964. Pesticides — in our ecosystem. *American Scientist.* 52:110–136.

Egler, F. E. 1979. Polarization of the herbicide problem. *BioScience.* 29:339.

Egler, F. E. 1986. "Physics envy" in ecology. *Bulletin of the Ecological Society of America.* 67:233–235.

Egler, F. E. and S. R. Foote. 1975. *The Plight of the Rightofway Domain: Victim of Vandalism*, Part 1. Futura Media Services, Mt. Kisco, NY.

Ehrenfeld, D. 1981. *The Arrogance of Humanism.* Oxford University Press, New York.

Ehrlich, P. R. 1978. Preface. pp. vii–viii. In: R. van den Bosch. *The Pesticide Conspiracy.* Doubleday, Garden City, NY.

Ehrlich, P. R. 1986. Which animal will invade? pp. 79–95. In: *Ecology of Biological Invasions of North America and Hawaii*. H. A. Mooney and J. A. Drake (eds.). Springer-Verlag, New York.

Ehrlich, P. R. 1989. Attributes of invaders and the invading processes: vertebrates. pp. 315–328. In: *Biological Invasions: A Global Perspective*. J. A. Drake, H. A. Mooney, F. di Castri, R. H. Groves, F. J. Kruger, M. Rejmanek, and M. Williamson (eds.). John Wiley & Sons, Chichester, U.K.

Ehrlich, P. R. and H. A. Mooney. 1983. Extinction, substitution, and ecosystem services. *BioScience*. 33:248–254.

Eigen, M. and P. Schuster. 1979. *The Hypercycle — A Principle of Natural Self-Organization*. Springer-Verlag, Berlin.

El-Fadel, M., A. N. Findikakis, and J. O. Leckie. 1997. Gas simulation models for solid waste landfills. *Critical Reviews in Environmental Science and Technology*. 27:237–283.

Ellison, L. 1960. Influence of grazing on plant succession of rangelands. *Botanical Review*. 26:1–66.

Ellison, J. C. 1993. Mangrove retreat with rising sea level, Bermuda, Estuarine, Coastal and Shelf. *Science*. 37:75–87.

Ellison, J. C. and D. R. Stoddart. 1991. Mangrove ecosystem collapse during predicted sea-level rise: Holocene analogues and implications. *Journal of Coastal Research*. 7:151–165.

Elser, J. J., D. R. Dobberfuhl, N. A. MacKay, and J. H. Schampel. 1996. Organism size, life history, and N:P stoichiometry: towards a unified view of cellular and ecosystem processes. *BioScience*. 46:674–684.

Elton, C. and M. Nicholson. 1942. Fluctuations in numbers of muskrat (*Ondatra zibethica*) in Canada. *Journal of Animal Ecology*. 11:96–126.

Elton, C. S. 1958. *The Ecology of Invasions by Animals and Plants*. Methuen, London.

Elton, C. S. 1966. *The Pattern of Animal Communities*. Methuen, London.

Elwood, J. W., J. D. Newbold, R. V. O'Neill, and W. Van Winkle. 1983. Resource spiraling: an operational paradigm for analyzing lotic ecosystems. pp. 3–27. In: *Dynamics of Lotic Ecosystems*. T. D. Fontaine III and S. M. Bartell (eds.). Ann Arbor Science, Ann Arbor, MI.

Engelberg, J. and L. L. Boyarsky. 1979. The noncybernetic nature of ecosystems. *American* Naturalist. 114:317–324.

Engineering Technology Associates and Biohabitats. 1993. Design Manual for Use of Bioretention in Stormwater Management. Report prepared for Watershed Protection Branch, Prince George's County, Landover, MD.

Enriquez, S., C. M. Duarte, and K. Sand-Jensen. 1993. Patterns in decomposition rates among photosynthetic organisms: the importance of detritus C:N:P content. *Oecologia*. 94:457–471.

Erickson, R. C. 1964. Planting and misplanting. pp. 579–591. In: *Waterfowl Tomorrow*. Fish and Wildlife Service, USDI, Washington, DC.

Errington, P. L. 1963. *Muskrat Populations*. The Iowa State University Press, Ames, IA.

Escheman, R. no date. Soil bioengineering: an introduction. Unpublished course materials. Soil Bioengineering for Streambank, Shoreline and Slope Stabilization Short Course 1996, Rutgers University, New Brunswick, NJ.

Estrada, M., M. Alcaraz, and C. Marrase. 1987. Effects of turbulence on the composition of phytoplankton assemblages in marine microcosms. *Marine Ecology Progress Series*. 38:267–281.

Etnier, C. and B. Guterstam (eds.). 1991. *Ecological Engineering for Wastewater Treatment*. Bokskogen, Gothenburg, Sweden.

Evans, F. C. 1956. Ecosystem as the basic unit in ecology. *Science*. 123:1127–1128.

Evans, F. C. 1976. A sack of uncut diamonds: the study of ecosystems and the future of resources of mankind. *Journal of Ecology*. 64:1–39.

Ewel, J. J. 1986a. Designing agricultural ecosystems for the humid tropics. *Annual Reviews of Ecology and Systematics.* 17:245–271.

Ewel, J. J. 1986b. Invasibility: lessons from South Florida. pp. 214–230. In: *Ecology of Biological Invasions of North America and Hawaii.* H. A. Mooney and J. A. Drake (eds.). Springer-Verlag, New York.

Ewel, J. J. 1987. Restoration is the ultimate test of ecological theory. pp. 31–33. In: *Restoration Ecology.* W. R. Jordan III, M. E. Gilpin, and J. D. Aber (eds.). Cambridge University Press, Cambridge, U.K.

Ewel, J. J. and P. Hogberg. 1995. Experimental studies on islands. pp. 227–232. In: *Islands.* P. M. Vitousek, L. L. Loope, and H. Adersen (eds.). Springer-Verlag, New York.

Ewel, J. J. and 20 other authors. 1999. Deliberate introductions of species: research needs. *BioScience.* 49:619–630.

Ewel, K. C. 1997. Water quality improvement by wetlands. pp. 329–344. In: *Nature's Services.* G. C. Daily (ed.). Island Press, Washington, DC.

Ewel, K. C. and H. T. Odum (eds.). 1984. *Cypress Swamps.* University Presses of Florida, Gainesville, FL.

Ewing, R. C. and A. Macfarlane. 2002. Yucca Mountain. *Science.* 296:659–660.

Faber, S. and R. Costanza. 1987. The economic value of wetlands systems. *Journal of Environmental Management.* 24:41–51.

Fairweather, P. G. 1991. Implications of "supply-side" ecology for environmental assessment and management. *Trends in Ecology and Evolution,* 6:60–63.

Falk, J. H. 1976. Energetics of a sururban lawn ecosystem. *Ecology.* 57:141–150.

Falk, J. H. 1980. The primary productivity of lawns in a temperate environment. *Journal of Applied Ecology.* 17:689–696.

Farnworth, E. G., T. H. Tidrick, C. F. Jordan, and W. M. Smathers, Jr. 1981. The value of natural ecosystems: an economic and ecological framework. *Environmental Conservation.* 8:275–282.

Fath, B. D. and B. C. Patten. 2000. Ecosystem theory: network environ analysis. pp. 345–360. in: *Handbook of Ecosystem Theories and Management.* S. E. Jorgensen and F. Muller (eds.). Lewis Publishers, Boca Raton, FL.

Feigenbaum, E. A. and J. Feldman (eds.). 1963. *Computers and Thought.* McGraw-Hill, New York.

Fenchel, T. 1982. Ecology of heterotrophic microflagellates. IV. Quantitative occurrence and importance as bacterial consumers. *Marine Ecology Progress Series.* 9:35–42.

Fenchel, T. and T. H. Blackburn. 1979. *Bacteria and Mineral Cycling.* Academic Press, London.

Ferguson, B. K. 1994. *Stormwater Infiltration.* Lewis Publishers, Boca Raton, FL.

Ferguson, B. K. 1998. *Introduction to Stormwater: Concept, Purpose, Design.* John Wiley & Sons. New York.

Ferguson, E. S. 1992. *Engineering and the Mind's Eye.* MIT Press, Cambridge, MA.

Fetter, C. W., Jr., W. E. Sloey, and F. L. Spangler. 1976. Potential replacement of septic tank drain fields by artificial marsh wastewater treatment systems. *Ground Water.* 14:396–401.

Field, C. D. 1995. Impact of expected climate change on mangroves. *Hydrobiologia.* 295:75–81.

Findlay, S. 1995. Importance of surface-subsurface exchange in stream ecosystems: the hyporheic zone. *Limnology and Oceanography.* 40:159–164.

Findlay, S. E. G., E. Kiviat, W. C. Nieder, and E. A. Blair. 2002. Functional assessment of a reference wetland set as a tool for science, management and restoration. *Aquatic Sciences.* 64:107–117.

Finley, W. L. 1937. The beaver — conserver of soil and water. *Transactions of the North American Wildlife Conference.* 2:295–297.

Finn, M. 1996. Comparison of Mangrove Forest Structure and Function in a Mesocosm and Florida. Ph.D. dissertation, Georgetown University, Washington, DC.

Finson, B. and K. E. Taylor. 1986. Steinbeck and Ricketts: Fishing in the mind. *Oceanus.* 29:86–91.

Finstein, M. S. 1972. *Pollution Microbiology: A Laboratory Manual.* Marcel Dekker, New York.

Fitz, H. C., E. B. DeBellevue, R. Costanza, R. Boumans, T. Maxwell, L. Wainger, and F. H. Sklar. 1996. Development of a general ecosystem model for a range of scales and ecosystems. *Ecological Modelling.* 88:263–295.

Flack, S. and E. Furlow. 1996. America's least wanted. *Nature Conservancy.* 46(6):17–23.

Flanagan, P. W. 1986. Genetically engineered organisms and ecology. *Bulletin of the Ecological Society of America.* 67:26–30.

Flecker, A. S. 1996. Ecosystem engineering by a dominant detritivore in a diverse tropical stream. *Ecology.* 77:1845–1854.

Flemer, D. A., J. R. Clark, R. S. Stanley, C. M. Burdrick, and G. R. Plaia. 1993. The importance of physical scaling factors to benthic marine invertebrate recolonization of laboratory microcosms. *International Journal of Environmental Studies.* 44:161–179.

Flemming, H.-C. 1993. Biofilms and environmental protection. *Water Science* Technology. 27:1–10.

Florman, S. C. 1976. *The Existential Pleasures of Engineering.* St. Martin's Press, New York.

Fogel, D. B. 1995. *Evolutionary Computation.* Institute of Electrical and Electronics Engineers, New York.

Fogel, L. J. 1999. *Intelligence Through Simulated Evolution: Forty Years of Evolutionary Programming.* John Wiley & Sons. New York.

Fogel, R. 1980. Mycorrhizae and nutrient cycling in natural forest ecosystems. *New Phytologist.* 86:199–212.

Fogg, M. J. 1995. *Terraforming, Engineering Planetary Environments.* Society of Automotive Engineers, Warrendale, PA.

Folsome, C. E. 1979. *The Origin of Life: A Warm Little Pond.* W. H. Freeman, San Francisco, CA.

Folsome, C. E. and J. A. Hanson. 1986. The emergence of materially-closed-system ecology. pp. 269–288. In: *Ecosystem Theory and Application.* N. Polunin (ed.). John Wiley & Sons, Chichester, U.K.

Foote, R. H. and D. R. Cook. 1959. Mosquitoes of medical importance. *Agriculture Handbook No. 152*, Agricultural Research Service, U.S. Department of Agriculture, Washington, DC.

Forbes, S. A. 1887. The lake as a microcosm. *Illinois Natural History Survey Bulletin* 15:537–550. Reprinted in: E. J. Kormondy (ed.). 1965. *Readings in Ecology.* Prentice Hall, Englewood Cliffs, NJ.

Forcella, F. 1984. Ecological Biotechnology. *Bulletin of the Ecological Society of America.* 65:434–436.

Ford, J. 1989. The effects of chemical stress on aquatic species composition and community structure. pp. 99–144. In: *Ecotoxicology: Problems and Approaches.* Springer-Verlag, New York.

Ford, M. A., D. R. Cahoon, and J. C. Lynch. 1999. Restoring marsh elevation in a rapidly subsiding saltmarsh by thin-layer deposition of dredged material. *Ecological Engineering.* 12:189–205.

Forthman-Quick, D. L. 1984. An integrative approach to environmental engineering in zoos. *Zoo Biology* 3:65–78.

Fosberg, F. R. 1987. Artificial diversity. *Environmental Conservation.* 14:74.

Foster, G. R. 1977. *Soil Erosion: Prediction and Control.* Soil Conservation Society of America, Ankeny, IA.

Foster, K. L., F. W. Steimle, W. C. Muir, R. K. Kropp, and B. E. Conlin. 1994. Mitigation potential of habitat replacement: concrete artificial reef in Delaware Bay — preliminary results. *Bulletin of Marine Sciences.* 55:783–795.

Fox, M. D. and B. J. Fox. 1986. The susceptibility of natural communities to invasion. pp. 57–66. In: *The Ecology of Biological Invasions.* R. H. Groves and J. Burdon (eds.). Australian Academy of Science, Canberra, Australia.

Fraisse, T., E. Muller and O. Decamps. 1997. Evaluation of spontaneous species for the revegetation of reservoir margins. *Ambio.* 26:375–381.

Francis, G. R., J. J. Magnuson, H. A. Regier, and D. R. Talhelm. 1979. Rehabilitating Great Lakes Ecosystems. Technical Report No. 37, Great Lakes Fishery Commission. Ann Arbor, MI.

Frank, S. 1994. *City Peregrines.* Hancock House, Surrey, BC, Canada.

Frankel, O. H. and M. E. Soule. 1981. *Conservation and Evolution.* Cambridge University Press, Cambridge, U.K.

Frankland, J. C. 1966. Succession of fungi on decaying petioles of *Pteridium aquilinum. Journal of Ecology.* 54:41–63.

Fraser, L. H. and P. Keddy. 1997. The role of experimental microcosms in ecological research. *TREE.* 12:478–481.

French, M. J. 1988. *Invention and Evolution, Design in Nature and Engineering.* Cambridge University Press, Cambridge, U.K.

Fretwell, S. D. 1987. Food chain dynamics: the central theory of ecology? *Oikos.* 50:291–301.

Friedlander, S. K. 1994. The two faces of technology: changing perspectives in design for environment. pp. 217–227. In: *The Greening of Industrial Ecosystems.* B. R. Allenby and D. J. Richards (eds.). National Academy Press, Washington, DC.

Fulfer, K. 2002. Fresh Kills redevelopment offers huge ecological promise. *Land and Water.* 46(6):21–25.

Fuller, R. B. 1963. A comprehensive anticipatory design science. pp. 75–104. In: *No More Secondhand God, and Other Writings.* Anchor Books, Garden City, NY.

Fuller, R. B. 1983. *Inventions: The Patented Works of R. Buckminster Fuller.* St. Martin's Press, New York.

Funk and Wagnalls. 1973. *Standard College Dictionary.* Funk & Wagnalls, New York.

Furley, P. A. and W. W. Newey. 1988. *Geography of the Biosphere: An Introduction to the Nature, Distribution and Evolution of the World's Life Zones.* Butterworths, London.

Futrell, A. W., Jr. 1961. *Orientation to Engineering.* Charles E. Merrill Books, Inc., Columbus, OH.

Futuyma, D. J. 1979. *Evolutionary Biology.* Sinauer Associates, Sunderland, MA.

Gadgil, M. and V. D. Vartak. 1976. The sacred groves of Western Ghats in India. *Economic Botany.* 30:152–160.

Galatowitsch, S. M. and A. G. van der Valk. 1994. *Restoring Prairie Wetlands: An Ecological Approach.* Iowa State University Press, Ames, IA.

Galatowitsch, S. M. and A. G. van der Valk. 1996. The vegetation of restored and natural prairie wetlands. *Ecological Applications.* 6:102–112.

Galston, A. W. 1992. Photosynthesis as a basis for life support on Earth and in space. *BioScience.* 42:490–493.

Gamble, J. C. 1990. Mesocosms: statistical and experimental design considerations. pp. 188–196. In: *Enclosed Experimental Marine Ecosystems: A Review and Recommendations*. C. M. Lalli (ed.). Springer-Verlag, New York.

Gamez, R., A. Piva, A. Sittenfeld, E. Leon, J. Jimenez, and G. Mirabelli. 1993. Costa Rica's conservation program and National Biodiversity Institute (INBio). pp. 53–67. In: *Biodiversity Prospecting*. W. V. Reid (ed.). World Resources Institute, Washington, DC.

Gano, L. 1917. A study in physiographic ecology in northern Florida. *Botanical Gazette.* 63:338–371.

Gardner, R. H., W. M. Kemp, V. S. Kennedy, and J. E. Petersen (eds.). 2001. *Scaling Relations in Experimental Ecology.* Columbia University Press, New York.

Garshelis, D. L. 200. Delusions in habitat evaluation: measuring use, selection, and importance. pp. 111–164. In: *Research Techniques in Animal Ecology: Controversies and Consequences*. L. Boitani and T. K. Fuller (eds.). Columbia University Press, New York.

Gaudy, A. F., Jr. 1972. Biochemical oxygen demand. pp. 305–332. In: *Water Pollution Microbiology.* R. Mitchell (ed.). John Wiley & Sons, New York.

Gaudy, A. F., Jr. and E. T. Gaudy. 1966. Microbiology of waste waters. *Annual Review of Microbiology.* 20:319–336.

Gaufin, A. R. 1973. Use of aquatic invertebrates in the assessment of water quality. pp. 96–116. In: *Biological Methods for the Assessment of Water Quality.* J. Cairns, Jr. and K. L. Dickson (eds.). American Society for Testing and Materials. Philadelphia, PA.

Gause, G. F. 1934. *The Struggle for Existence*. The Williams & Wilkens Co., Baltimore, MD.

Gearing, J. N. 1989. The role of aquatic microcosms in ecotoxicologic research as illustrated by large marine systems. pp. 411–470. In: *Ecotoxicology: Problems and Approaches.* S. A. Levin, M. A. Harwell, J. R. Kelly, and K. D. Kimball (eds.). Springer-Verlag, New York.

Gemmell, R. P. 1977. *Colonization of Industrial Wasteland.* Edward Arnold, London.

Georgescu-Roegen, N. 1971. *The Entropy Law and the Economic Process*. Harvard University Press, Cambridge, MA.

Georgescu-Roegen, N. 1977. The steady-state and ecological salvation: a thermodynamic analysis. *BioScience.* 27:266–270.

Gersberg, R. M., B. V. Elkins, S. R. Lyon, and C. R. Goldman. 1986. Role of aquatic plants in wastewater treatment by artificial wetlands. *Water Research.* 20:363–368.

Gibbs, J. P., M. L. Hunter, Jr., and E. J. Sterling. 1998. *Problem-Solving in Conservation Biology and Wildlife Management.* Blackwell Science, Malden, MA.

Giddings, J. M. and G. K. Eddlemon. 1977. The effects of microcosm size and substrate type on aquatic microcosm behavior and arsenic transport. *Archives of Environmental Contamination and Toxicology.* 6:491–505.

Giddings, J. M. and P. J. Franco. 1985. Calibration of laboratory bioassays with results from microcosms and ponds. pp. 104–119. In: *Contaminants in Aquatic Ecosystems*. T. P. Boyle (ed.). ASTM, Philadelphia, PA.

Giere, O. 1993. *Meiobenthology: The Microscopic Fauna in Aquatic Sediments*. Springer-Verlag, Berlin.

Giesy, J. P., Jr. (ed.). 1980. *Microcosms in Ecological Research.* CONF-781101, DOE Symposium Series, Technical Information Center, U.S. Dept. of Energy, Washington, DC.

Giesy, J. P., Jr. and P. M. Allred. 1985. Replicability of aquatic multispecies test systems. pp. 187–247. In: *Multispecies Toxicity Testing*. J. Cairns, Jr. (ed.). Pergamon Press, New York.

Giesy, J. R., Jr. and E. P. Odum. 1980. Microcosmology: introductory comments. pp. 1–13. In: *Microcosms in Ecological Research*. J. P. Giesy, Jr. (ed.). CONF-781101, DOE Symposium Series, Technical Information Center, U.S. Dept. of Energy, Washington, DC.

Gilbert, O. L. and P. Anderson. 1998. *Habitat Creation and Repair.* Oxford University Press, Oxford, U.K.

Gill, A. M. and P. B. Tomlinson. 1971. Studies on the growth of red mangrove (*Rhizophora mangle L.*). 2. Growth and differentiation of aerial roots. *Biotropica.* 3:63–77.

Gill, A. M. and P. B. Tomlinson. 1977. Studies on the growth of red mangrove (R*hizophora mangle L.*). 4. The adult root system. *Biotropica.* 9:145–155.

Gillett, J. D. 1973. The mosquito: still man's worst enemy. *American Scientist.* 61:430–436.

Gillett, J. W. and J. D. Gile. 1976. Pesticide fate in terrestrial laboratory ecosystems. *International Journal of Environmental Studies.* 10:15–22.

Gillett, J. W. (ed.). 1986. Potential impacts of environmental release of biotechnology products: assessment, regulation, and research needs. *Environmental Management.* 10:433–563.

Gilliland, M. W. and P. G. Risser. 1977. The use of systems diagrams for environmental impact assessment: procedures and an application. *Ecological Modelling.* 3:183–209.

Gilpin, M. E. and M. E. Soule. 1986. Minimum viable populations: processes of species extinction. pp. 19–34. In: *Conservation Biology.* M. E. Soule (ed.). Sinauer Assoc., Sunderland, MA.

Givnish, T. J., E. L. Burkhardt, R. E. Happel, and J. D. Weintraub. 1984. Carnivory in the bromeliad Brocchina reducta, with a cost/benefit model for the general restriction of carnivorous plants to sunny, moist, nutrient-poor habitats. *American Naturalist.* 124:479–497.

Gleason, H. A. 1922. On the relation between species and area. *Ecology.* 3:158–162.

Gliessman, S.. R. 1988. Local resource use systems in the tropics: taking pressure off the forests. pp. 53–70. In: *Tropical Rainforests: Diversity and Conservation.* F. Almeda and C. M. Pringle (eds.). California Academy of Sciences, San Francisco, CA.

Gliessman, S. R. 1991. Ecological basis of traditional management of wetlands in tropical Mexico: learning from agroecosystem models. pp. 211–229. In: *Biodiversity: Culture, Conservation, and Ecodevelopment.* M. L. Oldfield and J. B. Alcorn (eds.). Westview Press, Boulder, CO.

Gloyna, E. F., J. F. Malina, Jr., and E. M. Davis (eds.). 1976. *Ponds as a Wastewater Treatment Alternative.* Center for Research in Water Resources, University of Texas, Austin, TX.

Glynn, P. W. 1997. Bioerosion and coral-reef growth: a dynamic balance. pp. 68–95. In: *Life and Death of Coral Reefs.* C. Birkeland (ed.). Chapman & Hall, New York.

Godfray, H. C. J. and B. T. Grenfell. 1993. The continuing quest for chaos. *Trends in Ecology and Evolution.* 8:43–44.

Godfrey, P. J. and M. M. Godfrey. 1976. *Barrier Island Ecology of Cape Lookout National Seashore and Vicinity, North Carolina.* National Park Service Scientific Monograph Series, No. 9. Washington, DC.

Godfrey, P. J., E. R. Kaynor, S. Pelczarski, and J. Benforado (eds.). 1985. *Ecological Considerations in Wetlands Treatment of Municipal Wastewaters.* Van Nostrand Reinhold Co., New York.

Godfrey, P. J., S. P. Leatherman, and R. Zaremba. 1979. A geobotanical approach to classification of barrier beach systems. pp. 99–126. In: *Barrier Islands.* S. P. Leatherman (ed.). Academic Press, New York.

Godish, T. 2001. *Indoor Environmental Quality.* Lewis Publishers, Boca Raton, FL.

Goldberg, D. E. 1989. *Genetic Algorithms in Search, Optimization, and Machine Learning*. Addison-Wesley Publishing Co., Reading, MA.

Goldberg, L. H. and W. Middleton (eds.). 2000. *Green Electronics/Green Bottom Line: Environmentally Responsible Engineering*. Newnes, Boston, MA.

Goldman, J. C. 1984. Oceanic nutrient cycles. pp. 137–170. In: *Flows of Energy and Materials in Marine Ecosystems: Theory and Practice*. M. J. R. Fasham (ed.). Plenum Press, New York.

Goldman, J. C. and J. H. Ryther. 1976. Waste reclamation in an integrated food chain system. pp. 197–214. In: *Biological Control of Water Pollution*. J. Tourbier and R. W. Pierson, Jr. (eds.). University of Pennsylvania Press, Philadelphia, PA.

Goldman, J. C., K. R. Tenore, and H. I. Stanley. 1973. Inorganic nitrogen removal from wastewater: effect on phytoplankton growth in coastal marine waters. *Science*. 180:955–956.

Goldman, J. C., K. R. Tenore, J. H. Ryther, and N. Corwin. 1974a. Inorganic nitrogen removal in a combined tertiary treatment — marine aquaculture system — I. Removal efficiencies. *Water Research*. 8:45–54.

Goldman, J. C., K. R. Tenore, and H. I. Stanley. 1974b. Inorganic nitrogen removal in a combined tertiary treatment — marine aquaculture system — II. Algal bioassays. *Water Research*. 8:55–59.

Goldsmith, W. and L. Bestmann. 1992. An overview of bioengineering for shore protection. *Proceedings of the Conference of the International Erosion Control Association*. 23:267–272.

Golley, F. B. 1966. The variety of occupations in human communities compared with the variety of species in natural communities. *Bulletin of the Georgia Academy of* Sciences. 24:1–5.

Golley, F. B. (ed.). 1977. *Ecological Succession: Benchmark Paper in Ecology*. Vol. 5. Dowden, Hutchinson & Ross, Inc., Stroudsburg, PA.

Golley, F. B. 1993. *A History of the Ecosystem Concept in Ecology*. Yale University Press, New Haven, CT.

Golley, F. B., J. T. McGinnis, R. G. Clements, G. I. Child, and M. J. Duever. 1975. *Mineral Cycling in a Tropical Moist Forest Ecosystem*. University of Georgia Press, Athens, GA.

Golueke, C. G. 1977. *Biological Reclamation of Solid Wastes*. Rodale Press, Emmaus, PA.

Golueke, C. G. 1991. Early compost research. pp. 37–39. In: *The Biocycle Guide to the Art and Science of Composting*. The JG Press, Emmaus, PA.

Golueke, C. G. and W. J. Oswald. 1973. An algal regenerative system for single-family farms and villages. *Compost Science*. 14(3):12–15.

Gomez-Pompa, A., H. L. Morales, E. J. Avilla, and J. J. Avilla. 1982. Experiences in traditional hydraulic agriculture. pp. 327–342. In: *Maya Subsistence*. K. V. Flannery (ed.). Academic Press, New York.

Goodman, D. 1975. The theory of diversity-stability relationships in ecology. *Quarterly Review of Biology*. 50:237–266.

Goodwin, R. H. and W. A. Niering. 1975. *Inland Wetlands of the United States*. National Park Service, Natural History Theme Studies, No. 2, USDI, Washington, DC.

Gordon, M. S., D. J. Chapman, L. Y. Kawasaki, E. Tarifeno-Silva, and D. P. Yu. 1982. Aquacultural approaches to recycling of dissolved nutrients in secondarily treated domestic wastewater-IV. Conclusions, design and operational considerations for artificial food chains. *Water Research*. 16:67–71.

Gordon R. W. 1969. A proposed energy budget of a soybean field. *Bulletin of the Georgia Academy of Sciences*. 27:41–52.

Gordon, R. W., R. J. Beyers, E. P. Odum, and R. G. Eagon. 1969. Studies of a simple laboratory microecosystem: bacterial activities in a heterotrophic succession. *Ecology.* 50:86–100.

Gordon, T. T. and A. S. Cookfair. 1995. *Patent Fundamentals for Scientists and Engineers.* CRC Press, Boca Raton, FL.

Gore, J. A., F. L. Bryant, and D. J. Crawford. 1995. River and stream restoration. pp. 245–275. In: *Rehabilitating Damaged Ecosystems.* J. Cairns, Jr. (ed.). Lewis Publishers, Boca Raton, FL.

Gorman, M. 1979. *Island Ecology.* Chapman and Hall. London.

Gosselink, J. G., E. P. Odum, and R. M. Pope. 1973. The value of the tidal marsh. *Working Paper No. 3.* Urban and Regional Development Center, University of Florida, Gainesville, FL.

Gotaas, H. B., W. J. Oswald, and H. F. Ludwig. 1954. Photosynthetic reclamation of organic wastes. *Scientific Monthly New York.* 79:368–378.

Goto, E., K. Kurata, M. Hayashi, and S. Sase (eds.). 1997. *Plant Production in Closed Ecosystems.* Kluwer Academic Publ., Dordrecht, the Netherlands.

Gottlieb, S. J. and M. E. Schweighofer. 1996. Oysters and the Chesapeake Bay ecosystem: a case for exotic species introduction to improve environmental quality? *Estuaries.* 19:639–650.

Gould, F. 1991. The evolutionary potential of crop pests. *American Scientist.* 79:496–507.

Gould, S. J. 1988. On replacing the idea of progress with an operational notion of directionality. pp. 319–338. In: *Evolutionary Progress.* M. H. Nitecki (ed.). University of Chicago Press, Chicago, IL.

Gracias, D. H., J. Tien, T. L. Breen, C. Hsu, and G. M. Whitesides. 2000. Forming electrical networks in three dimensions by self-assembly. *Science.* 289:1170–1172.

Graedel, T. 1994. Industrial ecology: definition and implementation. pp. 23–41. In: *Industrial Ecology and Global Change.* R. Socolow, C. Andrews, F. Berkhout, and V. Thomas (eds.). Cambridge University Press, Cambridge, U.K.

Graedel, T. E. and B. R. Allenby. 2003. *Industrial Ecology*, 2nd ed. Pearson Education, Inc., Upper Saddle River, NJ.

Graney, R. L., J. H. Kennedy, and J. H. Rodgers, Jr. (eds.). 1994. *Aquatic Mesocosm Studies in Ecological Risk Assessment.* CRC Press, Inc., Boca Raton, FL.

Grant, R. R., Jr. and R. Patrick. 1970. Tinicum Marsh as a water purifier. pp. 105–123. In: *Two Studies of Tinicum Marsh, Delaware and Philadelphia Counties, PA.* The Conservation Foundation, Washington, DC.

Graf, W. L. 1988. Applications of catastrophe theory in fluvial geomorphology. pp. 33–47. In: M. G. Anderson (ed.). *Modelling Geomorphological Systems.* John Wiley & Sons, Chichester, U.K.

Grant, E. L. and W. G. Ireson. 1964. *Principles of Engineering Economy.* The Ronald Press Co., New York.

Grant, V. 1991. *The Evolutionary Process.* Columbia University Press, New York.

Gray, D. H. and A. T. Leiser. 1982. *Biotechnical Slope Protection and Erosion Control.* Krieger Publishing, Malabar, FL.

Gray, J. S. 1974. Animal-sediment relationships. *Annual Review of Oceanography and Marine* Biology. 12:223–261.

Gray, K. R. and A. J. Biddlestone. 1974. Decomposition of urban waste. pp. 743–775. In: *Biology of Plant Litter Decomposition.* C. H. Dickinson and G. J. F. Pugh. (eds.). Academic Press. London.

Gray, N. F. 1989. *Biology of Wastewater Treatment.* Oxford University Press, Oxford, U.K.

Gray, N. F. and C. A. Hunter. 1985. Heterotrophic slimes in Irish rivers. *Water Research.* 19:685–691.

Greaves, T. (ed.). 1994. *Intellectual Property Rights for Indigenous People: A Source Book.* Society for Applied Anthropology, Oklahoma City, OK.

Green, H. R. 1932. Financing sewage disposal. *Sewage Works Journal.* 4:288–295.

Greene, H. C. and J. T. Curtis. 1953. The re-establishment of an artificial prairie in the University of Wisconsin Arboretum. *Wild Flower.* 29:77–88.

Greenfield, V. 1986. *Making Do or Making Art: A Study of American Recycling.* UMI Research Press, Ann Arbor, MI.

Gregory, K. J. 1992. Vegetation and river channel process interactions. pp. 255–270. In: *River Conservation and Management.* P. J. Boon, P. Calow, and G. E. Petts (eds.). John Wiley & Sons, Chichester, U.K.

Gregory, S. V., F. J. Swanson, W. A. McKee, and K. W. Cummins. 1991. An ecosystem perspective of riparian zones. *BioScience.* 41:540–551.

Grice, G. D. and M. R. Reeve. (eds.). 1982. *Marine Mesocosms.* Springer-Verlag, New York.

Griffin, D. R. 1974. Preface. In: *Animal Engineering.* D. R. Griffin (ed.). W. H. Freeman, San Francisco, CA.

Grime, J. P. 1979. *Plant Strategies and Vegetation Processes.* John Wiley & Sons, Chichester, U.K.

Grime, J. P. 1974. Vegetation classification by reference to strategies. *Nature.* 250:26–31.

Grimm, V. and C. Wissel. 1997. Babel, or the ecological stability discussions: an inventory of terminology and a guide for avoiding confusion. *Oecologia.* 109:323–334.

Grmek, M. D. 1972. A survey of the mechanical interpretations of life from the Greek atomists to the followers of Descartes. pp. 181–195. In: *Biology, History, and Natural Philosophy.* A. D. Breck and W. Yourgrau (eds.). Plenum Press, New York.

Grodins, F. S. 1963. *Control Theory and Biological Systems.* Columbia University Press, New York.

Grossman, E. 2002. *Watershed: The Undamming of America.* Counterpoint, New York.

Grove, R. S. and C. J. Sonu. 1985. Fishing reef planning in Japan. pp. 187–251. In: *Artificial Reefs, Marine and Freshwater Applications.* F. M. D'Itri (ed.). Lewis Publishers, Chelsea, MI.

Grove, R. S., M. Nakamura, H. Kakimoto, and C. J. Sonu. 1994. Aquatic habitat technology innovation in Japan. *Bulletin of Marine Science.* 55:276–294.

Grove, R. S., C. J. Sonu, and M. Nakamura. 1991. Design and engineering of manufactured habitats for fisheries enhancement. pp. 109–152. In: *Artificial Habitats for Marine and Freshwater Fisheries.* W. Seaman, Jr. and L. M. Sprague (eds.). Academic Press, San Diego, CA.

Groves, R. H. 1989. Ecological control of invasive terrestrial plants. pp. 437–461. In: *Biological Invasions: A Global Perspective.* J. A. Drake, H. A. Mooney, F. di Castri, R. H. Groves, F. J. Kruger, M. Rejmanek, and M. Williamson (eds.). John Wiley & Sons, Chichester, U.K.

Gruber, D. and J. Diamond. 1988. *Automated Biomonitoring: Living Sensors as Environmental Monitors.* John Wiley & Sons, New York.

Gunderson, L. H., C. S. Holling, L. Pritchard, Jr., and G. D. Peterson. 2002. Resilience of large-scale resource systems. pp. 3–20. In: *Resilience and the Behavior of Large-Scale Systems.* L. H. Gunderson and L. Pritchard, Jr. (eds.). Island Press, Washington, DC.

Gunn, A. S. and P. A. Vesilind. 1986. *Environmental Ethics for Engineers.* Lewis Publishers, Chelsea, MI.

Gunn, C. R. and J. V. Dennis. 1976. *World Guide to Tropical Drift Seeds and Fruits.* The New York Times Book Co., New York.

Gunnison, D. and J. W. Barko. 1989. The rhizosphere ecology of submersed macrophytes. *Water Resources Bulletin.* 25:193–201.

Gurney, W. S. C. and J. H. Lawton. 1996. The population dynamics of ecosystem engineers. *Oikos.* 76:273–283.

Guterstam, B. 1996. Demonstrating ecological engineering for wastewater treatment in a Nordic climate using aquaculture principles in a greenhouse mesocosm. *Ecological Engineering.* 6:73–97.

Guterstam, B. and J. Todd. 1990. Ecological engineering for wastewater treatment and its application in New England and Sweden. *Ambio.* 19:173–175.

Gutierrez, L. T. and W. R. Fey. 1980. *Ecosystem Succession.* MIT Press, Cambridge, MA.

Hackney, C. T. and S. M. Adams. 1992. Aquatic communities of the southeastern United States: past, present, and future. pp. 747–762. In: *Biodiversity of the Southeastern United States, Aquatic Communities.* C. T. Hackney, S. M. Adams, and W. H. Martin (eds.). John Wiley & Sons, New York.

Haemig, P. D. 1978. Aztec Emperor Auitzotl and the great-tailed grackle. *Biotropica.* 10:11–17.

Haeuber, R. and J. Franklin. 1996. Perspectives on ecosystem management. *Ecological Applications.* 6:692–693.

Hagen, J. B. 1992. *An Entangled Bank: The Origins of Ecosystem Ecology.* Rutgers University Press, New Brunswick, NJ.

Haggett, P. and R. J. Chorley. 1969. *Network Analysis in Geography.* Edward Arnold, London.

Haila, Y. 1999. Islands and fragments. pp. 234–264. In: *Maintaining Biodiversity in Forest Ecosystems.* M. L. Hunter, Jr. (ed.). Cambridge University Press. Cambridge, U.K.

Haines, E. B. 1979. Growth dynamics of cordgrass, *Spartina alterniflora Loisel.*, on control and sewage sludge fertilized plots in a Georgia saltmarsh. *Estuaries.* 2:50–53.

Hairston, N. G., Sr. 1989. Hard choices in ecological experimentation. *Herpetologica.* 45:119–122.

Hairston, N. G., Sr. 1989. *Ecological Experiments.* Cambridge University Press, Cambridge, U.K.

Hairston, N. G., J. D. Allan, R. K. Colwell, D. J. Futuyma, J. Howell, M. D. Lubin, J. Mathias, and J. H. Vandermeer. 1968. The relationship between species diversity and stability: an experimental approach with protozoa and bacteria. *Ecology.* 49: 1091–1101.

Halacy, D. S., Jr. 1965. *Bionics: The Science of "Living" Machines.* Holiday House, New York.

Hales, L. 1995. Accomplishments of the Corps of Engineers dredging research program. *Journal of Coastal Research.* 11:68–88.

Hall, C. A. S. 1990. Sanctioning resource depletion: economic development and neo-classical economics. *The Ecologist.* 20 (3):99–104.

Hall, C. A. S. 1995a. Introduction: what is maximum power? pp. xiii–xvi. In: *Maximum Power.* C. A. S. Hall (ed.). University Press of Colorado, Niwot, CO.

Hall, C. A. S. 1995b. Introduction to the section on environmental management and engineering. p. 99. In: *Maximum Power: The Ideas and Applications of H. T. Odum.* University Press of Colorado, Niwot, CO.

Hall, C. A. S., J. W. Day, Jr., and H. T. Odum. 1977. A circuit language for energy and matter. pp. 38–48. In: *Ecosystem Modeling in Theory & Practice: An Introduction with Case Histories.* John Wiley & Sons, New York.

Hall, C. A. S., D. Lindenberger, R. Kummel, T. Kroeger, and W. Eichhorn. 2001. The need to reintegrate the natural sciences with economics. *BioScience.* 51:663–673.

Hall, D. J., W. E. Cooper, and E. E. Werner. 1970. An experimental approach to the production of animal communities. *Limnology and Oceanography.* 15:839–928.

Hall, L. S., P. R. Krausman, and M. L. Morrison. 1997. The habitat concept and a plea for standard terminology. *Wildlife Society Bulletin.* 25:173–182.

Hammer, D. A. (ed.). 1989. *Constructed Wetlands for Wastewater Treatment.* Lewis Publishers, Chelsea, MI.

Hammerschlag, R. 1999. Personal communication.

Hammons, A. S. (ed.). 1981. *Methods for Ecological Toxicology: A Critical Review of Laboratory Multispecies Tests.* Ann Arbor Science, Ann Arbor, MI.

Handel, S. N. 1997. The role of plant-animal mutualisms in the design and restoration of natural communities. pp. 111–132. In: *Restoration Ecology and Sustainable Development.* K. M. Urbanska, N. R. Webb, and P. J. Edwards (eds.). Cambridge University Press, Cambridge, U.K.

Hannon, B. 1986. Ecosystem control theory. *Journal of Theoretical Biology.* 121:417–437.

Hannon, B. and J. Bentsman. 1991. Control theory in the study of ecosystems: a summary view. pp. 240–260. In: *Theoretical Studies of Ecosystems: The Network Perspective.* M. Higashi and T. P. Burns (eds.). Cambridge University Press, Cambridge, U.K.

Hanski, I. and Y. Cambefort (eds.). 1991. *Dung Beetle Ecology.* Princeton University Press, Princeton, NJ.

Hanski, I. and D. Simberloff. 1997. The metapopulation approach, its history, conceptual domain, and application to conservation. pp. 5–26. In: *Metapopulation Biology: Ecology, Genetics, and Evolution.* I. Hanski and M. E. Gilpin (eds.). Academic Press, San Diego, CA.

Hardenbergh, W. A. 1942. *Sewerage and Sewage Treatment*, 2nd ed. International Textbook Co., Scranton, PA.

Hardenburg, W. E. 1922. *Mosquito Eradication.* McGraw-Hill, New York.

Hardesty, D. L. 1975. The niche concept: suggestions for its use in human ecology. *Human Ecology.* 3:71–84.

Hardin, G. 1960. The competitive exclusion principle. *Science.* 131:1292–1297.

Hardin, G. 1968. The tragedy of the commons. *Science.* 162:1243–1248.

Hardin, G. 1986. Cultural carrying capacity: a biological approach to human problems. *BioScience.* 36:599–606.

Hardin, G. 1993. *Living Within Limits: Ecology, Economics and Population Taboos.* Oxford University Press, New York.

Harlan, J. R. 1975. *Crops and Man.* American Society of Agronomy, Madison, WI.

Harlin, J. M. and G. M. Berardi (eds.). 1987. *Agricultural Soil Loss.* Westview Press, Boulder, CO.

Harman, P. M. 1998. *The Natural Philosophy of James Clerk Maxwell.* Cambridge University Press, Cambridge, U.K.

Harmer, R. and G. Kerr. 1995. Creating woodlands: to plant trees or not? pp. 113–128. In: *The Ecology of Woodland Creation.* R. Ferris-Kaan (ed.). John Wiley & Sons, Chichester, U.K.

Harper, J. L. 1974. Agricultural ecosystems. *Agro-Ecosystems.* 1:1–6.

Harris, A. S., I. Wengatz, M. Wortberg, S. B. Kreissig, S. J. Gee, and B. D. Hammock. 1998. Development and application of immunoassys for biological and environmental monitoring. pp. 135–154. In: *Multiple Stresses in Ecosystems.* J. J. Cech, Jr., B. W. Wilson, and D. G. Crosby (eds.). Lewis Publishers, Boca Raton, FL.

Harris, J. A., P. Birch, and J. Palmer. 1996. *Land Restoration and Reclamation: Principles and Practice.* Longman, Essex, U.K.

Harris, L. D. 1984. *The Fragmented Forest, Island Biogeography Theory and the Preservation of Biotic Diversity.* University of Chicago Press. Chicago, IL.

Harris, L. D. 1988. Edge effects and conservation of biotic diversity. *Conservation Biology.* 2:330–332.

Harris, L. D. and P. Kangas. 1989. Reconsideration of the habitat concept. *Transactions of the North American Wildlife and Natural Resources Conference.* 53:137–144.

Hart, C. W., Jr. and S. L. H. Fuller (eds.). 1974. *Pollution Ecology of Freshwater Invertebrates.* Academic Press, New York.

Hart, D. D. and N. L. Poff. 2002. A special section on dam removal and river restoration. *BioScience.* 52:653–655.

Hart, M. M., R. J. Reader, and J. N. Klironomos. 2001. Biodiversity and ecosystem function: alternate hypotheses or a single theory? *Bulletin of the Ecological Society of America.* 82:88–90.

Hart, R. D. 1980. A natural ecosystem analog approach to the design of a successional crop system for tropical forest environments. In: *Tropical Succession: Supplement to Biotropica.* J. Ewel (ed.). 12(2):73–82.

Hartenstein, R. 1986. Earthworm biotechnology and global biogeochemistry. *Advances in Ecological Research.* 15: 379–409.

Hartland-Rowe, R. and P. B. Wright. 1975. Effects of sewage effluent on a swampland stream. *Verhandlungen Internationale Vereinigung für Theoretische und Angewandte Limnologie.* 19:1575–1583.

Harvey, H. T. and M. N. Josselyn. 1986. Wetlands restoration and mitigation policies: comment. *Environmental Management.* 10:567–569.

Harvey, R. M. and J. L. Fox. 1973. Nutrient removal using Lemna minor. *Journal of the Water Pollution Control Federation.* 45:1928–1938.

Harwell, C. C. 1989. Regulatory framework for ecotoxicology. pp. 497–516. In: *Ecotoxicology: Problems and Approaches.* S. A. Levin, M. A. Harwell, J. R. Kelly, and K. D. Kimball (eds.). Springer-Verlag, New York.

Haselwandter, K. 1997. Soil micro-organisms, mycorrhiza, and restoration ecology. pp. 65–80. In: *Restoration Ecology and Sustainable Development.* K. M. Urbanska, N. R. Webb, and P. J. Edwards (eds.). Cambridge University Press, Cambridge, U.K.

Hashimoto, Y., G. P. A. Bot, W. Day, H.-J. Tantau, and H. Nonami (eds.). 1993. *The Computerized Greenhouse: Automatic Control Application in Plant Production.* Academic Press, San Diego, CA.

Hassall, M. and S. P. Rushton. 1985. The adaptive significance of coprophagous behavior in the terrestrial isopod, *Porcellio scaber. Pedobiologia.* 28:169–175.

Hastings, A. 1977. Spatial heterogeneity and the stability of predator prey systems. *Theoretical Population Biology.* 12:37–48.

Hastings, A. 1978. Spatial heterogeneity and the stability of predator-prey systems: predator-mediated coexistence. *Theoretical Population Biology.* 14:380–395.

Hastings, A. 1986. The invasion question. *Journal of Theoretical Biology.* 121:211–220.

Hastings, A., C. L. Hom, S. Ellner, P. Turchin, and H. C. J. Godfray. 1993. Chaos in ecology. *Annual Review of Ecology and Systematics.* 24:1–33.

Hauert, C., S. De Monte, J. Hofbauer, and K. Sigmund. 2002. Volunteering as red queen mechanism for cooperation in public goods games. *Science.* 296:1129–1132.

Haug, R. T. 1980. *Composting Engineering: Principles and Practice.* Technomic Publishing, Lancaster, PA.

Haven, K. J., L. M. Varnell, and J. G. Bradshaw. 1995. An asssessment of ecological conditions in a constructed tidal marsh and two natural reference tidal marshes in coastal Virginia. *Ecological Engineering.* 4:117–141.

Hawkes, H. A. 1963. *The Ecology of Waste Water Treatment.* Macmillan, New York.

Hawkes, H. A. 1965. The ecology of sewage bacteria beds. pp. 119–146. In: *Ecology and the Industrial Society.* G. T. Goodman, R. W. Edwards, and J. M. Lambert (eds.). John Wiley & Sons, New York.

Hawksworth, D. L. 1996. Microorganisms: the neglected rivets in ecosystem maintenance. pp. 130–138. In: *Biodiversity, Science and Development.* F. di Castri and T. Younes (eds.). CAB International, Wallingford, U.K.

Hayes, T. D., H. R. Isaacson, K. R. Reddy, D. P. Chynoweth, and R. Biljetina. 1987. Water hyacinth systems for water treatment. pp. 121–140. In: *Aquatic Plants for Water Treatment and Resource Recovery.* K. R. Reddy and W. H. Smith (eds.). Magnolia Publishing, Orlando, FL.

Haynes, D. L., R. L. Tummala, and T. L. Ellis. 1980. Ecosystem management for pest control. *BioScience* 30:690–696.

Haynes, R. H. and C. P. McKay. 1991. The implantation of life on Mars: feasibility and motivation. *Advances in Space Research.* 12:133–140.

Hedgpeth, J. W. 1978a. As blind men see the elephant: the dilemma of marine ecosystem research. pp. 3–15. In: *Estuarine Interactions.* M. L. Wiley (ed.). Academic Press, New York.

Hedgpeth, J. W. (ed.). 1978b. *The Outer Shores*, Parts 1 and 2. Mad River Press, Eureka, CA.

Heede, B. H. 1985. Channel adjustments to the removal of log steps: an experiment in a mountain stream. *Environmental Management.* 9:427–432.

Heimbach, F., J. Berndt, and W. Pflueger. 1994. Fate and biological effects of an herbicide on two artificial pond ecosystems of different size. pp. 303–320. In: *Aquatic Mesocosm Studies in Ecological Risk Assessment.* R. L. Graney, J. H. Kennedy, and J. H. Rodgers, Jr. (eds.). CRC Press, Boca Raton, FL.

Hendrix, P. F., C. L. Langner, and E. P. Odum. 1982. Cadmium in aquatic microcosms: implications for screening the ecological effects of toxic substances. *Environmental Management.* 6:543–553.

Hendrix, P. F., R. W. Parmelee, D. A. Crossley, Jr., D. C. Coleman, E. P. Odum, and P. M. Groffman. 1986. Detritus food webs in conventional and no-tillage agroecosystems. *BioScience.* 36:374–380.

Herbold, B. and P. B. Moyle. 1986. Introduced species and vacant niches. *American Naturalist.* 128:751–760.

Hergarten, S. 2002. Self-Organized Criticality in Earth Systems. Springer, Berlin.

Herrera, R., T. Merida, N. Stark, and C. F. Jordan. 1978. Direct phosphorus transfer from leaf litter to roots. *Naturwissenschaften.* 65:208–209.

Herricks, E. 1993. Integrating ecological and engineering design elements. pp. 111–117. In: *Symposium on Ecological Restoration.* Office of Policy, Planning, and Evaluation, U.S. Environmental Protection Agency, Washington, DC.

Hessen, D. O. 1997. Stoichiometry in food webs. Lotka revisited. *Oikos.* 79:195–200.

Heuvelmans, M. 1974. *The River Killers.* Stackpole Books, Harrisburg, PA.

Hewlett, J. D. and W. L. Nutter. 1969. *An Outline of Forest Hydrology.* University of Georgia Press, Athens, GA.

Hey, R. D. 1996. Environmentally sensitive river engineering. pp. 80–105. In: *River Restoration.* G. Petts and P. Calow (eds.). Blackwell Science, Cambridge, MA.

Heywood, V. H. 1996. The importance of urban environments in maintaining biodiversity. pp. 543–550. In: *Biodiversity, Science and Development: Towards a New Partnership.* F. di Castri and T. Younes (eds.). CAB International, Wallingford, U.K.

Hickey, C. W. 1988. Oxygen uptake kinetics and microbial biomass of river sewage fungus biofilms. *Water Research.* 22:1365–1373.

Hickman, H. L., Jr. 1999. *Principles of Integrated Solid Waste Management*. American Academy of Environmental Engineers, Annapolis, MD.

Higashi, M., B. C. Patten, and T.P. Burns. 1993. Network trophic dynamics: the nodes of energy utilization in ecosystems. Ecological Modelling. 66:1–42.

Higashi, M. and B. C. Patten 1989. Dominance of indirect causality in ecosystems. *American Naturalist.* 133:288–302.

Hergarten, S. 2002. *Self-Organized Criticality in Earth Systems*. Springer, Berlin.

Higgins, J. 1988. Development and applications of amperometric biosensors. Biotechnology. 2:3–8.

Higgins, S. I. and D. M. Richardson. 1996. A review of models of alien plant spread. *Ecological Modelling*. 87:249–265.

Higgs, E. S. 1997. What is good ecological restoration? *Conservation Biology.* 11:338–348.

Hild, A. and C.-P. Gunther. 1999. Ecosystem engineers: *Mytilus edulis* and *Lanice conchilega*. pp. 43–49. In: *The Wadden Sea Ecosystem, Stability Properties and Mechanisms*. S. Dittmann (ed.). Springer, Berlin.

Hill, I. R., F. Heimbach, P. Leeuwangh, and P. Matthiessen. (eds.). 1994. *Freshwater Field Tests for Hazard Assessment of Chemicals*. CRC Press, Boca Raton, FL.

Hill, J., IV and R. G. Wiegert. 1980. Microcosms in ecological modeling. pp. 138–163. In: *Microcosms in Ecological Research*. J. P. Giesy, Jr. (ed.). CONF-781101. U.S. Department of Energy, Technical Information Center, Washington, DC.

Hillman, W. S. and D. D. Culley, Jr. 1978. The uses of duckweed. *American Scientist.* 66: 442–451.

Hobbs, R. J. 1989. The nature and effects of disturbance relative to invasions. pp. 389–405. In: *Biological Invasions: A Global Perspective*. J. A. Drake, H. A. Mooney, F. di Castri, R. H. Groves, F. J. Kruger, M. Rejmanek, and M. Williamson (eds.). John Wiley & Sons, Chichester, U.K.

Hobbs, R. J. and L. F. Huenneke. 1992. Disturbance, diversity, and invasion: implications for conservation. *Conservation Biology.* 6:324–337.

Hodgson, R. G. 1930. *Successful Muskrat Farming*. Fur Trade Journal of Canada, Toronto.

Hoffmann, A. A. and P. A. Parsons. 1997. *Extreme Environmental Change and Evolution*. Cambridge University Press, Cambridge, U.K.

Holdridge, L. R. 1967. *Life Zone Ecology.* Tropical Research Center, San Jose, Costa Rica.

Holling, C. S. 1973. Resilience and stability of ecological systems. *Annual Reviews of Ecology and Systematics.* 4:1–23.

Holling, C. S. 1996. Engineering resilience versus ecological resilience. pp. 31–44. In: *Engineering Within Ecological Constraints*. P. C. Schulze (ed.). National Academy Press, Washington, DC.

Holm, L. G., D. L. Plucknett, J. V. Pancho, and J. P. Herberger. 1977. *The World's Worst Weeds*. East-West Center, University Press of Hawaii, Honolulu, HI.

Hopkinson, C. S., Jr. and J. W. Day, Jr. 1980. Modeling the relationship between development and storm water and nutrient runoff. *Environmental Management* 4:315–324.

Horenstein, M. N. 1999. *Design Concepts for Engineers*. Prentice Hall, Upper Saddle River, NJ.

Horikoshi, K. and W. D. Grant (eds.). 1991. *Superbugs, Microorganisms in Extreme Environments*. Springer-Verlag, Berlin.

Horn, H. S. and R. H. MacArthur. 1972. Competition among fugitive species in a harlequin environment. *Ecology.* 53:749–752.

Horvitz, C. C. 1997. The impact of natural disturbances. pp. 63–74. In: *Strangers in Paradise: Impact and Management of Nonindigenous Species in Florida*. D. Simberloff, D. C. Schmitz, and T. C. Brown (eds.). Island Press, Washington, DC.

Hoshino, S., R. Onodera, H. Minato, and H. Itabashi (eds.). 1990. *The Rumen Ecosystem: The Microbial Metabolism and Its Regulation*. Springer-Verlag, Berlin.

Hoskins, J. K. 1933. The oxygen demand test and its application to sewage treatment. *Sewage Works Journal.* 5:923–936.

Houston, D. B. 1971. Ecosystems of national parks. *Science.* 172:648–651.

Howard, J. A. and C. W. Mitchell. 1985. *Phytogeomorphology.* John Wiley & Sons, New York.

Howe, H. F. and J. Smallwood. 1982. Ecology of seed dispersal. *Annual Reviews of Ecology and Systematics.* 13:201–228.

Hubbs, C. L. and R. W. Eschmeyer. 1938. *The Improvement of Lakes for Fishing*. Institute for Fisheries Research, University of Michigan, Ann Arbor, MI.

Huffaker, C. B. 1958. Experimental studies on predation: dispersion factors and predator-prey oscillations. *Hilgardia.* 27:343–383.

Hughes, R. M. 1994. Defining acceptable biological status by comparing with reference conditions. pp. 31–47. In: *Biological Assessment and Criteria.* W. S. Davis and T. P. Simon (eds.). Lewis Publishers, Boca Raton, FL.

Hughes, R. M., D. P. Larsen and J.M. Omernik. 1986. Regional reference sites: a method for assessing stream potentials. *Environmental Management.* 10:629–635.

Hughes, S. A. 1993. *Physical Models and Laboratory Techniques in Coastal Engineering.* World Scientific Publishing, Singapore.

Hughes, T. P. 1994. Catastrophes, phase shifts, and large-scale degradation of a Caribbean coral reef. *Science.* 265:1547–1551.

Hungate, R. E. 1966. *The Rumen and Its Microbes.* Academic Press, New York.

Hunt, R. L. 1993. *Trout Stream Therapy.* The University of Wisconsin Press, Madison, WI.

Hunter, C. J. 1991. *Better Trout Habitat: A Guide to Stream Restoration and Management.* Island Press, Washington, DC.

Hunter, M. D. and P. W. Price. 1992. Playing chutes and ladders, bottom-up and top-down forces in natural communities. *Ecology.* 73:724–732.

Huntly, N. 1995. How important are consumer species to ecosystem functioning? pp. 72–83. In: *Linking Species & Ecosystems.* C. G. Jones and J. H. Lawton (eds.). Chapman & Hall, New York.

Hupp, C. R., W. R. Osterkamp, and A. D. Howard. (eds.). 1995. *Biogeomorphology, Terrestrial and Freshwater Systems.* Elsevier, Amsterdam.

Hurlbert, S. H. 1984. Pseudoreplication and the design of ecological field experiments. *Ecological Monographs.* 54:187–211.

Hurlbert, S. H. and M. S. Mulla. 1981. Impacts of mosquitofish (*Gambusia affinis*) predation on plankton communities. *Hydrobiologia.* 83:125–151.

Hurlbert, S. H., M. S. Mulla, and H. R. Wilson. 1972. Effects of an organophosphorus insecticide on the phytoplankton, zooplankton and insect populations of fresh-water ponds. *Ecological Monographs.* 42:269–299.

Hurlbert, S. H., J. Zedler, and D. Fairbanks. 1972. Ecosystem alteration by mosquitofish (*Gambusia affinis*) predation. *Science.* 175:639–641.

Hushak, L. J., D. O. Kelch, and S. J. Glenn. 1999. The economic value of the Lorain County, Ohio, artificial reef. *American Fisheries Society Symposium.* 22:348–362.

Hushon, J. M., R. J. Clerman, and B. O. Wagner. 1979. Tiered testing for chemical hazard assessment. *Environmental Science and Technology.* 13:1202–1207.

Huston, M. 1979. A general hypothesis of species diversity. *American Naturalist.* 113:81–101

Huston, M. and T. M. Smith. 1987. Plant succession — life history and competition. *American Naturalist* 130:168–198.

Huston, M. A. 1994. *Biological Diveristy: The Coexistence of Species on Changing Landscapes.* Cambridge University Press, Cambridge, U.K.

Hutchings, M. J., E. A. John, and A. J. A. Stewart (eds.). 2000. *The Ecological Consequences of Environmental Heterogeneity.* Blackwell Science Publishing, Malden, MA.

Hutchinson, G. E. 1951. Copepodology for the ornithologist. *Ecology.* 32:571–577.

Hutchinson, G. E. 1957. Concluding remarks. *Cold Spring Harbor Symposium on Quantitative Biology.* 22:415–427.

Hutchinson, G. E. 1961. The paradox of the plankton. *American Naturalist.* 95:137–145.

Hutchinson, G. E. 1964. The lacustrine microcosm reconsidered. *American Scientist.* 52:334–341.

Hutchinson, G. E. 1965. *The Ecological Theater and the Evolutionary Play.* Yale University Press, New Haven, CT

Hutchinson, G. E. 1971. Scale effects in ecology. pp. xvii–xxvi. In: *Statistical Ecology* Vol. 1. G. P. Patil, E. C. Pielou, and W. E. Waters (eds.). Pennsylvania State University Press, University Park, PA.

Hutchinson, G. E. 1973. Eutrophication. *American Scientist.* 61:269–279.

Hutchinson, G. E. 1978. *An Introduction to Population Ecology.* Yale University Press, New Haven, CT.

Hylleberg, J. 1975. Selective feeding by Abarenicola pacifica with notes on *Abarenicola vagabunda* and a concept of gardening in lugworms. *Ophelia.* 14:113–137.

Hynes, H. B. N. 1960. *The Biology of Polluted Waters.* Liverpool University Press, Liverpool, U.K.

Imaoka, T. and S. Teranishi. 1988. Rates of nutrient uptake and growth of the water hyacinth (*Eichhornia crassipes* (*Mart.*) *Solms*). *Water Research.* 22:943–951.

Ingle, K. A. 1994. *Reverse Engineering.* McGraw-Hill, New York.

Inman, D. L. and B. M. Brush. 1973. The coastal challenge. *Science.* 181:20–32.

Insam, H., N. Riddech, and S. Klammer (eds.). 2002. *Microbiology of Composting.* Springer-Verlag, Berlin, Germany.

Isensee, A. R. 1976. Variability of aquatic model ecosystem-derived data. *International Journal of Environmental Studies.* 10:35–41.

Israel, P. 1989. Telegraphy and Edison's invention factory. pp. 66–83. In: *Working at Inventing: Thomas A. Edison and the Menlo Park Experience.* W. S. Pretzer (ed.). Henry Ford Museum & Greenfield Village, Dearborn, MI.

Jackson, J. 2001. Engineering run II. *FERMI NEWS.* 24(8):4–9.

Jackson, L. L. 1992. The role of ecological restoration in conservation biology. pp. 438–452. In: *Conservation Biology.* P. L. Fiedler and S. K. Jain (eds.). Chapman and Hall, New York.

Jackson, L. L., N. Lopoukhine, and D. Hillyard. 1995. Ecological restoration: a definition and comments. *Restoration Ecology.* 3:71–75.

Jackson, W. 1980. *New Roots for Agriculture.* University of Nebraska Press, Lincoln, NE.

Jackson, W. 1999. Natural systems agriculture: the truly radical alternative. pp. 191–199. In: *Recovering the Prairie.* R. F. Sayre (ed.). University of Wisconsin Press, Madison, WI.

Jaeger, R. G. and S. C. Walls. 1989. On salamander guilds and ecological methodology. *Herpetologica.* 45:111–119.

James, A. 1964. The bacteriology of trickling filters. *Journal of Applied Bacteriology.* 27:197–207.

Janzen, D. H. 1973. Tropical agroecosystems. *Science.* 182:1212–1219.

Janzen, D. H. (ed.). 1983. *Costa Rican Natural History.* University of Chicago Press, Chicago, IL.

Janzen, D. H. 1985. On ecological fitting. *Oikos.* 45:308–310.

Janzen, D. H. 1988. Complexity is in the eye of the beholder. pp. 29–51. In: *Tropical Rainforests, Diversity and Conservation.* F. Almeda and C. M. Pringle (eds.). California Academy of Sciences, San Francisco, CA.

Jaynes, M. L. and S. R. Carpenter. 1986. Effects of vascular and nonvascular macrophytes on sediment redox and solute dynamics. *Ecology.* 67:875–882.

Jeffers, J. N. R. 1978. *An Introduction to Systems Analysis: With Ecological Applications.* University Park Press, Baltimore, MD.

Jenkins, J. C. 1994. *The Humanure Handbook: A Guide to Composting Human Manure.* Jenkins, Grove City, PA.

Jenny, H. 1941. *Factors of Soil Formation: A System of Quantitative Pedology.* McGraw-Hill, New York.

Jenny, H. 1958. The role of the plant factor in pedogenic functions. *Ecology.* 39:5–16.

Jenny, H. 1961. Derivation of state factor equations of soils and ecosystems. *Proceedings of the Soil Science Society.* 1961:385–388.

Jenny, H. 1980. *The Soil Resource: Origin and Behavior.* Springer-Verlag, New York.

Jenny, H., S. P. Gessel, and F. T. Bingham. 1949. Comparative study of decomposition rates of organic matter in temperate and tropical regions. *Soil Science.* 68:419–432.

Jensen, H. L. 1929. On the influence of the carbon nitrogen ratios of organic materials on the mineralization of nitrogen. *Journal of Agricultural Science.* 19:71–82.

Jingsong, Y. and Y. Honglu. 1989. Integrated fish culture management in China. pp. 375–408. In: *Ecological Engineering, An Introduction to Ecotechnology.* W. J. Mitsch and S. E. Jorgensen (eds.). John Wiley & Sons, New York.

Jobin, W. R. and A. T. Ippen. 1964. Ecological design of irrigation canals for snail control. *Science.* 145:1324–1326.

Johnson, A. T. and D. C. Davis. 1990. Biological engineering: a discipline whose time has come. *Engineering Education.* 80:15–18.

Johnson, A. T. and W. M. Phillips. 1995. Philosophical foundations of biological engineering. *Journal of Engineering Education.* 84:311–318.

Johnson, B. L., W. B. Richardson, and T. J. Naimo. 1995. Past, present, and future concepts in large river ecology. *BioScience.* 45:134–141.

Johnson, C. E. 1925. The muskrat in New York. *Roosevelt Wild Life Bulletin,* Vol. 3, No. 2. The New York State College of Forestry, Syracuse, NY.

Johnson, C. E. 1927. The beaver in the Adirondacks. *Roosevelt Wild Life Bulletin,* Vol. 4, No. 4. The New York State College of Forestry, Syracuse University, Syracuse, NY.

Johnson, K. H., K. A. Vogt, H. J. Clark, O. J. Schmitz, and D. J. Vogt. 1996. Biodiversity and the productivity and stability of ecosystems. *Trends in Ecology and Evolution.* 11:372–377.

Johnson, K. S. and D. M. Karl. 2002. Is ocean fertilization credible or creditable? *Science.* 296:467–468.

Johnson, L. E. and W. V. McGuinness, Jr. 1975. Guidelines for material placement in marsh creation. *Contract Report D-75-2*, Dredged Material Research Program, U.S. Army Engineer Waterways Experiment Station, Vicksburg, MS.

Johnson, M. S. and A. D. Bradshaw. 1979. Ecological principles for the restoration of disturbed and degraded land. pp. 142–200. In: *Applied Biology.* Vol. IV. T. H. Coaker (ed.). Academic Press, New York.

Johnson, P. C., J. H. Kennedy, R. G. Morris, F. E. Hambleton, and R. L. Graney. 1994. Fate and effects of cyfluthrin (pyrethroid insecticide) in pond mesocosms and concrete microcosms. pp. 337–371. In: *Aquatic Mesocosm Studies in Ecological Risk Assessment.* R. L. Graney, J. H. Kennedy, and J. H. Rodgers, Jr. (eds.). CRC Press, Inc., Boca Raton, FL.

Johnston, C. A. 1994. Ecological engineering of wetlands by beavers. pp. 379–384. In: *Global Wetlands: Old World and New.* W. J. Mitsch (ed.). Elsevier, Amsterdam, the Netherlands.

Jones, C. G. and J. H. Lawton. 1995. *Linking Species and Ecosystems*. Chapman & Hall, New York.

Jones, C. G., J. H. Lawton, and M. Shachak. 1994. Organisms as ecosystem engineers. *Oikos*. 69:373–386.

Jones, C. G., J. H. Lawton, and M. Shachak. 1997. Positive and negative effects of organisms as physical ecosystem engineers. *Ecology*. 78:1946–1957.

Jones, D. D. 1975. The application of catastrophe theory to ecological systems. pp. 133–148. In: *New Directions in the Analysis of Ecological Systems*, Part 2. G. S. Innis (ed.). The Society for Computer Simulation, La Jolla, CA.

Jones, P. 1989. Computerized greenhouse environmental controls. Circular 830. Florida Cooperative Extension Service, University of Florida, Gainesville, FL.

Jones, T. H. 1996. Biospherics, closed systems and life support. *Trends in Ecology and Evoluation*. 11:448–450.

Jordan, C. F. 1981. Do ecosystems exist? *American Naturalist*. 118:284–287.

Jordan, W. R. III. 1994. "Sunflower Forest": ecological restoration as the basis for a new environmental paradigm. pp. 17–34. In: *Beyond Preservation, Restoring and Inventing Landscapes*. A. D. Baldwin, Jr., J. de Luce, and C. Pletsch (eds.). University of Minnesota Press, Minneapolis, MN.

Jordan, W. R. III. 1995. Restoration ecology: a synthetic approach to ecological research. pp. 373–384. In: *Rehabilitating Damaged Ecosystems*. J. Cairns, Jr. (ed.). Lewis Publishers, Boca Raton, FL.

Jordan, W. R., III. and S. Packard. 1989. Just a few oddball species: restoration practice and ecological theory. pp. 18–26. In: *Biological Habitat Reconstruction*. G. P. Buckley (ed.). Belhaven Press, London.

Jordan, W. R. III., M. E. Gilpin, and J. D. Aber. 1987. Restoration ecology: ecological restoration as a technique for basic research. pp. 3–21. In: *Restoration Ecology: A Synthetic Approach To Ecological Research*. W. R. Jordan III, M. E. Gilpin, and J. D. Aber (eds.). Cambridge University Press, Cambridge, U.K.

Jordan, W. R., III., R. L. Peters, II., and E. B. Allen. 1988. Ecological restoration as a strategy for conserving biological diversity. *Environmental Management*. 12:55–72.

Jørgensen, S. E. 1982. Exergy and buffering capacity in ecological systems. pp. 61–72. In: *Energetics and Systems*. W. J. Mitsch, R. K. Ragade, R. W. Bosserman, and J. A. Dillon, Jr. (eds.). Ann Arbor Science, Ann Arbor, MI.

Jørgensen, S. E. 2000. The tentative fourth law of thermodynamics. pp. 161–175. In: *Handbook of Ecosystem Theories and Management*. S. E. Jørgensen and F. Muller (eds.). Lewis Publishers, Boca Raton, FL.

Jørgensen, S. E. (ed.). 2001. *Thermodynamics and Ecological Modelling*. Lewis Publishers, Boca Raton, FL.

Jørgensen, S. E. and F. Muller (eds.). 2000. *Handbook of Ecosystem Theories and Management*. Lewis Publishers, Boca Raton, FL.

Jørgensen, S. E., H. Mejer, and S. N. Nielsen. 1998. Ecosystem as self-organizing critical systems. *Ecological Modelling*. 111:261–268.

Josephson, B. 1995. The San Francisco living machine. *Annals of Earth*. 13(2):11–12.

Josephson, B., J. Todd, S. Serfling, A. Smith, L. Stuart, and K. Locke. 1996a. 1995 Report on the performance of the advance ecologically engineered system in Frederick, Maryland. *Annals of Earth*. 14(1):15–17.

Josephson, B., J. Todd, D. Austin, K. Locke, M. Shaw, and K. Little. 1996b. Performance report for the Frederick, Maryland living machine. *Annals of Earth*. 14(3):7–11.

Joy, B. 2000. Why the future doesn't need us. *Wired*. 8(4):238–262.

Judson, S. 1968. Erosion of the land. *American Scientist*. 56:356–374.

Jumars, P. A. 2000. Animal guts as ideal chemical reactors: maximizing absorption rates. *American Naturalist.* 155:527–543.

Junk, W. J., P. B. Bayley, and R. E. Sparks. 1989. The flood pulse concept in river-flood plain systems. pp. 110–127. In: Proceedings of the International Large River Symposium, D. P. Dodge (ed.). *Special Publication of Canadian Fisheries and Aquatic Sciences.* No. 106, Ottawa, Canada.

Kadlec, J. A. and W. A. Wentz. 1974. State-of-the-art Survey and Evaluation of Marsh Plant Establishment Techniques: Induced and Natural. *Contract Report DACW72-74-C-0010,* U.S. Army Engineer Waterways Experiment Station, Vicksburg, MS.

Kadlec, R. H. 1997. An autobiotic wetland phosphorus model. *Ecological Engineering.* 8:145–172.

Kadlec, R. H. and R. L. Knight. 1996. *Treatment Wetlands.* CRC Press, Boca Raton, FL.

Kahn, L., B. Allen, and J. Jones. 2000. *The Septic System Owner's Manual.* Shelter Publications, Bolinas, CA.

Kaiser, J. 1994. Wiping the slate clean at Biosphere 2. *Science.* 265:1027.

Kaiser, J. 2001a. Recreated wetlands no match for original. *Science.* 293:2.

Kaiser, J. 2001b. Breeding a hardier weed. *Science.* 293:1425–1426.

Kangas, P. 1983a. Energy Analysis of Landforms, Succession and Reclamation. Ph.D. dissertation, University of Florida, Gainesville, FL.

Kangas, P. 1983b. Simulating succession as a reclamation alternative. pp. 451–455. In: *Analysis of Ecological Systems: State-of-the-Art in Ecological Modelling.* W. K. Lauenroth, G. V. Skogerboe, and M. Flug (eds.). Elsevier, Amsterdam, the Netherlands.

Kangas, P. 1988. A chess analogy: teaching the role of animals in ecosystems. *American Biology Teacher.* 50:160–162.

Kangas, P. 1990. An energy theory of landscape for classifying wetlands. pp. 15–23. In: *Forested Wetlands,* Vol. 15, *Ecosystems of the World.* A. E. Lugo, M. Brinson, and S. Brown (eds.). Elsevier, Amsterdam, the Netherlands.

Kangas, P. 1995. Contributions of H. T. Odum to ecosystem simulation modeling. pp. 11–18. In: *Maximum Power: The Ideas and Applications of H. T. Odum.* C. A. S. Hall (ed.). University Press of Colorado. Niwot, CO.

Kangas, P. 1997. Tropical sustainable development and biodiversity. pp. 389–409. In: M. L. Reaka-Kudla, D. E. Wilson, and E. O. Wilson (eds.). *Biodiversity II.* Joseph Henry Press, Washington, DC.

Kangas, P. and W. Adey. 1996. Mesocosms and ecological engineering. *Ecological Engineering.* 6:1–5.

Kangas, P. and A. Lugo. 1990. The distribution of mangroves and saltmarsh in Florida. *Tropical Ecology.* 31:32–39.

Kangas, P. and P. G. Risser. 1979. Species packing in the fast food restaurant guild. *Bulletin of the Ecological Society of America.* 60:143–148.

Kania, H. J. and R. J. Beyers. no date. Feedback control of light input to a microecosystem by the system. Unpublished report. Savannah River Ecology Laboratory, Aiken, SC.

Kaplan, O. B. 1991. *Septic Systems Handbook.* Lewis Publishers. Chelsea, MI.

Kareiva, P., I. M. Parker, and M. Pascual. 1996. Can we use experiments and models in predicting the invasiveness of genetically engineered organisms? *Ecology.* 77:1670–1675.

Karr, J. R. 1981. Assessment of biotic integrity using fish communities. *Fisheries.* 6(6):21–27.

Karr, J. R. 1991. Biological integrity: a long neglected aspect of water resource management. *Ecological Applications.* 1:66–84.

Karr, J. R. 1996. Ecological integrity and ecological health are not the same. pp. 97–110. In: *Engineering within Ecological Constraints*. P. C. Schulze (ed.). National Academy Press, Washington, DC.

Karr, J. R. 2000. Health, integrity, and biological assessment: the importance of measuring whole things. pp. 209–225. In: *Ecological Integrity, Integrating Environment, Conservation, and Health*. D. Pimental, L. Westra, and R. F. Noss (eds.). Island Press, Washington, DC.

Karr, J. R. and E. W. Chu. 1999. *Restoring Life in Running Waters: Better Biological Monitoring*. Island Press, Washington, DC.

Karr, J. R. and D. R. Dudley. 1981. Ecological perspective on water quality. *Environmental Management* 5:55–68.

Kassner, S. L. 2001. Soil Development as a Functional Indicator of a Reconstructed Freshwater Tidal Marsh. M.S. thesis. University of Maryland, College Park, MD.

Kauffman, S. 1995. *At Home in the Universe: The Search for the Laws of Self-Organization and Complexity*. Oxford University Press, New York.

Kautsky, L. 1988. Life strategies of aquatic soft bottom macrophytes. *Oikos*. 53:126–135.

Kay, J. J. 2000. Ecosystems as self-organising holarchic open systems: narratives and the Second Law of Thermodynamics. pp. 135–160. In: *Handbook of Ecosystem Theories and Management*. S. E. Jørgensen and F. Muller (eds.). Lewis Publishers, Boca Raton, FL.

Kay, J. J. and H. A. Regier. 1999. An ecosystemic two-phase attractor approach to Lake Erie's ecology. pp. 511–533. In: *State of Lake Erie — Past, Present and Future*. M. Munawar, T. Edsall, and I. F. Munawar (eds.). Backhuys Publishers, Leiden, the Netherlands.

Kearns, E. A. and C. E. Folsome. 1981. Measurement of biological activity in materially closed microbial ecosystems. *BioSystems*. 14:205–209.

Keddy, P. A. 1976. Lakes as islands: the distributional ecology of two aquatic plants, *Lemna minor* L. and *L. trisulca* L. *Ecology*. 57:353–359.

Keiper, R. R. 1985. *The Assateague Ponies*. Tidewater Publishers, Centreville, MD.

Kelcey, J. G. 1975. Industrial development and wildlife conservation. *Environmental Conservation*. 2:99–108.

Keller, D. R. and E. C. Brummer. 2002. Putting food production in context: toward a post-mechanistic agricultural ethic. *BioScience*. 52:264–271.

Kelley, J. C. 1997. John Steinbeck, and Ed Ricketts: understanding life in the great tide pool. pp. 27–42. In: *Steinbeck and the Environment: Interdisciplinary Approaches*. S. F. Beegel, S. Shillinglaw, and W. N. Tiffney, Jr. (eds.). The University of Alabama Press, Tuscaloosa, AL.

Kellman, M. 1974. Preliminary seed budgets for two plant communities in coastal British Columbia. *Journal of Biogeography*. 1:123–133.

Kelly, K. 1994. *Out of Control: The New Biology of Machines, Social Systems, and the Economic World*. Addison-Wesley, Reading, MA.

Kemp, W. M. 1977. Energy Analysis and Ecological Evaluation of a Coastal Power Plant. Ph.D. dissertation, University of Florida, Gainesville, FL.

Kemper, J. D. 1982. *Engineers and Their Profession*. Holt, Rinehart and Winston, New York.

Kendeigh, S. C. 1944. Measurement of bird populations. *Ecological Monographs*. 14:67–106.

Kendeigh, S. C. 1961. *Animal Ecology*. Prentice Hall, Englewood Cliffs, NJ.

Kennedy, J. H., Z. B. Johnson, P. D. Wise, and P. C. Johnson. 1995a. Model aquatic ecosystems in ecotoxicological research: Considerations of design, implementation, and analysis. pp. 117–162. In: *Handbook of Ecotoxicology*. D. J. Hoffman, B. A. Rattner, G. A. Burton, Jr., and J. Cairns, Jr. (eds.). CRC Press, Boca Raton, FL.

Kennedy, J. H., P. C. Johnson, and Z. B. Johnson. 1995b. The use of constructed or artificial ponds in simulated field studies. pp. 149–167. *Ecological Toxicity Testing*. J. Cairns, Jr. and B. R. Niederlehner (eds.). Lewis Publishers, Boca Raton, FL.

Kennedy, T. A., S. Naeem, K. M. Howe, J. M. H. Knops, D. Tilman, and P. Reich. 2002. Biodiversity as a barrier to ecological invasion. *Nature*. 417:636–638.

Kentula, M. E., R. P. Brooks, S. E. Gwin, C. C. Holland, A. D. Sherman, and J. C. Sifneos. 1992. *Wetlands: An Approach to Improving Decision Making in Wetland Restoration and Creation*. Island Press, Washington, DC.

Kerfoot, W. C. and A. Sih. (eds.). 1987. *Predation, Direct and Indirect Impacts on Aquatic Communities*. University Press of New England, Hanover, NH.

Kersting, K. 1984. Normalized ecosystem strain: a system parameter for the analysis of toxic stress in (micro-)ecosystems. *Ecological Bulletin*. 36:150–153.

Keup, L. E., W. M. Ingram, and K. M. Mackenthum (eds.). 1967. *Biology of Water Pollution*, a collection of selected papers on stream pollution, waste water, and water treatment. Federal Water Pollution Control Administration, USDI, Washington, DC.

Kibert, C. J., J. Sendzimir, and G. Bradley Guy. (eds.). 2002. *Construction Ecology: Nature as the Basis for Green Buildings*. Spon Press, London.

Kimball, K. D. and S. A. Levin. 1985. Limitations of laboratory bioassays: The need for ecosystem-level testing. *BioScience*. 35:165–171.

King, W. V. 1952. Mosquitoes and DDT. pp. 327–330. In: *The Yearbook of Agriculture: Insects*. U.S. Department of Agriculture, Washington, DC.

King, W. V., M. Hay, and J. Charbonneau. 1979. Valuation of riparian habitats. pp. 161–165. In: *Strategies for Protection and Management of Floodplain Wetlands and Other Riparian Ecosystems*. R. R. Johnson and J. F. McCormick (eds.). General Technical Report. WO-12. USDA, Forest Service, Washington, DC.

Kingsland, S. E. 1995. *Modelling Nature*. 2nd ed. University of Chicago Press, Chicago, IL.

Kinkel, L. L., J. H. Andrews, F. M. Berbee, and E. V. Nordheim. 1987. Leaves as islands for microbes. *Oecologia*. 71:405–408.

Kinner, N. E. and C. R. Curds. 1987. Development of protozoan and metazoan communities in rotating biological contactor biofilms. *Water Research*. 21:481–490.

Kirby, C. J., J. W. Keeley, and J. Harrison. 1975. An overview of the technical aspects of the Corps of Engineers National Dredged Material Research Program. pp. 523–535. In: *Estuarine Research, Vol. II: Geology and Engineering*. L. E. Cronin (ed.). Academic Press, New York.

Kitchell, J. F. (ed.). 1992. *Food Web Management: A Case Study of Lake Mendota*. Springer-Verlag, New York.

Kitchell, J. F., S. M. Bartell, S. R. Carpenter, D. J. Hall, D. J. McQueen, W. E. Neill, D. Scavia, and E. E. Werner. 1988. Epistemology, experiments, and pragmatism. pp. 263–280. In: *Complex Interactions in Lake Communities*. S. R. Carpenter (ed.). Springer-Verlag, New York.

Kitchell, J. F., R. V. O'Neill, D. Webb, G. W. Gallepp, S. M. Bartell, J. F. Koonce, and B. S. Ausmus. 1979. Consumer regulation of nutrient cycling. *BioScience*. 29:28–34.

Kitchens, W. M., Jr., J. M. Dean, L. H. Stevenson, and J. H. Cooper. 1975. The Santee Swamp as a nutrient sink. pp. 349–366. In: *Mineral Cycling in Southeastern Ecosystems*. U.S. Energy Research and Development Administration Symposium CONF-740513, U.S. Atomic Energy Commission, Washington, DC.

Kitching, R. L. 2000. *Food Webs and Container Habitats: The Natural History and Ecology of Phytotelmata*. Cambridge University Press, Cambridge, U.K.

Kittredge, J. 1948. *Forest Influences*. Dover Publications, New York.

Klausner, S. D. no date. Manure management. pp. 43–48. In: *Cornell Field Crops Handbook.* New York State College of Agriculture and Life Sciences, Cornell University, Ithaca, NY.

Klein, M. J. 1970. Maxwell, his demon, and the Second Law of Thermodynamics. *American Scientist.* 58:84–97.

Klimisch, R. L. 1994. Designing the modern automobile for recycling. pp. 165–170. In: *The Greening of Industrial Ecosystems.* B. R. Allenby and D. J. Richards (eds.). National Academy Press, Washington, DC.

Klopfer, P. H. 1981. Islands as models. *BioScience.* 31:838–839.

Knabe, W. 1965. Observations on world-wide efforts to reclaim industrial waste land. pp. 263–296. In: *Ecology and the Industrial Society.* G. T. Goodman, R. W. Edwards, and J. M. Lambert (eds.). John Wiley & Sons, New York.

Knight, R. L. 1983. Energy basis of ecosystem control at Silver Springs, Florida. pp. 161–179. In: *Dynamics of Lotic Ecosystems.* T. D. Fontaine, III and S. M. Bartell (eds.). Ann Arbor Science, Ann Arbor, MI.

Knight, R. L. 1992. Natural land treatment with Carolina Bays. *Water, Environment, and Technology.* 4:13–16.

Knight, R. L. 1995. Wetland systems for wastewater management: implementation. pp. 123–131. In: *Maximum Power: The Ideas and Applications of H. T. Odum.* C. A. S. Hall (ed.). University Press of Colorado, Niwot, CO.

Knight, R. L. and D. P. Swaney. 1981. In defense of ecosystems. *American Naturalist.* 117:991–992.

Knowlton, N. 1992. Thresholds and multiple stable states in coral reef community dynamics. *American Zoologist.* 32:674–682.

Knutson, R. M. 1974. Heat production and temperature regulation in eastern skunk cabbage. *Science.* 186:746–747.

Ko, J.-Y., J. Martin, and J. W. Day. 2000. Embodied energy and emergy analysis of wastewater treatment using wetlands. pp. 197–210. In: *Emergy Synthesis: Theory and Applications of the Emergy Methodology.* M. T. Brown (ed.). Center for Environmental Policy, University of Florida, Gainesville, FL.

Koerner, R. M. 1986. *Designing with Geosynthetics.* Prentice Hall, Englewood Cliffs, NJ.

Kohler, M. and M. Schmidt. 1990. The importance of roofs covered with vegetation for the urban ecology: biotic factors. pp. 153–161. In: *Terrestrial and Aquatic Ecosystems, Perturbation and Recovery.* O. Ravera (ed.). Ellis Horwood Publishing, New York.

Kohnke, R. E. and A. K. Boller. 1989. Soil bioengineering for streambank protection. *Journal of Soil and Water* Conservation. 44:286–287.

Kok, R. and R. Lacroix. 1993. An analytical framework for the design of autonomous, enclosed agro-ecosystems. *Agricultural Systems.* 43:235–260.

Kolasa, J. and S. T. A. Pickett (eds.). 1991. *Ecological Heterogeneity.* Springer-Verlag, New York.

Kolata, G. B. 1977. Catastrophe theory: the emperor has no clothes. *Science.* 196:287, 350–351.

Komar, P. D. 1998. *Beach Processes and Sedimentation,* 2nd ed. Prentice Hall, Upper Saddle River, NJ.

Kormondy, E. J. (ed.). 1965. *Readings in Ecology.* Prentice Hall, Englewood Cliffs, NJ.

Korner, C. and J. A. Arnone III. 1992. Responses to elevated carbon dioxide in artificial tropical ecosystems. *Science.* 257:1672–1675.

Korringa, P. 1976. Safeguards in the exploitation of domestic effluents for aquaculture. pp. 201–209. In: *Harvesting Polluted Waters.* O. Devik (ed.). Plenum Press, New York.

Koshland, D. E., Jr. 1989. The engineering of species. *Science.* 244:1233.

Kountz, R. R. and J. B. Nesbitt. 1958. A bacterium looks at anaerobic digestion. pp. 44–47. In: *Biological Treatment of Sewage and Industrial Wastes*, Vol. II. J. McCabe and W. W. Eckenfelder, Jr. (eds.). Reinhold Publishing, New York.

Kovalevsky, A. L. 1987. *Biogeochemical Exploration for Mineral Deposits*, 2nd ed. Vnu-science Press, Utrecht, the Netherlands.

Krantz, W. B. 1990. Self-organization manifest as patterned ground in recurrently frozen soils. *Earth Science Review.* 29:117–130.

Krebs, C. J. 1995. Population regulation. pp. 183–202. In: *Encyclopedia of Environmental Biology*, Vol. 3. W. A. Nierenberg (ed.). Academic Press, San Diego, CA.

Kremer, J. N. and J. S. Lansing. 1995. Modeling water temples and rice irrigation in Bali: A lesson in socio-ecological communication. pp. 100–108. In: *Maximum Power: The Ideas and Applications of H. T. Odum.* C. A. S. Hall (ed.). University Press of Colorado, Niwot, CO.

Kruczynski, W. L. 1990. Mitigation and the Section 404 Program: A perspective. pp. 549–554. In: *Wetland Creation and Restoration: The Status of the Science.* J. A. Kusler and M. E. Kentula (eds.). Island Press, Washington, DC.

Kurten, B. 1968. *Pleistocene Mammals of Europe.* Aldine Publishing, Chicago, IL.

Kurtz, C. 2001. *A Practical Guide to Prairie Reconstruction.* University of Iowa Press, Iowa City, IA.

Kusler, J. A. and M. E. Kentula (eds.). 1990. *Wetland Creation and Restoration: The Status of the Science.* Island Press, Washington, DC.

la Riviere, J. W. M. 1977. Microbial ecology of liquid wastewater treatment. pp. 215–259. In: *Advances in Microbial Ecology*, Vol. 1. M. Alexander (ed.). Plenum Press, New York.

Lack, D. 1954. *The Natural Regulation of Animal Numbers.* Oxford University Press, Oxford, U.K.

Lakshman, G. 1979. An ecosystem approach to the treatment of waste waters. *Journal of Environmental Quality.* 8:353–361.

Lakshman, G. 1994. Design and operational limitations of engineered wetlands in cold climates — Canadian experience. pp. 399–409. In: *Global Wetlands: Old World and New.* W. J. Mitsch (ed.). Elsevier, Amsterdam, the Netherlands.

Laliberte, G., D. Proulx, N. DePauw, and J. De La Noue. 1994. Algal technology in wastewater treatment. pp. 283–299. In: *Algae and Water Pollution.* L. C. Lai and J. P. Gaur (eds.). *Advances in Limnology.* Heft 42, E. Schweizerbartsche Verlagsbuchhandlung, Stuttgart, Germany.

Lalli, C. M. (ed.). 1990. *Enclosed Experimental Marine Ecosystems: A Review and Recommendations.* Springer-Verlag, New York.

Lambou, V. W., W. D. Taylor, S. C. Hern, and L. P. Williams. 1983. Comparisons of trophic state measurements. *Water Research.* 27:1619–1626.

Lampert, W. 1999. Nutrient ratios. *Archives of Hydrobiology.* 146:1.

Lampton, C. 1993. *Nanotechnology Playhouse: Building Machines from Atoms.* Waite Group Press, Corte Madera, CA.

Landin, M. C. 1986. Wetland beneficial use applications of dredged material disposal sites. pp. 118–129. In: *Proceedings of the 13th Annual Conference on Wetlands Restoration and Creation.* F. J. Webb, Jr. (ed.). Hillsborough Community College, Plant City, FL.

Landis, R. 1992. An introduction to engineering. p. 26. In: *1992 Directory of Engineering and Engineering Technology*, Undergraduate Programs. American Society of Engineering Education, Washington, DC.

Langbein, W. B. and S. A. Schumm. 1958. Yield of sediment in relation to mean annual precipitation. *Transactions of the American Geophysical Union.* 39:1076–1084.

Langbein, W. B. and L. B. Leopold. 1966. River meanders and the theory of minimum variance. USGS Prof. Paper 442-H.

Lange, L., P. Kangas, G. Robbins, and W. Adey. 1994. A mesocosm model of the everglades: an extreme example of wetland creation. pp. 95–105. In: *Proceedings of the 21st Annual Conference on Wetlands Restoration and Creation*. F. J. Webb, Jr. (ed.). Hillsborough Community College, Tampa, FL.

Lange, L. E. 1998. An Analysis of the Hydrology and Fish Community Strucutre of the Florida Everglades Mesocosm. M.S. thesis, University of Maryland, College Park, MD.

Langewiesche, W. 2000. The shipbreakers. *The Atlantic Monthly.* 286(2):31–49.

Langley, L. L. 1965. *Homeostasis*. Reinhold Publishing, New York.

Langley, L. L. (ed.). 1973. *Homeostasis: Origins of the Concept*. Benchmark Papers in Human Physiology. Dowden, Hutchinson & Ross, Stroudsburg, PA.

Langton, C. G. (ed.). 1989. *Artificial Life*. Addison-Wesley, Redwood City, CA.

Lansing, J. S. 1987. Balinese water temples and the management of irrigation. *American Anthropologist.* 89:326–341.

Lansing, J. S. 1991. *Priests and Programmers: Technologies of Power in the Engineered Landscape of Bali*. Princeton University Press, Princeton, NJ.

Lappin-Scott, H. M. and J. W. Costerton. (ed.). 1995. *Microbial Biofilms*. Cambridge University Press, Cambridge, U.K.

Larsen, D. P., F. deNoyelles, Jr., F. Stay, and T. Shiroyama. 1986. Comparisons of single-species, microcosm and experimental pond responses to atrazine exposure. *Environment Toxicology Chemistry.* 5:179–190.

Lassen, H. H. 1975. The diversity of freshwater snails in view of the equilibrium theory of island biogeography. *Oecologia.* 19:1–8.

Lasserre, P. 1990. Marine microcosms: small-scale controlled ecosystems. pp. 20–60. In: *Enclosed Experimental Marine Ecosystems: A Review and Recommendations*. C. M. Lalli (ed.). Springer-Verlag, New York.

Latchum, J. A. 1996. Ecological Engineering Factors of a Constructed Wastewater Treatment Wetland. M.S. thesis, University of Maryland, College Park, MD.

Latchum, J. and P. Kangas. 1996. The economic basis of a wetland wastewater treatment plant in Maryland. *Constructed Wetlands in Cold Climates Conference Proceedings*, Niagara-on-the-Lake, Ontario, Canada.

Lavelle, P. 1997. Faunal activities and soil processes: adaptive strategies that determine ecosystem function. *Advances in Ecological Research.* 27:93–132.

Lavies, B. 1993. *Compost Critters*. Dutton Children's Books, New York.

Lavine, M. J. and T. J. Butler. 1981. Energy analysis and economic analysis: a comparison of concepts. pp. 757–765. In: *Energy and Ecological Modelling*. W. J. Mitsch, R. W. Bosserman, and J. M. Klopatek (eds.). Elsevier, Amsterdam, the Netherlands.

Law, R. and R. D. Morton. 1993. Alternative permanent states of ecological communities. *Ecology.* 74:1347–1361.

Lawler, S. P. 1998. Ecology in a bottle, Using microcosms to test theory. pp. 236–253. In: *Experimental Ecology*. W. J. Resetarits, Jr. and J. Bernardo. (eds.). Oxford University Press, New York.

Lawrence, M. G. 2002. Side effects of oceanic iron fertilization. *Science.* 297:1993.

Lawrey, J. D. 1977. The relative decomposition potential of habitats variously affected by surface coal mining. *Canadian Journal of Botany.* 55:1544–1552.

Lawton, J. H. 1994. What do species do in ecosystems? *Oikos.* 71:367–374.

Lawton, J. H. 1995. Ecological experiments with model systems. *Science.* 269:328–331.

Lawton, J. H. 1997. The role of species in ecosystems: Aspects of ecological complexity and biological diversity. pp. 215–228. In: *Biodiversity: An Ecological Perspective*. T. Abe, S. A. Levin, and M. Higashi (eds.). Springer. New York.

Lawton, J. H. 1998. Ecological experiments with model systems: the Ecotron facility in context. pp. 170–182. In: W. J. Resetarits, Jr. and J. Bernardo (eds.). *Experimental Ecology.* Oxford University Press, New York.

Lawton, J. H. and C. G. Jones. 1995. Linking species and ecosystems: Organisms as ecosystem engineers. pp. 141–150. In: *Linking Species and Ecosystems*. C. G. Jones and J. H. Lawton (eds.). Chapman & Hall, New York.

Laycock, W. A. 1991. Stable states and thresholds of range conditions on North America rangelands: a viewpoint. *Journal of Range Management.* 44:427–433.

Layton, E. T., Jr. 1976. American ideologies of science and engineering. *Technology and Culture.* 17:688–701.

Leavitt, P. R. 1992. An analogy for plankton interactions. pp. 483–492. In: *Food Web Management: A Case Study of Lake Mendota*. J. F. Kitchell (ed.). Springer-Verlag, New York.

Leck, M. A., V. T. Parker, and R. L. Simpson (eds.). 1989. *Ecology of Seed Banks*. Academic Press, San Diego, CA.

Lee, G. F., E. Bentley, and R. Amundson. 1975. Effects of marshes on water quality. pp. 105–127. In: *Coupling of Land and Water Systems*. A. D. Hasler (ed.). Springer-Verlag, New York.

Lee, J. J. 1995. Living sands. *BioScience.* 45:252–261.

Lee, J. J. and D. L. Inman. 1975. The ecological role of consumers — an aggregated systems view. *Ecology.* 56:1455–1458.

Lee, K. E. and T. G. Wood. 1971. *Termites and Soils*. Academic Press, London.

Leentvaar, P. 1967. Observations in guanotrophic environments. *Hydrobiologia.* 29:441–489.

Leffler, J. W. 1978. Ecosystem responses to stress in aquatic microcosms. pp. 102–119. In: *Energy and Environmental Stress in Aquaatic Systems*. J. H. Thorp and J. W. Gibbons (eds.). CONF-771114, U.S. Dept. of Energy, Washington, DC.

Leffler, J. W. 1980. Microcosmology: theoretical applications of biological models. pp. 14–29. In: *Microcosms in Ecological Research*, J. P. Giesy, Jr. (ed.). CONF-781101, U.S. Dept. of Energy, Washington, DC.

Leffler, J. W. 1984. The use of self-selected, generic aquatic microcosms for pollution effects assessment. pp. 139–158. In: *Concepts in Marine Pollution Measurements*. H. H. White (ed.). Sea Grant Publication, University of Maryland, College Park, MD.

Leffler, M. no date. *Restoring Oysters to U. S. Coastal Waters*. UM-SG-TS-98-03. Maryland Sea Grant Program, University of Maryland, College Park, MD.

Leopold, A. 1933. *Game Management*. Charles Scribner and Sons, New York.

Leopold, A. 1953. *Round River.* Oxford University Press, Oxford, U.K.

Leopold, A. S. 1966. Adaptability of animals to habitat change. pp. 66–75. In: *Future Environments of North America*. F. F. Darling and J. P. Milton (eds.). The Natural History Press, Garden City, NY.

Leopold, L. B. 1994. *A View of the River.* Harvard University Press, Cambridge, MA.

Leopold, L. B. and T. Maddock, Jr. 1954. *The Flood Control Controversy.* Ronald Press Co., New York.

Lepart, J. and M. Debussche. 1991. Invasion processes as related to succession and disturbance. pp. 159–177. In: *Biogeography of Mediterranean Invasions*. R. H. Groves and F. Di Castri (eds.). Cambridge University Press, Cambridge, U.K.

Lesinski, J. M. 1996. *Exotic Invaders*. Walker & Co., New York.

Leson, G. and A. M. Winer. 1991. Biofiltration: an innovative air pollution control technology for VOC emissions. *Journal of Air and Waste Management Association.* 41:1045–1054.

Lessard, P. and M. B. Beck. 1991. Dynamic modeling of wastewater treatment processes. *Environment Science Technology.* 25:30–39.

Levin, S. A. 1992. The problem of pattern and scale in ecology. *Ecology.* 73:1943–1967.

Levin, S. A. 1994. Frontiers in ecosystem science. pp. 381–389. In: *Frontiers in Mathematical Biology.* S. A. Levin (ed.). Springer-Verlag, Berlin.

Levin, S. A. 1998. Extrapolation and scaling in ecotoxicology. pp. 9–11. In: *Multiple Stresses in Ecosystems.* J. J. Cech, Jr., B. W. Wilson, and D. G. Crosby (eds.). Lewis Publishers, Boca Raton, FL.

Levin, S. A. and K. D. Kimball (eds.). 1984. New perspectives in ecotoxicology. *Environmental Management.* 8:375–442.

Levin, S. A. and R. T. Paine. 1974. Disturbance, patch formation, and community structure. *Proceedings of the National Academy of Science USA.* 72:2744–2747.

Levin, S. A., M. A. Harwell, J. R. Kelly, and K. D. Kimball (eds.). 1989. *Ecotoxicology: Problems and Approaches.* Springer-Verlag, New York.

Levins, R. 1998. Qualitative mathematics for understanding, prediction, and intervention in complex ecosystems. pp. 178–204. In: *Ecosystem Health.* D. Rapport, R. Costanza, P. R. Epstein, C. Gaudet, and R. Levins (eds.). Blackwell Science, Malden, MA.

Levinton, J. 1995. Bioturbators as ecosystem engineers: control of the sediment fabric, inter-individual interactions, and material fluxes. pp. 29–36. In: *Linking Species and Ecosystems.* C. G. Jones and J. H. Lawton (eds.). Chapman & Hall, New York.

Levinton, J. S., J. E. Ward, S. E. Shumway, and S. M. Baker. 2001. Feeding processes of bivalves: connecting the gut to the ecosystem. pp. 385–400. In: *Organisms-Sediment Interactions.* J. Y. Aller, S. A. Woodin, and R. C. Aller (eds.). University of South Carolina Press, Columbia, SC.

Levy, D., G. Lockett, J. Oldfather, J. Rees, E. Saegebarth, R. Schneider, and J. Harte. 1985. Realism and replicability of lentic freshwater microcosms. pp. 43–56. In: *Validation and Predictability of Laboratory Methods for Assessing the Fate and Effects of Contaminants in Aquatic Ecosystems.* T. P. Boyle (ed.). ASTM, Philadelphia, PA.

Levy, S. 1984. *Hackers: Heroes of the Computer Revolution.* Dell Publishing, New York.

Levy, S. 1992. *Artificial Life.* Pantheon Books, New York.

Lewin, R. 1986. Supply-side ecology. *Science.* 234:25–27.

Lewis, R. R., III. 1990. Wetlands restoration/creation/enhancement terminology: suggestions for standardization. pp. 417–419. In: *Wetland Creation and Restoration, The Status of the Science.* J. A. Kusler and M. E. Kentula (eds.). Island Press, Washington, DC.

Lichtenberg, E. R. and W. Getz. 1985. Economics of rice-field mosquito control in California. *BioScience.* 35:292–297.

Lifset, R. J. 2000. Full accounting. *The Sciences.* 40(3):32–37.

Likens, G. E. (ed.). 1972. *Nutrients and Eutrophication: The Limiting-Nutrient Controversy. Special Symposia*, Vol. 1, American Society of Limnology and Oceanography, Inc., Lawrence, KS.

Likens, G. E. (ed.). 1989. *Long-Term Studies in Ecology, Approaches and Alternatives.* Springer-Verlag, New York.

Likens, G. E. 1992. *The Ecosystem Approach: Its Use and Abuse.* Ecology Institute, Oldendorf/Luhe, Germany.

Likens, G. E., F. H. Bormann, R. S. Pierce, J. S. Eaton, and N. M. Johnson. 1977. *Biogeochemistry of a Forested Ecosytem.* Springer-Verlag, New York.

Limburg, P. R. 1980. *Farming the Waters.* Beaufort Books, Inc., New York.

Lin, H. R. 1982. Polycultural system of freshwater fish in China. *Canadian Journal of Fisheries and Aquatic Science*. 39:143–150.

Lipson, H. and J. B. Pollack. 2000. Automatic design and manufacture of robotic lifeforms. *Nature*. 406:974–978.

Lipton, R. J. and E. B. Baum (eds.). 1996. *DNA Based Computers*. American Mathematical Society, Princeton University Press, Princeton, NJ.

Little, C. E. 1987. *Green Fields Forever: The Conservation Tillage Revolution in America*. Island Press, Washington, DC.

Lloyd, C., D. Rapport, and J. E. Turner. 1975. The market adaptation of the firm. pp. 119–135. In: *Adaptive Economic Models*. R. H. Day and T. Groves (eds.). Academic Press, New York.

Lockaby, B. G. and W. H. Conner. 1999. N:P balance in wetland forests: productivity across a biogeochemical continuum. *Botanical Review*. 65:171–185.

Lockwood, J. L. and S. L. Pimm. 1999. When does restoration succeed? pp. 363–392. In: *Ecological Assembly Rules: Perspectives, Advances, Retreats*. Cambridge University Press, Cambridge, U.K.

Lockwood, J. L. and M. L. McKinney (eds.). 2001. *Biotic Homogenization*. Kluwer Academic/Plenum Publishers, New York.

Lodge, D. J., D. L. Hawksworth, and B. J. Ritchie. 1996. Microbial diversity and tropical forest functioning. pp. 69–100. In: *Biodiversity and Ecosystem Processes in Tropical Forests*. Springer-Verlag, Berlin.

Loehle, C. 1987. Hypothesis testing in ecology: psychological aspects and the importance of theory maturation. *The Quarterly Review of Biology*. 62:397–409.

Loehle, C. 1989. Catastrophe theory in ecology: a critical review and an example of the butterfly catastrophe. *Ecological Modelling*. 49:125–152.

Loehle, C. 1990. Indirect effects: a critique and alternative methods. *Ecology*. 71:2382–2386.

Loehle, C. 1988. Philosophical tools: potential contributions to ecology. *Oikos*. 51:97–104.

Limburg, P. R. 1980. *Farming the Waters*. Beaufort Books, Inc., New York.

Logsdon, G. 1978. *Getting Food from Water: A Guide to Backyard Aquaculture*. Rodale Press, Emmaus, PA.

Loomis, J. B. and R. G. Walsh. 1986. Assessing wildlife and environmental values in cost-benefit analysis: state of the art. *Journal of Environmental Management*. 22:125–132.

Lorsch, H. G., M. G. Koesterer, and E. Fried. 1968. *Biological Handbook for Engineers*. NASA CR-61237. Marshall Space Flight Center, Huntsville, AL.

Lotka, A. J. 1925. *Physical Biology*. Williams and Wilkins, Baltimore, MD.

Loucks, O. L. 1977. Emergence of research on agro-ecosystems. *Annual Reviews of Ecology and Systematics*. 8:173–192.

Loucks, O. L. 1985. Looking for surprise in managing stressed ecosystems. *BioScience*. 35:428–432.

Lowe, J. C. and S. Moryadas. 1975. *The Geography of Movement*. Houghton Mifflin, Boston, MA.

Lowes, A. L. and C. C. Blackwell, Jr. 1974. Applications of modern control theory to ecological systems. pp. 299–305. In: *Ecosystem Analysis and Prediction*. S. A. Levin (ed.). Society of Industrial and Applied Mathematics, Philadelphia, PA.

Lowrance, R. 1998. Riparian forest ecosystems as filters for nonpoint-source pollution. pp. 113–141. In: *Successes, Limitations, and Frontiers in Ecosystem Science*. M. L. Pace and P. M. Groffman (eds.). Springer, New York.

Lowrance, R. and S. R. Crow. 2002. Implementation of riparian buffer systems for landscape management. pp. 145–158. In: *Landscape Ecology in Agroecosystems Management*. L. Ryszkowski (ed.). CRC Press, Boca Raton, FL.

Luckett, C., W. H. Adey, J. Morrissey, and D. M. Spoon. 1996. Coral reef mesocosms and microcosms — successes, problems, and the future of laboratory models. *Ecological Engineering.* 6:57–72.

Luckinbill, L. S. 1979. Regulation, stability, and diversity in a model experimental microcosm. *Ecology.* 60:1098–1102.

Ludwig, D., D. D. Jones, and C. S. Holling. 1978. Qualitative analysis of insect outbreak systems: the spruce budworm and forest. *Journal of Animal Ecology.* 47:315–332.

Ludwig, D. F. 1989. Anthropic ecosystems. *Bulletin of the Ecological Society of America.* 70:12–14.

Lugo, A. E. 1978. Stress and ecosystems. pp. 62–101. In: *Energy and Environmental Stress in Aquatic Systems.* J. H. Thorp and J. W. Gibbons (eds.). DOE Symposium Series CONF-71114, National Technical Information Service, Springfield, VA.

Lugo, A. E. 1980. Mangrove ecosystems: successional or steady state? pp. 65–72. In: *Tropical Succession.* J. Ewel (ed.). Supplement to *Biotropica.* 12(2).

Lugo, A. E. 1982. Some aspects of the interactions among nutrient cycling, hydrology, and soils in wetlands. *Water International.* 7:178–184.

Lugo, A. E. 1988. Estimating reductions in the diversity of tropical forest species. pp. 58–70. In: *Biodiversity.* E. O. Wilson (ed.). National Academy Press, Washington, DC.

Lugo, A. E. 1990. Removal of exotic organisms. *Conservation Biology.* 4:345.

Lugo, A. E. 1994. Maintaining an open mind on exotic species. pp. 218–220. In: *Principles of Conservation Biology.* G. K. Meffe and C. R. Carroll (eds.). Sinauer Assoc., Sunderland, MA.

Lugo, A. E. and M. M. Brinson. 1979. Calculations of the value of salt water wetlands. pp. 120–130. In: *Wetland Functions and Values: The State of Our Understanding.* P. E. Greeson, J. R. Clark, and J. E. Clark (eds.). American Water Resources Association, Minneapolis, MN.

Lugo, A. E. and S. C. Snedaker. 1974. The ecology of mangroves. *Annual Reviews of Ecology and Systematics.* 5:39–64.

Lugo, A. E., S. Brown, and M. M. Brinson. 1990. Concepts in wetland ecology. pp. 53–85. In: *Forested Wetlands: Ecosystems of the World,* Vol. 15. A. E. Lugo, M. Brinson, and S. Brown (eds.). Elsevier, Amsterdam, the Netherlands.

Lugo, A. E., E.G., Farnworth, D. J. Pool, P. Jerez, and G. Kaufman. 1973. The impact of the leaf cutter ant *Atta colombica* on the energy flow of a tropical wet forest. *Ecology.* 54:1292–1301.

Luken, J. O. 1996. *Directing Ecological Succession.* Chapman & Hall, London.

Lund, H. F. (ed.). 2001.*The McGraw-Hill Recycling Handbook,* 2nd ed. McGraw-Hill, New York.

Lussenhop, J. 1992. Mechanisms of microarthropod-microbial interactions in soil. *Advances in Ecological Research.* 23:1–34.

Maas, D. and A. Schopp-Guth. 1995. Seed banks in fen areas and their potential use in restoration ecology. pp. 189–200. In: *Restoration of Temperate Wetlands.* B. D. Wheller, S. C. Shaw, W. J. Fojt, and R. A. Robertson (eds.). John Wiley & Sons, Chichester, U.K.

Maass, A. 1951. *Muddy Water: The Army Engineers and the Nation's Rivers.* Harvard University Press, Cambridge, MA.

MacArthur, R. H. 1955. Fluctuations of animal populations and a measure of community stability. *Ecology.* 36:533–536.

MacArthur, R. H. 1968. The theory of the niche. pp. 159–176. In: *Population Biology and Evolution.* R. C. Lewontin (ed.). Syracuse University Press, Syracuse, NY.

MacArthur, R. H. and E. O. Wilson. 1963. An equilibrium theory of insular biogeography. *Evolution.* 17:373–387.

MacArthur, R. H. and E. O. Wilson. 1967. *The Theory of Island Biogeography.* Princeton University Press, Princeton, NJ.

MacCready, P. 1997. The inventive process. pp. 44–50. In: *Design Outlaws on the Ecological Frontier.* C. Zelov and P. Cousineau (eds.). Knossus Publishing, Philadelphia, PA.

MacDonald, D. W. and M. D. Thom. 2001. Alien carnivores: unwelcome experiments in ecological theory. pp. 93–122. In: *Carnivore Conservation.* J. L. Gittleman, S. M. Funk, D. W. MacDonald, and R. K. Wayne (eds.). Cambridge University Press, Cambridge, U.K.

MacDonald, I. A. W. and J. Cooper. 1995. Insular lessons for global biodiversity conservation with particular reference to alien invasions. pp. 189–203. In: *Islands.* P. M. Vitousek, L. L. Loope, and H. Adsersen (eds.). Springer, New York.

MacIsaac, H. J. 1996. Potential abiotic and biotic impacts of zebra mussels on the inland waters of North America. *American Zoologist.* 36:287–299.

Mack, R. N. 1985. Invading plants: their potential contribution to population biology. pp. 127–142. In: *Studies on Plant Demography.* J. White (ed.). Academic Press, London.

Mack, R. N. 1992. Characteristics of invading plant species. pp. 42–46. In: *Alien Plant Invasions in Native Ecosystems of Hawaii.* C. P. Stone, C. W. Smith, and J. T. Tunison (eds.). University of Hawaii Cooperative National Park Resources Studies Unit, Honolulu, HI.

Mack, R. N., D. Simberloff, W. M. Lonsdale, H. Evans, M. Clout, and F. A. Bazzaz. 2000. Biotic invasions: causes, epidemiology, global consequences and control. *Issues in Ecology,* No. 5. Ecological Society of America, Washington, DC.

Mackin, J. H. 1948. Concept of the graded river. *Bulletin of the Geological Society of America. 59:463–512.*

MacLean, D. 1996. Evaluation of a Wetland Mitigation Site Using Self-Organization to Establish the Vegetative Community. M.S. thesis, University of Maryland, College Park, MD.

MacLean, D. and P. Kangas. 1997. Self-organization and planting strategies at a wetland mitigation site in Central Maryland. pp. 207–222. In: *Proceedings of the 24th Annual Conference on Ecosystems Restoration and Creation,* P. J. Cannizzaro (ed.). Hillsborough Community College, Plant City, FL.

MacMahon, J. A. 1979. Ecosystems over time: succession and other types of change. pp. 21–58. In: *Forests: Fresh Perspectives from Ecosystem Analysis.* R. H. Waring (ed.). Oregon State University Press, Corvallis, OR.

MacMahon, J. A. 1998. Empirical and theoretical ecology as a basis for restoration: an ecological success story. pp. 220–246. In: *Successes, Limitations, and Frontiers in Ecosystem Science.* M. L. Pace and P. M. Groffman (eds.). Springer, New York.

Mader, P., A. Fliebach, D. Dubois, L. Gunst, P. Fried, and U. Niggli. 2002. Soil fertility and biodiversity in organic farming. *Science.* 296:1694–1697.

Madsen, B. J. 1989. Biogeomorphology and spatial structure of northern patterned peatlands. pp. 235–247. In: *Freshwater Wetlands and Wildlife.* R. R. Sharitz and J. W. Gibbons (eds.). CONF-8603101, U.S. Department of Energy, Oak Ridge, TN.

Maguire, B. 1971. Phytotelmata: biota and community structure determinations in plant-held waters. *Annual Reviews of Ecology and Systematics.* 2:439–464.

Maguire, B., Jr., L. B. Slobodkin, H. J. Morowitz, B. Moore III, and D. B. Botkin. 1980. A new paradigm for the examination of closed ecosystems. pp. 30–68. In: *Microcosms in Ecological Research.* J. P. Giesy, Jr. (ed.). CONF-781101, DOE Symposium Series, National Technical Information Service, Springfield, VA.

Majer, J. D. 1983. Ants: bioindicators of minesite rehabilitation, land-use, and land conservation. *Environmental Management.* 7:375–383.

Majer, J. D. 1997. Invertebrates assist the restoration process: an Australian perspective. pp. 212–237. In: *Restoration Ecology and Sustainable Development.* K. M. Urbanska, N. R., Webb, and P. J. Edwards (eds.). Cambridge University Press, Cambridge, U.K.

Major, J. 1969. Historical development of the ecosystem concept. pp. 9–22. In: *The Ecosystem Concept in Natural Resource Management.* G. M. Van Dyne (ed.). Academic Press, New York.

Makepeace, D. K., D. W. Smith, and S. J. Stanley. 1995. Urban stormwater quality: summary of contaminant data. *Critical Reviews in Environmental Science and Technology.* 25:93–139.

Malakoff, D. 1998. Restored wetlands flunk real-world test. *Science.* 280:371–372.

Malakoff, D. 1999. Biological invaders sweep in. *Science.* 285:1834–1843.

Maltby, L. 1992. Detritus processing. pp. 331–353. In: *The Rivers Handbook, Hydrological and Ecological Principles.* P. Calow and G. E. Petts (eds.). Blackwell Science, London.

Malyuga, D. P. 1964. *Biogeochemical Methods of Prospecting.* Consultants Bureau, New York.

Mandt, M. G. and B. A. Bell. 1982. *Oxidation Ditches in Wastewater Treatment.* Ann Arbor Science, Ann Arbor, MI.

Margalef, R. 1967. Laboratory analogues of estuarine plankton systems. pp. 515–524. In: *Estuaries.* G. H. Lauff (ed.). American Association for the Advancement of Science, Washington, DC.

Margalef, R. 1968. *Perspectives in Ecological Theory.* University of Chicago Press, Chicago, IL.

Margalef, R. 1969. Diversity and stability: A practical proposal and a model of interdependence. pp. 25–37. In: *Diversity and Stability in Ecological Systems.* BNL-50175, Brookhaven National Laboratory, Upton, NY.

Margalef, R. 1984. Simple facts about life and the environment not to forget in preparing schoolbooks for our grandchildren. pp. 299–320. In: *Trends in Ecological Research for the 1980s.* J. H. Cooley and F. B. Golley (eds.). Plenum Press, New York.

Margalef, R. 1985. From hydrodynamic processes to structure (information) and from information to process. pp. 200–220. In: *Ecosystem Theory for Biological Oceanography.* R. E. Ulanowicz and T. Platt (eds.). Canadian Bulletin of Fisheries and Aquatic Sciences 213, Department of Fisheries and Oceans, Ottawa, Canada.

Marglin, F. A. and P. C. Mishra. 1993. Sacred Groves, regenerating the body, the land, the community. pp. 197–207. In: *Global Ecology: A New Arena of Political Conflict.* W. Sachs (ed.). Zed Books, London.

Margulis, L., D. Chase, and R. Guerrero. 1986. Microbial communities. *BioScience.* 36:160–170.

Margulis, L. and J. E. Lovelock. 1989. Gaia and geognosy. pp. 1–30. In: *Global Ecology: Towards a Science of the Biosphere.* M. B. Rambler, L. Margulis, and R. Fester (eds.). Academic Press, Boston, MA.

Mark, J., G. M. Chapman, and T. Gibson. 1985. Bioeconomics and the theory of niches. *Futures.* 17:632–651.

Markowitz, H. 1982. *Behavioral Enrichment in the Zoo.* Van Nostrand Reinhold, New York.

Martin, C. V. and B. F. Eldridge. 1989. California's experience with mosquitoes in aquatic wastewater treatment systems. pp. 393–398. In: *Constructed Wetlands for Wastewater Treatment.* D. A. Hammer (ed.). Lewis Publishers, Chelsea, MI.

Martin, D. L. and G. Gershuny (eds.). 1992. *The Rodale Book of Composting.* Rodale Press, Emmaus, PA.

Martin, J. H. 1992. Iron as a limiting factor in oceanic productivity. pp. 123–136. In: *Primary Productivity and Biogeochemical Cycles in the Sea*. P. G. Falkowski and A. D. Woodhead (eds.). Plenum Press, New York.

Maser, C. and J. R. Sedell. 1994. *From the Forest to the Sea, The Ecology of Wood in Streams, Rivers, Estuaries, and Oceans*. St. Lucie, Press, Delray Beach, FL.

Matsuda, H. and M. Shimada. 1993. Cost-benefit model for the evolution of symbiosis. pp. 228–238. In: *Mutualism and Community Organization: Behavioural, Theoretical, and Food-Web Approaches*. H. Kawanabe, J. E. Cohen, and K. Iwasaki (eds.). Oxford University Press, Oxford, U.K.

Matthews, G. A. and T. J. Minello. 1994. Technology and Success in Restoration, Creation, and Enhancement of *Spartina alterniflora* Marshes in the United States. Decision Analysis Series No. 2, National Oceanic and Atmospheric Administration, Coastal Ocean Office, Washington, DC.

Mattson, W. J. and N. D. Addy. 1975. Phytophagous insects as regulators of forest primary production. *Science*. 190:515–522.

Maxwell, J. and R. Costanza. 1989. An ecological economics for ecological engineering. pp. 57–77. In: *Ecological Engineering*. W. J. Mitsch and S. E. Jørgensen (eds.). John Wiley & Sons, New York.

May, P. I. 2000. How (and why) to build a tidal freshwater mudflat mesocosm. pp. 45–58. In: *Proceedings of the Annual Ecosystems Restoration and Creation Conference*, P. J. Cannizzaro (ed.). Hillsborough Community College, Plant City, FL.

May, P. I., Experimental Ecology and Restoration of Anacostia Freshwater Tidal Marshes, Ph.D. dissertation, University of Maryland, College Park, MD, in preparation.

May, R. M. 1973. *Stability and Complexity in Model Ecosystems*. Princeton University Press, Princeton, NJ.

May, R. M. 1974a. Scaling in ecology. *Science*. 184:1131.

May, R. M. 1974b. Biological populations with nonoverlapping generations: stable points, stable cycles, and chaos. *Science*. 186:645.

May, R. M. 1975. Island biogeography and the design of wildlife preserves. *Nature*. 254:177–178.

May, R. M. 1976. Simple mathematical models with very complicated dynamics. *Nature*. 261:459–467.

May, R. M. 1977. Thresholds and breakpoints in ecosystems with a multiplicity of stable states. *Nature*. 269:471–477.

May, R. M. 1988. How many species are there on Earth? *Science*. 241:1441–1449.

May, R. M. and G. F. Oster. 1976. Bifurcations and dynamic complexity in simple ecological models. *American Naturalist*. 110:573–599.

Mayr, O. 1970. *The Origins of Feedback Control*. MIT Press, Cambridge, MA.

McAllister, C. D., R. J. LeBrasseus, and T. R. Parsons. 1972. Stability of enriched aquatic ecosystems. *Science*. 175:562–564.

McAllister, D. M. 1980. *Evaluation in Environmental Planning*. MIT Press, Cambridge, MA.

McCabe, J. and W. W. Eckenfelder, Jr. 1958. Preface. pp. iii–iv. In: *Biological Treatment of Sewage and Industrial Wastes*, Vol. II. J. McCabe and W. W. Eckenfelder, Jr. (eds.). Reinhold Publishing New York.

McClanahan, T. R. and R. W. Wolfe. 1993. Accelerating forest succession in a fragmented landscape: the role of birds and perches. *Conservation Biology*. 7:279–288.

McCormick, J. 1970. The natural features of Tinicum Marsh, with particular emphasis on the vegetation. pp. 1–104. In: *Two Studies of Tinicum Marsh, Delaware and Philadelphia Counties, PA*. The Conservation Foundation, Washington, DC.

McCormick, J. 1971. Tinicum Marsh. *Bulletin of the Ecological Society of America.* 52(3):8, 23.

McCoy, J. H. 1971. Sewage pollution of natural waters. pp. 33–50. In: *Microbial Aspects of Pollution.* G. Sykes, and F. A. Skinner (eds.). Academic Press, New York.

McCulloch, W. S. 1962. The imitation of one form of life by another — Biomimesis. pp. 393–397. In: *Biological Prototypes and Synthetic Systems*, Vol. 1. Plenum Press, New York.

McCullough, D. R. 1979. *The George Reserve Deer Herd: Population Ecology of a K-selected Species.* University of Michigan Press, Ann Arbor, MI.

McCutcheon, S. C. and T. M. Walski. 1994. Ecological engineers: friend or foe? *Ecological Engineering.* 3:109–112.

McCutcheon, S. C. and W. J. Mitsch. 1994. Ecological and environmental engineering: potential for progress. *Ecological Engineering.* 3:107–109. (Also published in *Journal of Environmental Engineering.* 120:479–480)

McDonough, W. and M. Braungart. 2002. *Cradle to Cradle, Remaking the Way We Make Things.* North Point Press, New York.

McGuinness, K. A. 1984. Equations and explanations in the study of species-area curves. *Biological* Reviews. 59:423–440.

McHarg, I. L. 1969. *Design with Nature.* Doubleday, Garden City, NY.

McIntosh, R. P. 1981. Succession and ecological theory. pp. 10–23. In: *Forest Succession, Concepts and Application.* D. C. West, H. H. Shugart, and D. B. Botkin (eds.). Springer-Verlag, New York.

McIntosh, R. P. 1985. *The Background of Ecology.* Cambridge University Press, Cambridge, U.K.

McIntyre, R. 1996. *A Society of Wolves.* Voyageur Press, Inc., Stillwater, MN.

McKay, C. P. 1999. Bringing life to Mars. *Scientific American Presents: The Future of Space Exploration.* 10(1):52–57.

McKinney, R. E. and A. Gram. 1956. Protozoa and activated sludge. *Sewage Industrial Wastes.* 28:1219–1231.

McLachlan, A. 1980. Exposed sandy beaches as semi-enclosed ecosystems. *Marine Environmental Research.* 4:59–63.

McLachlan, A. and T. Erasmus (eds.). 1983. *Sandy Beaches as Ecosystems.* Junk Publishers, The Hague, the Netherlands.

McLachlan, A., T. Erasmus, A. H. Dye, T. Wooldridge, G. Van der Horst, G. Rossouw, T. A. Lasiak, and L. McGwynne. 1981. Sand beach energetics: an ecosystem approach towards a high energy interface. *Estuarine, Coastal and Shelf Science.* 13:11–25.

McLarney, W. O. and J. Todd. 1977. Walton two: a complete guide to backyard fish farming. pp. 74–107. In: *The Book of the New Alchemists.* N. J. Todd (ed.). E. P. Dutton, New York.

McNaughton, S. J. and M. B. Coughenour. 1981. The cybernetic nature of ecosystems. *American Naturalist.* 117:985–990.

McPhee, J. 1989. *The Control of Nature.* Farrar, Straus and Giroux Publishing, New York.

Mead, R. 1971. A note on the use and misuse of regression models in ecology. *Journal of Ecology.* 59:215–219.

Meadows, P. S. and A. Meadows (eds.). 1991. *The Environmental Impact of Burrowing Animals and Animal Burrows.* Clarendon Press, Oxford, U.K.

Meckler, M. (ed.). 1991. *Indoor Air Quality: Design Guidebook.* The Fairmont Press, Inc., Lilburn, GA.

Meffe, G. K. 1992. Techno-arrogance and halfway technologies: Salmon hatcheries on the Pacific Coast of North America. *Conservation Biology.* 6:350–354.

Meffe, G. K., L. A. Nielsen, R. L. Knight, and D. A. Schenborn. 2002. *Ecosystem Management: Adaptive, Community-Based Conservation.* Island Press, Washington, DC.

Meine, C. 1999. Reimagining the prairie: Aldo Leopold and the origins of prairie restoration. pp. 144–160. In: *Recovering the Prairie.* R. F. Sayre (ed.). University of Wisconsin Press, Madison, WI.

Melchiorri-Santolini, U. and J. W. Hopton (eds.). 1972. *Detritus and Its Role in Aquatic Ecosystems.* Memorie, Del'Instituto Italiano di Idrobiologia Vol. 29 Suppl. Pallanza, Italy.

Menduno, M. 1998. Reefer madness. *Wired.* 6(9):47.

Metcalf, L. and H. P. Eddy. 1916. *American Sewerage Practice.* Vol. III*: Disposal of Sewage.* McGraw-Hill, New York.

Metcalf, L. and H. P. Eddy. 1930. *Sewerage and Sewage Disposal, A Textbook.* McGraw-Hill, New York.

Metcalf & Eddy, Inc., revised by G. Tchobanoglous. 1979. *Wastewater Engineering: Treatment: Disposal and Reuse.* McGraw-Hill, New York.

Metcalf, R. L. 1977a. Model ecosystem approach to insecticide degradation: a critique. *Annual Review of Entomology.* 22:241–261.

Metcalf, R. L. 1977b. Model ecosystem studies of bioconcentration and biodegradation of pesticides. pp. 127–144. In: *Pesticides in Aquatic Environments.* M. A. Quddus Khan (ed.). Plenum Press, New York.

Metcalf, R. L., G. K. Sangha, and I. P. Kapoor. 1971. Model ecosystem for the evaluation of pesticide biodegradability and ecological magnification. *Environmental Science and Technology.* 5:709–713.

Metzker, K. D. and W. J. Mitsch. 1997. Modelling self-design of the aquatic community in a newly created freshwater wetland. *Ecological Modelling.* 100:61–86.

Michener, W. K., T. J. Baerwald, P. Firth, M. A. Palmer, J. L. Rosenberger, E. A. Sandlin, and H. Zimmerman. 2001. Defining and unraveling biocomplexity. *BioScience.* 51:1018–1023.

Middlebrooks, E. J., C. H. Middlebrooks, J. H. Reynolds, G. Z. Watters, S. C. Reed, and D. B. George. 1982. *Wastewater Stabilization Lagoon Design, Performance, and Upgrading.* Macmillan, New York.

Middleton, B. 1999. *Wetland Restoration: Flood Pulsing and Disturbance Dynamics.* John Wiley & Sons, New York.

Middleton, B. A. (ed.). 2002. *Flood Pulsing in Wetlands: Restoring the Natural Hydrological Balance.* John Wiley & Sons, New York.

Mikkelsen, L. 1993. Soft engineering repairs ailing Montana stream. *Land and Water.* 37(2):6–9.

Mikkola, D. E. 1993. Accreditation and engineering design: ABET and TMS. pp. 21–36. In: *Design Education in Metallurgical and Material Engineering.* The Minerals, Metals and Materials Society, Warrendale, PA.

Milhorn, H. T., Jr. 1966. *The Application of Control Theory to Physiological Systems.* W. B. Saunders Co., Philadelphia, PA.

Millard, A. 1990. *Edison and the Business of Innovation.* Johns Hopkins University Press, Baltimore, MD.

Miller, G. T., Jr. 1991. *Environmental Science.* Wadsworth, Belmont, CA.

Miller, J. A. 1980. Reef alive. *Science News.* 118:250–252.

Miller, J. W. and I. G. Koblick. 1995. *Living and Working in the Sea.* Five Corners Publications, Plymouth, VT.

Miller, R. M. 1987. Mycorrhizae and succession. pp. 205–220. In: *Restoration Ecology: A Synthetic Approach to Ecological Research.* W. R. Jordan, III, M. E. Gilpin, and J. D. Aber (eds.). Cambridge University Press, Cambridge, U.K.

Miller, R. M. and J. D. Jastrow. 1992. The application of VA mycorrhizae to ecosystem restoration and reclamation. pp. 438–467. In: *Mycorrhizal Functioning.* M. A. Allen (ed.). Chapman & Hall, New York.

Miller, S. L. 1953. A production of amino acids under possible primitive earth conditions. *Science.* 117:528–529.

Miller, S. L. 1955. Production of some organic compounds under possible primitive earth conditions. *Journal of the American Chemical Society.* 77:2351–2361.

Miller, S. L. and H. C. Urey. 1959. Organic compound synthesis on the primitive Earth. *Science.* 130:245–251.

Miller, T. E. and W. C. Kerfoot. 1987. Redefining indirect effects. pp. 33–37. In: *Predation: Direct and Indirect Impacts on Aquatic Communities.* W. C. Kerfoot and A. Sih (eds.). University Press of New England, Hanover, NH.

Mills, L. S., M. E. Soule, and D. F. Doak. 1993. The keystone-species concept in ecology and conservation. *BioScience.* 43:219–224.

Mills, S. 1997. *Turning Away From Technology.* Sierra Club Books, San Francisco, CA.

Milsum, J. H. 1966. *Biological Control Systems Analysis.* McGraw-Hill, New York.

Minckley, W. L. 1973. *Fishes of Arizona.* Publication of the Arizona Game and Fish Dept., Phoenix, AZ.

Minshall, G. W. 1967. Role of *Allochthonous detritus* in the trophic structure of a woodland springbrook community. *Ecology.* 48:139–149.

Minshall, G. W., D. A. Andrews, and C. Y. Manuel-Faler. 1983. Application of island biogeographic theory to streams: macroinvertebrate recolonization of the Teton River, Idaho. pp. 279–298. In: *Stream Ecology, Application and Testing of General Ecological Theory.* J. R. Barnes and G. W. Minshall (eds.). Plenum Press, New York.

Mitchell, M. 1996. *An Introduction to Genetic Algorithms.* MIT Press, Cambridge, MA.

Mitsch, W. J. 1977. Energy conservation through interface ecosystems. pp. 875–881. In: *Proceedings of the International Conference on Energy Use Management.* R. Fazzolari and C. B. Smith (eds.). Pergamon Press, Oxford, U.K.

Mitsch, W. J. 1990. Review of: Constructed Wetlands for Wastewater Treatment: Municipal, Industrial, and Agricultural by D. A. Hammer. *Journal of Environmental Quality.* 19:784–785.

Mitsch, W. J. 1991. Ecological engineering: approaches to sustainability and biodiversity in the U.S. and China. pp. 428–448. In: *Ecological Economics: The Science and Management of Sustainability.* R. Costanza (ed.). Columbia University Press, New York.

Mitsch, W. J. 1992. Wetlands, ecological engineering, and self-design. *Great Lakes Wetlands Newsletter.* 3:1–3,7.

Mitsch, W. J. 1993. Ecological engineering — a cooperative role with the planetary life-support systems. *Environmental Science and Technology.* 27:438–445.

Mitsch, W. J. 1994. Ecological engineering: a friend. *Ecological Engineering.* 3:112–115.

Mitsch, W. J. 1995a. Ecological engineering: from Gainesville to Beijing — a comparison of approaches in the United States and China. pp. 109–122. In: *Maximum Power: The Ideas and Applications of H. T. Odum.* C. A. S. Hall (ed.). University Press of Colorado, Niwott, CO.

Mitsch, W. J. 1995b. Restoration and creation of wetlands — providing the science and engineering basis and measuring success. *Ecological Engineering.* 4:61–64.

Mitsch, W. J. 1996. Ecological engineering: a new paradigm for engineers and ecologists. pp. 111–128. In: *Engineering within Ecological Constrainsts*. P. C. Schulze (ed.). National Academy Press, Washington, DC.

Mitsch, W. J. 1998a. Self-design and wetland creation: early results of a freshwater marsh experiment. pp. 635–655. In: *Wetlands for the Future*. A. J. McComb and J. A. Davis (eds.). Gleneagles Publishing, Adelaide, Australia.

Mitsch, W. J. 1998b. Ecological engineering — the 7-year itch. *Ecological Engineering*. 10:119–130.

Mitsch, W. J. 1998c. Olentangy River wetland research park. *Land and Water*, 42(5):22–26.

Mitsch, W. J. 2000. Self-design applied to coastal restoration. pp. 554–564. In: *Concepts and Controversies in Tidal Marsh Ecology*. M. P. Weinstein and D. A. Kreeger (eds.). Kluwer Academic, Dordrecht, the Netherlands.

Mitsch, W. J. and J. K. Cronk. 1992. Creation and restoration of wetlands: some design consideration for ecological engineering. *Advances in Soil Science*. 17:217–259.

Mitsch, W. J. and S. E. Jørgensen. (eds.). 1989. *Ecological Engineering*. John Wiley & Sons, New York.

Mitsch, W. J. and R. F. Wilson. 1996. Improving the success of wetland creation and restoration with know-how, time, and self-design. *Ecological Applications*. 6:77–83.

Mitsch, W. J., X. Wu, R. W. Nairn, P. E. Weihe, N. Wang, R. Deal, and C. E. Boucher. 1998. Creating and restoring wetlands. *BioScience*. 48:1019–1030.

Mitsch, W. J. and P. Kangas. 1999. Academic programs in ecological engineering: proceedings of a workshop held at the Ohio State University, March 15–16, 1999. Unpublished Report. Natural Resources Management Program, University of Maryland, College Park, MD.

Mohan, P. C. and C. Aruna. 1994. The biology of Serpulid worms in relation to biofouling. pp. 59–64. In: *Recent Developments in Biofouling Control*. A. A. Balkema/Rotterdam, Amsterdam, the Netherlands.

Mohr, C. O. 1943. Cattle droppings as ecological units. *Ecological Monographs*. 13:275–298.

Molles, M. C., Jr. 1978. Fish species diversity on model and natural patch reefs: experimental insular biogeography. *Ecological Monographs*. 48:289–305.

Montague, C. L. 1980. A natural history of temperate western Atlantic fiddler crabs (genus *Uca*) with reference to their impact on the saltmarsh. *Publications of the Marine Science Institute of Texas*. 23:25–55.

Montague, C. L. 1982. The influence of fiddler crab burrows and burrowing on metabolic processes in saltmarsh sediments. pp. 283–301. In: *Estuarine Comparisons*. V. S. Kennedy (ed.). Academic Press, New York.

Montague, C. L. 1993. Ecological engineering of inlets in southeastern Florida: Design criteria for sea turtle nesting beaches. *Journal of Coastal Research*. 18:267–276.

Montague, C. L. and H. T. Odum. 1997. Setting and functions. pp. 9–33. In: *Ecology and Management of Tidal Marshes: A Model from the Gulf of Mexico*. C. L. Coultas and Y.-P. Hsieh (eds.). St. Lucie Press, Delray Beach, FL.

Montague, C. L., A. V. Zale, and H. F. Percival. 1985. A conceptual model of saltmarsh management on Merritt Island National Wildlife Refuge, Florida. Technical Report No. 17. Florida Cooperative Fish and Wildlife Research Unit, University of Florida. Gainesville, FL.

Mooney, H. A. and J. A. Drake. 1989. Biological invasions: a SCOPE program overview. pp. 491–506. In: *Biological Invasions: A Global Perspective*. J. A. Drake, H. A. Mooney, F. di Castri, R. H. Groves, F. J. Kruger, M. Rejmanek, and M. Williamson (eds.). John Wiley & Sons, Chichester, U.K.

Morgan, A. E. 1971. *Dams and Other Disasters, A Century of the Army Corps of Engineers in Civil Works*. Porter Sargent Publishers, Boston, MA.

Morgan, L. H. 1868. *The American Beaver*. J. B. Lippincott, Philadephia, PA.

Morgan, R. P. C. and R. J. Rickson (eds.). 1995. *Slope Stabilization and Erosion Control: A Bioengineerinig Approach*. E & FN SPON, London.

Morin, P. J. 1989. New directions in amphibian community ecology. *Herpetologica*. 45:124–128.

Morin, P. J. 1998. Realism, precision, and generality in experimental ecology. pp. 50–70. In: *Experimental Ecology*. W. J. Resetarits, Jr. and J. Bernardo (eds.). Oxford University Press, New York.

Morisawa, M. 1968. *Streams: Their Dynamics and Morphologies*. McGraw-Hill, New York.

Morowitz, H. J. 1970. *Entropy for Biologists: An Introduction to Thermodynamics*. Academic Press, New York.

Morowitz, H. J. 1996. Complexity and epistemology. pp. 188–198. In: *Boundaries and Barriers on the Limits of Scientific Knowledge*. J. L. Casti and A. Karlqvist (eds.). Addison-Wesley, Reading, MA.

Morrison, M. L., B. G. Marcot, and R. W. Mannan. 1992. *Wildlife-Habitat Relationships*. University of Wisconsin Press, Madison, WI.

Moshiri, G. A. (ed.). 1993. *Constructed Wetlands for Water Quality Improvement*. Lewis Publishers, Boca Raton, FL.

Mottet, M. G. 1985. Enhancement of the marine environment for fisheries and aquaculture in Japan. pp. 13–112. In: *Artificial Reefs, Marine and Freshwater Applications*. F. M. D'Itri (ed.). Lewis Publishers, Chelsea, MI.

Moy, L. D. and L. A. Levin. 1991. Are Spartina marshes a replaceable resource? A functional approach to evaluation of marsh creation efforts. *Estuaries*. 14:1–16.

Moyle, P. B., H. W. Li, and B. A. Barton. 1986. The Frankenstein effect: impact of introduced fishes on native fishes in North America. pp. 415–426. In: *Fish Culture in Fisheries Management*. R. H. Stroud (ed.). American Fisheries Society, Bethesda, MD.

Mulamoottil, G., E. A. McBean, and F. Rovers (eds.). 1999. *Constructed Wetlands for the Treatment of Landfill Leachates*. Lewis Publishers, Boca Raton, FL.

Mulholland, R. J. and C. S. Sims. 1976. Control theory and regulation of ecosystems. pp. 373–388. In: *Systems Analysis and Simulation in Ecology*, Vol. IV. B. C. Patten (ed.). Academic Press, New York.

Muller, R. N. and F. H. Bormann. 1976. Role of *Erythronium americanum* Ker. in energy flow and nutrient dynamics of a northern hardwood forest ecosystem. *Science*. 193:1126–1128.

Murden, W. R. 1984. The national dredging program in relationship to the excavation of suspended and settled sediments. pp. 40–47. In: *Fate and Effects of Sediment-Bound Chemicals in Aquatic Systems*. K. L. Dickson, A. W. Maki, and W. A. Brungs (eds.). Pergamon Press, New York.

Murdoch, A. and J. A. Capobianco. 1979. Effects of treated effluent on a natural marsh. *Journal of the Water Pollution Control Federation*. 51:2243–2256.

Murdoch, W. W. and C. J. Briggs. 1996. Theory for biological control: recent developments. *Ecology*. 77:2001–2013.

Murphy, D. W. and L. E. Carr. 1991. Composting dead birds, Fact Sheet 537, Cooperative Extension Service, University of Maryland, College Park, MD.

Murphy, G. J. 1959. *Control Engineering*. D. Van Nostrand Co., Princeton, NJ.

Myers, J. 1963. Space biology: ecological aspects, introductory remarks. *American Biology Teacher*. 26:409–411.

Myers J. and L. B. Clark 1944. Culture conditions and the development of the photosynthetic mechanism. II. An apparatus for the continuous culture of Chlorella. *Journal of General Physiology.* 28:103–112.

Naeem, S., F. S. Chapin III, R. Costanza, P. R. Ehrlich, F. B. Golley, D. U. Hooper, J. H. Lawton, R. V. O'Neill, H. A. Mooney, O. E. Sala, A. J. Symstad, and D. Tilman. 1999. Biodiversity and ecosystem functioning: maintaining natural life support processes. *Issues in Ecology,* No. 4. Ecological Society of America, Washington, DC.

Naiman, R. J. 1988. Animal influences on ecosystem dynamics. *BioScience.* 38:750–752.

Naiman, R. J., C. A. Johnston, and J. C. Kelley. 1988. Alteration of North American streams by beaver. *BioScience.* 38:753–762.

Nakamura, M. 1985. Evolution of artificial fishing reef concepts in Japan. *Bulletin of Marine Sciences.* 37:271–278.

National Research Council (NRC). 1976. Appendix A, Duckweeds and their uses. pp. 148–154. In: *Making Aquatic Weeds Useful: Some Perspectives for Developing Countries.* National Academy of Sciences, Washington, DC.

National Research Council (NRC). 1981. *Food, Fuel, and Fertilizer from Organic Wastes.* National Academy Press, Washington, DC.

National Research Council (NRC). 1989a. *Alternative Agriculture.* National Academy Press, Washington, DC.

National Research Council (NRC). 1989b. *Field Testing Genetically Modified Organisms.* National Academy Press, Washington, DC.

National Research Council (NRC). 1992. *Restoration of Aquatic Ecosystems: Science, Technology, and Public Policy.* National Academy Press, Washington, DC.

National Research Council (NRC). 1994. Restoring and Protecting Marine Habitat. National Academy Press, Washington, DC.

National Research Council (NRC). 1996a. *Stemming the Tide.* National Academy Press, Washington, DC.

National Research Council (NRC). 1996b. *Ecologically Based Pest Management.* National Academy Press, Washington, DC.

National Science and Technology Council (NSTC). 1995. *Bridge to a Sustainable Future: National Environmental Technology Strategy.* Interagency Environmental Technologies Office, Washington, DC.

Nava, A. 1999. Personal communication.

Naylor, R. L., R. J. Goldburg, J. Primavera, N. Kautsky, M. C. M. Beveridge, J. Clay, C. Folke, J. Lubchenco, H. Mooney, and M. Troell. 2001. Effects of aquaculture on world fish supplies. *Issues in Ecology,* No. 8. Ecological Society of America, Washington, DC.

Nelson, D. J. and D. C. Scott. 1962. Role of detritus in the productivity of a rock-outcrop community in a Piedmont stream. *Limnology and Oceanography.* 7:396–413.

Nelson, M. 1992. 3001 species. *Restoration and Management Notes.* 10:167.

Nelson, M. and W. F. Dempster. 1993. Biosphere 2 — A new approach to experimental ecology. *Environmental Conservation.* 20:74–75.

Nelson, M., T. L. Burgess, A. Alling, N. Alvarez-Romo, W. F. Dempster, R. L. Walford, and J. P. Allen. 1993. Using a closed ecological system to study Earth's biosphere: initial results from Biosphere 2. *BioScience.* 43:225–236.

Nepstad, D. C. and S. Schwartzman (eds.). 1992. *Non-Timber Products from Tropical Forests.* The New York Botanical Garden, Bronx, NY.

Newbold, J. D. 1992. Cycles and spirals of nutrients. pp. 370–408. In: *The Rivers Handbook: Hydrological and Ecological Principles.* Vol. I. P. Calow and G. E. Petts (eds.). Blackwell Scientific Publications, Oxford, U.K.

Newbold, J. D., J. W. Elwood, R. V. O'Neill, and W. Van Winkle. 1981. Measuring nutrient spiralling in stream ecosystems. *Canadian Journal of the Fisheries and Aquatic Science.* 38:860–863.

Newbold, J. D., R. V. O'Neill, J. W. Elwood, and W. Van Winkle. 1982. Nutrient spiraling in streams: implications for nutrient limitation and invertebrate activity. *American Naturalist.* 120:628–652.

Nichols, F. H., J. E. Cloern, S. N. Luoma, and D. H. Peterson. 1986. The modification of an estuary. *Science.* 231:567–571.

Niering, W. A. 1958. Principles of sound right-of-way vegetation management. *Economic Botany.* 12:140–144.

Niering, W. A. 1997. Tidal wetlands restoration and creation along the east coast of North America. pp. 259–285. In: *Restoration Ecology and Sustainable Development.* K. M. Urbanska, N. R. Webb, and P. J. Edwards (eds.). Cambridge University Press, Cambridge, U.K.

Niering, W. A. and R. H. Goodwin. 1974. Creation of relatively stable shrublands with herbicides: arresting "succession" on rights-of-way and pastureland. *Ecology.* 55:784–795.

Nixon, S. W. 1969. A synthetic microcosm. *Limnology and Oceanography.* 14:142–145.

Nixon, S. W. 2001. Some reluctant ruminations on scales (and claws and teeth) in marine mesocosms. pp. 178–190. In: *Scaling Relations in Experimental Ecology.* R. H. Gardner, W. M. Kemp, V. S. Kennedy, and J. E. Petersen (eds.). Columbia University Press, New York.

Nixon, S. W., D. Alonso, M. E. Q. Pilson, and B. A. Buckley. 1980. Turbulent mixing in aquatic microcosms. pp. 818–849. In: *Microcosms in Ecological Research.* J. P. Giesy (ed.). DOE Symposium Series 52, U.S. Dept. Energy, Washington, DC.

Nixon, S. W., C. A. Oviatt, J. N. Kremer, and K. Perez. 1979. The use of numerical models and laboratory microcosms in estuarine ecosystem analysis — simulations of a winter phytoplankton bloom. pp. 165–206. In: *Marsh-Estuarine Systems Simulation.* R. F. Dame (ed.). University of South Carolina Press, Columbia, SC.

Noble, I. R. 1989. Attributes of invaders and the invading process: terrestrial and vascular plants. pp. 301–314. In: *Biological Invasions: A Global Perspective.* J. A. Drake, H. A. Mooney, F. di Castri, R. H. Groves, F. J. Kruger, M. Rejmanek, and M. Williamson (eds.). John Wiley & Sons, Chichester, U.K.

Noble, I. R. and R. O. Slatyer. 1980. The use of vital attributes to predict successional changes in plant communities subject to periodic disturbances. *Vegetation.* 43:5–21.

Nolfi, S. and D. Floreano. 2000. *Evolutionary Robotics: The Biology, Intelligence, and Technology of Self-Organizing Machines.* MIT Press, Cambridge, MA.

Norgaard, R. B. 1987. The epistemological basis of agroecology. pp. 21–27. In: *Agroecology, The Scientific Basis of Alternative Agriculture.* M. A. Altieri (ed.). Westview Press, Boulder, CO.

Norton, S. B., D. J. Rodier, J. H. Gentile, M. E. Troyer, R. B. Landy, and W. van der Schalie, 1995. The EPA's framework for ecological risk assessment. pp. 703–737. In: *Handbook of Ecotoxicology.* D. J. Hoffman, B. A. Rattner, G. A. Burton, Jr., and J. Cairns, Jr. Lewis Publishers, Boca Raton, FL.

Novak, I. D. 1973. Predicting coarse sediment transport: the Hjulstrom curve revisited. pp. 14–25. In *Fluvial Geomorphology.* M. Morisawa (ed.). State University of New York, Binghamton, NY.

Novick, A. 1955. Growth of bacteria. *Annual Reviews of Microbiology.* 9:97–110.

Obenhuber, D. C. and C. E. Folsome. 1984. Eucaryote/procaryote ratio as an indicator of stability for closed ecological systems. *BioSystems.* 16:291–296.

Obenhuber, D. C. and C. E. Folsome. 1988. Carbon recycling in materially closed ecological life support systems. *BioSystems.* 21:165–173.

O'Brien, R. 1964. *Machines.* Time Incorporated, New York.

Odum, E. P. 1959. The macroscopic organism and its environment. pp. 53–61. In: *Proceedings of the Semicentennial Celebration.* The University of Michigan Biological Station, Pellston, MI.

Odum, E. P. 1962. Relationships between structure and function in the ecosystem. *Japanese Journal of Ecology* 12:108–118. (Also published as pp. 6–20 In: G. W. Cox (ed.). 1969. *Readings in Conservation Ecology.* Meredith Corp., New York.)

Odum, E. P. 1969. The strategy of ecosystem development. *Science.* 164:262–270.

Odum, E. P. 1971. *Fundamentals of Ecology,* 3rd ed. W. B. Saunders Co., Philadelphia, PA.

Odum, E. P. 1972. Ecosystem theory in relation to man. pp. 11–24. In: *Ecosystem Structure and Function.* J. A. Wiens (ed.). Oregon State University Press, Corvallis, OR.

Odum, E. P. 1979a. The value of wetlands: a hierarchical approach. pp. 16–25. In: *Wetland Function and Values: The State of Our Understanding.* P. E. Greeson, J. R. Clark, and J. E. Clark (eds.). American Water Resources Association, Minneapolis, MN.

Odum, E. P. 1979b. Rebuttal of "Economic value of natural coastal wetlands: a critique." *Coastal Zone Management Journal.* 5:231–237.

Odum, E. P. 1984. The mesocosm. *BioScience.* 34:558–562.

Odum, E. P. 1986. Introductory review: Perspective on ecosystem theory and application. pp. 1–11. In: *Ecosystem Theory and Application.* N. Polunin (ed.). John Wiley & Sons, Chichester, U.K.

Odum, E. P. 1992. Great ideas in ecology for the 1990s. *BioScience.* 42:542–545.

Odum, E. P. 1993. Biosphere 2: a new kind of science. *Science.* 260:878–879.

Odum, E. P. 1997. *Ecology: A Bridge Between Science and Society.* Sinauer Associates, Inc., Sunderland, MA.

Odum, E. P. and A. A. de la Cruz. 1963. Detritus as a major component of ecosystems. *AIBS Bulletin (BioScience).* 13:39–40.

Odum, E. P., E. H. Franz, and J. T. Finn. 1979. Perturbation theory and the subsidy-stress gradient. *BioScience* 29:349–352.

Odum, H. T. 1950. The Biogeochemistry of Strontium. Ph.D. dissertation, Yale University, New Haven, CT.

Odum, H. T. 1960. Ecological potential and analogue circuits for the ecosystem. *American Scientist.* 48:1–8.

Odum, H. T. 1963. Limits of remote ecosystems containing man. *American Biology Teacher.* 26:429–443.

Odum, H. T. 1967. Biological circuits and the marine systems of Texas. pp. 99–157. In: *Pollution and Marine Ecology.* T. A. Olson and F. J. Burgess (eds.). John Wiley & Sons, New York.

Odum, H. T. 1970. Summary: an emerging view of the ecological system at El Verde. pp. I-191–I-289. In: *A Tropical Rain Forest.* H. T. Odum and R. F. Pigeon (eds.). TID-24270, National Technical Information Service, U.S. Department of Commerce, Springfield, VA.

Odum, H. T. 1971. *Environment, Power, and Society.* John Wiley & Sons, New York.

Odum, H. T. 1972. An energy circuit language for ecological and social systems: its physical basis. pp. 140–211. In: *Systems Analysis and Simulation in Ecology,* Vol. II. B. C. Patten (ed.). Academic Press, New York.

Odum, H. T. 1975. Combining energy laws and corollaries of the maximum power principle with visual systems mathematics. pp. 239–263. In: *Ecosystem Analysis and Prediction*. S. A. Levin (ed.). Society for Industrial and Applied Mathematics. Philadelphia, PA.

Odum, H. T. 1976. Energy quality and carrying capacity of the Earth. *Tropical Ecology.* 16:1–8.

Odum, H. T. 1978a. Value of wetlands as domestic ecosystems. Center for Wetlands, University of Florida, Gainesville, FL.

Odum, H. T. 1978b. Energy analysis, energy quality, and environment. pp. 55–87. In: *Energy Analysis: A New Public Policy Tool.* M. W. Gilliland (ed.). AAAS Selected Symposium 9. Westview Press, Boulder, CO.

Odum, H. T. 1979. Principle of environmental energy matching for estimating potential economic value: a rebuttal. *Coastal Zone Management Journal.* 5:239–241.

Odum, H. T. 1982. Pulsing, power, and hierarchy. pp. 33–59. In: *Energetics and Systems*. W. J. Mitsch, R. K. Ragade, R. W. Bosserman, and J. A. Dillon, Jr. (eds.). Ann Arbor Science, Ann Arbor, MI.

Odum, H. T. 1983. *Systems Ecology: An Introduction*. John Wiley & Sons, New York.

Odum, H. T. 1985. Self-Organization of Ecosystems in Marine Ponds Receiving Treated Sewage. UNC Sea Grant Publication UNC-SG-85-04, North Carolina State University, Raleigh.

Odum, H. T. 1989a. Ecological engineering and self-organization. pp. 79–101. In: *Ecological Engineering*. W. J. Mitsch and S. E. Jørgensen (eds.). John Wiley & Sons, New York.

Odum, H. T. 1989b. Experimental study of self-organization in estuarine ponds. pp. 291–340. In: *Ecological Engineering*. W. J. Mitsch and S. E. Jørgensen (eds.). John Wiley & Sons, New York.

Odum, H. T. 1994a. Ecological engineering: the necessary use of ecological self-design. *Ecological Engineering.* 3:115–118.

Odum, H. T. 1994b "Emergy" evaluation of biodiversity for ecological engineering. pp. 339–359. In: *Biodiversity and Landscapes*. K. C. Kim and R. D. Weaver (eds.). Cambridge University Press, New York.

Odum, H. T. 1995. Self-organization and maximum empower. pp. 311–330. In: *Maximum Power: The Ideas and Applications of H. T. Odum*. C. A. S. Hall (ed.). University Press of Colorado, Niwot, CO.

Odum, H. T. 1996. *Environmental Accounting, Emergy and Environmental Decision Making*. John Wiley & Sons, New York.

Odum, H. T. 1996. Scales of ecological engineering. *Ecological Engineering*. 6:7–19.

Odum, H. T., W. McConnell, and W. Abbott. 1958. The chlorophyll a of communities. *Publications of the Institute of Marine Science*. University of Texas 5:65–96.

Odum, H. T. and B. J. Copeland. 1972. Functional classification of coastal ecological systems of the United States. pp. 9–28. In: *Environmental Framework of Coastal Plain Estuaries*. B. W. Nelson (ed.). The Geological Society of America. Boulder, CO.

Odum, H. T. and D. A. Hornbeck. 1997. Emergy evaluation of Florida saltmarsh and its contribution to economic wealth. pp. 209–230. In: *Ecology and Management of Tidal Marshes*. C. L. Coultas and Y.-P. Hsieh (eds.). St. Lucie Press, Delray Beach, FL.

Odum, H. T. and C. M. Hoskin. 1957. Metabolism of a laboratory stream microcosm. *Institute of Marine Science Publication, Texas.* 4:115–133.

Odum, H. T., W. Kemp, M. Sell, W. Boynton, and M. Lehman. 1977. Energy analysis and the coupling of man and estuaries. *Environmental Management.* 1:297–315.

Odum, H. T., G. A. Knox, and D. E. Campbell. 1983. *Organization of a New Ecosystem: Exotic Spartina saltmarsh in New Zealand.* Center for Wetlands, University of Florida, Gainesville, FL.

Odum, H. T. and A. E. Lugo. 1970. Metabolism of forest floor microcosms. pp. I-35–I-56. In: *A Tropical Rain Forest*. H. T. Odum and R. Pigeon (eds.). TID-24270 (PRNC-138) U.S. Atomic Energy Commission, Oak Ridge, TN.

Odum, H. T. and E. C. Odum. 1976. *Energy Basis for Man and Nature*, First Edition. McGraw-Hill, New York.

Odum, H. T. and E. C. Odum. 2000. *Modeling for All Scales: An Introduction to System Simulation*. Academic Press, San Diego, CA.

Odum, H. T. and R. C. Pinkerton. 1955. Time's speed regulator: the optimum efficiency for maximum power output in physical and biological systems. *American Scientist.* 43:331–343.

Odum, H. T. and J. Ruiz-Reyes. 1970. Holes in leaves and the grazing control mechanism. pp. I-69–I-80. In: *A Tropical Rain Forest*, H. T. Odum (ed.). U. S. Atomic Energy Commission, Oak Ridge, TN.

Odum, H. T., R. P. Cuzon du Rest, R. J. Beyers, and C. Allbaugh. 1963. Diurnal metabolism, total phosphorus, Ohle anomaly, and zooplankton diversity of abnormal marine ecosystems of Texas. *Institute of Marine Science Publications, Texas*. 9:404–453.

Odum, H. T., K. C. Ewel, W. J. Mitsch, and J. W. Ordway. 1977. Recycling treated sewage through cypress wetlands in Florida. pp. 35–67. In: *Wastewater Renovation and Reuse*. F. M. D'Itri (ed.). Marcel Dekker. New York.

Odum, H. T., W. Kemp, M. Sell, W. Boynton, and M. Lehman. 1977. Energy analysis and the coupling of man and estuaries. *Environmental Management.* 1:297–315.

Odum, H. T., W. L. Siler, R. J. Beyers, and N. Armstrong. 1963. Experiments with engineering of marine ecosystems. *Institute of Marine Science Publication, Texas.* 9:373–403.

Odum, W. E. 1974. Potential effects of aquaculture on inshore coastal waters. *Environmental Conservation.* 1:225–230.

Odum, W. E., E. P. Odum, and H. T. Odum. 1995. Nature's pulsing paradigm. *Estuaries.* 18:547–555.

Oertel, B. 2001. Root wads to stabilize streambanks. *Land and Water.* 45(3):34–35.

Office of Technology Assessment. 1991. Bioremediation of Marine Oil Spills — Background Paper. OTA-BP-O-70. U.S. Government Printing Office, Washington, DC.

Office of Technology Assessment. 1993. *Harmful Non-Indigenous Species in the United States*. OTA-F-565. U.S. Government Printing Office. Washington, DC.

Officer, C. B., T. J. Smayda, and R. Mann. 1982. Benthic filter feeding: a natural eutrophication control. *Marine Ecology–Progress Series.* 9:203–210.

Offner, D. 1995. *Design Homology: An Introduction to Bionics*. PDQ Printing Services, Urbana, IL.

Okun, D. A. 1991. The evolution of the environmental engineer. pp. 23–32. In: *Environmental Concerns*. J. A. Hansen (ed.). Elsevier Applied Science, London.

Oldeman, R. A. A. and J. van Dijk. 1991. Diagnosis of the temperament of tropical rain forest trees. pp. 21–66. In: *Rain Forest Regeneration and Management: Man and the Biosphere Series,* Vol. 6. A. Gomez-Pompa, T. C. Whitmore, and M. Hadley (eds.). UNESCO, Paris.

Olgyay, V. 1963. *Design with Climate: Bioclimatic Approach to Architectural Regionalism.* Princeton University Press, Princeton, NJ.

Olson, J. S. 1963. Energy storage and the balance of producers and decomposers in ecological systems. *Ecology.* 44:322–331.

O'Neil, T. 1949. *The Muskrat in the Louisiana Coastal Marshes*. Louisiana Dept. of Wild Life and Fisheries, New Orleans, LA.

O'Neill, R. V. 1976. Ecosystem persistence and heterotrophic regulation. *Ecology.* 57:1244–1253.

O'Neill, R. V. 1989. Perspectives in hierarchy and scale. pp. 140–156. In: *Perspectives in Ecological Theory.* J. Roughgarden, R. M. May, and S. A. Levin (eds.). Princeton University Press, Princeton, NJ.

O'Neill, R. V. and A. W. King. 1998. Homage to St. Michael: or, why are there so many books on scale? pp. 3–15. In: *Ecological Scale, Theory and Applications.* D. L. Peterson and V. T. Parker (eds.). Columbia University Press, New York.

O'Neill, R. V. and J. B. Waide. 1981. Ecosystem theory and the unexpected: implications for environmental toxicology. pp. 43–73. In: *Management of Toxic Substances in Our Ecosystems.* B. W. Cornaby (ed.). Ann Arbor Science, Ann Arbor, MI.

O'Neill, R. V., D. L. DeAngelis, J. B. Waide, and T. F. H. Allen. 1986. *A Hierarchical Concept of Ecosystems.* Princeton University Press, Princeton, NJ.

O'Neill, R. V., R. H. Gardner, and D. E. Weller. 1982. Chaotic models as representations of ecological systems. *American Naturalist.* 120:259–263.

Ordish, G. 1981. *The Living American House: The 350-Year Story of a Home — An Ecological History.* William Morrow & Co., New York.

Orians, G. H. 1986. Site characteristics favoring invasions. pp. 133–148. In: *Ecology of Biological Invasions of North America and Hawaii.* H. A. Mooney and J. A. Drake (eds.). Springer-Verlag, New York.

Orians, G. H. and O. T. Solbrig. 1977. A cost income model of leaves and roots with special reference to arid and semi-arid areas. *American Naturalist.* 111:677–690.

Orr, D. W. 1992a. *Ecological Literacy.* State University of New York Press, Albany, NY.

Orr, D. W. 1992b. Education and the ecological design arts. *Conservation Biology.* 6:162–164.

Orr, D. W. 2002. *The Nature of Design.* Oxford University Press, Oxford, U.K.

Osborne, J. M. 1975. Tertiary treatment of campground wastes using a native Minnesota peat. *Journal of Soil Water Conservation.* 30:235–236.

Oster, G. F. 1974. A simple analog for teaching demographic concepts. *BioScience.* 24:212–213.

Oswald, W. J. 1963. Fundamental factors in stabilization pond design. pp. 357–393. In: *Advances in Biological Waste Treatment.* W. W. Eckenfelder, Jr. and J. McCabe (eds.). Macmillan, New York.

Oswald, W. J. 1988. The role of microalgae in liquid waste treatment and reclamation. pp. 255–281. In: *Algae and Human Affairs.* C. A. Lembi and J. R. Waaland (eds.). Cambridge University Press, Cambridge, U.K.

Oswald, W. J., H. B. Gotaas, C. G. Golucke, and W. R. Kellen. 1957. Algae in waste treatment. *Sewage Industry Wastes.* 29:437–455.

Oviatt, C. A., S. W. Nixon, K. T. Perez, and B. Buckley. 1979. On the season and nature of perturbations in microcosm experiments. pp. 143–164. In: *Marsh-Estuarine Systems Simulation.* R. F. Dame (ed.). University of South Carolina Press, Columbia, SC.

Oviatt, C. A., C. D. Hunt, G. A. Vargo, and K. W. Kopchynski. 1981. Simulation of a storm event in marine microcosms. *Journal of Marine Research.* 39:605–626.

Owen, D. F. and R. G. Wiegert. 1976. Do consumers maximize plant fitness? *Oikos.* 27:488–492.

Owens, O. D. 1994. "Porcupines" help rescue trout. *Garbage.* 6(1):60–61.

Pace, M. L. and P. M. Groffman (eds.). 1998. *Successes, Limitations, and Frontiers in Ecosystem Science.* Springer-Verlag, New York.

Packard, S. and C. F. Mutel. 1997a. Perspective. pp. xix–xxvii. In: *The Tallgrass Restoration Handbook.* S. Packard and C. F. Mutel (eds.). Island Press, Washington, DC.

Packard, S. and C. F. Mutel (eds.). 199b7. The *Tallgrass Restoration Handbook.* Island Press, Washington, DC.

Paine, R. T. 1966. Food web complexity and species diversity. *American Naturalist.* 100:65–75.

Paine, R. T. 1971. The ecologist's Oedipus complex: community structure. *Ecology.* 52:376–377.

Paine, R. T. 1995. A conversation on refining the concept of keystone species. *Conservation Biology.* 9:962–964.

Palermo, M. R. 1992. Wetlands restoration and establishment projects to benefit from systematic engineering approach. *The Wetlands Research Program Bulletin.* 2(4):5–7.

Palmer, M. A. and 12 other authors. 1997. Biodiversity and ecosystem processes in freshwater sediments. *Ambio.* 26:571–577.

Pandolfi, J. M. and D. R. Robertson. 1998. Roles for worms in reef-building. *Coral Reefs.* 17:120.

Papanek, V. 1971. *Design for the Real World.* Bantam Books, Toronto, Canada.

Park, E. 1993. A special treat awaits zoophiles in Washington. *Smithsonian.* 23(11):54–59.

Park, T. 1948. Experimental studies of interspecies competition, I. Competition between populations of the flour beetles, *Tribolium confusum* Duval and *Tribolium castaneum* Herbst. *Ecological Monographs.* 18:265–307.

Park, T. 1954. Competition: an experimental and statistical study. pp. 175–195. In: *Statistics and Mathematics in Biology.* O. Kempthorne, T. A. Bancroft, J. W. Gowen, and J. L. Lush (eds.). Hafner Publishing, New York.

Park, T. 1962. Beetles, competition and populations. *Science.* 138:1369–1375.

Parker, C. D. 1962. Microbiological aspects of lagoon treatment. *Journal of the Water Pollution Control Federation.* 34:149–161.

Parlange, M. 1998. The city as ecosystem. *BioScience.* 48:581–585.

Parrish, S. A. 2000. Patterns of Invasiveness of an Exotic Tree Species, *Ailanthus altissima.* M.S. thesis, University of Maryland, College Park, MD.

Parrott, L., R. Kok, G. Clark, and R. Molenaar. 1996. Crafting new life-engines. *Resource.* 3(3):8–10.

Parsons, T. R. 1990. The use of mathematical models in conjunction with mesocosm ecosystem research. pp. 197–210. In: *Enclosed Experimental Marine Ecosystems: A Review and Recommendations.* C. M. Lalli (ed.). Springer-Verlag, New York.

Parsons, W. F. J., D. M. Durall, and D. Parkinson. 1986. Evaluation of the actual evapotranspiration model of decomposition at a subalpine coal mine. *Canadian Journal of Botany.* 64:35–38.

Patrick, R. (ed.). 1983. Diversity. *Benchmark Papers in Ecology,* Vol. 13. Hutchinson Ross Publishing Co., Stroudsburg, PA.

Patten, B. C. 1983. On the quantitative dominance of indirect effects in ecosystems. pp. 27–37. In: *Analysis of Ecological Systems, State-of-the-Art in Ecological Modelling.* Elsevier, Amsterdam, the Netherlands.

Patten, B. C. and S. E. Jørgensen. (eds.). 1995. *Complex Ecology: The Part–Whole Relation in Ecosystems.* Prentice Hall, Englewood Cliffs, NJ.

Patten, B. C. and E. P. Odum. 1981. The cybernetic nature of ecosystems. *American Naturalist.* 118:886–895.

Patten, B. C. 1985. Energy cycling in the ecosystem. *Ecological Modelling.* 28:1–71.

Patten, B. C. 1991. Network ecology: indirect determination of the life-environment relationship in ecosystems. pp. 288–351. In: *Theoretical Studies of Ecosystems: The Network Perspective.* M. Higashi and T. P. Burns (eds.). Cambridge University Press, Cambridge, U.K.

Patten, B. C., R. W. Bosserman, J. T. Finn, and W. G. Cale. 1976. Propagation of cause in ecosystems. pp. 457–579. In: *Systems Analysis and Simulation in Ecology.* Vol. IV.

B. C. Patten (ed.). Academic Press, New York.

Paul, E. 2002. SPACE bill looks to control alien invaders. *BioScience.* 52:472.

Paul, M. J. and J. L. Meyer. 2001. Streams in the urban landscape. *Annual Reviews of Ecology and Systematics.* 32:333–365.

Payne, J. A. 1965. A summer carrion study of the baby pig *Sus scrofa* Linnaeus. *Ecology.* 46:592–602.

Pearse, A. S., H. J. Humm, and G. W. Wharton. 1942. Ecology of sand beaches at Beaufort, NC. *Ecological Monographs.* 12:35–190.

Pearson, O. P. 1960. A mechanical model for the study of population dynamics. *Ecology.* 41:494–508.

Pedros-Alio, C. and R. Guerrero. 1994. Prokaryotology for the limnologist. pp. 37–57. In: *Limnology: A Paradigm of Planetary Problems*. R. Margalef (ed.). Elsevier, Amsterdam, the Netherlands.

Peek, J. M., D. G. Miquelle, and R. G. Wright. 1987. Are bison exotic in the Wrangell-St. Elias National Park and Preserve? *Environmental Management.* 11:149–153.

Pennak, R. W. 1989. *Fresh-Water Invertebrates of the United States: Protozoa to Mollusca*, 3rd ed. John Wiley & Sons. New York.

Penrose, L. S. 1959. Self-reproducing machines. *Scientific American.* 200(6):105–114.

Penry, D. L. and P. A. Jumars. 1986. Chemical reactor analysis and optimal digestion. *BioScience.* 36:310–315.

Penry, D. L. and P. A. Jumars. 1987. Modeling animal guts as chemical reactors. *American Naturalist.* 129:69–96.

Perez, K. T. 1995. Role and significance of scale to ecotoxicology. pp. 49–72. In: *Ecological Toxicity Testing*. J. Cairns, Jr. and B. R. Niederlehner (eds.). Lewis Publishers, Boca Raton, FL.

Perez, K. T., G. M. Morrison, N. F. Lackie, C. A. Oviatt, S. W. Nixon, B. A. Buckley, and J. F. Heltshe. 1977. The importance of physical and biotic scaling to a experimental simulation of coastal marine ecosystem. *Helgolaender Wissensehaften Meeresunters.* 30:144–162.

Perez, K. T., G. E. Morrison, E. W. Davey, N. F. Lackie, A. E. Soper, R. J. Blasco, and D. L. Winslow. 1991. Influence of size on fate and ecological effects of kepone in physical models. *Ecological Applications.* 1:237–248.

Perfecto, I., R. A. Rice, R. Greenberg, and M. E. Van der Voort. 1996. Shade coffee: a disappearing refuge for biodiversity. *BioScience.* 46:598–608.

Perry, D. A. 1995. Self-organizing systems across scales. *Trends in Ecology and Evolution.* 10:241–244.

Persson, L., J. Bengtsson, B. A. Menge, and M. E. Power. 1996. Productivity and consumer regulation: concepts, patterns, and mechanisms. pp. 396–434. In: *Food Webs, Integration of Patterns and Dynamics*. G. A. Polis and K. O. Winemiller (eds.). Chapman & Hall, New York.

Pestrong, R. 1974. Unnatural shoreline. *Environment.* 16(9):27–35.

Peters, C. M., A. H. Gentry, and R. O Mendelsohn. 1989. Valuation of an Amazonian rainforest. *Nature.* 339:655–656.

Peters, R. H. 1991. *A Critique for Ecology.* Cambridge University Press, Cambridge, U.K.

Petersen, J. E. 1998. Scale and Energy Input in the Dynamics of Experimental Estuarine Ecosystems. Ph.D. dissertation, University of Maryland, College Park, MD.

Petersen, J. E. 1999. Personal communication.

Petersen, J. E. 2001. Adding artificial feedback to a simple aquatic ecosystem: the cybernetic nature of ecosystems revisited. *Oikos.* 94:533–547.

Petersen, J. E., C.-C. Chen, and W. M. Kemp. 1997. Spatial scaling of aquatic primary productivity: experiments under nutrient and light-limited conditions. *Ecology.* 78:2326–2338.

Petersen, J. E., J. Cornwell, and W. M. Kemp. 1999. Implicit scaling in the design of experimental aquatic ecosystems. *Oikos.* 85:3–18.

Petersen, R. C., Jr. and K. W. Cummins. 1974. Leaf processing in a woodland stream ecosystem. *Freshwater Biology.* 4:343–368.

Petersen, R. C., Jr., K. W. Cummins, and G. M. Ward. 1989. Microbial and animal processing of detritus in a woodland stream. *Ecological Monographs.* 59:21–39.

Peterson, D. L. and V. T. Parker (eds.). 1998. *Ecological Scale, Theory and Applications.* Columbia University Press, New York.

Peterson, S. 1991. Commercial prospects for ecological engineering. pp. 121–124. In: *Ecological Engineering for Wastewater Treatment.* C. Etnier and B. Guterstam (eds.). Stensund Folk College, Trosa, Sweden.

Peterson, S. B. and J. M. Teal. 1996. The role of plants in ecologically engineered wastewater treatment systems. *Ecological Engineering.* 6:137–148.

Petraitis, P. S., R. E. Latham, and R. A. Niesenbaum. 1989. The maintenance of species diversity by disturbance. *Quarterly Review of Biology.* 64:393–418.

Petroski, H. 1982. *To Engineer Is Human:The Role of Failure in Successful Design.* Vintage Books, New York.

Petroski, H. 1989. *The Pencil.* Alfred A. Knopf, Inc., New York.

Petroski, H. 1992. *The Evolution of Useful Things.* Alfred A. Knopf, Inc., New York.

Petroski, H. 1994. *Design Paradigms, Case Histories of Error and Judgement in Engineering.* Cambridge University Press, New York.

Petroski, H. 1995. Learning from paper clips. *American Scientist.* 83:313–316.

Petroski, H. 1996. *Invention by Design: How Engineers Get from Thought to Thing.* Harvard University Press, Cambridge, MA.

Petroski, H. 1997a. *Remaking the World: Adventures in Engineering.* Vintage Books, New York.

Petroski, H. 1997b. Designed to fail. *American Scientist.* 85:412–416.

Petrusewicz, K. and W. L. Grodzinski. 1975. The role of herbivore consumers in various ecosystems. pp. 64–70. In: *Productivity of World Ecosystems.* National Academy of Sciences, Washington, DC.

Pfafflin, J. R. and E. N. Ziegler (eds.). 1979. *Advances in Environmental Science and Engineering*, Vol. 2. Gordon & Breach, New York.

Phelps, E. B. 1944. *Stream Sanitation.* John Wiley & Sons, New York.

Phelps, H. L. 1994. The Asiatic clam (*Corbicula fluminea*) invasion and system-level ecological change in the Potomac River Estuary near Washington DC. *Estuaries.* 17:614–621.

Phillips, J. D. 1989. An evaluation of the state factor model of soil ecosystems. *Ecological Modelling.* 45:165–177.

Phillips, J. D. 1992. Deterministic chaos in surface runoff. pp. 177–197. In: *Overland Flow.* A. J. Parsons and A. D. Abrahams (eds.). Chapman & Hall, New York.

Phillips, J. D. 1995. Self-organization and landscape evolution. *Progress in Physical Geography.* 19:309–321.

Phillips, J. D. and W. H. Renwick (eds.). 1992. *Geomorphic Systems.* Elsevier, Amsterdam, the Netherlands.

Phillips, R. E., R. L. Blevins, G. W. Thomas, W. W. Frye, and S. H. Phillips. 1980. No-tillage agriculture. *Science.* 208:1108–1113.

Pianka, E. R. 1970. On r- and K-selection. *American Naturalist.* 104:592–597.

Pianka, E. R. 1983. *Evolutionary Ecology*, 3rd ed. Harper & Row, New York.

Piburn, M. D. 1972. New beaches from old bottles. *Natural History.* 81(4):48–50.

Pickett, S. T. A. and P. S. White (eds.). 1985. *The Ecology of Natural Disturbance and Patch Dynamics*. Academic Press, Orlando, FL.

Pickett, S. T. A., M. L. Cadenasso, J. M. Grove, C. H. Nilon, R. V. Pouyat, W. C. Zipperer, and R. Costanza. 2001. Urban ecological systems: linking terrestrial ecological, physical, and socioeconomic components of metropolitan areas. *Annual Reviews of Ecology and Systematics.* 32:127–157.

Pickett, S. T. A., J. Kolasa, J. J. Armesto, and S. L. Collins. 1989. The ecological concept of disturbance and its expression at various hierarchical levels. *Oikos.* 54:129–136.

Pielou, E. C. 1969. *Introduction to Mathematical Ecology.* Wiley Interscience, New York.

Pilette, R. 1989. Evaluating direct and indirect effects in ecosystems. *American Naturalist.* 133:303–307.

Pilkey, O. H. and K. L. Dixon. 1996. *The Corps and the Shore*. Island Press, Washington, DC.

Pilson, M. E. Q. and S. W. Nixon. 1980. Marine microcosms in ecological research. pp. 724–741. In: *Microcosms in Ecological Research.* J. P. Giesy, Jr. (ed.). DOE Symposium Series 52, U.S. Dept. of Energy, Washington, DC.

Pimentel, D. and C. A. Edwards. 1982. Pesticides and ecosystems. *BioScience.* 32:595–600.

Pimentel, D., H. Acquay, M. Biltonen, P. Rice, M. Silva, J. Nelson, V. Lipner, S. Giordano, A. Horowitz, and M. D'Amore. 1992. Environmental and economic costs of pesticide use. *BioScience.* 42:750–760.

Pimentel, D., J. Allen, A. Beers, L. Guinand, R. Linder, P. McLaughlin, B. Meer, D. Musonda, D. Perdue, S. Poisson, S. Siebert, K. Stoner, R. Salazar, and A. Hawkins. 1987. World agriculture and soil erosion. *BioScience.* 37:277–283.

Pimentel, D., D. Andow, R. Dyson-Hudson, D. Gallahan, S. Jacobson, M. Irish, S. Kroop, A. Moss, I. Schreiner, M. Shepard, T. Thompson, and B. Vinzant. 1980. Environmental and social costs of pesticides: a preliminary assessment. *Oikos.* 34:126–140.

Pimentel, D., M. S. Hunter, J. A. LaGro, R. A. Efroymson, J. C. Landers, F. T. Mervis, C. A. McCarthy, and A. E. Boyd. 1989. Benefits and risks of genetic engineering in agriculture. *BioScience.* 39:606–614.

Pimentel, D., L. Lach, R. Zuniga, and D. Morrison. 2000. Environmental and economic costs of nonindigenous species in the United States. *BioScience.* 50:53–65.

Pimm, S. L. 1982. *Food Webs*. Chapman & Hall Publishers, London.

Pimm, S. L. 1989. Theories of predicting success and impact of introduced species. pp. 351–367. In: *Biological Invasions, A Global Perspective*. J. A. Drake, H. A. Mooney, F. di Castri, R. H. Groves, F. J. Kruger, M. Rejmanek, and M. Williamson (eds.). John Wiley & Sons, Chichester, U.K.

Pinchot, G. B. 1966. Whale culture — a proposal. *Perspectives in Biology and Medicine.* 10:33–43.

Pinchot, G. B. 1974. Ecological aquaculture. *BioScience.* 24:265.

Pirsig, R. M. 1974. *Zen and the Art of Motorcycle Maintenance: An Inquiry into Values*. Quill, William, and Morrow, New York.

Pittendrigh, C. S., W. Vishniac, and J. P. T. Pearman. 1966. *Biology and the Exploration of Mars*. National Academy of Sciences, Washington, DC.

Planty-Tabacchi, A.-M., R. Tabacchi, R. J. Naiman, C. Deferrari, and H. Decamps. 1996. Invasibility of species-rich communities in riparian zones. *Conservation Biology.* 10:598–607.

Platt, R. B. and J. F. McCormick. 1964. Manipulatable terrestrial ecosystems. *Ecology.* 45:649–650.

Plotkin, M. and L. Famolare. (eds.). 1992. *Sustainable Harvest and Marketing of Rain Forest Products*. Island Press, Washington, DC.

Poincelot, R. P. 1974. A scientific examination of the principles and practice of composting. *Compost Science*. 15(3):24–31.

Pollock, M. M., R. J. Naiman, H. E. Erickson, C. A. Johnston, J. Pastor, and G. Pinay. 1995. Beaver as engineers: influences on biotic and abiotic characteristics of drainage basins. pp. 117–126. In: *Linking Species and Ecosystems*. C. G. Jones and J. H. Lawton (eds.). Chapman & Hall, New York.

Pomeroy, L. R. (ed.). 1974a. *Cycles of Essential Elements: Benchmark Papers in Ecology*. Dowden, Hutchinson & Ross, Stroudsburg, PA.

Pomeroy, L. R. 1974b. The ocean's food web, a changing paradigm. *BioScience*. 24:499–504.

Pomeroy, L. R. 1980. Detritus and its role as a food source. pp. 84–102. In: *Fundamentals of Aquatic Ecosystems*. R.S.K. Barnes and K. H. Mann (eds.). Blackwell Scientific Publications, Oxford, U.K.

Pomeroy, L. R. and J. J. Alberts. (eds.). 1988. *Concepts of Ecosystem Ecology*. Springer-Verlag, New York.

Poulson, T. L. 1972. Bat guano ecosystems. *Bulletin of the National Speleological and Society*. 34:55–59.

Pound, C. E. and F. E. Egler. 1953. Brush control in southeastern New York: fifteen years of stable tree-less communities. *Ecology*. 34:63–73.

Poundstone, W. 1999. *Carl Sagan: A Life in the Cosmos*. Henry Holt and Co., New York.

Power, M. E., D. Tilman, J. A. Estes, B. A. Menge, W. J. Bond, L. S. Mills, G. Daily, J. C. Castilla, J. Lubchenco, and R. T. Paine. 1996. Challenges in the quest for keystones. *BioScience*. 46:609–620.

Prach, K. and P. Pysek. 2001. Using spontaneous succession for restoration of human-disturbed habitats: experience from Central Europe. *Ecological Engineering*. 17:55–62.

Pratt, J. R. 1994. Artificial habitats and ecosystem restoraton: managing for the future. *Bulletin of Marine Sciences*. 55:268–275.

Pretzer, W. S. (ed.). 1989. *Working at Inventing: Thomas A. Edison and the Menlo Park Experience*. Henry Ford Museum & Greenfield Village, Dearborn, MI.

Price, M. R. S. 1989. Reconstructing ecosystems. pp. 210–218. In: *Conservation for the Twenty-First Century*. D. Western and M. Pearl (eds.). Oxford University Press, New York.

Prigogine, I. 1980. *From Being to Becoming: Time and Complexity in the Physical Sciences*. W. H. Freeman, San Francisco, CA.

Pritchard, P. H. and A. W. Bourquin. 1984. The use of microcosms for evaluation of interactions between pollutants and microorganisms. *Advances in Microbial Ecology*. 7:133–215.

Provost, M. W. 1974. saltmarsh management in Florida. pp. 5–17. In: *Proceedings of the Tall Timbers Conference on Ecological Animal Control by Habitat Management*. Tall Timbers Research Station, Tallahassee, FL.

Prugh, T. with R. Costanza, J. H. Cumberland, H. Daly, R. Goodland, and R. B. Norgaard. 1995. *Natural Capital and Human Economic Survival*. International Society for Ecological Economics Press, Solomons, MD.

Public Affairs Office. No date. *Summary of Capabilities*. Waterways Experiment Station. U.S. Army Corps of Engineers, Vicksburg, MS.

Pullin, B. P. and D. A. Hammer. 1991. Aquatic plants improve wastewater treatment. *Water Environment and Technology*. 3(3):37–40.

Purrington, C. B. and J. Bergelson. 1995. Assessing weediness of transgenic crops: industry plays plant ecologist. *Trends in Ecology and Evolution.* 10:340–342.

Puschnig, M., J. Schettler-Wiegel, and V. Schulz-Berendt. 1990. Can soil animals be used to improve decontamination of oil residues in polluted soils? pp. 485–492. In: *Terrestrial and Aquatic Ecosystems: Perturbation and Recovery.* O. Ravera (ed.). Ellis Horwood Publishing, New York.

Rabenhorst, M. C. 1997. The chrono-continuum: an approach to modeling pedogenesis in marsh soils along transgressive coastlines. *Soil Science.* 162:2–9.

Race, M. S. 1985. Critique of present wetlands mitigation policies in the United States based on an analysis of past restoration projects in San Francisco Bay. *Environmental Management.* 9:71–82.

Race, M. S. 1986. Wetlands restoration and mitigation policies: reply. *Environmental Management.* 10:571–572.

Race, M. S. and D. R. Christie, 1982. Coastal zone development: mitigation, marsh creation and decision-making. *Environmental Management.* 6:317–328.

Randall, J. M. 1997. Defining weeds of natural areas. pp. 18–25. In: *Assessment and Management of Plant Invasions.* J. O. Luken and J. W. Thieret (eds.). Springer, New York.

Raphael, C. N. and E. Jaworski. 1979. Economic value of fish, wildlife, and recreation in Michigan's coastal wetlands. *Coastal Zone Management Journal.* 5:181–194.

Rapoport, A. 1969. *House Form and Culture.* Prentice Hall, Englewood Cliffs, NJ.

Rapport, D. J. and J. E. Turner. 1977. Economic models in ecology. *Science.* 195:367–373.

Rapport, D. J. and H. A. Regier. 1995. Disturbance and stress effects on ecological systems. pp. 397–414. In: *Complex Ecology: The Part–Whole Relation in Ecosystems.* B. C. Patten and S. E. Jørgensen (eds.). Prentice Hall, Englewood Cliffs, NJ.

Rapport, D. J. and W. G. Whitford. 1999. How ecosystems respond to stress. *BioScience.* 49:193–203.

Rapport, D. J., H. A. Regier, and T. C. Hutchinson. 1985. Ecosystem behavior under stress. *American Naturalist.* 125:617–640.

Raskin, I., A. Ehmann, W. R. Melander, and B. J. D. Meeuse. 1987. Salicylic acid: a natural inducer of heat production in *Arum* lilies. *Science.* 237:1601–1602.

Rathje, W. L. and L. Psihoyos. 1991. Once and future landfills. *National Geographic.* 179(5):116–134.

Rathje, W. L. and C. Murphy. 1992. *Rubbish! The Archaeology of Garbage.* Harper Collins Publishers, New York.

Rathje, W. and C. Murphy. 1992. Five major myths about garbage, and why they're wrong. *Smithsonian.* 23(4):113–122.

Rau, E. H. et al. 2000. Minimization and management of wastes from biomedical research. *Environmental Health Perspectives.* 108:953–977.

Rawson, D. M. 1993. Bioprobes and biosensors. pp. 428–437. In: *Handbook of Ecotoxicology.* P. Calow (ed.). Blackwell Scientific, London.

Ray, D. L. (ed.). 1959. *Marine Boring and Fouling Organisms.* University of Washington Press, Seattle, WA.

Rebele, F. 1994. Urban ecology and special feature of urban ecosystems. *Global Ecology and Biogeography Letters.* 4:173–187.

Rechcigl, J. E. and H. C. MacKinnon (eds.). 1997. *Agricultural Uses of By-Products and Wastes.* American Chemical Society, Washington, DC.

Reddy, A. S. 1998. Indian ethos for restoration of ecosystem. pp. 301–313. In: *Damaged Ecosystems and Restoration.* B. C. Rana (ed.). World Scientific Publishing, Singapore.

Reddy, K. R. and W. H. Smith (eds.). 1987. *Aquatic Plants for Water Treatment and Resource Recovery.* Magnolia Publishing, Orlando, FL.

Redfield, A. C. 1958. The biological control of chemical factors in the environment. *American Scientist.* 46:205–221.

Redfield, A. C., B. H. Ketchum, and F. A. Richards. 1963. The influence of organisms on the composition of sea-water. pp. 26–77. In: *The Sea*, Vol. 2. M. N. Hill (ed.). Wiley-Interscience Publishing, New York.

Reed, D. J. 2000. Coastal biogeomorphology. pp. 347–361. In: *Estuarine Science: A Synthetic Approach to Research and Practice*. J. E. Hobbie (ed.). Island Press, Washington, DC.

Reed, S. C. 1991. Constructed wetlands for wastewater treatment. *BioCycle*, January :44–49.

Reed, S. C. and D. S. Brown. 1992. Constructed wetland design — the first generation. *Water Environment Research.* 64:776–781.

Reed, S. C., R. W. Crites, and E. J. Middlebrooks. 1995. *Natural Systems for Waste Management and Treatment.* McGraw-Hill, New York.

Reeves, G. H., L. E. Benda, K. M. Burnett, P. A. Bisson, and J. R. Sedell. 1995. A disturbance-based ecosystem approach to maintaining and restoring freshwater habitats of evolutionary significant units of anadromous salmonids in the Pacific Northwest. pp. 334–349. In: *Evolution and the Aquatic Ecosystem: Defining Unique Units in Population Conservation.* American Fisheries Society, Bethesda, MD.

Reganold, J. P., L. F. Elliott, and Y. L. Unger. 1987. Long-term effects of organic and conventional farming on soil erosion. *Nature.* 330:370–372.

Regier, H. A., S. A. Bocking, and H. F. Henderson. 1995. Sustainability of temperate zone fisheries: biophysical foundations for its definition and measurement. pp. 335–354. In: *Defining and Measuring Sustainability: The Biogeophysical Foundations.* M. Munasinghe and W. Shearer (eds.). The United Nations University and The World Bank, Washington, DC.

Rehbock, P. F. 1980. The Victorian aquarium in ecological and social perspective. pp. 522–539. In: *Oceanography: The Past.* M. Sears and D. Merriman (eds.). Springer-Verlag, New York.

Reice, S. R. 1994. Nonequilibrium determinants of biological community structure. *American Scientist.* 82:424–435.

Reice, S. R. 2001. *The Silver Lining: The Benefits of Natural Disasters.* Princeton University Press, Princeton, NJ.

Reichard, S. E. 1997. Prevention of invasive plant introductions on national and local levels. pp. 215–227. In: *Assessment and Management of Plant Invasions.* J. O. Luken and J. W. Thieret (eds.). Springer, New York.

Reichel–Dolmatoff, G. 1976. Cosmology as ecological analysis: a view from the rain forest. *Man.* 11:307–318.

Reid, G. W. and J. R. Assenzo. 1963. Biological slimes. pp. 347–355. In: *Advances in Biological Waste Treatment.* W. W. Eckenfelder, Jr. and J. McCabe (eds.). Macmillan, New York.

Reid, W. V. 1993. Bioprospecting. *Environmental Science and Technology.* 27:1730–1732.

Reid, W. V., S. A. Laird, C. A. Meyer, R. Gamez, A. Sittenfeld, D. H. Janzen, M. A. Gollin, and C. Juma. 1993. *Biodiversity Prospecting: Using Genetic Resources for Sustainable Development.* World Resources Institute, Washington, DC.

Reimers, H. and K. Branden. 1994. Algal colonization of a tire reef — influence of placement date. *Bulletin of Marine Sciences.* 55:460–469.

Reise, K. 1985. *Tidal Flat Ecology.* Springer-Verlag, Berlin.

Rejmanek, M. 1989. Invasibility of plant communities. pp. 369–388. In: *Biological Invasions: A Global Perspective.* J. A. Drake, H. A. Mooney, F. di Castri, R. H. Groves, F. J. Kruger, M. Rejmanek, and M. Williamson (eds.). John Wiley & Sons, Chichester, U.K.

Repetto, R., W. Magrath, M. Wells, C. Beer, and F. Rossini. 1989. Wasting Assets: Natural Resources in the National Income Accounts. World Resources Institute, Washington, DC.

Repetto, R. C., P. Faeth, and J. Westra. 1999. Accounting for natural resources in income and productivity measurements. pp. 70–88. In: *Measures of Environmental Performance and Ecosystem Condition*. P. C. Schulze (ed.). National Academy Press, Washington, DC.

Resetaritis, W. J., Jr., and J. Bernardo (eds.). 1998. *Experimental Ecology: Issues and Perspectives*. Oxford University Press, New York.

Resetarits, W. J., Jr., and J. E. Fauth. 1998. From cattle tanks to Carolina bays. pp. 133–151. In: *Experimental Ecology*. W. J. Resetarits, Jr. and J. Bernardo (eds.). Oxford University Press, New York.

Resh, V. H. 2001. Mosquito control and habitat modification: case history studies of San Francisco Bay wetlands. pp. 413–428. In: *Bioassessment and Management of North American Freshwater Wetlands*. R. B. Rader, D. P. Batzer, and S. A. Wissinger (eds.). John Wiley & Sons, New York.

Resnick, M. 1994. *Turtles, Termites, and Traffic Jams*. MIT Press, Cambridge, MA.

Reynolds, C. S. 1994. The ecological basis for the successful biomanipulation of aquatic communities. *Archives of Hydrobiology*. 130:1–33.

Rhoads, D. C. and D. K. Young. 1970. The influence of deposit-feeding organisms on sediment stability and community trophic structure. *Journal of Marine Research*. 28:150–178.

Rhoads, D. C., P. L. McCall, and J. Y. Yingst. 1978. Disturbance and production of the estuarine seafloor. *American Scientist*. 66:577–586.

Ricciardi, A. and H. J. MacIsaac. 2000. Recent mass invasion of the North American Great Lakes by Ponto-Caspian species. *Trends in Ecology and Evolution*. 15:62–65.

Rich, L. G. 1963. *Unit Processes of Sanitary Engineering*. John Wiley & Sons, New York.

Rich, P. H. and R. G. Wetzel. 1978. Detritus in lake ecosystems. *American Naturalist*. 112:57–71.

Richards, D. J. (ed.). 1997. *The Industrial Green Game: Implications for Environmental Design and Management*. National Academy Press, Washington, DC.

Richards, D. J. and G. Pearson (eds.). 1998. *The Ecology of Industry*. National Academy Press, Washington, DC.

Richards, D. J., B. R. Allenby, and R. A. Frosch. 1994. The greening of industrial ecosystems: overview and perspective. pp. 1–19. In: *The Greening of Industrial Ecosystems*. B. R. Allenby and D. J. Richards (eds.). National Academy Press, Washington, DC.

Richardson, A. M. M. 1983. The effect of the burrows of a crayfish on the respiration of the surrounding soil. *Soil Biology and Biochemistry*. 15:239–242.

Richardson, C. J. 1985. Mechanisms controlling phosphorus retention capacity in freshwater wetlands. *Science*. 228:1424–1427.

Richardson, C. J. 1989. Freshwater wetlands: transformers, filters or sinks? pp. 25–46. In: *Freshwater Wetlands and Wildlife*. R. R. Sharitz and J. W. Gibbons (eds.). USDOE Office of Scientific and Technical Information, Oak Ridge, TN.

Richardson, C. J. and C. B. Craft. 1993. Effective phosphorus retention in wetlands: fact or fiction? pp. 271–282. In: *Constructed Wetlands for Water Quality Improvement*. G. A. Moshiri (ed.). Lewis Publishers, Boca Raton, FL.

Richardson, J. R. and H. T. Odum. 1981. Power and a pulsing production model. pp. 641–648. In: *Energy and Ecological Modelling*. W. J. Mitsch (ed.). Elsevier, Amsterdam, the Netherlands.

Richey, J. E. 1970. The Role of Disordering Energy in Microcosms. M.S. thesis, Department of Environmental Science and Engineering, School of Public Health, University of North Carolina, Chapel Hill, NC.

Richey, J. E. et al. 1978. Carbon flow in four lake ecosystems: a structural approach. *Science.* 202:1183–1186.

Ricketts, E. F. and J. Calvin. 1939. *Between Pacific Tides*. Stanford University Press, Stanford, CA.

Riedel, K. 1998. Application of biosensors to environmental samples. pp. 267–294. In: *Commercial Biosensors*. G. Ramsay (ed.). John Wiley & Sons, New York.

Riessen, H. P. 1992. Cost-benefit model for the induction of an antipredator defense. *American Naturalist.* 140:349–362.

Rietkerk, M. and J. Van de Koppel. 1997. Alternative stable states and threshold effects in semi-arid grazing systems. *Oikos.* 79:69–76.

Rietman, E. A. 2001. *Molecular Engineering of Nanosystems*. Springer, New York.

Riley, A. L. 1998. *Restoring Streams in Cities*. Island Press, Washington, DC.

Ringold, P. 1979. Burrowing, root mat density, and the distribution of fiddler crabs in the eastern United States. *Journal of Experimental Marine Biology and Ecology.* 36:11–21.

Risser, P. G. 1995a. The status of the science examining ecotones. *BioScience.* 45:318–325.

Risser, P. G. 1995b. Biodiversity and ecosystem function. *Conservation Biology.* 9:742–746.

Rissler, J. and M. Mellon. 1996. *The Ecological Risks of Engineered Crops*. MIT Press, Cambridge, MA.

Roberts, H. A. 1981. Seed banks in soils. *Advances in Applied Biology.* 6:1–55.

Robertson, D. W. 1974. *The Mind's Eye of Buckminster Fuller.* St. Martin's Press, New York.

Robertson, G. P. and E. A. Paul. 1998. Ecological research in agricultural ecosystems: contributions to ecosystem science and to the management of agronomic resources. pp. 142–164. In: *Successes, Limitations, and Frontiers in Ecosystem Science*. M. L. Pace and P. M. Groffman (eds.). Springer, New York.

Robinson, G. R. and S. N. Handel. 1993. Forest restoration on a closed landfill: rapid addition of new species by bird dispersal. *Conservation Biology.* 7:271–278.

Robinson, J. V. and J. E. Dickerson, Jr. 1984. Testing the invulnerability of laboratory island communities to invasion. *Oecologia.* 61:169–174.

Robison, E. G., and R. L. Beschta. 1990. Coarse woody debris and channel morphology interactions for undisturbed streams in southeast Alaska, U.S. *Earth Surface Processes and Landforms.* 15:149–156.

Robles-Diaz-de-Leon, L. F. and P. Kangas. 1999. Evaluation of potential gross income from non-timber products in a model riparian forest for the Chesapeake Bay watershed. *Agroforestry Systems*. 44:215–225.

Rodriguez-Iturbe, I. and A. Rinaldo. 1997. *Fractal River Basins, Chance and Self-Organization*. Cambridge University Press, Cambridge, U.K.

Roels, O. A., B. A. Sharfstein, and V. M. Harris. 1978. An evaluation of the feasibility of a temperate climate effluent-aquaculture-tertiary treatment system in New York City. pp. 145–156. In: *Estuarine Interactions*. M. L. Wiley (ed.). Academic Press, New York.

Roff, D. A. 1974. Spatial heterogeneity and the persistence of populations. *Oecologia.* 15:245–258.

Rogers, H. H. and D. E. Davis. 1972. Nutrient removal by water-hyacinths. *Weed Science.* 20:423–428.

Rooth, J. E. and L. Windham. 2000. Phragmites on death row: is biocontrol really warranted? *Wetland Journal.* 12:29–37.

Rosenberg, D. M. and V. H. Resh (eds.). 1993. *Freshwater Biomonitoring and Benthic Macroinvertebrates*. Chapman & Hall, New York.

Rosenzweig, M. L. 1971. Paradox of enrichment: destabilization of exploitation ecosystems in ecological time. *Science*. 171:385–387.

Rosenzweig, M. L. 1995. *Species Diversity in Space and Time*. Cambridge University Press, Cambridge, U.K.

Rosgen, D. 1996. *Applied River Morphology*. Wildland Hydrology, Pagosa Springs, CO.

Ross, S. T. 1991. Mechanisms structuring stream fish assemblages: are there lessons from introduced species? *Environmental Biology of Fishes*. 30:359–368.

Rossi, L. and G. Vitagliano-Tadini. 1978. Role of adult faeces in the nutrition of larvae of *Asellus aquaticus* (Isopoda). *Oikos*. 30:109–113.

Roughgarden, J. 1975. Evolution of marine symbiosis — a simple cost-benefit model. *Ecology*. 56:1201–1208.

Roughgarden, J. 1995. Vertebrate patterns on islands. pp. 51–56. In: *Islands*. P. M. Vitousek, L. L. Loope, and H. Adersen (eds.). Springer, New York.

Roughgarden, J. 2001. Guide to diplomatic relations with economics. *Bulletin of the Ecological Society of America*. 82:85–88.

Roughgarden, J., S. Gaines, and S. W. Pacala. 1986. Supply side ecology: the role of physical transport processes. pp. 491–518. In: *Organization of Communities, Past and Present*. J. H. R. Gee and P. S. Giller (eds.). Blackwell Scientific Publications, Oxford, U.K.

Roush, W. 1995. When rigor meets reality. *Science*. 269:313–315.

Rubin, S. M. and S. C. Fish. 1994. Biodiversity prospecting: using innovative contractual provisions to foster ethnobotanical knowledge, technology, and conservation. pp. 23–58. In: *Endangered Peoples, Indigenous Rights and the Environment*. University Press of Colorado, Niwot, CO.

Ruesink, J. L., I. M. Parker, M. J. Groom, and P. M. Kareiva. 1995. Reducing the risks of nonindigenous species introductions. *BioScience*. 45:465–477.

Ruggiero, L. F., G. D. Hayward, and J. R. Squires. 1994. Viability analysis in biological evaluations: concepts of population viability analysis, biological population, and ecological scale. *Conservation Biology* 8:364–372.

Russell, R. C. 1999. Constructed wetlands and mosquitoes: health hazards and management options — an Australian perspective. *Ecological Engineering*. 12:107–124.

Russell-Hunter, W. D. 1970. *Aquatic Productivity: An Introduction to Some Basic Aspects of Biological Oceanography and Limnology*. Macmillan, New York.

Ruth, B. F., D. A. Flemer, and C. M. Bundrick. 1994. Recolonization of estuarine sediments by macroinvertebrates: does microcosm size matter? *Estuaries*. 17:606–613.

Ruttenber, A. J., Jr. 1979. Urban Areas as Energy Flow Systems. Ph.D. dissertation, Emory University, Atlanta, GA.

Rybczynski, W. 2000. *One Good Turn: A Natural History of the Screwdriver and the Screw*. Simon & Schuster, New York.

Ryther, J. H. 1969. The potential of the estuary for shellfish production. *Proceedings of the National Shellfisheries Association*. 59:18–22.

Ryther, J. H., W. M. Dunstan, K. R. Tenore, and J. E. Huguenin. 1972. Controlled eutrophication — increasing food production from the sea by recycling human wastes. *BioScience*. 22:144–152.

Sagoff, M. 1995. Carrying capacity and ecological economics. *BioScience*. 45:610–620.

Sakai, A. K. and 14 coauthors. 2001. The population biology of invasive species. *Annual Reviews of Ecology and Systematics*. 32:305–332.

Sale, P. F. 1977. Maintenance of high diversity in coral reef fish communities. *American Naturalist*. 111:337–359.

Sale, P. F. 1989. Diversity of the tropics: causes of high diversity in reef fish systems. pp. 1–20. In: *Vertebrates in Complex Tropical Systems*. M. L. Harmelin-Vivien and F. Bourliere (eds.). Springer-Verlag, New York.

Salisbury, F. B., J. I. Gitelson, and G. M. Lisovsky. 1997. Bios-3: Siberian experiments in bioregenerative life support. *BioScience*. 47:575–585.

Salvato, J. A., Jr. 1992. *Environmental Engineering and Sanitation*. John Wiley & Sons, New York.

Sanford, L. P. 1997. Turbulent mixing in experimental ecosystem studies. *Marine Ecology Progress* Series. 161:265–293.

Sanks, R. L. and T. Asano. (eds.). 1976. *Land Treatment and Disposal of Municipal and Industrial Wastewater*. Ann Arbor Science, Ann Arbor, MI.

Sarai, D. S. 1975. Insects on trickling filter rocks and their role in sewage treatment. *Environmental Entomology* 4:238–240.

Sarikaya, M. and I. A. Aksay (eds.). 1995. *Biomimetics*. American Institute of Physics, Woodbury, NY.

Satchell, J. E. (ed.). 1983. *Earthworm Ecology: From Darwin to Vermiculture*. Chapman & Hall, London.

Saunders, D. A., R. J. Hobbs, and C. R. Margules. 1991. Biological consequences of ecosystem fragmentation: a review. *Conservation Biology*. 5:18–32.

Saunders, P. T. 1983. Catastrophe theory. pp. 105–139. In: *Mathematics in Microbiology*. M. Bazin (ed.). Academic Press, London.

Savage, C. 1995. *Bird Brains: The Intelligence of Crows, Ravens, Magpies, and Jays*. Sierra Club Books, San Francisco, CA.

Savage, N. 1986. The mitigation predicament. *Environmental Management*. 10:319–320.

Saward, D. 1975. An experimental approach to the study of pollution effects in aquatic ecosystems. pp. 86–90. In: *Ecological Toxicology Research*. A. D. McIntyre and C. F. Mills. (eds.). Plenum Press, New York.

Sawyer, C. N. 1944. Biological engineering in sewage treatment. *Sewage Works Journal*. 16:925–935.

Scarsbrook, E. and D. E. Davis. 1971. Effect of sewage effluent on the growth of five vascular aquatic species. *Hyacinth Control Journal*. 9:26–30.

Schaffer, W. M. 1985. Order and chaos in ecological systems. *Ecology*. 66:93–106.

Schaffer, W. M. 1988. Perceiving order in the chaos of nature. pp. 313–350. In: *Evolution of Life Histories of Mammals, Theory and Pattern*. M. S. Boyce (ed.). Yale University Press, New Haven, CT.

Schaffer, W. M. and M. Kot. 1985. Do strange attractors govern ecological systems? *BioScience*. 35:342–350.

Schaller, F. 1968. *Soil Animals*. University of Michigan Press, Ann Arbor, MI.

Schardt, J. D. 1997. Maintenance control. pp. 229–243. In: *Strangers in Paradise: Impact and Management of Nonindigenous Species in Florida*. D. Simberloff, D. C. Schmitz, and T. C. Brown (eds.). Island Press, Washington, DC.

Scheffer, M. and R. J. De Boer. 1995. Implications of spatial heterogeneity for the paradox of enrichment. *Ecology*. 76:2270–2277.

Scheffer, M. and E. Jeppesen. 1998. Alternative stable states. pp. 397–406. In: *The Structuring Role of Submerged Macrophytes in Lakes*. E. Jeppesen, M. Sondergaard, M. Sondergaard, and K. Christoffersen (eds.). Springer, New York.

Scheiner, S. M., A. J. Hudson, and M. A. VanderMeulen. 1993. An epistemology for ecology. *Bulletin of the Ecological Society of America*. 74:17–21.

Schelske, C. L. 1984. In situ and natural phytoplankton assemblage bioassays. pp. 15–47. In: *Algae as Ecological Indicators*. Academic Press, London.

Schiaparelli, S. and R. Cattaneo-Vietti. 1999. Functional morphology of vermetid feeding-tubes. *Lethaia.* 32:41–46.

Schiechtl, H. M. and R. Stern. 1997. *Water Bioengineering Techniques for Watercourse: Bank and Shoreline Protection.* Blackwell Science, Cambridge, MA.

Schlesinger, W. H. 1977. Carbon balance in terrestrial detritus. *Annual Reviews of Ecology and Systematics.* 8:51–81.

Schlesinger, W. H. 1997. *Biogeochemistry — An Analysis of Global Change.* 2nd ed. Academic Press, San Diego, CA.

Schmink, M., K. H. Redford, and C. Padoch. 1992. Traditional peoples and the biosphere: framing the issues and defining the terms. pp. 3–20. In: *Conservation of Neotropical Forests: Working from Traditional Resource Use.* K. H. Redford and C. Padoch (eds.). Columbia University Press, New York.

Schmitz, D. C. and D. Simberloff. 1997. Biological invasions: a growing threat. *Issues in Science and Technology.* 13:33–40.

Schmitz, J. P. 2000. Meso-scale Community Organization and Response to Burning in Mesocosms and a Field saltmarsh. M.S. thesis, University of Maryland, College Park, MD.

Schneider, D. C. 2001. The rise of the concept of scale in ecology. *BioScience.* 51:545–553.

Schoener, A. 1974. Experimental zoogeography: colonization of marine mini-islands. *American Naturalist.* 108:715–738.

Schoener, T. W. 1988. The ecological niche. pp. 79–113. In: *Ecological Concepts: The Contribution of Ecology to an Understanding of the Natural World.* J. M. Cherrett (ed.). Blackwell Scientific, Oxford, U.K.

Schueler, T. R. 1987. *Controlling Urban Runoff.* Washington Metropolitan Water Resources Planning Board, Washington, DC.

Schueler, T. R. 1995. The peculiarities of perviousness. *Watershed Protection Techniques.* 2:233–238.

Schueler, T. R. 1995. *Site Planning for Urban Stream Protection.* Center for Watershed Protection, Silver Spring, MD.

Schulenberg, R. 1969. Summary of Morton Arboretum prairie restoration work, 1963–1968. pp. 45–46. In: *Proceedings of a Symposium on Prairie and Prairie Restoration.* P. Schramm (ed.). Knox College, Galesburg, IL.

Schultz, A. M. 1964. The nutrient-recovery hypothesis for Arctic microtine cycles. II. Ecosystem variables in relation to Arctic microtine cycles. pp. 57–68. In: *Grazing in Terrestrial and Marine Environments.* D. J. Crisp (ed.). Blackwell Scientific, Oxford, U.K.

Schultz, A. M. 1969. A study of an ecosystem: the Arctic tundra. pp. 77–93. In: *The Ecosystem Concept in Natural Resource Management.* G. M. Van Dyne (ed.). Academic Press, New York.

Schultz, J. S. 1991. Biosensors. *Scientific American.* 265(2):64–69.

Schulze, P. C. 1991. Incorporation of environmental damages into cost-benefit analyses: an introduction. *Bulletin of the Ecological Society of America.* 72:15–19.

Schulze, P. C. (ed.). 1996. *Engineering Within Ecological Constraints.* National Academy Press, Washington, DC.

Schumacher, E. F. 1968. Buddhist economics. *Resurgence* 1, January–February.

Schumacher, E. F. 1973. *Small Is Beautiful: Economics as if People Mattered.* Harper & Row, New York.

Schwartz, E. L. (ed.). 1990. *Computational Neuroscience.* MIT Press, Cambridge, MA.

Schwartz, G. E. 1975. Biofeedback, self-regulation, and the patterning of physiological processes. *American Scientist.* 63:314–324.

Schwemmler, W. 1984. *Reconstruction of Cell Evolution: A Periodic System*. CRC Press, Boca Raton, FL.

Scoffin, T. P. 1970. The trapping and binding of subtidal carbonate sediments by marine vegetation in Bimini Lagoon, Bahamas. *Journal of Sedimentary Petrology*. 40:249–273.

Seastedt, T. R. 1984. The role of microarthropods in decomposition and mineralization processes. *Annual Review of Entomology*. 29:25–46.

Sedell, J. R. and R. L. Beschta. 1991. Bringing back the "bio" in bioengineering. pp. 160–175. In: *Fisheries Bioengineering Symposium*. J. Colt and R. J. White (eds.). American Fisheries Society, Bethesda, MD.

Seidel, K. 1976. Macrophytes and water purification. pp. 109–121. In: *Biological Control of Water Pollution*. J. Tourbier and R. W. Pierson, Jr. (eds.). University of Pennsylvania Press, Philadelphia, PA.

Seidel, K. 1966. Purification of water by higher plants. *Naturwissenschaften* (Ger.). 53:289

Seneca, E. D. 1974. Stabilization of coastal dredge spoil with *Spartina alterniflora*. pp. 525–529. In: *Ecology of Halophytes*. R. J. Reimold and W. H. Queen (eds.). Academic Press, New York.

Seneca, E. D. and S. W. Broome. 1992. Restoring tidal marshes in North Carolina and France. pp. 53–78. In: *Restoring the Nation's Marine Environment*. G. W. Thayer (ed.). Maryland Sea Grant Program, University of Maryland, College Park, MD.

Seneca, E. D., W. W. Woodhouse, Jr., and S. W. Broome. 1975. Salt-water marsh creation. pp. 427–437. In: *Estuarine Research Vol. II. Geology and Engineering*. L. E. Cronin (ed.). Academic Press, New York.

Seneca, E. D., S. W. Broome, W. W. Woodhouse, Jr., L. M. Cammen, and J. T. Lyon. 1976. Establishing *Spartina alterniflora* marsh in North Carolina. *Environmental Conservation*. 3:185–188.

Sensabaugh, W. M. 1975. The beach — a natural protection from the sea. Sea Grant Marine Advisory Program, University of Florida, Gainesville, FL.

Sepkoski, J. J., Jr. and M. A. Rex. 1974. Distribution of freshwater mussels: coastal rivers as biogeographic islands. *Systematic Zoology*. 23:165–188.

Sepulveda, J. A., W. E. Souder, and B. S. Gottfried. 1984. *Theory and Problems of Engineering Economics*. McGraw-Hill, New York.

Service, R. F. 1995. Prompting complex patterns to form themselves. *Science*. 270:1299–1300.

Service, R. F. 2001. Arson strikes research labs and tree farm in Pacific Northwest. *Science*. 292:1622–1623.

Severinghaus, J. P., W. S. Broecker, W. F. Dempster, T. MacCallum, and M. Wahlen. 1994. Oxygen loss in Biosphere 2. EOS. *Transactions of the American Geophysical Union*. 75:33–40.

Shabman, L. A. and S. S. Batie. 1978. Economic value of natural coastal wetlands: a critique. *Coastal Zone Management Journal*. 4:231–247.

Shabman, L. A. and S. S. Batie. 1980. Estimating the economic value of coastal wetlands: conceptual issues and research needs. pp. 3–16. In: *Estuarine Perspectives*. V. S. Kennedy (ed.). Academic Press, New York.

Shafer, C. L. 1990. *Nature Reserves: Island Theory and Conservation Practice*. Smithsonian Institution Press, Washington, DC.

Shanks, R. E. and J. S. Olson. 1961. First-year breakdown of leaf litter in southern Appalachian forests. *Science*. 134:194–195.

Shapiro, J., V. Lamarra, and M. Lynch. 1975. Biomanipulation: an ecosystem approach to lake restoration. pp. 85–96. In: *Water Quality Management Through Biological Control*. P. L. Brezonik and J. L. Fox (eds.). University of Florida, Gainesville, FL.

Shaw, M. 1999. Industrial ecology and living machines. *Annals of Earth.* 17(3):7–10.

Shaw, M. 2001. Restorers for treating industrial wastewater at the Tyson Foods Plant. *Annals of Earth.* 19(2):12.

Shea, C. 1997. Terns hit the roof. *Audubon.* 99(6):22.

Sheehan, P. J. 1989. Statistical and nonstatistical considerations in quantifying pollutant-induced changes in microcosms. pp. 178–188. In: *Aquatic Toxicology and Hazard Assessment,* Vol. 12. U. M. Cowgill and L. R. Williams (eds.). ASTM, Philadelphia, PA.

Sheehy, D. J. and S. F. Vik. 1992. Developing prefabricated reefs: an ecological and engineering approach. pp. 543–582. In: *Restoring the Nation's Marine Environment.* G. W. Thayer (ed.). Maryland Sea Grant, College Park, MD.

Sheffield, C. W. 1967. Water hyacinth for nutrient removal. *Hyacinth Control Journal.* 6:27–30.

Sheldon, A. L. 1968. Species diversity and longitudinal succession in stream fishes. *Ecology.* 49:194–198.

Shelley, C. 1999. Preadaptation and the explanation of human evolution. *Biology and Philosophy.* 14:65–82.

Sheng, Y. P. 2000. Physical characteristics and engineering at reef sites. pp. 53–94. In: *Artificial Reef Evaluation.* W. Seaman, Jr. (ed.). CRC Press, Boca Raton, FL.

Shenot, J. 1993. An Analysis of Wetland Planting Success at Three Stormwater Management Ponds in Montgomery County, Maryland. M.S. thesis, University of Maryland, College Park, MD.

Shenot, J. and P. Kangas. 1993. Evaluation of wetland plantings in three stormwater retention ponds in Maryland. pp. 187–195. In: *Proceedings of the 20th Annual Conference on Wetlands Restoration and Creation.* F. J. Webb, Jr. (ed.). Hillsborough Community College, Plant City, FL.

Shepherdson, D. J. 1998. Tracing the path of environmental enrichment in zoos. pp. 1–12. In: *Second Nature: Environmental Enrichment for Captive Animals.* D. J. Shepherdson, J. D. Mellen, and M. Hutchins (eds.). Smithsonian Institution Press, Washington, DC.

Shepherdson, D. J., J. D. Mellen, and M. Hutchins (eds.). 1998. *Second Nature, Environmental Enrichment for Captive Animals.* Smithsonian Institution Press, Washington, DC.

Sheppard, S. C. 1997. Toxicity testing using microcosms. pp. 346–373. In: *Soil Ecotoxicology.* J. Tarradellas, G. Bitton, and D. Rossel (eds.). Lewis Publishers, Boca Raton, FL.

Shields, F. D., Jr., A. J. Bowie, and C. M. Cooper. 1995. Control of streambank erosion due to bed degradation with vegetation and structure. *Water Resources Bulletin.* 31:475–489.

Shigesada, N. and K. Kawasaki. 1997. *Biological Invasions: Theory and Practice.* Oxford University Press, Oxford, U.K.

Shirley, S. 1994. *Restoring the Tallgrass Prairie.* University of Iowa Press, Iowa City, IA.

Shorrocks, B. and I. R. Swingland (eds.). 1990. *Living in a Patchy Environment.* Oxford University Press, Oxford, U.K.

Shrader-Frechette, K. S. 1993. *Burying Uncertainty: Risk and the Case Against Geological Disposal of Nuclear Waste.* University of California Press, Berkeley, CA.

Shugart, H. H. 1984. *A Theory of Forest Dynamics: The Ecological Implications of Forest Succession Models.* Springer-Verlag. New York.

Shulman, M. J., J. C. Ogden, J. P. Ebersole, W. N. McFarland, S. L. Miller, and N. G. Wolf. 1983. Priority effects in the recruitment of coral reef fishes. *Ecology.* 64:1508–1513.

Sibert, J. R. and R. J. Naiman. 1980. The role of detritus and the nature of estuarine ecosystems. pp. 311–323. In: *Marine Benthic Dynamics.* K. R. Tenore and B. C. Coull (eds.). University of South Carolina Press. Columbia, SC.

Sieburth, J. M. 1976. Bacterial substrates and productivity in marine ecosystems. *Annual Reviews of Ecology and Systematics.* 7:259–285.

Siever, R. 1968. Sedimentological consequences of a steady-state ocean-atmosphere. *Sedimentology.* 11:5–29.

Silver, M. W., A. L. Shanks, and J. D. Trent. 1978. Marine snow: microplankton habitat and source of small-scale patchiness in pelagic populations. *Science.* 201:371–373.

Simberloff, D. 1974. Equilibrium theory of island biogeography and ecology. *Annual Reviews of Ecology and Systematics.* 5:161–182.

Simberloff, D. 1981. Community effects of introduced species. pp. 53–81. In: *Biotic Crises in Evolutionary Time.* M. H. Nitecki (ed.). Academic Press, New York.

Simberloff, D. 1984. A succession of paradigms in ecology: essentialism to materialism and probabilism. pp. 63–99. In: *Conceptual Issues in Ecology.* E. Saarinen (ed.). D. Reidel Publishing, Dordrecht, Holland.

Simberloff, D. 1989. Which insect introductions succeed and which fail? pp. 61–76. In: *Biological Invasions: A Global Perspective.* J. A. Drake, H. A. Mooney, F. di Castri, R. H. Groves, F. J. Kruger, M. Rejmanek, and M. Williamson (eds.). John Wiley & Sons, Chichester, U.K.

Simberloff, D. and B. Von Holle. 1999. Positive interactions of nonindigenous species: invasional meltdown? *Biological Invasions.* 1:21–32.

Simberloff, D. 1995. Introduced species. pp. 323–336. In: *Encyclopedia of Environmental Biology,* Vol. 2. W. A. Nierenberg (ed.). Academic Press, San Diego, CA.

Simberloff, D. 1997. Biogeographic approaches and the new conservation biology. pp. 274–284. In: *The Ecological Basis of Conservation,* S. T. A. Pickett, R. S. Ostfeld, M. Shachak, and G. E. Likens (eds.). Chapman & Hall, New York.

Simberloff, D. 1997. Eradication. pp. 221–228. In: *Strangers in Paradise: Impact and Management of Nonindigenous Species in Florida.* D. Simberloff, D. C. Schmitz, and T. C. Brown (eds.). Island Press, Washington, DC.

Simberloff, D. S. and L. G. Abele. 1976. Island biogeography theory and conservation practice. *Science* 191:285–286.

Simberloff, D., and P. Stiling. 1996. How risky is biological control? *Ecology.* 77:1965–1974.

Simberloff, D. S. and E. O. Wilson. 1969. Experimental zoogeography of islands: the colonization of empty islands. *Ecology.* 50:278–296.

Simberloff, D. S. and E. O. Wilson. 1970. Experimental zoogeography of islands: a two-year record of colonization. *Ecology.* 51:934–937.

Simmons, M. 1997. Unstoppable seaweed becomes monster of the deep. *The Cleveland Plain Dealer,* Cleveland, OH. August 17, p. 8-A.

Singhvi, R., A. Kumar, G. P. Lopez, G. N. Stephanopoulos, D. I. C. Wang, G. M. Whitesides, and D. E. Ingber. 1994. Engineering cell shape and function. *Science.* 264:696–698.

Sinha, R. N. 1991. Storage ecosystems. pp. 17–30. In: *Ecology and Management of Food-Industry Pests.* J. R. Gorham (ed.). FDA Technical Bull. 4, Association of Official Analytical Chemists, Arlington, VA.

Sipper, M., G. Tempesti, D. Mange, and E. Sanchez. 1998. Guest Editors' Introduction: Von Neumann's legacy — special issue on self-replication. *Artificial Life.* 4: iii–iv.

Slattery, B. E. 1991. *WOW!: The Wonders of Wetlands.* Environmental Concern, Inc., St. Michaels, MD.

Slobodkin, L. B. 1964. The strategy of evolution. *American Scienctist.* 52:342–357.

Slobodkin, L. B. 1968. Animal populations and ecologies. pp. 149–163. In: *Positive Feedback: A General Systems Approach to Positive/Negative Feedback and Mutual Causality.* J. H. Milsum (ed.). Pergamon Press, Oxford, U.K.

Slobodkin, L. 1993. Gaia: hoke and substance (review of Scientists on Gaia). *BioScience.* 43:255–256.

Slobodkin, L. B. and D. E. Dykhuizen. 1991. Applied ecology, its practice and philosophy. pp. 63–70. In: *Integrated Environmental Management.* J. Cairns, Jr. and T. V. Crawford (eds.). Lewis Publishers, Chelsea, MI.

Slobodkin, L. B., D. B. Botkin, B. Maguire, Jr., B. Moore, III, and H. Morowitz. 1980. On the epistemology of ecosystem analysis. pp. 497–507. In: *Estuarine Perspectives.* V. S. Kennedy (ed.). Academic Press, New York.

Small, A. M., W. H. Adey, and D. Spoon. 1998. Are current estimates of coral reef biodiversity too low? The view through the window of a microcosm. *Atoll Research Bulletin.* 458:1–20.

Small, M. M. 1975. Brookhaven's two sewage treatment systems. *Compost Science.* 16(5):7–9.

Smith, C. C. and O. J. Reichman. 1984. The evolution of food caching by birds and mammals. *Annual Reviews of Ecology and Systematics.* 15:329–351.

Simberloff, D. and B. Von Holle. 1999. Positive interactions of nonindigenous species: invasional meltdown? *Biological Invasions.* 1:21–32.

Smith, D. S. and P. C. Hellmund (eds.). 1993. *Ecology of Greenways.* University of Minnesota Press, Minneapolis, MN.

Smith, F. E. 1954. Quantitative aspects of population growth. pp. 277–294. In: *Dynamics of Growth Processes.* E. Boell (ed.). Princeton University Press, Princeton, NJ.

Smith, F. E. 1972. Spatial heterogeneity, stability, and diversity in ecosystems. pp. 309–335. In: *Growth by Intussusception.* E. S. Deevey (ed.). *Transactions of Connecticut Academy of Arts and* Sciences. The Connecticut Academy of Arts and Sciences, New Haven, CT.

Smith, G. B. 1979. Relationship of eastern Gulf of Mexico reef-fish communities to the species equilibrium theory of insular biogeography. *Journal of Biogeography.* 6:49–61.

Smith, S. R. 1996. *Agricultural Recycling of Sewage Sludge and the Environment.* CAB International, Wallingford, U.K.

Smith, W., V. R. Gibson, and J. F. Grassle. 1982. Replication in controlled marine systems: presenting the evidence. pp. 217–225. In: *Marine Mesocosms.* G. D. Grice and M. R. Reeve (eds.). Springer-Verlag, New York.

Smock, L. A., G. M. Metzler, and J. E. Gladden. 1989. Role of debris dams in the structure and functioning of low-gradient headwater streams. *Ecology.* 70:764–775.

Snelgrove, P. V. R. and 10 other authors. 1997. The importance of marine sediment biodiversity in ecosystem processes. *Ambio.* 26:578–583.

Snow, A. A. and P. M. Palma. 1997. Commercialization of transgenic plants: potential ecological risks. *BioScience.* 47:86–96.

Soares, A. M. V. M. and P. Calow. 1993. Seeking standardization in ecotoxicology. pp. 1–6. In: *Progress in Standardization of Aquatic Toxicity Tests.* A. M. V. M. Soares and P. Calow (eds.). Lewis Publishers, Boca Raton, FL.

Socolow, R., C. Andrews, F. Berkhout, and V. Thomas (eds.). 1994. *Industrial Ecology and Global Change.* Cambridge University Press, Cambridge, U.K.

Solbrig, O. T. and G. H. Orians. 1977. The adaptive characteristics of desert plants. *American Scientist.* 65:412–421.

Soleri, P. 1973. *Matter Becoming Spirit.* Anchor Books, Garden City, NY.

Soleri, P. 1993. *Arcosanti: An Urban Laboratory?* 3rd ed. Cosanti Press, Scottsdale, AZ.

Solomon, K. R., G. L. Stephenson, and N. K. Kaushik. 1989. Effects of methoxychlor on zooplankton in freshwater enclosures: influence of enclosure size and number of applications. *Environment Toxicology Chemistry.* 8:659–669.

Sommer, U. 1991. Convergent succession of phytoplankton in microcosms with different inoculum species composition. *Oecologia.* 87:171–179.

Sonntag, N. C. and T. R. Parsons. 1979. Mixing an enclosed 1300 m^3 water column: effects on the planktonic food web. *Journal for Plankton Research.* 1:85–102.

Sopper, W. E. and S. N. Kerr (eds.). 1979. *Utilization of Municipal Sewage Effluent and Sludge on Forest and Disturbed Land.* Pennsylvania State University Press, University Park, PA.

Sopper, W. E. and L. T. Kardos (eds.). 1973. *Recycling Treated Municipal Wastewater and Sludge through Forest and Cropland.* Pennsylvania State University Press, University Park, PA.

Soule, M. E. 1980. Thresholds for survival: maintaining fitness and evolutionary potential. pp. 151–169. In: *Conservation Biology: An Evolutionary-Ecological Perspective.* M. E. Soule and B. A. Wilcox (eds.). Sinauer Associates, Sunderland, MA.

Soule, M. E. 1990. The onslaught of alien species, and other challenges in the coming decades. *Conservation Biology.* 4:233–239.

Sousa, W. P. 1984. The role of disturbance in natural communities. *Annual Reviews of Ecology and Systematics.* 15:535–591.

Space Studies Board. 1997. *Mars Sample Return.* National Academy Press, Washington, DC.

Space Studies Board. 1998. *Evaluating the Biological Potential in Samples Returned from Planetary Satellites and Small Solar System Bodies.* National Academy Press, Washington, DC.

Spangler, F., W. Sloey, and C. W. Fetter. 1976. Experimental use of emergent vegetation for the biological treatment of municipal wastewater in Wisconsin. pp. 161–171. In: *Biological Control of Water Pollution.* J. Tourbier and R. W. Pierson (eds.). University of Pennsylvania Press, Philadelphia, PA.

Spieler, R. E., D. S. Gilliam, and R. L. Sherman. 2001. Artificial substrate and coral reef restoration: what do we need to know to know what we need. *Bulletin of Marine Science.* 69:1013–1030.

Sponsel, L. E. and P. Natadecha. 1988. Buddhism, ecology, and forests in Thailand: past, present, and future. pp. 305–325. In: *Changing Tropical Forests,* J. Dargavel, K. Dixon, and N. Semple (eds.). Centre for Resource and Environmental Studies, Australian National University, Canberra, Australia.

Sponsel, L. E. and P. Natadecha-Sponsel. 1993. The potential contribution of Buddhism in developing an environmental ethic for the conservation of biodiversity. pp. 75–97. In: *Ethics, Religion and Biodiversity: Relations Between Conservation and Cultural Values.* L. S. Hamilton (ed.). The White Horse Press, Cambridge, U.K.

Spotte, S. 1974. Aquarium techniques: closed-system marine aquariums. pp. 2–21. In: *Experimental Marine Biology.* R. N. Mariscal (ed.). Academic Press, New York.

Spurgeon, D. 1997. Guiliani, New York City Declare War on Rats — All 28 Million of Them. *The Washington Post.* October 4: A3.

St. John, T. V. 1983. Response of tree roots to decomposing organic matter in two lowland Amazonian rain forests. *Canadian Journal of Forest* Research. 13:346–349.

Stachowicz, J. J. 2001. Mutualism, facilitation, and the structure of ecological communities. *BioScience.* 51:235–246.

Stahel, W. R. 1994. The utilization-focused service economy: resource efficiency and product-life extension. pp. 178–190. In: *The Greening of Industrial Ecosystems.* B. R. Allenby and D. J. Richards (eds.). National Academy Press, Washington, DC.

Stanford, J. A. and J. V. Ward. 1988. The hyporheic habitat of river ecosystems. *Nature.* 335:64–66.

Stanton, N. L. 1988. The underground in grasslands. *Annual Reviews of Ecology and Systematics.* 19:573–589.

Stark, N. M. and C. F. Jordan. 1978. Nutrient retention by the root mat of an Amazonian rain forest. *Ecology.* 59:434–437.

Stearns, F. and T. Montag. 1974. *The Urban Ecosystem: A Holistic Approach.* Dowden, Hutchinson & Ross, Stroudsburg, PA.

Stearns, S. C. 1976. Life-history tactics: a review of the ideas. *Quarterly Reviews of Biology.* 5:3–47.

Steavenson, H. A., H. E. Gearhart, and R. L. Curtis. 1943. Living fences and supplies of fence posts. *Journal of Wildlife Management.* 7:257–261.

Steele, J. 1997. *Sustainable Architecture: Principles, Paradigms, and Case Studies.* McGraw-Hill, New York.

Steele, J. H. 1979. The uses of experimental ecosystems. *Phil. Trans. R. Soc. Lond. B.* 286:583–595.

Steinbeck, J. 1937. *Cannery Row.* Penguin, New York.

Steinbeck, J. 1939. *Grapes of Wrath.* Penguin, New York.

Steinbeck, J. and E. Ricketts. 1941. *The Sea of Cortez: A Leisurely Journal of Travel and Research.* Viking Publishers, New York.

Stenseth, N. C. 1979. Where have all the species gone? On the nature of extinction and the Red Queen Hypothesis. *Oikos.* 33:196–227.

Stephenson, G. L., P. Hamilton, N. K. Kaushik, J. B. Robinson, and K. R. Solomon. 1984. Spatial distribution of plankton in enclosures of three sizes. *Canadian Journal of the Fisheries and Aquatic Sciences.* 41:1048–1054.

Sterner, R. W. 1995. Elemental stoichiometry of species in ecosystems. pp. 240–252. In: *Linking Species and Ecosystems.* C. G. Jones and J. H. Lawton (eds.). Chapman & Hall, New York.

Stevens, E. S. 2002. *Green Plastics: An Introduction to the New Science of Biodegradable Plastics.* Princeton University Press, Princeton, NJ.

Stevenson, J. C. 1999. Personal communication.

Stevenson, J. C., J. I. Marusic, B. Ozretic, A. Marson, G. Cecconii, and M. S. Kearney. 1999. Shallow water and shoreline ecosystems of the Chesapeake Bay compared to the Northern Adriatic Sea: transformation of habitat at the land-sea margin. pp. 29–79. In: *Ecosystems at the Land–Sea Margin: Drainage Basin to Coastal Sea.* T. C. Malone, A. Malej, L. W. Harding, Jr., N. Smodlaka, and R. E. Turner (eds.). American Geophysical Union, Washington, DC.

Stevenson, J. C., J. E. Rooth, M. S. Kearney, and K. L. Sundberg. 2000. The health and long-term stability of natural and restored marshes in Chesapeake Bay. pp. 709–735. In: *Concepts and Controversies in Tidal Marsh Ecology.* M. P. Weinstein and D. A. Kreeger (eds.). Kluwer Academic, Dordrecht, the Netherlands.

Steward, K. K. 1970. Nutrient removal potentials of various aquatic plants. *Hyacinth Control Journal.* 8:34–35.

Steward, K. K. and W. H. Ornes. 1975. Assessing a marsh environment for wastewater renovation. *Journal of the Water Pollution Control Federation.* 47:1880–1891.

Stinner, B. R., D. A. Crossley, Jr., E. P. Odum, and R. L. Todd. 1984. Nutrient budgets and internal cycling of N, P, K, Ca and Mg in conventional tillage, no-tillage, and old-field ecosystems on the Georgia piedmont. *Ecology.* 65:354–369.

Stitt, F. A. (ed.). 1999. *Ecological Design Handbook: Sustainable Strategies for Architecture, Landscape Architecture, Interior Design, and Planning.* McGraw-Hill, New York.

Stix, G. 1996. Waiting for breakthroughs. *Scientific American,* 274(4):94–99.

Stoeckeler, J. H. and R. A. Williams. 1949. Windbreaks and shelterbelts. pp. 191–199. In: *Trees. The Yearbook of Agriculture*. U.S. Department of Agriculture, Washington, DC.

Stolum, H.-H. 1996. River meandering as a self-organization process. *Science*. 271:1710–1713.

Stommel, H. 1963. Varieties of oceanographic experience. *Science*. 139:572–576.

Stone, R. 1993. Biosphere loses its advisers. *Science*. 259:1111.

Stone, R. B. 1985. History of artificial reef use in the United States. pp. 3–11. In: *Artificial Reefs, Marine and Freshwater Applications*. F. M. D'Itri (ed.). Lewis Publishers, Chelsea, MI.

Stone, R. B., L. M. Sprague, J. M. McGurrin, and W. Seaman, Jr. 1991. Artificial habitats of the world: synopsis and major trends. pp. 31–60. In: *Artificial Habitats for Marine and Freshwater Fisheries*. W. Seaman, Jr. and L. M. Sprague (eds.). Academic Press, San Diego, CA.

Stoner, P. M. (ed.). 1977. *Goodbye to the Flush Toilet*. Rodale Press, Emmaus, PA.

Stout, J. D. 1980. The role of protozoa in nutrient cycling and energy flow. pp. 1–50. In: *Advances in Microbial Ecology*, Vol. 4. M. Alexander (ed.). Plenum Press, New York.

Stowell, R., S. Weber, G. Tchobanoglous, B. A. Wilson, and K. R. Townzen. 1985. Mosquito considerations in the design of wetland systems for the treatment of wastewater. pp. 38–47. In: *Ecological Considerations in Wetlands Treatment of Municipal Wastewaters*. P. J. Godfrey, E. R. Kaynor, and S. Pelczarski (eds.). Van Nostrand Reinhold, New York.

Strahler, A. N. 1950. Equilibrium theory of erosional slopes approached by frequency distribution analysis. *American Journal of Science*. 248:673–696.

Straskraba, M. 1999. Self-organization, direct and indirect effects. pp. 29–51. In: *Theoretical Reservoir Ecology and Its Applications*. Backhuys Publishers, Leiden, the Netherlands.

Strauss, S. Y. 1991. Indirect effects in community: their definition, study and importance. *Trends in Ecology and Evolution (TREE)*. 6:206–210.

Strassmann, B. I. 1986. Rangelands. pp. 497–514. In: *Energy and Resource Quality: The Ecology of the Economic Process*. C. A. S. Hall, C. J. Cleveland, and R. Kaufmann (eds.). John Wiley & Sons. New York.

Strayer, D. L., N. F. Caraco, J. J. Cole, S. Findlay, and M. L. Pace. 1999. Transformation of freshwater ecosystems by bivalves. *BioScience*. 49:19–27.

Streb, C. 2001. Woody Debris Jams: Exploring the Principles of Ecological Engineering and Self-Design to Restore Streams. M.S. thesis, University of Maryland, College Park, MD.

Streb, C., E. Biermann, P. Kangas, and W. Adey. in press. The energy basis of a subtropical wetland mesocosm. In: *Emergy Analysis II*. M. Brown, H. T. Odum, and D. Tilley (eds.). Center for Environmental Policy, University of Florida, Gainesville, FL.

Streeter, H. W. and E. B. Phelps. 1925. A study of pollution and natural purification of the Ohio River. *USPHS Bulletin* 146:1–75.

Strong, D. R. 1992. Are trophic cascades all wet? Differentiation and donor-control in speciose ecosystems. *Ecology*. 73:747–754.

Sullivan, A. L. and M. L. Shaffer. 1975. Biogeography of the megazoo. *Science*. 189:13–17.

Susarla, S., V. F. Medina, and S. C. McCutcheon. 2002. Phytoremediation: an ecological solution to organic chemical contamination. *Ecological Engineering*. 18:647–658.

Sutherland, J. P. 1974. Multiple stable points in natural communities. *American Naturalist*. 108:859–873.

Sutton, D. B. and N. P. Harmon. 1973. *Ecology: Selected Concepts*. John Wiley & Sons. New York.

Svensson, B. M. 1995. Competition between *Sphagnum fuscum* and *Drosera rotundifolia*: a case of ecosystem engineering. *Oikos*. 74:205–212.

Swanson, F. J. 1979. Geomorphology and ecosystems. pp. 159–170. In: *Forests: Fresh Perspectives from Ecosystem Analysis*. R. H. Waring (ed.). Oregon State University Press, Corvallis, OR.

Swanson, F. J., T. K. Kratz, N. Caine, and R. G. Woodmansee. 1988. Landform effects on ecosystem patterns and processes. *BioScience*. 38:92–98.

Swartwood, S. in preparation. Use of Snail Populations as Indicators of Ecological Development of Restored Wetlands and Wetland Mesocosms. M.S. thesis, University of Maryland, College Park, MD.

Swartzmann, G. L. and T. M. Zaret. 1983. Modeling fish species introduction and prey extermination: the invasion of *Cichla ocellaris* to Gatun Lake, Panama. pp. 361–371. In: *Analysis of Ecological Systems: State-of-the-Art in Ecological Modelling*. W. K. Lauenroth, G. V. Skogerboe, and M. Flug (eds.). Elsevier, Amsterdam.

Swingle, H. S. 1950. *Relationships and Dynamics of Balanced and Unbalanced Fish Populations*. Agricultural Experiment Station, Alabama Polytechnic Inst. Birmingham, AL.

Swingle, H. S. and E. V. Smith. 1941. The management of ponds for the production of game and pan fish. pp. 218–226. In: *A Symposium on Hydrobiology*. University of Wisconsin Press, Madison, WI.

Takahashi, M. and F. A. Whitney. 1977. Temperature, salinity, and light penetration structures: controlled ecosystem pollution experiment. *Bulletin of Marine Sciences*. 27:8–16.

Takahashi, M., W. H. Thomas, D. L. R. Seibert, J. Beers, P. Koeller, and T. R. Parsons. 1975. The replication of biological events in enclosed water columns. *Archives of Hydrobiology*. 76:5–23.

Takano, C. T., C. E. Folsome, and D. M. Karl. 1983. ATP as a biomass indicator for closed ecosystems. *BioSystems*. 16:75–78.

Takayasu, H. and H. Inaoka. 1992. New type of self-organized criticality in a model of erosion. *Physical Review Letters*. 68:966–969.

Tallamy, D. W. 1983. Equilibrium biogeography and its application to insect host-parasite systems. *American Naturalist*. 121:244–254.

Talling, J. F. 1958. The longitudinal succession of water characteristics in the White Nile. *Hydrobiologia*. 11:73–89.

Tamarin, R. H. (ed.). 1978. *Population Regulation: Benchmark Papers in Ecology*, Vol. 7. Dowden, Hutchinson & Ross, Stroudsburg, PA.

Tammemagi, H. 1999. *The Waste Crisis: Landfills, Incinerators, and the Search for a Sustainable Future*. Oxford University Press, New York.

Tangley, L. 1986. The urban ecologist. *BioScience*. 36:68–71.

Tanner, W. F. 1958. The equilibrium beach. *Transactions of the American Geophysical Union*. 39:889–891.

Tansley, A. G. 1935. The use and abuse of vegetational concepts and terms. *Ecology*. 16:284–307.

Taub, F. B. 1969a. Gnotobiotic models of freshwater communities. *Verhandelingen International Vereins Limnology*. 17:485–496.

Taub, F. B. 1969b. A biological model of a freshwater community: a gnotobiotic ecosystem. *Limnology and Oceanography*. 14:136–142.

Taub, F. B. 1969c. A continuous gnotobiotic (species defined) ecosystem. pp. 101–120. In: *The Structure and Function of Fresh-water Microbial Communities*. J. Cairns, Jr. (ed.). Virginia Polytechnic Institute and State University, Blacksburg, VA.

Taub, F. B. 1974. Closed ecological systems. *Annual Review of Ecology and Systematies*. 5:139–160.

Taub, F. B. 1984. Introduction. pp. 113–116. In: *Concepts in Marine Pollution Measurements.* H. H. White (ed.). Sea Grant Publication, University of Maryland, College Park, MD.

Taub, F. B. 1989. Standardized aquatic microcosms. *Environment Science and Technology.* 23:1064–1066.

Taub, F. B. 1993. Standardizing an aquatic microcosm test. pp. 159–188. In: *Progress in Standardization of Aquatic Toxicity Tests.* A. M. V. M. Soares and P. Calow (eds.). Lewis Publishers, Boca Raton, FL.

Taub, F. B. 1997. Unique information contributed by multispecies systems: examples from the standardized aquatic microcosm. *Ecological Applications.* 7:1103–1110.

Taub, F. B. and A. M. Dollar. 1964. A Chlorella-Daphnia food-chain study: the design of a compatible chemically defined culture medium. *Limnology and Oceanography.* 9:61–74.

Taub, F. B. and A. M. Dollar. 1968. The nutritional inadequacy of Chlorella and Chlamydomonas as food for *Daphnia pulex. Limnology and Oceanography.* 13:607–617.

Taylor, J. S. and E. A. Stewart. 1978. Hyacinths. pp. 143–179. In: *Biological Nutrient Removal.* M. P. Wanielista and W. W. Eckenfelder, Jr. (eds.). Ann Arbor Science, Ann Arbor, MI.

Tchobanoglous, G. 1991. Land-based systems, constructed wetlands, and aquatic plant systems in the United States: An overview. pp. 110–120. In: *Ecological Engineering for Wastewater Treatment.* C. Etnier and B. Guterstam (eds.). Bokskogen, Gothenburg, Sweden.

Teal, J. M. 1991. Contributions of marshes and saltmarshes to ecological engineering. pp. 55–62. In: *Ecological Engineering for Wastewater Treatment.* C. Etnier and B. Guterstam (eds.). Bokskogen, Gothenburg, Sweden.

Teal, J. M. and S. B. Peterson. 1991. The next generation of septage treatment. *Research Journal of the Water Pollution Control Federation.* 63:84–89.

Teal, J. M. and S. B. Peterson. 1993. A solar aquatic system septage treatment plant. *Environmental Science and Technology.* 27:34–37.

Teal, J. M., B. L. Howes, S. B. Peterson, J. E. Petersen, and A. Armstrong. 1994. Nutrient processing in an artificial wetland engineered for high loading: a septage treatment example. pp. 421–428. In: *Global Wetlands Old World and New.* W. J. Mitsch and R. E. Turner (eds.). Elsevier, Amsterdam, the Netherlands.

Temple, S. A. 1990. The nasty necessity: eradicating exotics. *Conservation Biology.* 4:113–115.

Tenner, E. 1997. *Why Things Bite Back.* Vintage Books, New York.

Tenney, M. W., W. F. Echelberger, Jr., K. J. Guter, and J. B. Carberry. 1972. Nutrient removal from wastewater by biological treatment methods. pp. 391–419. In: *Nutrients in Natural Waters.* H. E. Allen and J. R. Kramer (eds.). John Wiley & Sons. New York.

Tenore, K. R., J. C. Goldman, and J. P. Clarner. 1973. The food chain dynamics of the oyster, clam, and mussel in an aquaculture food chain. *J. Exp. Mar. Biol.* Ecol. 12:157–165.

Terborgh, J. 1975. Faunal equilibria and the design of wildlife preserves. pp. 369–380. In: *Tropical Ecological Systems: Trends in Terrestrial and Aquatic Research.* F. Golley and E. Medina (eds.). Springer-Verlag, New York.

Thienemann, A. 1926. *Limnologie.* Jedermanns Bucherei, Breslau, Poland.

Thom, B. G. 1967. Mangrove ecology and deltaic geomorphology: Tabasco, Mexico. *Journal of Ecology.* 55:301–343.

Thom, B. G. 1984. Coastal landforms and geomorphic processes. pp. 3–17. In: S. C. Snedaker and J. G. Snedaker (eds.). *The Mangrove Ecosystem: Research Methods.* Unesco. Paris.

Thom, B. G., L. D. Wright, and J. M. Coleman. 1975. Mangrove ecology and deltaic estuarine geomorphology: Cambridge Gulf-Ord River, Western Australia. *Journal of Ecology.* 63:203–232.

Thom, R. 1975. *Structural Stability and Morphogenesis: An Outline of a General Theory of Models.* Benjamin/Cummings, Reading, MA.

Thomas, D. J. 1995. Biological aspects of the ecopoeisis and terraformation of Mars: current perspectives and research. *Journal of the British Interplanetary Society.* 48:415–418.

Thomas, J. W. and R. A. Dixon. 1973. Cemetery ecology. *Natural History.* March: 61–67.

Thomsen, D. E. 1982. The lone prairie. *Science News.* 122:250–251.

Thulesius, O. 1997. *Edison in Florida: The Green Laboratory.* University Press of Florida, Gainesville, FL.

Thunhorst, G. A. 1993. *Wetland Planting Guide for the Northeastern United States: Plants for Wetland Creation, Restoration, and Enhancement.* Environmental Concern, Inc., St. Michaels, MD.

Tiedje, J. M., R. K. Colwell, Y. L. Grossman, R. E. Hodson, R. E. Lenski, R. N. Mack, and P. J. Regal. 1989. The planned introduction of genetically engineered organisms: ecological considerations and recommendations. *Ecology.* 70:298–315.

Tilley, D. 2001. Personal communication.

Tilman, D., J. Knops, D. Wedin, P. Reich, M. Richie, and E. Siemann. 1997. The influence of functional diversity and composition on ecosystem processes. *Science.* 277:1300–1302.

Tilton, D. L. and R. H. Kadlec. 1979. The utilization of a fresh-water wetland for nutrient removal from secondarily treated waste water effluent. *Journal of Environmental Quality.* 8:328–334.

Timmons, M. B. and T. M. Losordo (eds.). 1994. *Aquaculture Water Reuse Systems: Engineering Design and Management.* Elsevier, Amsterdam, the Netherlands.

Toates, F. M. 1975. *Control Theory in Biology and Experimental Psychology.* Hutchinson Educational Publishing, London.

Todd, J. 1977. pp. 48–49. In: *Space Colonies.* S. Brand (ed.). Penguin Books, New York.

Todd, J. 1988a. Solar aquatic wastewater treatment. *BioCycle.* 29(2):38–40.

Todd, J. 1988b. Restoring diversity, the search for a social and economic context. pp. 344–352. In: *Biodiversity.* E. O. Wilson (ed.). National Academy Press. Washington, DC.

Todd, J. 1990. Solar aquatics. p. 85. In: *Whole Earth* Ecolog. J. Baldwin (ed.). Harmony Books, New York.

Todd, J. 1991. Ecological engineering, living machines and the visionary landscape. pp. 335–343. In: *Ecological Engineering for Wastewater Treatment.* C. Etnier and B. Guterstam (eds.). Bokskogen, Gothenburg, Sweden.

Todd, J. 1996a. Why lake restorer living machines work. *Annals of Earth.* 14(2):12–14.

Todd, J. 1996b. The ocean ark. *Annals of Earth.* 14(1):13–14.

Todd, J. 1998. Aikido aquaculture. *Annals of Earth.* 16(2):22–23.

Todd, J. and B. Josephson. 1996. The design of living technologies for waste treatment. *Ecological Engineering.* 6:109–136.

Todd, J. and N. J. Todd. 1980. *Tomorrow Is Our Permanent Address.* Harper & Row, New York.

Todd, J. and N. J. Todd. 1991. Biology as a basis for design. pp. 154–170. In: *Gaia 2, Emergence.* W. I. Thompson (ed.). Lindisfarne Press, Hudson, NY.

Todd, N. J. and J. Todd. 1984. *Bioshelters, Ocean Arks, City Farming: Ecology as the Basis of Design.* Sierra Club Books, San Francisco, CA.

Todd, N. J. and J. Todd. 1994. *From Eco-Cities to Living Machines: Principles of Ecological Design.* North Atlantic Books, Berkeley, CA.

Tomilson, P. B. 1986. *The Botany of Mangroves.* Cambridge University Press, Cambridge, U.K.

Trachtman, P. 2000. Redefining robots. *Smithsonian*, 30(11):97–110.

Transeau, E. N. 1926. The accumulation of energy by plants. *Ohio Journal of Science.* 26:1–10.

Tschirhart, J. 2000. General equilibrium of an ecosystem. *Journal of Theoretical Biology.* 203:13–32.

Tullock, G. 1971. The coal tit as a careful shopper. *American Naturalist.* 105:77–80.

Turing, A. M. 1950. Computing machinery and intelligence. *Mind.* 59:433–460.

Turner, F. 1994. The invented landscape. pp. 35–66. In: *Beyond Preservation.* A. D. Baldwin, Jr., J. de Luce, and C. Pletsch (eds.). University of Minnesota Press, Minneapolis, MN.

Turner, R. E., A. M. Redmond, and J. B. Zedler. 2001. Count it by acre or function — mitigation adds up to net loss of wetlands. *National Wetlands Newsletter.* 23(6):5–6, 14–16.

Tuska, C. D. 1947. *Patent Notes for Engineers.* Radio Corp. of American, Princeton, NJ.

Twinch, A. J. and C. M. Breen. 1978. Enrichment studies using isolation columns. I. The effects of isolation. *Aquatic Botany.* 4:151–160.

Ugolini, F. C. 1972. Ornithogenic soils of Antarctica. pp. 181–193. In: *Antarctic Terrestrial Biology.* G. A. Llano (ed.). American Geophysical Union. Washington, DC.

Uhl, C. 1988. Restoration of degraded lands in the Amazon Basin. pp. 326–332. In: *Biodiversity.* E. O. Wilson (ed.). National Academy of Sciences, Washington, DC.

Ulanowicz, R. E. 1981. A unified theory of self-organization. pp. 649–652. In: *Energy and Ecological Modelling.* W. J. Mitsch, R. W. Bosserman, and J. M. Klopatek (eds.). Elsevier Scientific Publishing, Amsterdam, the Netherlands.

Ulanowicz, R. E. 1993. Oecologia ex machine? *ECOMOD (International Society of Ecological Modelling Bulletin).* 11(2):1, 9.

Ulanowicz, R. E. 1997. *Ecology: The Ascendent Perspective.* Columbia University Press, New York.

Ulanowicz, R. E. and J. H. Tuttle. 1992. The trophic consequences of oyster stock rehabilitation in Chesapeake Bay. *Estuaries.* 15:298–306.

Underwood, A. J. and P. G. Fairweather. 1989. Supply-side ecology and benthic marine assemblages. *Trends in Ecology and Evolution.* 4:16–20.

Urbonas, B. and P. Stahre. 1993. *Stormwater: Best Management Practices and Detention.* PTR Prentice Hall, Englewood Cliffs, NJ.

Usinger, R. L. and W. R. Kellen. 1955. The role of insects in sewage disposal beds. *Hilgardia.* 23:263–321.

Valiela, I., J. M. Teal, and W. Sass. 1973. Nutrient retention in saltmarsh plots experimentally fertilized with sewage sludge. *Estuarine and Coastal Marine Science.* 1:261–269.

van de Koppel, J., P. M. J. Herman, P. Thoolen, and C. H. R. Heip. 2001. Do alternative stable states occur in natural ecosystems? Evidence from a tidal flat. *Ecology.* 82:3449–3461.

van den Bosch, R. 1978. *The Pesticide Conspiracy.* Doubleday, Garden City, NY.

van der Pijl, L. 1972. *Principles of Dispersal in Higher Plants.* Springer-Verlag, Berlin.

Van der Ryn, S. 1995. *The Toilet Papers: Recycling Waste and Conserving Water.* Ecological Design Press, Sausalito, CA.

Van der Ryn, S. and S. Cowan. 1996. *Ecological Design.* Island Press, Washington, DC.

van der Valk, A. G. 1981. Succession in wetlands: a Gleasonian approach. *Ecology.* 62:688–696.

van der Valk, A. G. 1988. From community ecology to vegetation management: providing a scientific basis for management. *Transactions of 53rd North American Wildlife and Nature Research Conference*. pp. 463–470.

van der Valk, A. G. 1998. Succession theory and restoration of wetland vegetation. pp. 657–668. In: *Wetlands for the Future*. A. J. McComb and J. A. Davis (eds.). Gleneagles Publishing, Adelaide, Australia.

van der Valk, A. G., R. L. Pederson, and C. B. Davis. 1992. Restoration and creation of freshwater wetlands using seed banks. *Wetlands Ecology Management*. 1:191–197.

van Ierland, E. C. and N. Y. H. de Man. 1996. Ecological engineering: first steps towards economic analysis. *Ecological Engineering*. 7:351–371.

Van Noordwijk, M. 1999. Nutrient cycling in ecosystems versus nutrient budgets of agricultural systems. pp. 1–26. In: *Nutrient Disequilibria in Agroecosystems*. E. M. A. Smaling, O. Oenema, and L. O. Fresco (eds.). CABI Publishing, Wallingford, U.K.

Van Valen, L. 1973. A new evolutionary law. *Evolution Theory*. 1:1–30.

Van Valen, L. 1977. The red queen. *American Naturalist*. 111:809–810.

Van Voris, P., R. V. O'Neill, W. R. Emanuel, and H. H. Shugart, Jr. 1980. Functional complexity and ecosystem stability. *Ecology*. 61:1352–1360.

Van Wilgen, B. W., R. M. Cowling, and C. J. Burgers. 1996. Valuation of ecosystem services. *BioScience*. 46:184–189.

Vander Wall, S. B. 1990. *Food Hoarding in Animals*. University of Chicago Press, Chicago, IL.

Vandermeer, J. H. 1972. Niche theory. *Annual Reviews of Ecology and Systematics*. 3:107–132.

Vandermeer, J. and I. Perfecto. 1997. The agroecosystem: a need for the conservation biologist's lens. *Conservation Biology*. 11: 591–592.

Varela, F. G., H. R. Maturana, and R. Uribe. 1974. Autopoiesis: The organization of living systems, Its characterization and a model. *Biosystems*. 5:187–196.

Velz, C. J. 1970. *Applied Stream Sanitation*. Wiley-Interscience, New York.

Verduin, J. 1969. Critique of research methods involving plastic bags in aquatic environments. *Transactions of American Fishery Society*. 98:355–356.

Vermeij, G. J. 1986. The biology of human-caused extinction. pp. 28–49. In: *The Preservation of Species*. B. G. Norton (ed.). Princeton University Press, Princeton, NJ.

Viles, H. A. (ed.). 1988. *Biogeomorphology*. Basil Blackwell, Oxford, U.K.

Vincenti, W. G. 1990. *What Engineers Know and How They Know It: Analytical Studies from Aeronautical History*. Johns Hopkins University Press, Baltimore, MD.

Visser, S. 1985. Role of the soil invertebrates in determining the composition of soil microbial communities. pp. 297–317. In: *Ecological Interactions in Soil*. A. H. Fitter (ed.). Blackwell Science, Oxford, U.K.

Visser, S. and D. Parkinson. 1975. Fungal succession on aspen popular leaf litter. *Canadian Journal of Botany*. 53:1640–1651.

Vitousek, P. M. 1986. Biological invasions and ecosystem properties: can species make a difference? pp. 163–174. In: *Ecology of Biological Invasions of North America and Hawaii*. H. A. Mooney and J. A. Drake (eds.). Springer-Verlag, New York.

Vitousek, P. M. 1988. Diversity and biological invasions of oceanic islands. pp. 181–189. In: *Biodiversity*. E. O. Wilson (ed.). National Academy Press, Washington, DC.

Vitousek, P. M. 1990. Biological invasions and ecosystem processes: towards an integration of population biology and ecosystem studies. *Oikos*. 57:7–13.

Vitousek, P. M. 1991. Can planted forests counteract increasing atmospheric carbon dioxide? *Journal of Environmental Quality*. 20:348–354.

Vitousek, P. M., C. M. D'Antonio, L. L. Loope, and R. Westbrooks. 1996. Biological invasions as global environmental change. *American Scientist*. 84:468–478.

Vitousek, P. M., L. R. Walker, L. D. Whiteaker, D. Mueller-Dombois, and P. A. Matson. 1987. Biological invasion by *Myrica faya* alters ecosystem development. *Science.* 802–804.

Vogel, S. 1998. *Cats' Paws and Catapults: Mechanical Worlds of Nature and People.* W. W. Norton, New York.

Vogt, K. A., C. C. Grier, and D. J. Vogt. 1986. Production, turnover, and nutrient dynamics of above- and belowground detritus of world forests. *Advances in Ecological Research.* 15:303–377.

von Frisch, K. and O. von Frisch. 1974. *Animal Architecture.* Harcourt Brace Jovanovich, New York.

Von Neumann, J. 1958. *The Computer and the Brain.* Yale University Press, New Haven, CT.

Von Neumann, J. 1966. *Theory of Self-Reproducing Automata.* Edited and completed by A. W. Burks. University of Illinois Press, Urbana, IL.

Voshell, J. R., Jr. (ed.). 1989. *Using Mesocosms to Assess the Aquatic Ecological Risk of Pesticides: Theory and Practice.* Misc. Publ. No. 75, Entomological Society of America.

Wackernagel, M., L. Lewan, and C. B. Hahsson. 1999. Evaluating the use of natural capital with the ecological foodprint. *Ambio.* 28:604–612.

Wackernagel, M. and W. Rees. 1996. *Our Ecological Footprint: Reducing Human Impact on the Earth.* New Society Publishers, Gabriola Island, British Columbia.

Wagener, S. M., M. W. Oswood, and J. P. Schimel. 1998. Rivers and soils: parallels in carbon and nutrient processing. *BioScience.* 48:104–108.

Wainscott, V. J., C. Bartley, and P. Kangas. 1990. Effect of muskrat mounds on microbial density on plant litter. *American Midland Naturalist.* 123:399–401.

Wajnberg, E., J. K. Scott, and P. C. Quimby (eds.). 2001. *Evaluating Indirect Ecological Effects of Biological Control.* CABI Publishing, Wallingford, U.K.

Waksman, S. A. 1952. *Soil Microbiology.* John Wiley & Sons, New York.

Walford, R. L. 2002. Biosphere 2 as voyage of discovery: the serendipity from inside. *BioScience.* 52:259–263.

Wali, M. (ed.). 1992. *Ecosytem Rehabilitation.* SPB Academic Publishing, the Hague, the Netherlands.

Walker, L. R. (ed.). 1999. *Ecosystems of Disturbed Ground: Ecosystems of the World.* Vol. 16. Elsevier, Amsterdam, the Netherlands.

Wallace, B. 1974. The biogeography of laboratory islands. *Evolution.* 29:622–635.

Wallace, J. B., J. R. Webster, and T. F. Cuffney. 1982. Stream detritus dymanics: regulation by invertebrate consumers. *Oecologia.* 53:197–200.

Walsh, W. J. 1985. Reef fish community dynamics on small artificial reefs: the influence of isolation, habitat structure, and biogeography. *Bulletin of Marine Sciences.* 36:357–376.

Walton, S. 1980. Smithsonian transplants coral system from tropics to tank. *BioScience.* 30:805–808.

Wann, D. 1990. *Biologic, Environmental Protection by Design.* Johnson Books, Boulder, CO.

Wann, D. 1996. *Deep Design: Pathways to a Livable Future.* Island Press, Washington, DC.

Wardle, D. A. 2002. *Communities and Ecosystems: Linking the Aboveground and Below-ground Components.* Princeton University Press, Princeton, NJ.

Waring, R. H. and W. H. Schlesinger. 1985. *Forest Ecosystems, Concepts and Management.* Academic Press, Orlando, FL.

Warren, C. E. and P. Doudoroff. 1971. *Biology and Water Pollution Control.* W. B. Saunders, Philadelphia, PA.

Warren, E. R. 1927. *The Beaver, Its Work and Its Ways.* Williams & Wilkins, Baltimore, MD.

Waterhouse, D. F. 1974. The biological control of dung. *Scientific American.* 230(4):100–109.

Watson, C. C., S. R. Abt, and D. Derrick. 1997. Willow posts bank stabilization. *Journal of American Water Resources Association.* 33:293–300.

Watson, E. S., DC. McCluricin, and M. B. Huneycutt. 1974. Fungal succession on loblolly pine and upland hardwood foliage and litter in north Mississippi. *Ecology.* 55:1128–1134.

Watson, T. 1993. Can basic research ever find a good home in Biosphere 2? *Science.* 259:1688–1689.

Weaver, J. E. 1919. *The ecological relations of roots.* Carnegie Institute Washington Publication 286., Washington, DC.

Weaver, J. E. 1954. *North American Prairie.* Johnsen Publishing, Lincoln, NE.

Weaver, J. E. 1958. Classification of root systems of forbs of grassland and a consideration of their significance. *Ecology.* 39:393–401.

Weaver, J. E. 1961. The living network in prairie soils. *Botanical Gazette.* 123:16–28.

Weaver, J. E. 1968. *Prairie Plants and Their Environment.* University of Nebraska Press, Lincoln, NE.

Weaver, J. E. and F. E. Clements. 1938. *Plant Ecology*, 2nd ed. McGraw-Hill, New York.

Weaver, W. 1947. Science and complexity. *American Scientist.* 36:536–544.

Webb, B. and T. R. Consi (eds.). 2001. *Biorobotics: Methods and Applications.* MIT Press, Menlo Park, CA.

Webb, R. H., J. C. Schmidt, G. R. Marzolf, and R. A. Valdez (eds.). 1999. *The Controlled Flood in Grand Canyon.* American Geophysical Union, Washington, DC.

Weber, I. P. and S. D. Wiltshire. 1985. *The Nuclear Waste Primer.* Nick Lyons Books, New York.

Weber, N. A. 1972. Gardening ants: the attines. *Memoir of American Philosophical Society.* 92:1–146.

Weigert, R. G. (ed.). 1976. *Ecological Energetics: Benchmark Papers in Ecology*, Vol. 4. Dowden, Hutchinson and Ross, Stroudsburg, PA.

Weigert, R. G. and J. Kozlowski. 1984. Indirect causality in ecosystems. *American Naturalist.* 124:293–298.

Weiher, E. and P. A. Keddy. 1995. The assembly of experimental wetland plant communities. *Oikos.* 73:323–335.

Weiher, E. and P. Keddy (eds.). 1999. *Ecological Assembly Rules: Perspectives, Advances, Retreats.* Cambridge University Press, Cambridge, U.K.

Weins, J. A. 1976. Population responses to patchy environments. *Annual Reviews of Ecology and Systematics.* 7:81–120.

Weir, J. S. 1977. Exotics: past, present and future. pp. 4–14. In: *Exotic Species in Australia — Their Establishment and Success.* D. Anderson (ed.). Proceedings of the Ecological Society of Australia, Vol. 10.

Welch, E. B. 1980. *Ecological Effects of Waste Water.* Cambridge University Press, Cambridge, U.K.

Welcomme, R. L. 1984. International transfers of inland fish species. pp. 22–40. In: *Distribution, Biology, and Management of Exotic Fishes.* W. R. Courtenay, Jr. and J. R. Stauffer, Jr. (eds.). John Hopkins University Press, Baltimore, MD.

Went, F. W. and N. Stark. 1968a. The biological and mechanical role of soil fungi. *Proceedings of the National Academy of Sciences, USA.* 60:497–504.

Went, F. W. and N. Stark. 1968b. Mycorrhiza. *BioScience.* 18:1035–1039.

Werker, A. G., J. M. Dougherty, J. L. McHenry, and W. A. Van Loon. 2002. Treatment variability for wetland wastewater treatment design in cold climates. *Ecological Engineering.* 19:1–11.

Werner, B. T. and T. M. Fink. 1993. Beach cusps as self-organized patterns. *Science.* 260:968–971.

Westman, W. E. 1977. How much are nature's services worth? *Science.* 197:960–964.

Westman, W. E. 1990. Park management of exotic plant species: problems and issues. *Conservation Biology.* 4:251–260.

Wetzel, R. G. 1995. Death, detritus, and energy flow in aquatic ecosystems. *Freshwater Biology.* 33:83–89.

Wharton, R. A., Jr., D. T. Smeroff, and M. M. Averner. 1988. Algae in space. pp. 485–507. In: *Algae and Human Affairs.* C. A. Lembi and J. R. Waaland (eds.). Cambridge University Press. Cambridge, U.K.

Wheaton, F. W. 1977. *Aquacultural Engineering.* John Wiley & Sons. New York.

Whigham, D. F. 1985. Vegetation in wetlands receiving sewage effluenct: the importance of the seed bank. pp. 231–240. In: *Ecological Considerations in Wetlands Treatment of Municipal Wastewaters.* P. J. Godfrey, E. R. Kaynor, S. Pelczarski, and J. Benforado. (eds.). Van Nostrand Reinhold Co., New York.

Whigham, D. F. and R. L. Simpson. 1976. The potential use of freshwater tidal marshes in the management of water quality in the Delaware River. pp. 173–186. In: *Biological Control of Water Pollution.* J. Tourbier and R. W. Pierson (eds.). University of Pennsylvania Press, Philadelphia, PA.

Whisenant, S. G. 1999. *Repairing Damaged Wildlands.* Cambridge University Press, Cambridge, U.K.

White, P. S. and J. L. Walker. 1997. Approximating nature's variation: selecting and using reference information in restoration ecology. *Restoration Ecology.* 5:338–349.

Whitesides, G. M. 1995. Self-assembling materials. *Scientific American.* 273(3):146–149.

Whitesides, G. M., J. P. Mathias, and C. T. Seto. 1991. Molecular self-assembly and nanochemistry: a chemical strategy for the synthesis of nanostructures. *Science.* 254:1312–1319.

Whitford, W. G. and N. Z. Elkins. 1986. The importance of soil ecology and the ecosystem perspective in surface-mine reclamation. pp. 151–187. In: *Principles and Methods of Reclamation Science: With Case Studies from the Arid Southwest.* C. C. Reith and L. D. Potter (eds.). University of New Mexico Press, Albuquerque, NM.

Whittaker, R. H. 1961. Experiments with radiophosphorus tracer in aquarium microcosms. *Ecological Monographs.* 31:157–188.

Whittaker, R. H. 1967. Gradient analysis of vegetation. *Biological Review.* 42:207–264.

Whittaker, R. H. 1977. Evolution of species diversity in land communities. *Evolutionary Biology.* 10:1–67.

Whittaker, R. H. and D. Goodman. 1979. Classifying species according to their demographic strategy. *American Naturalist.* 113:185–200.

Whittaker, R. H. and S. A. Levin. (eds.). 1975. *Niche: Theory and Application. Benchmark Papers in Ecology,* Vol. 3. Dowden, Hutchinson and Ross, Stroudsburg, PA.

Whittaker, R. H. and G. M. Woodwell. 1972. Evolution of natural communities. pp. 137–156. In: *Ecosystem Structure and Function.* J. A. Wiens (ed.). Oregon State University Press, Corvalis, OR.

Wieder, R. K. and G. E. Lang. 1982. A critique of the analytical methods used in examining decomposition data obtained from litter bags. *Ecology.* 63:1636–1642.

Wiener, N. 1948. *Cybernetics or Control and Communication in the Animal and the Machine.* MIT Press, Cambridge, MA.

Wiens, J. A. 1984. On understanding a non-equilibrium world: myth and reality in community patterns and processes. pp. 439–457. In: *Ecological Communities*. D. R. Strong, Jr., D. Simberloff, L. G. Abele, and A. B. Thistle (eds.). Princeton University Press, Princeton, NJ.

Wik, R. M. 1972. *Henry Ford and Grass-roots America*. University of Michigan Press, Ann Arbor, MI.

Wilbur, H. M. 1987. Regulation of structure in complex systems: experimental temporary pond communities. *Ecology*. 68:1437–1452.

Wilbur, H. M. 1989. In defense of tanks. *Herpetologica*. 45:122–123.

Wilbur, H. M. 1997. Experimental ecology of food webs: complex systems in temporary ponds. *Ecology*. 78:2279–2302.

Wilbur, H. M. and R. A. Alford. 1985. Priority effects in experimental pond communities: responses of *Hyla* to *Bufo* and *Rana*. *Ecology*. 66:1106–1114.

Wile, I., G. Miller, and S. Black. 1985. Design and use of artificial wetlands. pp. 26–37. In: *Ecological Considerations in Wetlands Treatment of Municipal Wastewaters*. P. J. Godfrey, E. R. Kaynor, S. Pelczarski, and J. Benforado (eds.). Van Nostrand Reinhold, New York.

Wilhm, J. L. and T. C. Dorris. 1968. Biological parameters for water quality criteria. *BioScience*. 18:477–481.

Wilkinson, S. 2000. "Gastrobots" — benefits and challenges of microbial fuel cells in food powered robot applications. *Autonomous Robots*. 9:99–111.

Williams, R. J., F. B. Griffiths, E. J. Van der Wal, and J. Kelly. 1988. Cargo vessel ballast water as a vector for the transport of nonindigenous marine species. *Estuarine and Coastal Shelf Science*. 26:409–420.

Williamson, M. 1989. The MacArthur and Wilson theory today: true but trivial. *Journal of Biogeography*. 16:3–4.

Willis, D. 1995. *The Sand Dollar and the Slide Rule: Drawing Blueprints from Nature*. Addison-Wesley, Reading, MA.

Wilson, E. O. 1988. *Biophilia*. Harvard University Press, Cambridge, MA.

Wilson, E. O. 1988. The current state of biological diversity. pp. 3–18. In: *Biodiversity*. E. O. Wilson (ed.). National Academy Press, Washington, DC.

Wilson, E. O. and W. H. Bossert. 1971. *A Primer of Population Biology*. Sinauer Associates, Stamford, CN.

Wilson, E. O. and E. O. Willis. 1975. Applied biogeography. pp. 522–534. In: *Ecology and Evolution of Communities*. M. L. Cody and J. M. Diamond (eds.). Harvard University Press, Cambridge, MA.

Wilson, R. F. and W. J. Mitsch. 1996. Functional assessment of five wetlands constructed to mitigate wetland loss in Ohio, USA. *Wetlands*. 16:436–451.

Wilsson, L. 1971. Observations and experiments on the ethology of the European beaver (*Castor fiber* L.). pp. 374–403. In: *External Construction by Animals*. N. E. Collias, and E. C. Collias (eds.). Dowden, Hutchinson & Ross, Stroudsburg, PA.

Winemiller, K. O. 1990. Spatial and temporal variation in tropical fish trophic networks. *Ecological Monographs*. 60:331–367.

Winner, L. 1977. *Autonomous Technology: Technics-out-of-control as a Theme in Political Thought*. MIT Press, Cambridge, MA.

Winston, M. L. 1997. *Nature Wars: People vs. Pests*. Harvard University Press, Cambridge, MA.

Winter, S. G., Jr. 1964. Economic "natural selection" and the theory of the firm. *Yale Economics Essay*. 4:225–275.

Wischmeier, W. H. 1976. Use and misuse of the universal soil loss equation. *Journal of Soil and Water Conservation*. 31:5–9.

Wisniewski, R. 1990. Shoals of *Dreissena polymorpha* as bio-processor of seston. *Hydrobiologia*. 200/201:451–458.

Wolf, L. 1996. Bioregeneration with maltose excreting Chlorella: system concept, technological development, and experiments. *Advances in Space Biology and Medicine*. 6:255–274.

Wolf, R. B., L. C. Lee, and R. R. Sharitz. 1986. Wetland creation and restoration in the United States from 1970 to 1985: An annotated bibliography. *Wetlands*. 6:1–88.

Wolfenbarger, D. O. 1975. *Factors Affecting Dispersal Distances of Small Organisms*. Exposition Press, Hicksville, New York.

Wolfgang, L. 1995. Biosphere 2 turned over to Columbia. *Science*. 270:1111.

Wolman, M. G. 1995. Personal communication.

Wolverton, B. C. 1996. *How to Grow Fresh Air: 50 Houseplants That Purify Your Home or Office*. Penguin Books, New York.

Wolverton, B. C. and R. C. McDonald. 1979a. Water hyacinth (*Eichornia crassipes*) productivity and harvesting studies. *Economic Botany*. 33:1–10.

Wolverton, B. C. and R. C. McDonald. 1979b. The water hyacinth: from prolific pest to potential provider. *Ambio*. 8(1):2–9.

Wolverton, B. C. and J. D. Wolverton. 2001. *Growing Clean Water: Nature's Solution to Water Pollution*. WES, Picayune, MS.

Wolverton, B. C., R. M. Barlow, and R. C. McDonald. 1976. Application of vascular aquatic plants for pollution removal, energy, and food production in a biological system. pp. 141–149. In: *Biological Control of Water Pollution*. J. Tourbier and R. W. Pierson, Jr. (eds.) University of Pennsylvania Press, Philadelphia, PA.

Wong, Y.-S. and N. F. Y. Tam (eds.). 1998. *Wastewater Treatment with Algae*. Springer, Berlin.

Wood, R. A., R. L. Orwell, J. Tarran, F. Torpy, and M. Burchett. 2002. Potted-plant/growth media interactions and capacities for removal of volatiles from indoor air. *Journal of Horticultural Science & Biotechnology*. 77:120–129.

Wood, A. E. 1980. The Oligocene rodents of north america. *Transactions of the American Philosophical Society*, Vol. 70, part 5. Philadelphia, PA.

Woodhouse, W. W., Jr. and P. L. Knutson. 1982. Atlantic coastal marshes. pp. 46–70. In: *Creation and Restoration of Coastal Plant Communities*. R. R. Lewis III (ed.). CRC Press, Boca Raton, FL.

Woodhouse, W. W., Jr., E. D. Seneca, and S. W. Broome. 1976. Propagation and use of *Spartina alterniflora* for shoreline erosion abatement. Technical Report No. 76-2, U. S. Army Corps of Engineers Coastal Engineering Research Center, Fort Belvoir, VA.

Woodroffe, C. D. 1990. The impact of sea-level rise on mangrove shorelines. *Progress in Physical Geography*. 14:483–520.

Woodruff, L. L. 1912. Observations on the origin and sequence of protozoan fauna of hay infusions. *Journal of Experimental Zoology*. 1:205–264.

Woodwell, G. M. 1967. Radiation and the patterns of nature. *Science*. 156:461–470.

Woodwell, G. M. 1970. Effects of pollution on the structure and physiology of ecosystems. *Science*. 168:429–433.

Woodwell, G. M. and R. A. Houghton. 1990. The experimental impoverishment of natural communities: effects of ionizing radiation on plant communities, 1961–1976. pp. 9–24. In: *The Earth in Transition: Patterns and Processes of Biotic Impoverishment*. G. M. Woodwell (ed.). Cambridge University Press, Cambridge, U.K.

Woodwell, G. M. and F. T. Mackenzie (eds.). 1995. *Biotic Feedbacks in the Global Climatic System*. Oxford University Press, New York.

Wootton, J. T. 1994. The nature and consequences of indirect effects in ecological communities. *Annual Reviews of Ecology and Systematics*. 25:443–466.

Wooten, J. W. and J. D. Dodd. 1976. Growth of water hyacinths in treated sewage effluent. *Economic Botany.* 30:29–37.

Wotton, R. S. 1980. Coprophagy as an economic feeding tactic in blackfly larvae. *Oikos.* 34:282–286.

Wotton, R. S. and K. Hirabayashi. 1999. Midge larvae (*Diptera: Chironomidae*) as engineers in slow sand filter beds. *Water Research.* 33:1509–1515.

Wotton, R. S. and B. Malmqvist. 2001. Feces in aquatic ecosystems. *BioScience.* 51:537–544.

Wright, D. H., C. E. Folsome, and D. Obenhuber. 1985. Competition and efficiency in closed freshwater algal systems: tests of ecosystem design principles. *BioSystems.* 17:233–239.

Wu, J. and O. L. Loucks. 1995. From balance of nature to hierarchical patch dynamics: a paradigm shift in ecology. *Quarterly Review of Biology.* 70:439–466.

Wuhrmann, K. 1972. Stream purification. pp. 119–151. In: *Water Pollution Microbiology.* R. Mitchell (ed.). Wiley-Interscience, New York.

Wulff, F., J. G. Field, and K. H. Mann (eds.). 1989. *Network Analysis in Marine Ecology: Methods and Applications.* Springer-Verlag, Berlin.

Wynne-Edwards, V. C. 1970. Feedback from food resources to population regulation. pp. 413–426. In: *Animal Populations in Relation to Their Food Resources.* A. Watson (ed.). Blackwell Scientific Publications, Oxford, U.K.

Yamane, T. 1989. Status and future plans of artificial reef projects in Japan. *Bulletin of Marine Sciences.* 44:1038–1040.

Yan, J. and Y. Zhang. 1992. Ecological techniques and their application with some case studies in China. *Ecological Engineering.* 1:261–285.

Yang, C. T. 1971a. On river meanders. *Journal of Hydrology.* 13:231–253.

Yang, C. T. 1971b. Formation of riffles and pools. *Water Resources Research.* 7:1567–1574.

Yaron, P., M. Walsh, C. Sazama, R. Bozek, C. Burdette, A. Farrand, C. King, J. Vignola, and P. Kangas. 2000. Design and construction of a floating living machine. pp. 92–101. In: *Proceedings of the 27th Annual Conference on Ecosystems Restoration and Creation.* P. J. Cannizzaro (ed.). Hillsborough Community College, Plant City, FL.

Yodzis, P. 1996. Food webs and perturbation experiments: theory and practice. pp. 192–200. In: *Food Webs, Integration of Patterns and Dynamics.* G. A. Polis and K. O. Winemiller (eds.). Chapman & Hall. New York.

Young, C. M. 1987. Novelty of "supply-side ecology." *Science.* 235:415–416.

Young, P. 1996. The "new science" of wetland restoration. *Environmental Science and Technology.* 30:292A–296A.

Zakin, S. 1993. *Coyotes and Town Dogs: Earth First! and the Environmental Movement.* Penguin Books, New York.

Zaret, T. M. 1975. The ecology of introductions — a case-study from a Central American lake. *Environmental Conservation.* 1:308–309.

Zaret, T. M. 1984. Ecology and epistemology. *Bulletin of the Ecological Society of America.* 65:4–7.

Zaret, T. M. and R. T. Paine. 1973. Species introduction in a tropical lake. *Science.* 182:449–455.

Zedler, J. B. 1988. Restoring diversity in saltmarshes, can we do it? pp. 317–325. In: *Biodiversity.* E. O. Wilson and F. M. Peter (eds.). National Academy Press, Washington, DC.

Zedler, J. B. 1995. saltmarsh restoration: lessons from California. pp. 75–95. In: *Rehabilitating Damaged Ecosystems.* J. Cairns, Jr. (ed.). Lewis Publishers, Boca Raton, FL.

Zedler, J. B. 1996a. Ecological issues in wetland mitigation: an introduction to the forum. *Ecological Applications.* 6:33–37.

Zedler, J. B. 1996b. Ecological function and sustainability in created wetlands. pp. 331–342. In: *Restoring Diversity: Strategies for Reintroduction of Endangered Species.* D. A. Falk, C. I. Millar, and M. Olwell (eds.). Island Press, Washington, DC.

Zedler, J. B. 1999. The ecological restoration spectrum. pp. 301–318. In: *An International Perspective on Wetland Rehabilitation.* W. Streever (ed.). Kluwer Academic, Dordrecht, the Netherlands.

Zedler, J. B. 2000. Progress in wetland restoration ecology. *Trends in Ecology and Evolution.* 15:402–407.

Zedler, J. B. (ed.). 2001. *Handbook for Restoring Tidal Wetlands.* CRC Press, Boca Raton, FL.

Zedler, J. B., G. D. Williams, and J. S. Desmond. 1997. Can fishes distinguish between natural and constructed wetlands? *Fisheries.* 22(3):26–28.

Zeiher, L. C. 1996. The Ecology of Architecture: *A Complete Guide to Creating the Environmentally Conscious Building.* Whitney Library of Design, New York.

Zelov, C. and P. Cousineau. (eds.). 1997. *Design Outlaws on the Ecological Frontier.* Knossus, Philadelphia, PA.

Zentner, J. J. 1999. The consulting industry in wetland rehabilitation. pp. 243–249. In: *An International Perspective on Wetland Rehabilitation.* W. Streever (ed.). Kluwer Academic, Dordrecht, the Netherlands.

Zinecker, E. K. 1997. Exotic Species and Riparian Plant Communities in Three Watersheds of Varying Levels of Urbanization in Prince William County, Virginia. M.S. thesis, University of Maryland, College Park, MD.

Zlotin, R. I. and K. S. Khodashova. 1980. *The Role of Animals in Biological Cycling of Forest-Steppe Ecosystems.* Dowden, Hutchinson & Ross, Stroudsburg, PA.

Zubrin, R. 1996. *The Case for Mars: The Plan to Settle the Red Planet and Why We Must.* Touchstone Publishing, New York.

Zweig, R. D. 1986. An integrated fish culture hydroponic vegetable production system. *Aquaculture Magazine.* 12(3):34–40.

Zweig, R. D., J. R. Wolfe, J. H. Todd, D. G. Engstrom, and A. M. Doolittle. 1981. Solar aquaculture: an ecological approach to human food production. pp. 210–226. In: *Proceedings of the Bio-Engineering Symposium for Fish Culture.* L. J. Allen and E. C. Kinney (eds.). American Fisheries Society, Bethesda, MD.

Zwillich, T. 2000. A tentative comeback for bioremediation. *Science.* 289:2266–2267.

Index

A

B

C

D

E

F

G

H

I

J

K

L

M

Mangroves *(Rhizophora),* 111–115

N

O

P

R

S

U

V

W

X

Y

Z